18th International Conference on Non-Conventional Materials and Technologies (IC-NOCMAT 2019)

Nairobi, Kenya
24-26 July 2019

Editors:

Khosrow Ghavami
Esther Adhiambo Obonyo
Normando Perazzo Barbosa
José Augusto Gomes Neto

ISBN: 978-1-7138-1158-9

Printed from e-media with permission by:

Curran Associates, Inc.
57 Morehouse Lane
Red Hook, NY 12571

CURRAN ASSOCIATES INC. proceedings .com

Published by IC-NOCMAT 2019

Printed with permission by Curran Associates, Inc. (2020)

For permission requests, please contact IC-NOCMAT 2019

Additional copies of this publication are available from:

Curran Associates, Inc.
57 Morehouse Lane
Red Hook, NY 12571 USA
Phone: 845-758-0400
Fax: 845-758-2633
Email: curran@proceedings.com
Web: www.proceedings.com

IC-NOCMAT 2019

18th International Conference on Non-Conventional Materials and Technologies

July 24th - 26th, 2019
University of Nairobi Towers
Nairobi, Kenya

PROCEEDINGS

Design José Augusto Gomes Neto

PPCEM
POST-GRADUATION PROGRAMM ON SCIENCE AND MATERIALS ENGINEERING FEDERAL UNIVERSITY OF PARAÍBA, BRAZIL

PUC RIO

SCIENTEC

abmtenc

PennState College of Engineering
ARCHITECTURAL ENGINEERING

Khosrow Ghavami

Esther Adhiambo Obonyo

Normando Perazzo Barbosa

José Augusto Gomes Neto

PROCEEDINGS of 18th INTERNATIONAL CONFERENCE ON NON-CONVENTIONAL MATERIALS AND TECHNOLOGIES – IC NOCMAT 2019

Nairobi, Kenya

July, 24th – 26th, 2019

University of Nairobi

IC-NOCMAT 2019

1st Edition

João Pessoa – Paraíba – Brazil
2019

Associação Para o Desenvolvimento da Ciência e da Tecnologia, 1st edition, João Pessoa, Paraíba, Brazil, 2019.

Review

Normando Perazzo Barbosa
José Augusto Gomes Neto

Barbosa, N. P.

Proceedings of 18th international conference on non-conventional materials and technologies – IC NOCMAT 2019 Nairobi, Kenya, 2019

533 p.

1. IC-NOCMAT 2019. 2. Proceedings.

Associação Para o Desenvolvimento da Ciência e da Tecnologia, 1st edition, João Pessoa, Paraíba, Brazil, 2019.

Conference Chairs / Co-chairs

- Dr. Khosrow Ghavami – ABMTENC – Rio, PUC-Rio
- Dr. Esther Obonyo – The Pennsylvania State University

Local Organizing Committee Chair / Co-chairs

- Patts Odira – University of Nairobi
- Dr. Titus Kivaa – Jomo Kenyatta University of Agriculture & Technology
- Q.S. James Okaka – Jomo Kenyatta University of Agriculture & Technology
- Eng. Moses Opiyo – Kenytta University

Scientific Committee

- Prof. Said Kenai, University Saad Dahlab
- Prof. Beatriz Garzon, University of Tucuman
- Dr. Conrado de Sousa Rodrigues, Federal Institute of Minas Gerais
- Prof. Holmer Savastano Jr., University of São Paulo / USP - Pirassununga
- Assoc. Prof. Luis Eustáquio Moreira, Federal University of Minas Gerais
- Prof. Normando Perazzo Barbosa, Federal University of Paraíba
- Prof. Pierre Ohayon, Federal University of Rio de Janeiro
- Prof. Romildo Dias Toledo Filho, Federal University of Rio de Janeiro
- Dr. Ulisses Targino Bezerra, Paraíba Federal Institute of Paraíba
- Dr. Aluisio Braz Melo, Federal University of Paraíba
- Prof. Milton Bezerra das Chagas Filho, Federal University of Campina Grande
- Prof. Antonio Beraldo, University of Campinas.
- Prof. Jianzhuang Xiao, Tongji University
- Prof. Yan Xiao (China/USA) Zhejiang University/U. of Southern California
- Prof. Jianzhuang Xiao, Tongji University
- Prof. Eduardo Moreira, Central University of las Villas- Cuba,
- Dra. Lena Mora, Central University of las Villas.
- Prof. Hanan El Nouhy, Housing and Building National Research Center / HBRC
- Prof. Jean-Claude Morel, École Nationale des Travaux Publics/Coventry University
- Prof. B.V. Venkatarama Reddy, Indian Institute of Science, Bangalore
- Assoc. Prof. Kambiz Pourtahmasi, University of Tehran
- Prof. Mohamad Zaman Kabir, Amirkabir University of Technology, Tehran
- Prof. Giorgio Monti, Sapienza University of Rome
- Prof. Manuela Mattone, Polytechnic of Turin
- Prof. Alejandro Manzano-Ramírez, CINVESTAV-IPN
- Prof. Pedro Jesús Herrera, Yucatan´s Center for Scientific Research
- Prof. Piet Stroeven, University of Delft
- Prof Antônio Tadeu – University of Coimbra
- Prof. José Barroso Aguiar, University of Minho
- Prof. Humberto Varum – University of Porto
- Prof. G. I. Yakoviev, Kalashnikov Izhevsk State Technical University Studencheskaya
- Dr. Jorge Juan Payá Bernabeu, Universidad Politécnica de Valencia
- Dr. José Maria Monzó Balbuena, Universidad Politécnica de Valencia
- Dr. Angel Palomo, Instituto Eduardo Toroja
- Prof. Mitchell Gohnert, University of the Witwatersrand / Johannesburg.
- Prof. G. Habert, Zurich Polytechnic
- Prof. J. V. Tesha, University of Dar es Salaam
- Prof. Pete Walker, University of Bath
- Dan Maskell, University of Bath
- Dr. Bruce King, Sustainable Design and Engineering
- Prof. Kent A. Harries, University of Pittsburgh

PREFACE

The IC-NOCMAT series provides an international forum for information dissemination and exchange, discussions, and debates on research and practice in a range of sustainable and innovative construction materials and technologies. The first NOCMAT held in 1984 in Rio de Janeiro, organized by Prof. Khosrow Ghavami, who already was preoccupied at that time with the impact of industrialized construction materials on the environment. In 1990, Prof. Ghavami created the Brazilian Association of Non-conventional Materials and Technologies, ABMTENC. Since then, the association has held NOCMAT conferences around the world.

The tropical countries are extremely rich in natural materials and appropriate local technologies. In those countries, with the increase of industrialization and the domination of big polluting industries, and through the implementation of corrupt administration the ecological local industries forgotten. Now, when environmental pollution problems involve the whole world, the interest to study the denominated non-conventional materials and Technologies, NOCMAT, becomes a necessity.

The NOCMAT 2019, that is the 18th edition of the NOCMAT series, in Nairobi, Kenya, attracted a wide range of academics, scientists, researchers, students, builders, designers, NGO field-workers, besides other industrialists from a wide variety of backgrounds, including fields of architecture, engineering, materials, sustainable and ecological technologies, biomaterials, materials sciences, and environmental engineering. The conference also was of interest to specialists in geography, economists and administrators of natural resources.

Khosrow Ghavami
Pontifical Catholic University of Rio de Janeiro

Acknowledgements

INTRODUCTION

The new generation faces two graves challenges on the Earth: the environmental and climate problems and the social inequality, in an overpopulated planet with already more than 7.7 billion souls. We consider that our actual generations has been irresponsible in these aspects, because we are seen how much the Men is attacking the Planet and how he has been incapable to solve basic problems to significant part of mankind, as hunger, housing and sanitation, although the extraordinary technological development we have had. Dramatic are the life conditions of more than two billion of people and, despite that, we see investments in buildings that will exceed 1 km in height, very luxurious houses, high tech cars and many excesses that are not a necessity of a human being. Let´s remember a summarized definition of sustainability, from the Brundtland Report: Sustainability is the ability of present generations to meet their needs without compromising future generations to meet their own needs. But with these excesses, knowing that the Earth is a finite planet, how justify that the new generations will find enough raw material to satisfy their needs? If we think about it, only part of the present generation is having their needs met. So, sustainability will be a utopia if mankind continues to be what it is! Nowadays, the world is dominated by a cruel capitalism that allows news like the following: 'World's richest 1% get 82% of the wealth', says Oxfam. The gap between the super rich and the rest of the world widened last year, as wealth continued to be owned by a small minority, Oxfam has claimed in Davos, 22 de jan de 2018. (Oxfam is an international confederation of 20 ngos working with partners in over 90 countries to end the injustices that cause poverty). The world military expenditure grows to $1.8 trillion in 2018! How can this be possible when hunger attaint hundreds of millions of people? This wrong system is based on increasing consumption on a finite planet! Something must be done! As the industrialized materials are high energy consumers, utilizes a lot of raw materials, generates a lot of pollutants and wastes, NOCMAT points for the necessary change. Diffusing the natural, bio and renewables materials, looking for reuse of wastes, encouraging research on low energy construction products, new cementitious materials and appropriate technologies, NOCMAT 2019 was held in Nairobi, Kenya. In this e-book you can find the contributions sent to the Congress by people from many countries. The order of the papers is that given by the organizers. We hope you diffuse it to yours colleagues. Knowledge must belong to all mankind, especially that favorable to change something on our not well-organized planet.

Prof. Normando Perazzo Barbosa
Federal University of Paraíba, Brazil

P. S. Thanks to José Augusto Gomes Neto, our Ph.D. student that organized the book.

SUMMARY

BIO-MATERIALS

IC-NOCMAT 2019

18th International Conference on
Non-Conventional Materials and Technologies
"Construction Materials & Technologies for Sustainability"
(18th NOCMAT 2019)
24th – 26th July 2019
Nairobi, Kenya

MONOLITHIC MYCELIUM: GROWING VAULT STRUCTURES

Jonathan Dessi-Olive[1]

[1] Georgia Institute of Technology School of Architecture, Atlanta, GA, USA;

ABSTRACT

This paper presents advancements in mycelium construction, demonstrated through a large-scale monolithic mycelium vault structure. In the burgeoning field of bio-materials, mycelium has emerged as a potentially viable building material because it is naturally occurring, demands little energy for its production, is lightweight, and completely bio-degradable. Whereas, previous architecture-scale mycelium constructions used component-based techniques, this paper presents novel research on grown-in-place and self-supporting *monolithic mycelium structures*, achieved using techniques resembling common practices for cast-in-place concrete.

Keywords: mycelium construction, vault construction, structural design, sustainable materials, bio-materials.

INTRODUCTION

There is an eminent need to address issues of dwindling material resources by exploring alternate construction materials and methods. Mycelium-composite materials have emerged as a contender to become a commercially viable lightweight and renewable building material. Whereas the mushroom is the fruit part of fungus, mycelium [1] is the plant part. These web-like root structures are everywhere underground and play an integral role in the way our natural environments grow and thrive; building networks between our farms, forests, and water-ways. Recently, scientists, artists, designers, and students have experimented with mycelium as a stabilizing compound for agricultural waste fibers to produce a variety of objects. At architecture-scale, several highly imaginative pavilion structures have been constructed using bricks or components made from mycelium materials. This paper presents novel advancements in the field of mycelium construction through a series of experiments that culminated in a large-scale monolithic mycelium vault prototype and demonstrates the possibility of growing building structures. In addition to reviewing recent design work with mycelium, this paper will summarize the monolithic construction methodology, outline initial small-scale experiments, and offer a detailed description of the "Monolito Micelio" pavilion, which was conceived of and grown in the context of an architecture seminar at the Georgia Institute of Technology School of Architecture.

BACKGROUND TO MYCELIUM CONSTRUCTION

Recently, scientists, artists, designers, and students have experimented with mycelium as a stabilizing compound for agricultural waste fibers to produce objects at many scales. This section reviews the motivation for using mycelium, its material properties, recent design and research work with mycelium-composites, and identifies a common approach, particularly among the architectural-scale examples.

Motivation for using Mycelium as a Building Material

The primary motivation for testing mycelium-composites in load-bearing material applications is the excessive amounts of waste produced by current construction practices. More than ever, the global construction industry has relied on materials that are neither renewable nor re-usable – often mined and manufactured with high energy processes. Buildings today are often constructed for increasingly shorter expected lifespans as it is often more financially economical to demolish and rebuild. In the United States, 548 million tons of construction and demolition waste to landfill sites in 2015 [2], mostly from demolition waste. Such figures highlight the eminent need to find ways of offsetting the current state through sustainable building methods and material economies that redirect resources away from landfills and recirculates them back into production. The ability for mycelium composites to become the source of new cultivation places them in a class of materials that support circular economies. In the United States, companies such as MycoWorks [3] and Ecovative [4] hold patents on the production of mycelium-stabilized materials, and offer highly engineered products that are reliable and easy to grow.

Mycelium Material Properties and Precautions

Mycelium-based building materials are made with agricultural waste products, replace energy-intensive manufacturing processes of common construction materials with organic growth, and are completely compostable. Mycelium-composites are unique because they require growing living mushroom cultures in very specific conditions. To thrive, mycelium needs a sterile environment with ample food supply, access to air, darkness, warmth and humidity. When combined with cellulosic materials (food), mycelium-stabilized materials have the ability to form dense composite matrices that when dried, have a density and strength that resembles polyurethane foam. Several different fibrous substrates (hemp, corn husk, or wood sawdust, for example) and mycelium strains can be combined to make mycelium bio-composites, with varying structural integrity, density, and visual quality. Given fibers that are inoculated with live mushroom spores and the appropriate formwork, nearly any shape can be made by packing the inoculated fibers into the formwork. Typically, the packed form is left to grow in favorable environmental conditions for several days, after which the mycelium part is baked to dry and kill off the mycelia, preventing unwanted growth or infection from other organisms. The limitations for growth are biological and relatively easy to manage: a growing environment that is inadequately sterile leads to several types of mold. The environmental conditions for mushrooms and several types of mold are identical, so sterility and proper access to air (via thickness control and breathable formwork) are of greatest importance. At certain thicknesses, mycelium may not grow sufficiently to bind to the fibers and die too soon due to lack of air, also presenting an opportunity for mold to develop. With proper management of these environmental requirements, making mycelium-composite parts is very easy.

Applications of Mycelium in Art and Academia

In recent years, several exhibitions, experiments, and pavilions have helped expose a broader audience to the promise of mycelium-stabilized composites. Furniture pieces and several other domestic-scale objects have been proposed by artists such as Philip Ross [5], Gediminas and Nomeda Urbonas [6], and Officina Corpuscoli [7]. Philip Ross coined the term "mycotecture" with his 2012 installation, which consisted of bricks grown from mushroom cultures, stacked into a catenary-form barrel vault. His installation inspired several small-scale experiments in schools of architecture in Spain [8], Austria [9], and Canada [10, 11, 12], and the United Kingdom [13], among others. Common areas of focus for these projects include developing strategies for producing mycelium materials, including testing different mushroom spores and fibrous substrates; performing compression tests on samples; growing mycelium materials different brick shapes and sizes and building small prototype structures; and proposing pavilion-scale structures.

Applications of Mycelium in Architecture and Engineering

Mycelium composite materials are advantageous because they are naturally occurring, demand very little energy for their production, and are completely bio-degradable. A major disadvantage to mycelium composite materials is they have a very low stress capacity compared to traditional building materials. As architects and engineers have sought to build with mycelium at architectural scale, rigorous structurally informed design approaches have been needed. Three pavilions demonstrate the range of approaches taken for mycelium constructions so far. A temporary pavilion called "Shell Mycelium" was built in Keria, India [14, 15] in 2017. The designers used mycelium panels to fill spaces in a wooden grid-shell structure – serving as a sufficient example of large-scale use, but in a definitively non-structural application. The largest, and perhaps most widely publicized mycelium structure was "Hi-Fi" [16], a 40-foot tower installation for the MOMA PS1 Young Architects Program by David Benjamin and The Living in 2014, engineered by ARUP [17] with the Lafarge-Holcim Foundation [18]. The structure was hand-stacked atop of a wood and steel supporting structure using mycelium bricks sourced from Ecovative. This ambitious temporary installation was fundamental in popularizing experimentation with mycelium in architecture and design research. The most structurally accomplished and formally evocative mycelium spatial structure to-date is the "MycoTree" [19, 20]. Installed at the 2017 Seoul Biennale of Architecture, the structure demonstrates how the structural capacity of mycelium can be exploited maximally by placing in compression-only configurations. The MycoTree benefits from well-established computational structural design techniques developed by the Block Research Group (BRG) at ETH-Zurich.

TOWARD MONOLITHC MYCELIUM CONSTRUCTION

From Growing Components to Growing Monoliths

Among the mycelium-composite spatial structures described above, all have taken a common tectonic approach: using modular bricks, custom components, or other discrete-element system that resemble historical masonry and stone construction logics. This paper seeks to contribute to the field of large-scale mycelium structures by identifying a gap in the literature and proposing a novel stereotomic approach: monolithic mycelium. Rather than growing components or bricks from mycelium, this paper claims it is possible to grow-in-place monolithic mycelium structures. The techniques proposed here address the need to propose alternative construction logics and uses for mycelium-stabilized materials by testing, validating, and showcasing the potential of grow-in-place structures. At small scale, artist Erik Klarenbeek has achieved numerous inspiring ways of structural monolithic growth [21] in combination with adapted 3D printing technologies. In 2016, a Master of Science in Civil Engineering thesis was written by University of Miami graduate student, Riccardo Andrea Baricci [22] suggesting analytical methods for mycelium based monolithic domes, but did not validate the methods through physical means. Though much weaker and lighter than concrete, grow-in-place monolithic mycelium techniques inherit many of the advantages (and also challenges) of cast-in-place concrete techniques, including the use of traditional board, plank, or sheeting formwork, as well as use flexible fabric formwork techniques similar to those perfected by Mark West [23].

Mycoarch

The origins of this monolithic mycelium construction research stem from a collaborative, experimental project called "Tactical Mycelium" [24]. One of the outcomes of the project was the "Mycoarch", a monolithically grown, self-supporting mycelium-composite arch [Figure 1]. A volume of about 0.4 m3 of mycelium was sourced from Ecovative and grown in a plastic sheeting membrane, cast over the top of an active-bent PVC pipe. The structure was grown outside over five days, after which it was unwrapped, and let to air-dry for several more days. Due to inaccuracies to the shape, and after sustaining rain, the arch collapsed prematurely. As a preliminary experiment, the Mycoarch experiment demonstrated that growing monolithically was possible, and motivated additional research into such techniques.

Figure 1: Mycoarch structure; and detail view showing unique qualities resulting from flexible formwork.

Thick and Thin

Thanks to its partial success, the Mycoarch experiment brought to light several issues that were addressed by a second prototype grown as a part of an experimental architecture course at the Georgia Tech School of Architecture, titled "Thick and Thin". The seminar focused on developing analytical skills, geometric and constructive intuition for funicular (compression-only) structures. Participants included undergraduate, graduate, and post-graduate architecture students the Georgia Tech School of Architecture. Monolithic mycelium was proposed to the seminar as a possible means of constructing funicular shell structures with distinct construction logics and material approaches, compared to previous component-based constructions. Following, a review of the observations made during the Mycoarch experiment, students were tasked with designing a small monolithic mycelium shape, its internal formwork, and outer soft membrane schemes, for a volume of approximately 0.2 m3 of material sourced from Ecovative.

The three primary areas of needed improvement observed from the Mycoarch were: using a more stucturally informed form-finding process, using an internal formwork system, and taking more care in controlling the growing environment. Due to the weakness of the material, a three-dimensional, doubly-curved, compression-only form was designed to make a self-supporting structure that would be both stiffer and more stable compared to the two-dimensional Mycoarch. The students used RhinoVault [25] to assist them in exploring compression-only shapes, and designed a custom-made internal reinforcing structure from laser-cut cardboard [Figures 2 and 3]. As a fiber-based sheet material, the cardboard acts as an additional food source for the mycelium, which causes the internal reinforcing to become integrated into the construction matrix with a far more sophisticated bond compared to concrete and steel reinforcing, which never fully integrate. The material was packed by hand over a few hours, let to grow for four full days, unwrapped and let to air-dry for a period of several weeks. Due to its small size, it was possible to grow the structure indoors, which brought a much higher level of control over the environmental conditions of the grow space. This minimized the possibility of the structure to prematurely terminate its growth due to infections from mold and other organisms. By addressing the issues observed with Mycoarch, the students managed to successfully grow a small monolithic mycelium vault structure [Figure 4]. Their approach to the external membrane portion of the formwork was identical to the Mycoarch, using thin polyethylene plastic sheeting, which replicated the unpredictable, strange, and evocative qualities on the surface of the structure. Although in some voids remained after the first growth period, the students managed to repair and patch those areas by growing fresh mycelium onto/into the already dried structure. Several months after the structure had grown and dried, it was stong enough to support a person weighing approximately 75 kg.

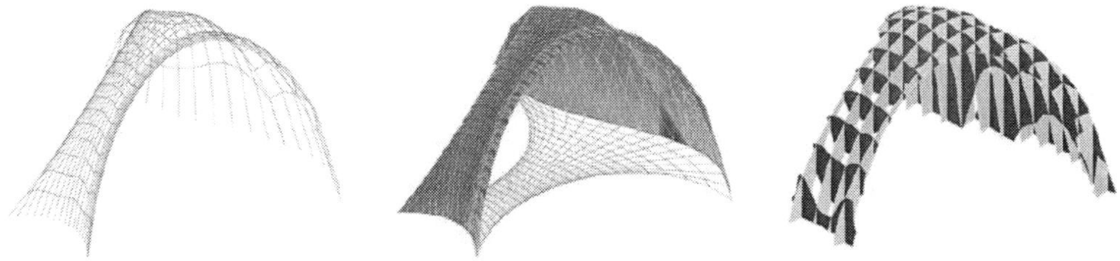

Figure 2: Form-finding for "Thick and Thin" prototype using Rhinovault; and internal formwork scheme.

Figure 3: Custom-made internal cardboard skeleton; packing live mycelium composite material and the cardboard skeleton into a membrane of plastic sheeting.

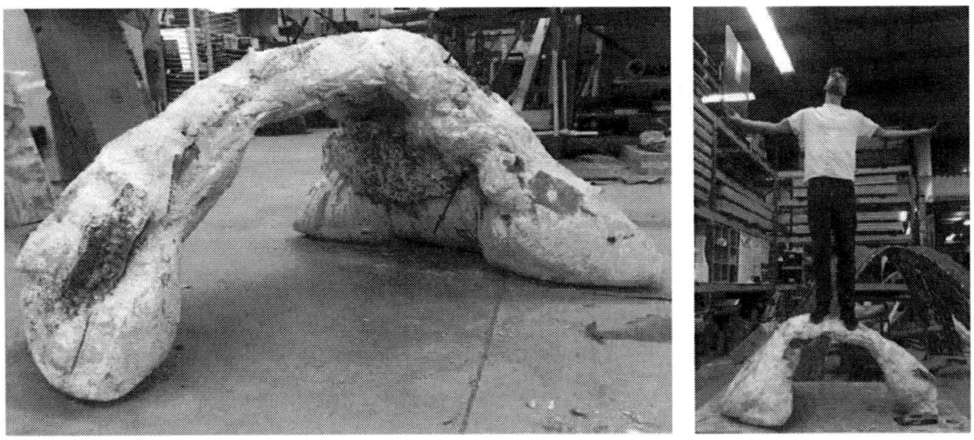

Figure 4: Formation diagram for the geometry of the Monolito Micelio.

THE MONOLITO MICELIO PAVILION

The "Monolito Micelio" was envisioned as the first large-scale monolithic mycelium vault structure. The success of the small "Thick and Thin" prototype inspired a collective mobilization in the seminar around attempting to scale-up our techniques. The self-supporting monolithic mycelium structure used structurally-informed computational design processes to ensure performative success, still allow for imprecisions inherent to flexible formwork systems, and produce the types of visual qualities in the material that are surprising, strange, and beautiful. The following section describes the design, production, and growth of the Monolito Micelio bio-pavilion which served as a performance structure for a barbershop quartet at the end-of-year exhibition at the School of Architecture at the Georgia Institute of Technology.

Design and Geometry

Mycelium materials present numerous ecological advantages over traditional engineered materials, but also have a much lower stress capacity. To build with such weak materials, compression-only forms that can produce equilibrated material configurations are preferred – achieving structural stability through geometry, rather than material strength. Assuming imperfections are inevitable at this scale, working toward an optimized form provides a safety factor to ensure performative success and allows for controlled imprecisions from fabric formwork to produce visual qualities in the material that are surprising, strange, and beautiful. To generate the geometry of the pavilion, a rule-based structurally-informed computational design process was developed [Figure 5]. The shape was computed by splitting a "mushroom column" into four equal parts and re-arranging them into a symmetrical, four-legged vault shell. To finalize the shape computation, a Boolean between the shell and a bounding cube make a thick, monolithic form. The resulting design, loosely resembling the shape transformations of a British-Gothic "fan vault", consists of flat, orthogonal exterior surfaces, contrasted with curved, vaulting interior surfaces. The overall dimensions of the structure are 2.5 x 2.5 x 2.5 meters. The monolithic volume of the vault is 2.75 m3, or around 800 kg of live mycelium material. The Monolito Micelio distinguishes itself from previous monolithic prototypes by strategically deploying a rigid internal "lost-work" reinforcing skeleton, and a hybrid removable formwork system that combines a highly controlled plywood exterior with a highly flexible geotextile interior.

Internal Lostwork

Our previous monolithic mycelium-composite prototypes demonstrated the importance of using fibrous internal "lostwork" reinforcing systems that can become fully integrated into the material matrix of the structure. When the internal reinforcing is based on a structurally-informed geometry and made precisely with digital fabrication tools, it can also serve as a target geometry to regulate the level of precision the mycelium structure takes as it is packed. At the scale of this pavilion, precautions had to be taken in order to safely vault material over peoples' heads, and to handle other external loads such as wind and rain. An internal reinforcing rib structure [Figure 6] was designed and fabricated from oriented strand-board (OSB) sheets cut on a CNC mill. OSB is made from wood strands, which if properly sanitized can be consumed by mycelium to make a strong bond with the mycelium. The rib pieces were assembled on top of a wooden platform, which was on wheels to facilitate moving the formwork structure to the final growing and exhibiting site.

Figure 5: Formation diagram for the geometry of the Monolito Micelio.

Figure 6: Construction sequence diagram for the internal reinforcing portion of the formwork system.

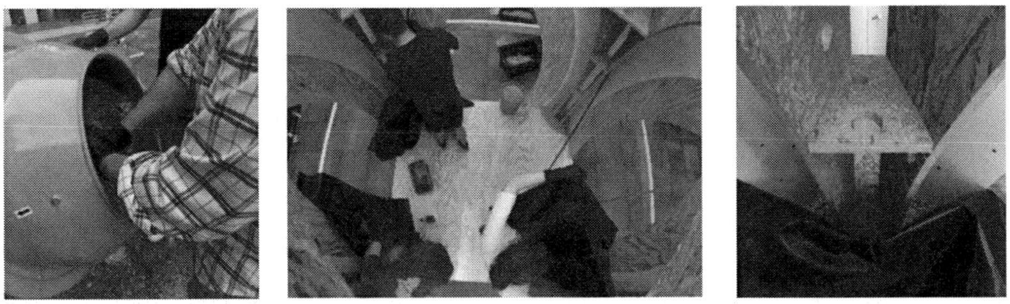

Figure 7: The geotextile formwork being installed as mycelium material is packed.

External Formwork

Formwork strategies for growing large monolithic mycelium structures must be as rigorous as with large, cast-in-place concrete pours. Applying different formwork logics to the interior and exterior parts of the shape presents an opportunity to demonstrate monolithic structural form with a multiplicity of different surface qualities. Mycelium require clean, dark, humid, and breathable environmental conditions for growth, which are relatively easy to achieve in laboratory settings. Growing outdoors at architectural scale poses several new challenges for designing and developing techniques for large grow-in-place mycelium formwork. Structurally, the formwork must be robust enough to resist forces from the wet and densely packed fibrous substrate. Large quantities of mycelium produce quantities of carbon dioxide that are damaging to its growth, so the formwork materials should be semi-porous and breathable. Whether rigid or flexible, formwork parts should be removed without damaging the structure or surface quality of the mycelium. Thus, in the areas where mycelium comes directly in contact with formwork, non-cellulosic materials should be used to ensure those formwork parts can be safely removed.

For the flat orthogonal exterior portions of the structure, the approach used for the Monolito Micelio was straight-forward: acrylic-laminated plywood sheets were milled and perforated (to let the mycelium breath) and installed around the perimeter of the formwork structure. For the curved vaulting interior portions of the structure, a flexible fabric formwork approach was taken. Previous monolithic prototypes demonstrated that plastic sheeting provided flexibility and favorable qualities, but did not sufficiently let the mycelium breath or ventilate. For the Monolito Micelio, a woven nylon geotextile was selected for its tensile strength and porosity – used as flexible and removable fabric formwork to expressively form the curving internal portions of the structure [Figure 7]. As a practical matter, the fabric was installed during the packing process using screws to keep it tight along edges of the laminated sheeting. Our goal was to have between five and ten centimeters of continuous mycelium material around the reinforcement rib structure. Between the fabric and the ribs, long screws were used to intuitively vary the thickness, in turn producing bulbous and rippled qualities.

Figure 8: The grow-site. Breakers, feeders and packers working late into the night to fill the formwork.

Mycelium Packing, Growing, and Drying

At architectural scale, packing mycelium-inoculated substrates is very labor intensive. The tools on-site included: several sizes of plastic mixing bins; impact drivers and screws; gloves, paper towels and bleach; a clean cement mixer; work-lights, a Bluetooth speaker, and other worksite sundries. As with previous monolithic prototypes, the material for the Monolito Micelio was sourced from Ecovative. The pre-inoculated hemp material, which arrived in 10kg bags, had both advantages and disadvantages. On one hand we saved time and effort typically put into the inoculation process. On the other hand, in the days it took to ship the bags from the factory, the material had already grown into hardened masses, had to be used almost immediately, and required some labor-intensive processing. Three working roles emerged on the grow-site [Figure 8]. Breakers took bags of live material and manually broke apart the fibers in large bins. Ecovative recommends their material be moistened and fed before it is packed into its form. Feeders transferred the fibers into a clean cement mixer. As the fibers were tumbled, flour and water were added to the fibers to kick-start mycelium growth. Packers installed the fabric formwork and packed the mycelium into the formwork. The pavilion was packed by an average of six people working for twelve hours. After letting the material grow for a full four days, the geotextile was progressively removed. Six days after the packing event all remaining formwork was removed, revealing the first large-scale, grown-in-place monolithic mycelium pavilion [Figure 9]. The structure was unveiled at the end-of-year exhibition at the School of Architecture at the Georgia Institute of Technology and served as a performance structure for a barbershop quartet [Figure 10].

Figure 9: The Monolito Micelio structure; after the external plywood and fabric formwork were removed.

Figure 10: The Monolito Micelio pavilion; unveiled at the end of year exhibition at the Georgia Tech School of Architecture. As a part of the event, a barbershop quartet performed under the structure.

CONCLUSION

This paper has presented advancements in mycelium construction, demonstrated through research that seeks to expand the possible construction logics that can be used with the emerging bio-material. Focusing specifically on developing techniques for large-scale compression-only structural forms, this paper has reviewed previous design work to highlight that at architecture-scale, several highly imaginative pavilion structures have been constructed using a similar construction logic: using bricks or components made from mycelium materials, assembled in a manner resembling stone and masonry construction. The research presented here distinguishes itself from previous work by physically validating through a series of prototypes that and demonstrates the possibility of growing building structures. Culminating in the Monolito Micelio pavilion, this research provides evidence that in addition to component-based approaches as used by previous mycelium structures, it is also possible to grow self-supporting monolithic mycelium structures in ways that resemble cast-in-place concrete construction. Much weaker and lighter than concrete, monolithic mycelium inherits many of the advantages and surface qualities possible with concrete techniques. At large scale, a combination of correct form, internal reinforcing systems, and robust yet breathable formwork materials are crucial to the success of grow structures with monolithic mycelium techniques. The Monolito Micelio used a wooden internal "lost-work" reinforcing skeleton as well as a hybrid removable formwork system that combines a highly controlled plywood exterior with a highly flexible and breathable geotextile interior.

In the face of material scarcity and insecurity, there is an increased need to explore and test low-energy and rapidly renewable building materials that contribute to circular material economies. Mycelium is among a class of bio-composite building materials that are grown and composted with very little external energy input can vastly offset the impact of the short lifespans of buildings in the modern era. The ability to rapidly grow structures cultivated from composites of bio-materials and agricultural waste has the potential could greatly impact current construction practices by promoting circular material economies. At a time when buildings are being constructed with very short expected lifespans that result in landfill disposal, mycelium can help challenge our traditional perceptions of the permanence of buildings, what materials we use, and where they go when we're done using them. As mycelium continues to demonstrate promise as a commercialized bio-material for building construction, it will not only be important to continue inventing new ways to use it, but also specify the contexts in which it is most appropriate. A major weakness of mycelium material used as a part of a superstructure, is how fragile it is in the face of external elements and organisms without proper treatment. Even if structural applications are not the best suited application for mycelium, the material still shows promise to have an impact on other design fields including, fashion which is already well established, and also in architectural acoustics, where despite the fact that nearly every new building receives acoustical treatment, very few sustainable acoustical products are commercially available.

Acknowledgements

Due to the highly collaborative nature of this project, there are several people to acknowledge – and without whom, the work presented here would not have been possible. Bridget Ayers Looby and Rebecca Ramsey led the first Mycoarch experiment and were fundamental in formulating the intellectual attitude around monolithic mycelium construction. Significant contributions were made by Georgia Tech architecture students: Roberto Bucheli, Bennet Crawford , Keyhan Khaki, Alex Knight, Anna McCuan, Sean Miller, Sounok Sarkar, Matt Singleton, and Justin Wilson. Special thanks to those who helped prep and pack the pavilion: Chao Dang, Julia Jones, and Jessica Bilgrad. In the Digital Fabrication Lab, Jake Thompkins provided space and facilitated production. Thanks to Professor Russell Gentry for his encouragement and for suggesting NOCMAT as a venue to publish this experimental work. Thanks to Shane Boland from Ecovative/Grow.bio for his commitment to customer service and helping us source mycelium material. Finally, this research would not have been possible without financial support from the Department of Architecture at Georgia Tech through the Ventulett NEXT Generation Fellowship and the departmental endowment for masonry research that supported the construction activities in the Thick and Thin Seminar, including purchasing materials for the Monolito Micelio Pavilion.

REFERENCES

1. Stamets, P. Mycelium Running: How Mushrooms Can Help Save the World. Berkeley, CA: 10 Speed Press, 2005.

2. Environmental Protection Agency. "Sustainable Management of Construction and Demolition Waste". <https://www.epa.gov/smm/sustainable-management-construction-and-demolition-materials> (accessed March 15, 2019)

3. MycoWorks. <https://www.mycoworks.com/> (accessed March 15, 2019)

4. Ecovative. <https://ecovativedesign.com/> (accessed March 15, 2019)

5. Mok, K. "Mycotecture: Building with Mushrooms? Inventor says yes." Treehugger, September 26, 2012. <https://www.treehugger.com/green-architecture/mycotecture-mushroom-bricks-philip-ross.html> (accessed March 15, 2019)

6. Studio Urbonas. < http://www.nugu.lt/us> (accessed March 15, 2019)

7. Officina Corpuscoli. < http://www.corpuscoli.com/projects/the-growing-lab/> (accessed March 15, 2019)

8. Alima. N. "Mycotecture: Growing in Form". Institute for Advanced Architecture of Catalonia, 2013. < http://www.iaacblog.com/programs/mycotecture-growing-into-form-2/>

9. Imhof, B. and Gruber, P. (eds.). Built to Grow: Blending Architecture and Biology. Basil, Switzerland: Birkhauser, 2016.

10. Dahmen, J. and Frid-Jimenez, A. "Research on Mycelium Construction Materials". University of British Columbia, 2015. (Accessed March 15, 2009)

11. Dahmen, J. "Soft Futures: Mushrooms and Regenerative Design". Journal of Architectural Education, 71:1 (2017), pg 57-64.

<

https://open.library.ubc.ca/cIRcle/collections/graduateresearch/66428/items/1.0343394>

12. Fatehi, P. Engaging Mycelium: explorations of a cultivated architecture. Master of Architecture Thesis. University of Waterloo, Ontario, Canada, 2018.

13. Morby, A. "Structure grown from 'mushroom sausages' shows potential for zero-waste architecture". Dezeen, June 20, 2017. < https://www.dezeen.com/2017/06/20/aleksi-vesaluoma-mushroom-mycelium-structure-shows-potential-zero-waste-architecture/> (accessed March 15, 2019)

14. Frearson, A. "Fungus used to build arching pavilion in Keria". Dezeen, August 26, 2017.

< https://www.dezeen.com/2017/08/26/shell-mycelium-fungus-pavilion-beetles-3-3-yassin-arredia-design-kerala-india/#disqus_thread> (accessed March 15, 2019)

15. Zeitoun, L. "Shell Mycelium: exploring fungus growth as a possible building block". Designboom, July 25, 2017. < https://www.designboom.com/architecture/shell-mycelium-degradation-movement-manifesto-07-25-2017/> (accessed March 15, 2019)

16. The Living. < http://www.thelivingnewyork.com/> (accessed March 15, 2019)

17. Clark, M. and Saporta, S. "How ARUP Engineered The Living's Mushroom Tower". Metropolis. <https://www.metropolismag.com/ideas/technology/how-arup-engineered-the-livings-mushroom-tower/> (accessed March 15, 2019)

18. LafargeHolcim Foundation for Sustainable Construction. "Hy-Fi: Zero Carbon Emissions Compostable Structure, New York, NY." LafargeHolcim Foundation. <https://www.lafargeholcim-foundation.org/projects/hy-fi> (accessed March 15, 2019).

19. Heisel, F. et al. "Design of a load-bearing mycelium structure through informed structural design". Proceedings of the World Congress on Sustainable Technologies, 2017 (December).

20. Heisel, F. et al. "Design, Cultivation, and Application of Load-Bearing Mycelium Components: The MycoTree at the 2017 Seoul Biennale of Architecture and Urbanism." International Journal of Sustainable Energy Development (IJSED), Volume 6, Issue 1, June 2017/2018, 296-303.

21. Pallister, J. "3D-printed mushroom roots "could be used to build houses". Dezeen, March 6, 2014. <https://www.dezeen.com/2014/03/06/movie-eric-klarenbeek-mushroom-roots-fungus-3d-printed-chair/> (accessed March 15, 2019).

22. Baricci, R. A. "Structural Analysis and Form-Finding of Mycelium-Based Monolithic Domes". Master of Science Thesis. University of Miami, Florida, USA, 2016.

23. West, M. The Fabric Formwork Book: methods for building new architectural and structural forms in concrete. New York: Routledge, 2017.

24. Ayers Looby, B. and Ramsey, R. "Tactical Mycelium: An Exploration of Mushroom Mycelium As Ephemeral Building Material". Perkins and Will Innovation Incubator, October 2017. < https://issuu.com/bridgetayerslooby/docs/tactical_mycelium_final> (accessed March 15, 2019).

25. M. Rippmann, L. Lachauer, P. Block, "Interactive Vault Design", International Journal of Space Structures Vol. 27 No. 4, 219-230. December 2012.

IC-NOCMAT 2019

18th International Conference on
Non-Conventional Materials and Technologies
"Construction Materials & Technologies for Sustainability"
(18th NOCMAT 2019)
24th – 26th July 2019
Nairobi, Kenya

EXTRACTION OF SILICA PRESENT IN RESIDUAL SUGAR CANE BAGASSE ASH

Primo Fernandes Filho[1], Sandro Marden[1], Normando Perazzo Barbosa[1], Kelly Cristiane Gomes da Silva[1], Paloma Alcântara[2]

[1] Prof. Center of Technology of Federal University of Paraíba, João Pessoa, Brazil;

[2] PhD Student, Federal University of Paraiba

ABSTRACT

After chemical characterization of the residual sugar cane bagasse ash (CBC), it was found that silica contributes with 71% of its composition. In this paper, we intend to determine the yield of soluble silica, contained in the residue, and the consequent potential for its use as a source of silica in the production of cement products. The methodology adopted for silica extraction takes into account that amorphous silica increases its solubility in high pH solutions and as a consequence, in the presence of acid solution, the silica present in the alkali silicate is precipitated. The physical characterization evaluated the density using the Pycnometer and the specific area using the Blaine equipment. The chemical characterization determined the composition in oxides through the technique of x-ray fluorescence spectroscopy and the mineralogical characterization resorted to the techniques of infrared spectroscopy and x-ray diffraction to identify the phases present in the samples. To give significance to the silica yield extracted from the residual ash, the result was compared to the silica yield extracted from samples of other materials, taken as reference, for example of metakaolin (META), and quartz (QP), using same method. The results were 2.89% (QP), 11.94% (CBC) and 34.17% (META). Considering the volume of available raw material, the result is encouraging and can be increased, from the amorphous character improvement, such as: more efficient washing of the sugar cane and controlled burning of the bagasse. A milling process aiming to increase the fineness of the material may also contribute to the increased yield of the extracted silica.
KEYWORDS: Sugar cane bagasse ash, extracted silica, cements products.

INTRODUCTION

The possibility of incorporating products from the agroindustry to Portland cement has been of interest to researchers and several works, with positive results, are under development worldwide. The efforts range from the incorporation of these residues as a small aggregate, to the partial replacement of cement by pozzolanic materials in concrete, as well as through a more radical alternative, namely the development of new materials such as alkaline activated cements. In Brazil, residual ash from sugarcane bagasse (CBC), due to the burning of bagasse, is worth mentioning, mainly due to the large amounts currently generated. Studies indicate that current production of sugarcane exceeds 612.2 million tons per year (Unica [1]) with 0.7% remaining in the form of residual ash, which is approximately 4 million tons. Each year the contribution of sugarcane biomass to the

Brazilian energy matrix is increasing (Tolmasquim et al [2]). Figure 6.1 shows this evolution.

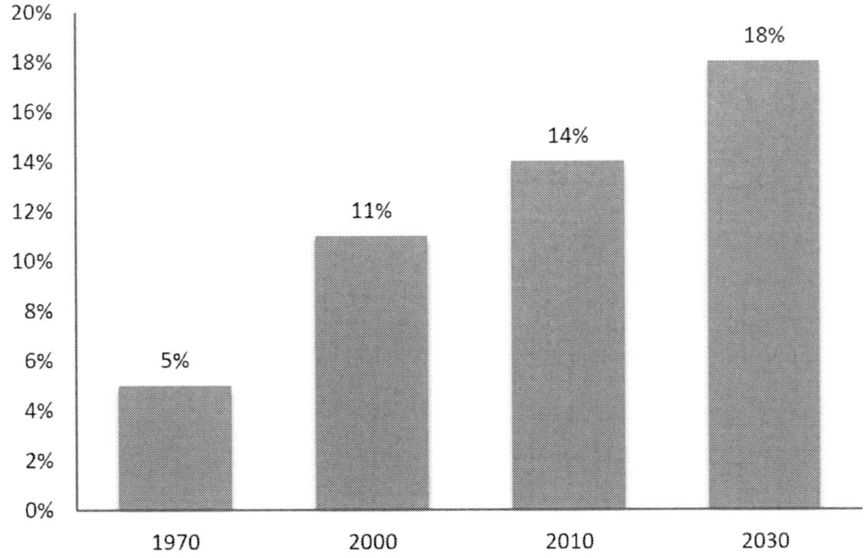

Figure 6.1 - Evolution of the contribution of sugarcane in the energy matrix of Brazil

Thus, the mountains of residual ashes from the burning of the bagasse that is used for the generation of energy that fires the boilers of the mills rise.

The chemical characterization indicates the silica as the predominant compound in the constitution of the residual ash of sugarcane bagasse (Payá [3]). The study of mineralogy indicates that it is a predominantly crystalline material. Improvement processes such as grinding and controlled burning can contribute to increase the reactivity of the material.

In this study, we intend to determine the yield of soluble silica contained in the in natura residue and the consequent potential for its use as a source of silica in obtaining alkaline activated products. In an attempt to solve the problem caused by solid waste, we are ready to meet two main advantages related to economic and environmental aspects; from an economic point of view, reducing the use of cement, if it will be, in the same proportion, reducing the high energy consumption during the process of clinquerization and also adding value to the residue; from an environmental point of view, in addition to providing a correct destination for waste, reducing the possibilities of the most varied forms of contamination, reducing the damaging effects caused by the emission of CO2 in nature during the cement manufacturing process.

EXPERIMENTAL PROCEDURE

Materials

The residual ash sample was collected at a plant located in the Santa Rita city, in the coastal region of Paraíba state. For analysis, it was considered the fraction passed through the 200 mesh sieve and the sample was denominated CBC. To give significance to the silica yield extracted from the residual ash, the result was compared to the silica yield extracted from samples of other materials, taken as reference, such as metakaolin (META) and quartz (QP), using the same method . In order to facilitate the solubility of the silica, alkaline solutions were produced from commercial sodium hydroxide in 97% purity and to favor the precipitation of the silica was used solution of hydrochloric acid.

Methods

The methodology adopted for silica extraction takes into account that amorphous silica increases its solubility in high pH solutions and as a consequence, in the presence of acid solution, the silica present in the alkali silicate is precipitated. The temperature and reaction time as well as the molarity of the alkaline solution and the sample mass / volume ratio of the solution are important variables for the solubility of silica (Bac et al [4]).

The ash sample was submitted to the mineralogical characterization through the techniques of X-ray, infrared spectroscopy, thermal analysis and Secondary Electron Scanning Electron Microscopy. The physical characterization of the samples evaluated the density using the pycnometer and the specific area using the blaine equipment. The chemical characterization determined the composition in oxides by the technique of x-ray fluorescence spectroscopy.

The silica extraction process was carried out under the following conditions: (25 °C), 24h reaction, alkaline solution 2M NaOH and sample mass / volume ratio of the solution (g/ml) equal to 1/30. These values were used by Bac et al. (2010), when studying the solubility of silica present in municipal solid waste. Thus, the following sequence of procedures was adopted:

- 3.3333g of each sample was immersed in 100 ml of 2M NaOH solution for 24 hours at room temperature.
- The reaction product was membrane-filtered with 0.45 μm pore diameter.
- The retained material was considered insoluble and discarded from the process.
- The material passing through the membrane was mixed with a solution of HCl, to neutralize the pH that was around 8.5, favoring the precipitation of the silica. Material was taken to the greenhouse for 24 hours at 110 ° C
- The solid residue was weighed and subjected to chemical and mineralogical characterization by x-ray fluorescence and x-ray diffraction to quantify and identify the nature of the silica present in the extracted material.

Results and Discussion

Physical characterization of the samples presented the result shown in table 1.

Table 1 - Physical characterization of the samples

Sample	Density (g/cm^3)	Specific area (cm^2/g)
CBC	2,60	2.892
META	2,39	19.391
QP	2,65	13.300

CBC and QP presented similar densities and META presented the largest specific area. The result of the chemical characterization is shown in table 2.

Table 2 - Chemical composition of the samples in (%)

Sample	SiO_2	Al_2O_3	K_2O	CaO	Fe_2O_3	SO_3	MgO	Na_2O	P_2O_5	Outros
CBC	71,07	4,13	7,53	4,72	2,59	1,75	3,04	0,74	2,83	1,60
META	60,86	35,1	2,78	0,08	0,62	-	0,25	0,15	0,07	0,09
QP	98,38	0,85	0,29	0,07	0,08	-	-	0,33	-	0,00

The high silica content found for CBC (71%) cooperates with the possibility of success for the extraction process. It is also observed in CBC, presence of alumina and some alkalis, being able to favor the alkaline activation.

The mineralogy study was restricted to the CBC sample since for META and QP the parameters of this analysis are already well established.

The X-ray diffraction presented the result shown in figure 2.

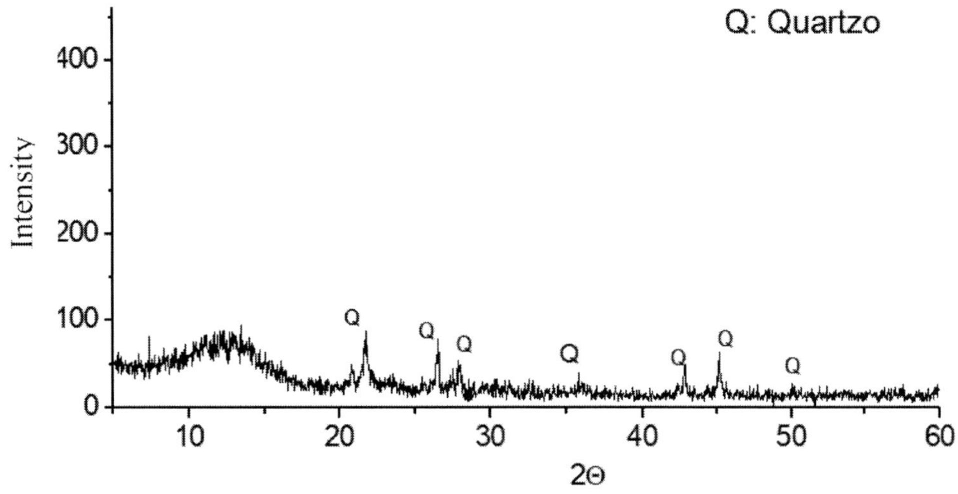

Figure 2 - X-ray diffraction of the CBC Sample

Crystalline phases of quartz are identified. The diffuse halo between the Bragg angles of 10° to 17° is indicative to the existence of amorphous material.

Figure 3 shows the infrared spectrum of the ash sample in the region between 1200 cm-1 and 400 cm-1 which is the region of interest for silica analysis, where reside the assignments for Si-O-Si bonds (Martin et al [5]).

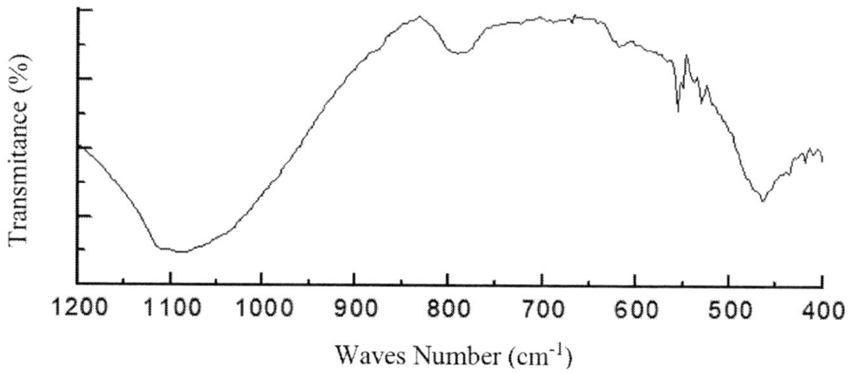

Figure 3 - Infrared spectra of the CBC Sample

Larger bands around 800 cm -1 indicate the presence of amorphous silica in the sample. The result of the thermal analysis to air presented the result shown in figure 4.

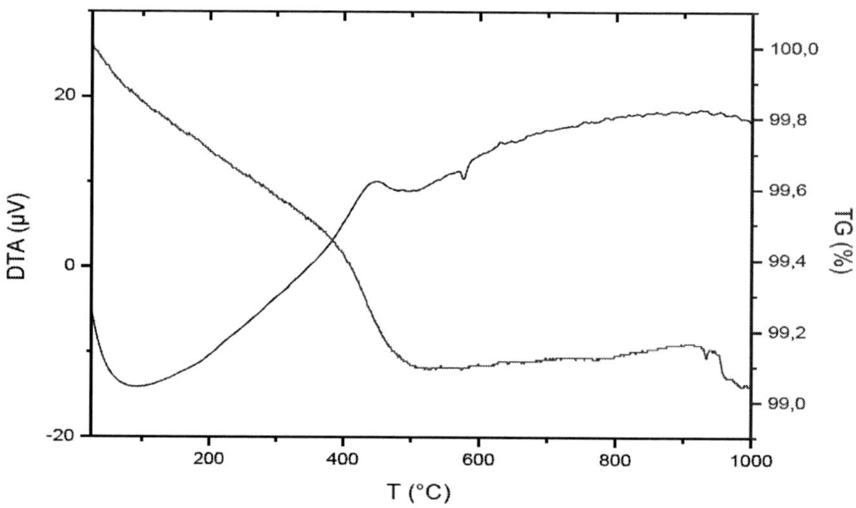

Figure 4 - TG and DTA of the CBC Sample

The low value for mass loss is justified because it is a previously calcined material. At 100°C, the endothermic peak in the DTA records the moisture loss of the sample. The exothermic peak at 420°C indicates combustion of the carbon remaining from the bagasse burning (Frias [6]). At 573°C, the peak is characteristic of the quartz alpha passage for beta quartz, indicating the presence of quartz in the sample. The Secondary Electron Scanning Electron Microscopy presented result shown in figure 5.

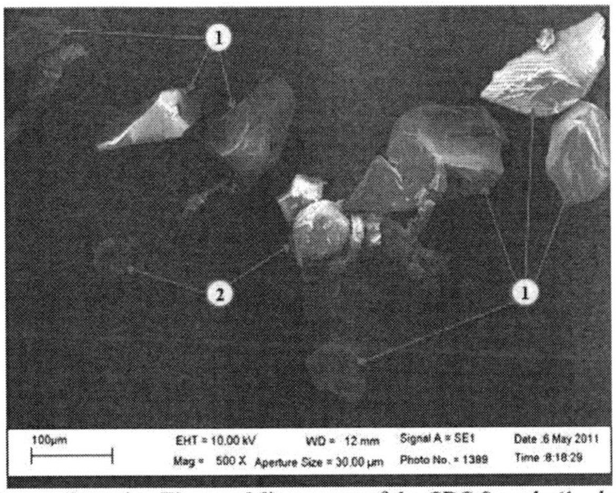

Figura 5 - Secondary Electron Scanning Electron Microscopy of the CBC Sample (1 – larger particles with sharp edges and 2 – particles of spherical shape)

The presence of particles of various shapes and sizes is observed. The larger particles have well defined edges, characteristic of fragile material, suggesting the presence of quartz in the sample (Faria [7]), being in agreement with the result presented by the other techniques of characterization. The smaller particles are porous and tend to agglomerate. The spherical particles may indicate amorphous phase in the sample.

In figure 6 shown precipitation of the silica after the extraction and neutralization processes of the CBC sample.

Figura 6 - Precipitation of the silica present in the CBC sample

After the silica extraction process, the samples were chemically characterized, providing elements to evaluate the yield of silica extracted and the result is shown in table 3.

Table 3 - Chemical composition of the samples in (%), after the extraction process

Sample	Cl	Na_2O	SiO_2	Al_2O_3	K_2O	SO_3	CaO	Fe_2O_3	Outros
CBC	50,54	45,48	2,51	0,42	0,41	0,47	0,04	0,02	0,11
META	46,92	42,13	6,18	4,50	0,1	0,03	0,08	0,04	0,02
QP	51,57	47,30	0,84	0,19	0,03	-	0,05	0,02	0,00

The higher values for chlorine and sodium are justified by the presence of the solutions of sodium hydroxide and hydrochloric acid used in the extraction process. As far as silica is concerned, the sample with the highest value was META and the smallest was the QP. The mineralogy of the CBC sample was characterized by the X-ray diffraction technique with the result shown in figure 7.

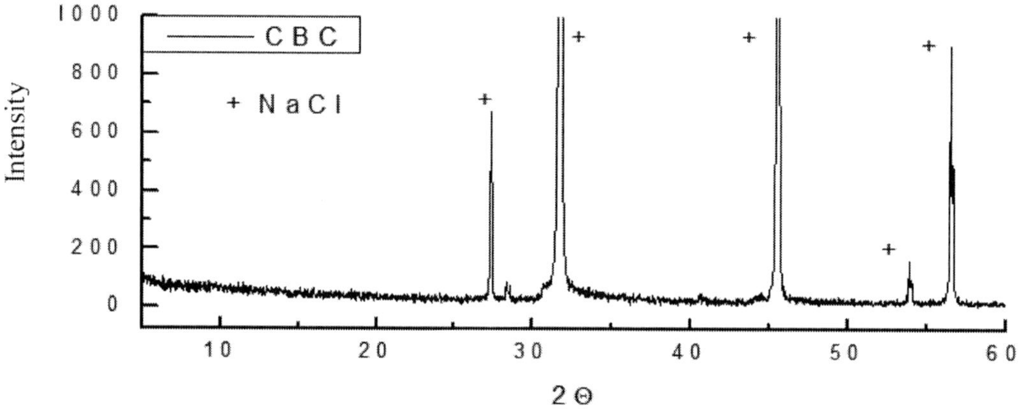

Figure 7 - X-ray diffraction of the samples after silica extraction

The absence of silica peaks testifies to the amorphous nature of the extracted silica. The yield of the extracted silica in % (g/g) was calculated using the expression:

$$R = M_2 \times C_2 / M_1 \times C_1$$

At where:

- M2 = sample mass after extraction procedure
- M1 = Sample mass before extraction
- C2 = Mass composition of the silica present in the sample after the extraction procedure
- C1 = Mass composition of the silica present in the sample before extraction
- M2 × C2 = mass of silica in the sample after extraction procedure
- M1 × C1 = mass of silica in the sample before extraction

The values of C1 and C2 for each sample were obtained by X-ray fluorescence spectroscopy and their values along with M1, M2 are presented in table 4.

Table 4 - Parameters used for silica extraction

Sample	M_1 (g)	M_2 (g)	C_1 (%)	C_2 (%)	V (ml)
CBC	3,3333	11,29	71,07	2,51	100
META	3,3333	11,22	60,86	6,18	100
QP	3,3333	11,37	98,38	0,83	100

The values obtained for R are shown in table 5.

Table 5- Yield of extracted silica (%)

Sample	Yield of extracted silica (%)
QP	2,89
CBC	11,94
META	34,17

Conclusions

- CBC is a residue rich in silica (71%).
- The silica present in the CBC sample closer to the crystalline nature than the amorphous nature.
- The methodology adopted showed that it is possible to extract silica the three samples. 2, 89% (QP), 11,94% (CBC) e 34,17% (META).
- The silica extracted from the CBC sample is in the amorphous state.
- Considering the large volume of available raw material, the values reached for the silica yield extracted from the CBC sample is significant.
- The silica yield extracted from the CBC sample can be increased by adopting the following beneficiation measures:
  - More efficient washing of sugarcane and controlled burning of bagasse; this would contribute to increase the amorphous character of the material.
  - Optimization of the grinding process aiming to increase the fineness of the material.
  - Optimization of the independent variables involved in the chemical reactions, highlighting temperature, time, molarity of the solutions and mass ratio of the sample (g) / volume of the alkaline solution (ml).

Bibliographic References

1. Única – União da Indústria da Cana-de-Açúcar, 2010.

2. Tolmasquim, M.T., et al., 2007, Matriz energética brasileira: uma prospectiva. São Paulo: Novos Estudos

3. Payá, J., Borrachero, M.V., Díaz-Pinzón, L., Ordóñez, L.M, 2002, Sugar-cane bagasse ash (SCBA): studies on its properties for reusing in concrete production. Journal of Chemical Technology and Biotechnology 77: pp. 321-325.

4. Martin, C.C., et al., 2007, Análise via espectroscopia no infravermelho da sílica proveniente da pirólise do bagaço de cana-de-açúcar. Sociedade Brasileira de Química (SBQ).

5. Bac, B. H., Song, Y., Moon, Y., Kim, M. H. e Kang, M, 2010, Effective utilization of incinerated municipal solid waste incineration ash: zeolitic material synthesis and silica extraction. Waste Management and Research.

6. Frias, M., et al., 2011, Brazilian sugarcane bagasse ashes from the cogeneration industry as active pozzolans for cement manufacture. Cement and concrete composites, v. 33, issue 4, pp. 490-496.

7. Faria, K. C. P., Gurgel, R. F., Holanda, J. N. F., 2012, Recycling of sugar cane bagasse ash waste in the production of clay bricks. Journal of environmental management, v. 101, pp. 7-12.

18th International Conference on
Non-Conventional Materials and Technologies
"Construction Materials & Technologies for Sustainability"
(18th NOCMAT 2019)
24th – 26th July 2019
Nairobi, Kenya

NATURAL FIBRES AS CEMENTITIOUS COMPOSITES REINFORCEMENT

Jerônimo Coura-Sobrinho[1], Camila Lacerda Gomes[1], Conrado de Souza Rodrigues[1], Mário Guimarães Júnior[1]

[1] Centro Federal de Educação Tecnológica de Minas Gerais (CEFET-MG), Brazil;

ABSTRACT

The ban on the use of asbestos as reinforcement of cementitious composites in some countries and the scarcity of asbestos in nature have stimulated research on the use of alternative fibres to replace that mineral. Unlike asbestos, natural fibers are eco-friendly and therefore present as a low cost alternative due to their availability. On the other hand, the production of natural fibre reinforced cementitious composites requires special care both to ensure the quality of the interface between the matrix and the reinforcement and to minimize the effects of alkaline attacks. The most tested natural fibres in composites are sisal, jute, curauá, cotton, flax, hemp, ramie, bamboo and plants such as eucalyptus and pine. Researches have assessed the performance and the durability of the composites with alternative fibres, in the search of compositions that guarantee the compatible characteristics to the applications of the cementitious composite. The chemical composition of natural fibres of vegetable origin (cellulose, hemicellulose and lignin), besides being responsible for the adhesion with the cement matrix, is also susceptible to alkaline attacks, causing the degradation of the reinforcement and consequent decrease of the durability of the composite. These factors are the major challenges for researchers in the field. This work has the objective of analyzing the effects of the addition of two reinforcements - jute mesh and microfiber of vegetal origin - in the tenacity of fiber cement. To minimize alkaline attack on the fibres, 15% of the cement was replaced by rice husk ash (RHA). In order to minimize the degradation of the reinforcements the jute mesh was submitted to bleaching in a solution of sodium hydroxide and hydrogen peroxide. The results of the mechanical tests showed that the combination of mesh jute and pinus microfiber consists of a promising reinforcement of cementitious composites.

Keywords: Asbestos-free cementitious composites. Jute fabric and cellulosic microfiber as reinforcement of fiber cement. Chemical bleaching of natural fibers.

INTRODUCTION

The prohibition of asbestos in fiber cements has motivated the use of alternative materials to be used as cementitious composite reinforcement. While the advance in asbestos-free fiber cement technology is undeniable, the current production costs are relatively high. A great advantage of the use of alternative fibers, such as vegetables or cellulosic, is the fact that they originate from renewable sources.

One of the greatest difficulties in achieving the same resistance and durability of asbestos fiber cements, with the use of other types of fibers, is the adherence of these alternative fibers with the cementitious matrix. In the case of plant fibers, the problem lies in their early deterioration, due to the alkaline medium of the cementitious matrix being quite aggressive. This aggressiveness of the environment occurs due to the natural porosity of the cementitious matrix, which favors the harmful effects of moisture on the fibers.

The composite strength is a function of variables such as the volumetric fractions of the constituents, which directly affect the mechanical properties of the composite. In other words, cementitious composites with different formulations in their constituents present a differentiated mechanical performance. These formulations include the use of pozzolanic materials in order to replace part of the cement and fibers or combination of fibers.

In this research, the objective is to evaluate the use of jute fibers combined with cellulosic microfibers as reinforcement of cementitious composites. In other words, the proposal consists in the use of a hybrid reinforcement. Although, according to Kureemun et al [1], in the literature of the composite area the term hybrid reinforcement refers to materials of different origins (cellulosic pulp and synthetic fiber, for example), in this research two types of reinforcement of plant origin were used – in form of fabric and in the form of microfibers.

Onuaguluchi and Bhantia [2] cite some strategies to improve the properties of cementitious composites reinforced by natural fibers of plant origin. One of these strategies concerns the partial substitution of the cement by silica, to reduce the alkaline content of the matrix. This substitution, combined with accelerated carbonation, has the potential to mitigate the deterioration of the fibers used in the composites. The pre-treatment of fibers is also cited as a way to protect the fibers, especially with regard to water absorption Li et al [3].

In this research, it was sought to protect the fibers by means of alkaline treatment in order to simultaneously eliminating undesirable impurities and also improving the bonding between reinforcement and matrix.

MATERIALS

The following materials were used in the composition of the samples of this research:

TABLE 1 – MATERIAILS USED IN THE SAMPLE COMPOSITION

Matrix	1) Cement CPV-ARI of the company Lafarge-Holcim.
	2) Rice husk ash (RHA) of Silcca Nobre, produced by Pilecco,
	3) Water provided by the Sanitation Company of Minas Gerais (COPASA) in Belo Horizonte city, Brazil.
	4) Limestone filler $CaCO_3$ (calcium carbonate)
Reinforcement	5) jute fiber in the form of fabric
	6) Microfiber of vegetable origin provided by Suzano Indústria de Papel e Celulose.

Half of the amount of cement was replaced by limestone filler in order to decreasing the alkalinity, permeability and capillarity of the mixture. In half of the samples, RHA was used, in the proportion of 15% of the weight relative to the cement and the filler.

METHODOLOGY

This research is applied in nature, since it seeks, based on knowledge already acquired and published about the use of alternative fibers as reinforcement of cementitious composites, applying the concept of hybridism in a way distinct from that identified in previous works. In this work we used fabric of jute fiber and pine microfiber.

As for the objectives, this research is simultaneously exploratory and explanatory. Exploratory, since it studies a new possibility of combining reinforcement in fiber cement; explanatory, because, by the experimental method, it seeks to identify the cause of the independent variable (use of RHA and alkaline treatment of the jute fibers) that has effects on the dependent variable (mechanical characteristics of the composite). Therefore, it seeks to mitigate the degradation of the fibers and, consequently, to improve the mechanical performance of the composite.

As for the procedures, this research is experimental, since it seeks to control and identify the impact of two variables on the resistance of the composite with hybrid natural reinforcement.

- Equipments used
- Bench scale (digital precision)
- Mixer
- Metallic mould
- Vacuum pump
- Digital caliper Mitutoyo Absolute
- Universal Testing Machine – EMIC model

Two groups of samples were molded: the first group had its matrix composed of cement, limestone filler and water (group called SRHA); in the composition of the second group,

15% of the cement and the limestone filler was replaced by rice husk ash (group called CRHA). Both groups of samples received as reinforcement a mesh of jute and microfiber of of vegetable origin in the proportion of 8%, taking into account its humidity, which at the time was 69%. The jute mesh received three different types of treatment, respectively A, B and C, for the removal of impurities. Therefore, for each group of samples, three subgroups (A, B and C) were formed, as a function of the chemical treatment applied to the jute fabric. Each piece was molded with an approximate size of 16x16x0,8 cm. Subsequently the pieces were divided into three parts with approximate dimensions of 4x16x0,8 cm, generating the total 90 pieces. Table 2 contains the identifications of the sample groups and subgroups.

TABLE 2 - GROUPS, SUBGROUPS AND QUANTITY OF SAMPLES

Groups	Subgroups with differentiated treatment of jute fabric			Number of pieces (approximate size 16x16x0,8 cm)	Number of pieces (approximate size 4x16x0,8 cm)
SRHA	A	B	C	15	45
CRHA	A	B	C	15	45
Total Samples				30	90

According to Table 2, 5 pieces of each subgroup were molded, totaling 30 pieces with an approximate size of 16x16x0,8 cm. Each of these 30 pieces, 7 days after the molding, was subdivided into 3 pieces with an approximate size of 4x16x0,8 cm. Consequently, 90 samples were finally obtained.

The subgroups of the samples are differentiated by the chemical treatment applied to the jut fabric. Subgroup A refers to samples pre-treated in a NaOH solution and subsequently washed in running water at room temperature. The subgroup B was chemically pre-treated in a NaOH solution, with a concentration of 5%, as recommended by Pires at al [4] at a temperature of 60°C with rotation, for 1 hour. After this procedure, the mesh was washed in running water until pH close to 7. For subgroup C, a NaOH solution with 5% (M/V) and H2O2 hydrogen peroxide 24% (V/V) was prepared, in the proportion of 1:1. Then the mesh of jute fiber was immersed for 2 hours at a temperature of 65°C under constant agitation; subsequently the mesh was washed in running water until pH close to 7. After treatments A, B and C, the meshes were dried at 80°C, for 24h.

TABLE 3 - COMPOSITION OF SAMPLES

Sample Group	Constituent materials of the samples						Treatment of Jute Mesh
	Cement (g)	Filler (g)	RHA (g)	Water (ml)	Pine Microfiber (g)	Jute Mesh	
1 - SRHA	115	115	0	900	31,10	Yes	A, B e C
2 - CRHA	97,75	97,75	34,5	900	31,10	Yes	A, B e C

According to table 3, Group 1 (SRHA) of the samples did not contain RHA in its composition, unlike Group 2 (CRHA), in which 15% of the cement and filler was replaced by 34,5g of RHA.

The mixture of cement, limestone filler, RHA (only for group 2 of the samples), microfiber and water was made in agitator, for 7 minutes. Then, one third of the mixture was inserted into a metallic mold with dimensions of 16x16 cm. At the base of the mold was placed a kind of filter to allow the passage of water and, at the same time, retain the other constituent materials of the sample. The excess water was removed by using a vacuum pump. Subsequently, the jute mesh, with the same shape size, was placed on the top of the mixture. Afterwards, the remainder of the mixture was inserted into the mold and again, the excess water was sucked by the pump. The mixture has been compacted with a metallic socket, in order to ensure uniformity in the consistency and thickness of the workpiece. Finally, excess water of the pieces was removed. This whole molding process is inspired in the manufacture of fibrocement consecrated by Austrian Ludvig Hastchek. The system called Hastchek consists in the filtration of a well-diluted cement suspension in a rotating cylinder and enables the manufacture of light cementitious plates with low consumption of materials.

The tests were performed after 21 days of molding the parts and the results are presented below.

RESULTS AND DISCUSSION

For all 90 samples, flexion assays were performed in 3 points, and the load was applied in the middle of the span of 12cm. From the data obtained in the flexural test, the rupture modules (Figure 1), the elasticity modules (Figure 2) and the specific energy (Figure 3) were calculated respectively. Of the 90 samples three were eliminated, because they presented an out-of-standards behavior (outliers).

All the following figures contain the results of the tests of the two groups of samples (CRHA and SRHA) and their subgroups (A, B and C) that contained in their composition jute mesh alkaline pre-treated in different forms.

Two main variables were considered in the following analysis: the presence or absence of RHA and the type of alkaline treatment applied to the jute mesh.

FIGURE 1 – RUPTURE MODULE

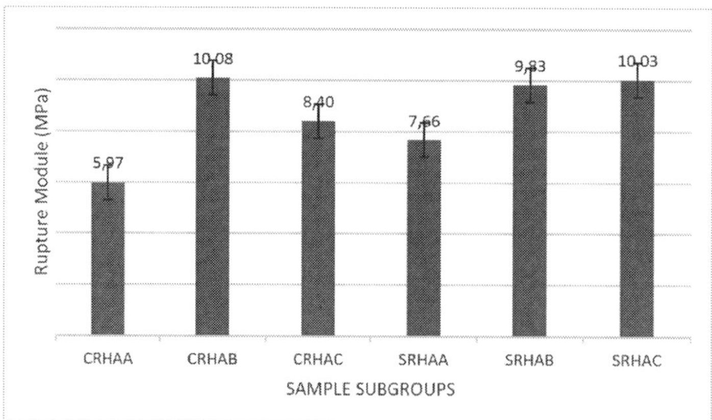

The presence of RHA in the samples in which jute mesh received chemical treatment a (only washing in NaOH solution, at room temperature) decreased the rupture modules (MOR) of the samples by 22%. On the other hand, the presence of RHA in the samples whose fiber meshes received chemical treatment B (immersion in NaOH solution, at controlled temperature of 60°C and in rotation for 1 hour) increased the MOR slightly by 2,5%. The presence of RHA in samples whose jute fabric received chemical treatment B (bleaching in NaOH and H2O2 solution at controlled temperature of 60°C and in rotation for 2 hours) decreased MOR by 16%. The MOR of the samples tends, therefore, to be greater in the subgroups in which there was no addition of RHA. Probably, the greatest effect of this addition is on the durability of the composite, a factor not analyzed in this study. It is also observed a tendency of increasing in the MOR as the treatment of the fibers passes from a simple immersion in NaOH solution (treatment A) to immersion in a solution heated to 60°C and in rotation (treatments B and C).

FIGURE 2 – TENSILE MODULUS

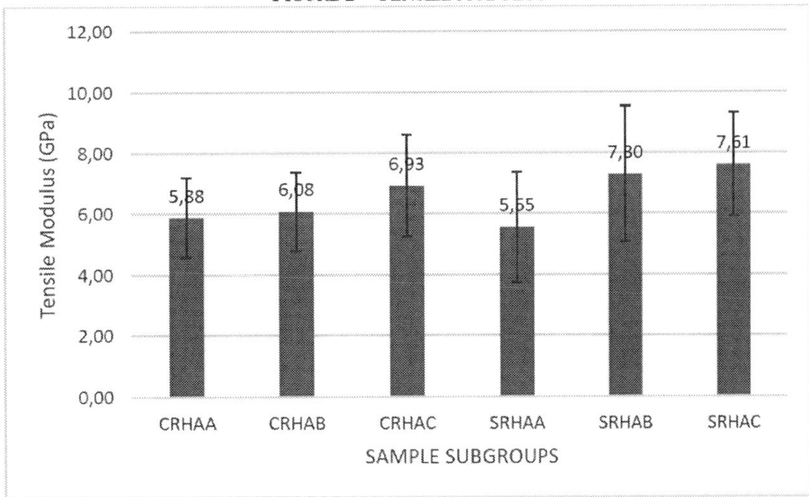

There is a tendency for the tensile modulus (MOE) of the fibrocements without RHA to be higher than those with RHA, especially in cases in which the chemical treatment of the fibers was made based on NaOH for 1 hour and in rotation (treatment B), in which an increment of 20% in the MOE was observed. When analyzing the two CRHA and SRHA groups separately, we observed an increment in the MOE as a function of the chemical treatment of jute fabric, as observed in the research by Fidelis et al [5]. In both groups, there was a gradual increase in the MOE as a function of the chemical treatment of the fibers. This fact indicates that there was an effective work of the reinforcement set (microfibers and mesh) in the resistance of the composite to the elastic deformation. In both groups, the chemical treatment of the hybrid fibers increased the rigidity of the material, in accordance with previous research by Pakravan et al [6].

FIGURE 3 – SPECIFIC ENERGY

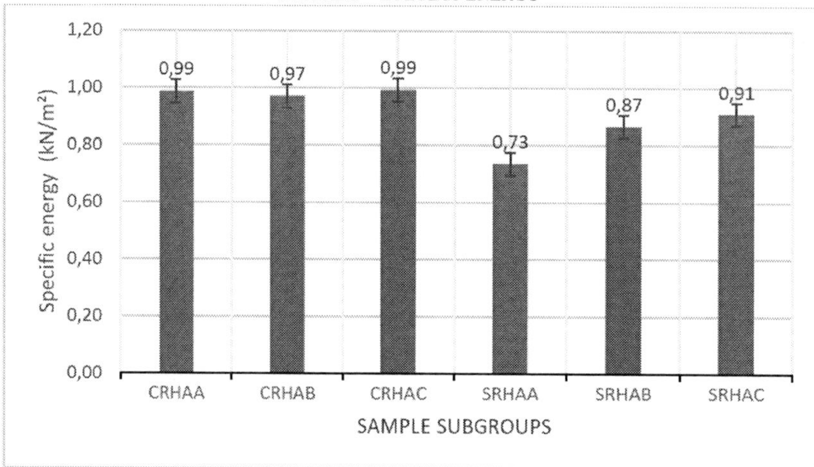

Figure 3 allows affirming that the samples with RHA presented superior ductility values for the three subgroups (A, B and C). The highest gain in ductility was observed in samples that received a chemical treatment a based on NaOH at room temperature (increase of 35%). As for treatments B and C, the increase was, respectively, 11,5% and 9%.

CONCLUSIONS

There is a tendency for the presence of RHA in composites with hybrid plant reinforcement (pine microfiber and jute fiber) to cause a decrease in the rupture modulus. There is probably an increment of reinforcement protection, which would be possible to be verified by durability tests, a factor not observed in this study. On the other hand, the type of alkaline treatment given to jute fibers tends to cause an increase in MOR, i.e., the more bleached the jute fibers, the greater the MOR of the composite.

The chemical treatment of jute fibers has a positive impact on the MOE of the cementitious composite. The chemical treatment of the fibers that increased the stiffness of the material was the treatment B (immersion in NaOH solution, for 1 hour, in a temperature of 60°C and constant rotation. The RHA tends to have no positive impact on the MOE at 21 days after molding. The need to verify the impact of the addition of RHA on the durability of the composite is emphasized.

The gain in ductility was expressive in the samples that received the addition of RHA in the proportion of 15% in substitution of cement and the limestone filler. In the case of samples without RHA, the chemical treatment of jute fibers had a direct impact on the ductility of the composite, and the chemical treatment in NaOH and H2O2 solution proved to be more effective.

REFERENCES

1. Umeyr Kureemun, M. Ravandi, L.Q.N. Tran, W.S. Teo, T.E. Tay, H.P. Lee. Effects of hybridization and hybrid fibre dispersion on the mechanical properties of woven flax-carbon epoxy at low carbon fibre volume fractions. In Composites Part B 134 (2018) 28-38.

2. Obinna Onuaguluchi e Nemkumar Banthia. Plant-based natural fibre reinforced cement composites: A review. In Cement and Concrete Composites 68 (2016) 96-108.

3. Xue Li, Lope G. Tabil, Satyanaraya Panigraphi. Chemical Treatments of natural fibre for use in natural Fiber-Reinforced Composites: A review. In Polyn Environ (2007) 15: 25-33.

4. Eduardo N. Pires, Claudia Merlini, Hazim Ali Al-Qureshi, Gean V. Salmória, Guilherme M. O. Barra. Efeito do Tratamento Alcalino de Fibras de Juta no Comportamento Mecânico de Compósitos de Matriz Epóxi. In Polímeros, vol. 22, n. 4, p. 339-344, 2012.

5. M.E.A. Fidelis, F.A. Silva, R.D. Toledo Filho, The influence of fiber treatment on the mechanical behavior of jute textile reinforced concrete. In Eng. Mater. 600 (2014) 469–474.

6. H. R. Pakravan; M. Latifi, J. Jamshidi. Hybrid short fiber reinforcement system in concrete: A review. In Construction and Building Materials 142 (2017) 280-294

IC-NOCMAT 2019

18th International Conference on
Non-Conventional Materials and Technologies
"Construction Materials & Technologies for Sustainability"
(18th NOCMAT 2019)
24th – 26th July 2019
Nairobi, Kenya

CHEMICAL COMPOSITION AND SELECTED PHYSICAL CHARACTERISTICS OF NEEM SEED HUSK ASH BLENDED CEMENT

A. A. Raheem[1], B. D. Ikotun[1], E. O Ibiwoye[2]

[1] Department of Civil and Chemical Engineering, University of South Africa;

[2] Department of Civil Engineering, Ladoke Akintola University of Technology (LAUTECH) Ogbomoso, Nigeria;

ABSTRACT

Neem tree (Azadirachtaindica) is a tropical evergreen tree scattered all over Nigeria. Production of neem oil from the seed generates large quantity of residue annually which constitutes a waste disposal problem. The residue can be converted into Neem Seed Husk Ash (NSHA) and used as partial substitution for cement in the production of concrete. This study investigated the characteristics of NSHA blended cement produced in the factory. Neem fruits were collected from Olooru town and Ilorin metropolis in Kwara State Nigeria and the seeds were burnt in a furnace at a temperature of 650 ± 50 oC and the chemical composition (SiO2, Al2O3 and Fe2O3) determined. Blended cement was produced by intergrinding 5, 10, 15, 20, 25, 30, 35, 40, 45 and 50% by weight of NSHA with Ordinary Portland Cement (OPC) clinker and gypsum. The blended cements produced were analyzed for chemical composition (SiO2, Al2O3, Fe2O3, CaO) and physical properties (fineness, initial and final setting time, heat of hydration). The total percentage of SiO2, Al2O3 and Fe2O3 for NSHA is 75.35%. The percentage of SiO2, Al2O3, Fe2O3 and CaO for blended cement with 5% - 50% ash replacements ranged from 20.21 – 20.81, 5.41 – 3.86, 3.21 – 1.92 and 62. 76 – 42.48, respectively. The corresponding fineness (m2/kg), initial and final setting time (min) and heat of hydration (cal/g) for the blended cement ranged 230 – 740, 45 – 72, 135 – 235, and 34 - 49, respectively. Factory produced NSHA – blended cement is suitable as it satisfied maximum requirements.

Keywords: NSHA, Chemical composition, Physical characteristics, Setting times, Heat of hydration.

INTRODUCTION

Growth in human population over the last five decades has doubled from 3 to 6 billion and it is expected to grow more in years to come [1]. According to Olusola and Adesanya [2], it is of great importance to make provision for availability and accessibility to building materials and decent housing for the teaming populace of the world, especially those in developing countries. Large quantity of materials is needed for the construction of infrastructures to cater for the increased population and human needs. Concrete is the most commonly used material in construction works. One of the constituent materials in making concrete is cement. The demand for cement is increasing as a result of rapid population growth. Thus, effort to find alternative materials which must be inexpensive, readily available in large quantity and require little energy to produce has to be undertaken. Nowadays, the current trend is to use industrial by – products or agricultural residue ashes that exhibit pozzolanic properties, to partially replace cement in the production of concrete.

According to Day [3], the usage of pozzolanic materials, some of which are industrial and agricultural by-products, preserves natural resources that are employed in the production of cement. Researchers have shown that rice husk ash [4-7]; corn cob ash [8, 9]; oil palm residue ash [10]; saw dust ash [11, 12]; groundnut husk ash [13, 14] and bagasse ash [15, 16] are pozzolanic in nature with high amount of silica in amorphous form. They further recorded that their inclusion in concrete enhanced the properties of blended cement concrete. However, there is paucity of literature on the use of Neem Seed Hush Ash (NSHA) as a pozzolanic material

Neem Seed Husk is a residue that is obtained during industrial processing of Neem Seed to extract oil and produce fertilizer [17]. Neem seeds are obtained from Neem tree (Azadirachta indica) which is a tropical evergreen tree scattered all over Nigeria. Neem tree is readily available in most temperate environment and is locally called "dongoyaro" in one of the Nigeria dialect (Yoruba), but it is predominantly present in the northern part of Nigeria and some parts of the South – West. NSHA is produced by burning the Neem seed husk which comes from the Neem tree. This study investigated the chemical composition and selected physical characteristics of NSHA blended cement produced in the controlled environment of a cement manufacturing company.

MATERIALS AND METHODS

Different materials were used in the production of NSHA – blended cement. These materials are discussed in detail highlighting their sources, usage and method of production.

Neem Seed Husk

The neem seed husk that was used in this study was collected from Neem Seed Pod, obtained from Neem trees at various locations in Olooru town and Ilorin metropolis, Kwara state, Nigeria. The The seed was removed with the use of Decorticator huskers, after which it was sun dried to reduce its moisture content and calcinated in a muffle furnace at temperature of 650 ± 50 oC to produce NSHA. The Neem tree, fruit, seed and husk are as shown in Figure 1.

Clinker and Gypsum

Ordinary Portland cement clinker (OPC clinker) and Gypsum used for this study was obtained from Lafarge, West Africa Portland Cement Company (WAPCO), Sagamu Plant, Ogun State, Nigeria. The gypsum was in granular form with particle size range of 2 to 8 mm which is suitable to achieve adequate blending of the cement [18].

Chemical Composition of NSHA and OPC-clinker

The determination of chemical composition of NSHA and OPC clinker was carried out at WAPCO, Ewekoro Plant in Ogun State using X-ray flourescence spectometer (Model: QS 1279) as specified in BS EN 196-2 [19].

Production of NSHA Blended cement

The production of NSHA blended cement was carried out at Lafarge, (WAPCO) Sagamu Plant, Ogun State, Nigeria. The production process comprised batching, preparation of cement mixture and milling process. The NSHA blended cement was produced by iintergrinding Portland cement clinker with 5 % - 50 % NSHA replacement during the cement manufacturing process.

RESULTS AND DISCUSSION

The results obtained from the study are as presented below.

Chemical Composition of Neem Seed Husk Ash (NSHA)

Table 1 shows the average percentage composition of each element present in NSHA. The main constituents are Silicon Oxide (SiO2) Aluminum Oxide (Al2O3) and Iron Oxide (Fe2O3). The combined percentages of SiO2, Al2O3 and Fe2O3 is 75.34 % which is above 70 % required by ASTM C 618[20] for a material to be suitable for use as a pozzolan. This result is similar to those of other pozzolans such as corn cob ash (CCA) and Rice husk ash (RHA) [21, 22].

a: Neem Tree
(indent: neem tree flowers)

b: Neem Fruit

c: Neem Seed

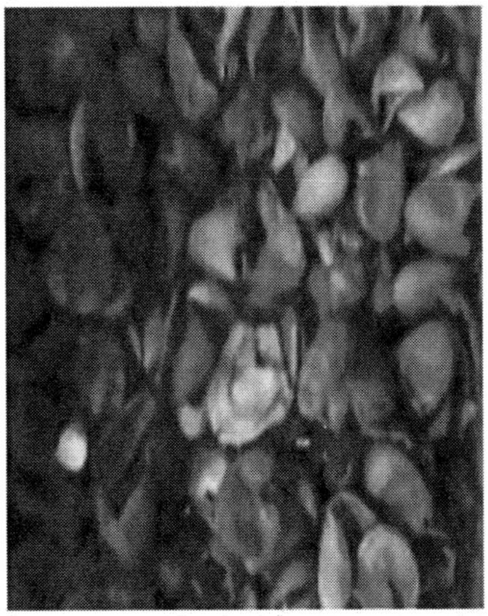

d: Neem Husk

FIGURE 1: NEEM TREE, POD, SEED, AND HUSK

TABLE 1- CHEMICAL PROPERTIES OF NEEM SEED HUSK ASH (NSHA)

Constituents	Sample 1 (%)	Sample 2 (%)	Sample 3 (%)	Average (%)
SiO_2	69.14	69.13	69.16	69.14
Al_2O_3	2.94	2.96	2.94	2.95
Fe_2O_3	3.25	3.24	3.27	3.25
CaO	2.75	2.77	2.76	2.76
MgO	0.54	0.56	0.55	0.55
SO_3	0.42	0.43	0.41	0.42
Na_2O	0.15	0.14	0.17	0.15
K_2O	15.01	15.02	15.04	15.02
TiO_2	0.24	0.26	0.24	0.25
P_2O_5	1.01	1.01	1.01	1.01
Mn_2O_3	0.04	0.04	0.05	0.04
$SiO_2+Al_2O_3+Fe_2O_3$	75.33	75.33	75.37	75.34

Chemical Composition of OPC – Clinker

Table 2 is the chemical composition of clinker used in the production of NSHA blended cements. The result shows that the mean value of silica (SiO2) and calcium oxide (CaO) contents of OPC-clinker samples were 20.30 % and 65.90 % respectively. These results show a slight reduction in SiO2 and CaO compared to the values of 22.56% and 66.54%) obtained by [21] for SiO2 and CaO respectively. The minimum ratio of 2:1 for CaO: SiO2 specified by BS EN 197-1:2000 [23] was satisfied by the OPC – clinker used.

The average values of Al2O3 (5.47%) and Fe2O3 (3.47%) in OPC-clinker used were slightly higher than those obtained by [9] which are (4.40% and 3.04%). The maximum contents of magnesium oxide (MgO) of 5.00% specified by BS 197 -1:2000 was satisfied by the OPC-clinker which has 2.02% MgO.

TABLE 2: CHEMICAL COMPOSITION OF OPC - CLINKER

Constituents	Sample 1 (%)	Sample 2 (%)	Sample 3 (%)	Average (%)
SiO_2	20.22	20.25	20.45	20.30
Al_2O_3	5.30	5.45	5.65	5.47
Fe_2O_3	3.43	3.26	3.72	3.47
CaO	66.00	64.85	66.85	65.90
MgO	2.28	1.98	1.80	2.02
SO_3	2.14	2.14	2.23	2.17
Na_2O	0.15	0.15	0.15	0.15
K_2O	0.29	0.32	0.23	0.28
P_2O_5	0.17	0.13	0.21	0.17
LOI	1.33	1.30	1.30	1.31
LSF	95.93	95.73	96.04	95.90
AR	1.55	1.64	1.45	1.55
SR	2.23	2.35	2.17	2.25
Free lime	1.11	1.05	0.94	1.03

Other minor oxides contents of SO3, K2O, Na2O, P2O5 are in the range of 0.17% – 2.17%. All the composition satisfied the requirements in ASTM C 150:2002 and NIS 439:2000. Thus, the clinker is suitable for cement production.

Chemical Composition of NSHA blended cement

The chemical compositions of NSHA blended cements produced are presented in Table 3. The silica content of the blended cement increased from 20.21% for 5% replacement to 20.81% for

50 % replacement. However, there was a different behavior with alumina (Al2O3) contents which decreased in percentage content from 5.43 for the control to 3.86 for 50% NSHA replacement.

TABLE 3- CHEMICAL COMPOSITION OF NSHA BLENDED CEMENT

Constituents	Percentage Composition (%)										
NSHA	0	5	10	15	20	25	30	35	40	45	50
SiO_2	20.17	20.21	20.23	20.27	20.32	20.36	20.42	20.48	20.57	20.79	20.81
Al_2O_3	5.43	5.41	5.32	5.22	5.01	4.78	4.65	4.53	4.51	4.32	3.86
Fe_2O_3	3.51	3.21	3.07	2.84	2.54	2.42	2.31	2.27	2.17	2.08	1.92
CaO	63.57	62.76	60.56	57.87	55.78	51.32	49.45	47.65	45.45	44.32	42.48
MgO	2.06	2.33	2.54	2.63	2.72	2.86	2.97	3.05	3.12	3.21	3.37
SO_3	2.21	2.34	2.44	2.51	2.74	2.82	2.85	2.89	2.92	3.08	3.11
Na_2O	0.15	0.12	0.12	0.06	0.04	0.03	0.02	0.02	0.01	0.02	0.03
K_2O	0.30	0.42	0.64	0.76	0.94	1.02	1.50	1.89	2.00	2.84	3.11
P_2O_5	0.11	0.16	0.19	0.21	0.24	0.27	0.32	0.58	0.62	0.69	0.73
Free lime	1.47	1.45	1.43	1.41	1.39	1.37	1.35	1.33	1.32	1.30	1.28
LSF	0.95	0.92	0.94	0.92	0.89	0.86	0.83	0.77	0.75	0.70	0.69
AR	1.70	1.63	1.62	1.65	1.63	1.53	1.55	1.61	1.66	1.62	1.64
SR	2.21	2.49	2.62	2.74	2.79	2.96	3.00	3.18	3.32	3.45	3.87
C_3S	65.79	67.00	65.71	68.02	68.87	70.02	71.31	72.11	73.02	72.22	73.01
C_3A	6.50	6.21	6.00	5.65	6.32	6.08	5.97	5.31	5.57	4.10	5.55
C_4AF	9.05	8.53	8.32	8.21	8.03	7.65	7.54	7.31	6.66	6.32	5.75

Ferric oxide (Fe2O3) content has a similar behavior with Alumina as it decreased from 3.51% to 1.92%. Lea (2005) reported that cement with high percentage content of Fe2O3 is responsible for impacting colour on cement product. All the NSHA blended cements produced satisfied the requirements for Type II and Type IIA Portland cement specified by ASTM C 150: 2000 [24].

Furthermore, all the NSHA blended cement produced have a lower percentage of calcium oxide (CaO) than the control and decreased from 5% to 50% replacement. These results are in consonance with the results reported by [9] showing a decrease in the values of calcium oxide from 2% to 25%. Calcium oxide is responsible for the formation of C3S (3CaO.SiO2) and C2S (2CaO.SiO2) in cement as well as all other clinker minerals which are also responsible for the strength in the early or later age of the concrete [25]. With increase in the percentage of NSHA replacement, more silicate and less alumina, ferric and calcium oxide are present in the blended cement to react with the lime produced during hydration of cement to produce additional cementitious products.

It could be observed from Table 3 that the magnesium oxide (MgO) contents of the NSHA blended cement is in the range of 2.33 to 3.37. These values are higher than that of the control (2.06). This is in line with the findings of [9, 25]. Due to the adverse effect of magnesia on cement's soundness, high content of magnesia in cement is not desirable. All the blended cement produced satisfied the maximum of 6.00% MgO content requirement for both Type II and Type IIA Portland cement specified by ASTM C 150:2000 [24].

The percentage of K2O for all the NSHA blended cements ranges from 0.42 to 3.11 and are higher than that of the control (0.30). The values are higher than those reported by [9] which ranged between 0.36% and 0.83% for 2% and 25% CCA replacement. The Na2O are lower ranging between 0.12% and 0.03% for 5% to 50% NSHA replacement, respectively. Both K2O and Na2O are known as alkalis. Higher alkalis in cement can cause damage to kiln, attack reinforced concrete [25] and enhanced alkali-silicates reaction [26].

The SO3 present in all the blended cement as presented in Table 3 are higher than that of the control and ranges from 2.34 to 3.11. This agrees with [9] which reported that SO3 content of CCA blended cements to be higher than the control. According to [27], high SO3 content in cement is from gypsum, which is one of the raw materials used in producing cements for controlling setting time. Neville [28], stated that cement produced with high content of SO3 are acid resistant while on the other hand, Taylor [25] opined that small quantity of SO3 present in cement prevents the formation of Tricalcium aluminate (C3A). Only blended cement with up to 40% NSHA replacement satisfied the maximum of 3.00% SO3 content required for Type II and Type IIA Portland cement specified by ASTM C 150:2000 [24].

The lime saturated factor (LSF) for NSHA blended cements with 5% – 50% ash replacements ranged from 0.92 to 0.69 making it lower than that of the control (0.95) indicating that as the percentage replacement of the ash is increasing, the LSF content of the blended cements decreases. The free lime content also decreases as the percentage replacements of ash increases. This is in line with the findings of [9] which reported a decrease in LSF for CCA blended cement. The soundness of cement will be enhanced due to the reduction of the free lime content. Consequently, all the blended cements produced satisfied the limiting value of 2 – 3% of free lime content specified by NIS 439: 2000 [29].

The silica ratio (SR) for all blended cements produced with 5 to 50% ash replacements ranged between 2.49 and 3.87. This corresponded with Adesanya and Raheem [9]) which reported increase in SR content for CCA blended cements from 2.69 to 2.96 between 2 – 25% CCA replacement. Burning becomes extremely easy for silica ratio below 2.1 but excessive liquid phase and low strength cement is obtained. All blended cements produced satisfied the recommended limit value of 5% by NIS 439: 2000 [29].

It can thus be affirmed that most of the NSHA blended cements produced satisfied the limiting values recommended by available standards in terms of their chemical compositions.

Fineness

The mean value of fineness of all NSHA blended cements is shown in Figure 2. The fineness ranged from 285 m2/kg for 5% to 740 m2/kg for 50% NSHA replacement. The fineness increases with increase in NSHA percentage replacement. This could be attributed to the lower density of NSHA compared to OPC – clinker and gypsum. All the blended cements produced satisfied the 250 m2/kg and 280 m2/kg minimum fineness specified by NIS 439:2000[29] and ASTM C 150: 2000 [24], respectively.

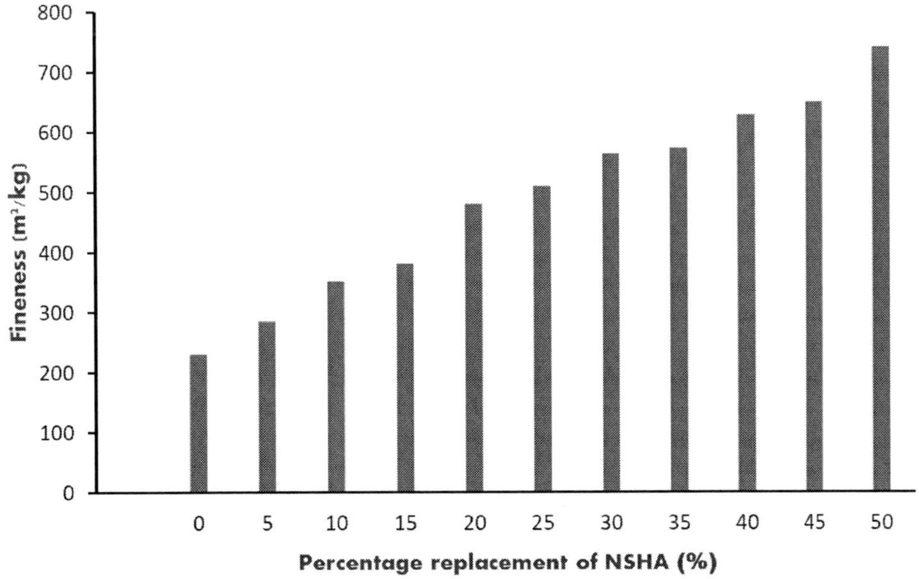

FIGURE 2: EFFECT OF NSHA REPLACEMENT ON FINENESS OF BLENDED CEMENTS

Setting times

Figure 3 shows the initial and final setting times for NSHA blended cements. It could be observed rom the figure that the initial and final setting time of NSHA blended cement increased from 45 to 72 min and 135 to 235 min, respectively, when the percentage replacement of NSHA increased from 5% to 50%. This agrees with the findings of [9, 30]. All the blended cements satisfied the NIS 439:2000 [29] requirement of 45 min minimum initial setting time and maximum of 10 hours final setting time. The variation of setting time with percentage replacement of NSHA blended cements shows that both the initial and final setting time increased as their NSHA contents increased, hence, the hydration process is slowed down. Pozzolanic materials exhibit low rate of hydration and heat development of concrete, which make them of great importance in mass concrete construction as it reduces thermal stresses. The effect of the NSHA on the cement paste setting times can be attributed to the slower rate of reaction and the reduced quantity of heat liberated, which eventually lead to late stiffening of the paste. However, all the

NSHA blended cement satisfied the maximum requirement of 375 min final setting time by ASTM C150:2002 [24].

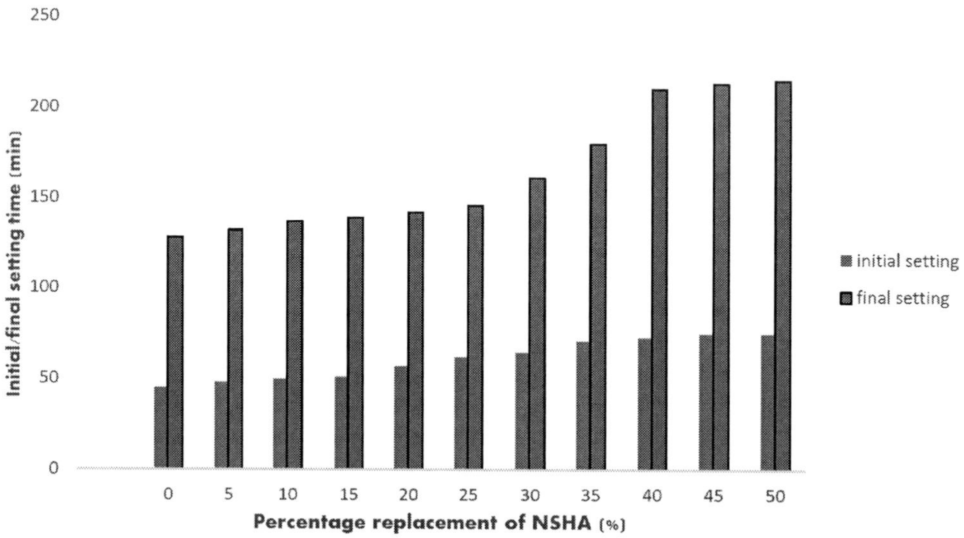

FIGURE 3: EFFECT OF NSHA REPLACEMENT ON INITIAL AND FINAL SETTING TIMES OF BLENDED CEMENT

Heat of hydration

The heat of hydration of NSHA blended cements for six-hour duration is as presented in Figure 4. It was discovered that the temperature of the control sample increased consistently while all others showed constant temperature for some hours before showing slight changes in temperature. This indicates that the control sample generated high heat of hydration at early stage while all other NSHA blended cements produced low heat of hydration. It was also discovered that the higher the percentage replacement of NSHA, the lower the heat of hydration of the blended cement. This is in line with the findings of [9]. Since heat produced by NSHA blended cement reduces with increase in ash replacement, it is most suitable for use in the production of mass concrete works.

CONCLUSIONS

From the findings in this study, the following conclusions were made.

(i). NSHA is a suitable material for use as a pozzolan, since it satisfied the requirement for such a material by having a combined SiO2, Fe2O3 and Al2O3 of more than 70%.

(ii). The addition of NSHA in blended cement increased marginally the percentage of SiO2 and FeO3 and decreased slightly that of Al2O3 and CaO.

(iii). NSHA blended cement has higher setting times than OPC cement, hence, it is most applicable where low rate of heat development is required such as in mass concrete work and dam construction.

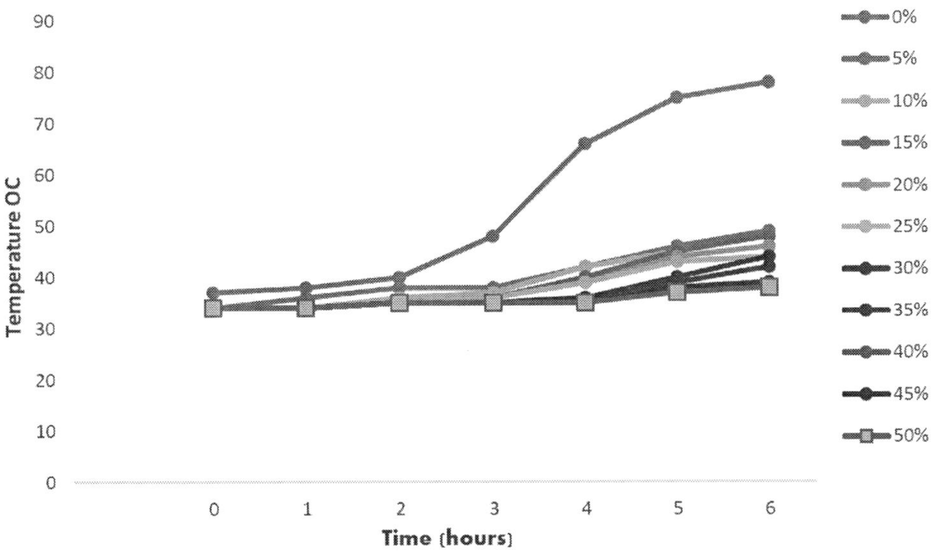

FIGURE 4: HEAT OF HYDRATION OF NSHA BLENDED CEMENT

REFERENCES

1. Ngab, AS. Concrete technology in developing world, Proceedings, 7th International Conference on Concrete Technology in Developing Countries. Kuala Lumpur, 2002. p. 19 – 32.

2. Olusola K, Adesanya DA. Public acceptability and evaluation of local building material housing construction in Nigeria. Build Environment 2004; 1: 83-89.

3. Day LR. Pozzolans, for use in Low- Cost Housing, A State-of-the-Art Report prepared for The International Development Research Centre Ottawa, Canada. 1990.

4. Bui DD, Stroeven P. Particle size effect on the strength of rice husk ash blended gap graded Portland cement concrete, Cem. Concr. Comp. 2005; 27: 357-366.

5. Chindasprasirt P, Rukzon S, Sirivivatnanon V. Resistance to chloride penetration of blended portland cement mortar containing palm oil fuel ash, rice husk ash and fly ash, Construct. Build Mater. 2008; 22 (5): 932-936.

6. Gastaldini AA, Isaia NS, Gomes GC, Sperb JEK. Chloride penetration and carbonation in concrete with rice husk ash and chemical activators, Cem. Concr. Comp. 2007; 25 (1): 176-180.

7. Nedhi M, Duquette A, Damatty EI. Performance of rice husk ash produced using a new technology as a mineral admixture, Cem. Concr. Res. 2003; 33: 1203-1210.

8. Adesanya DA. Evaluation of blended cement mortar, concrete and stabilized earth made from ordinary Portland cement and corn cob ash. Construct. Build. Mater. 1996; 10 (6): 451-6.

9. Adesanya DA, Raheem AA. Development of corn cob ash blended cement, Construct and Build. Mater. 2009a; 23: 347-352.

10. Nurdeen MA, Megat AM, Syed FS. Influence of calcination temperature on characteristics and pozzolanic activity of palm oil waste ash, Australian J Basic and Applied Sci. 2011; 5 (11): 1010-1018.

11.Marthong C. Sawdust Ash (SDA) as partial replacement of cement. Int. J. Eng. Res. and Applications. 2012; 2: 1980-1985.

12. Udoeye FF, Deshibi PU. Sawdust ash as concrete material, J. Mater in Civil Eng.2002; 14 (2): (2): 173 –17 6.

13. Elinwa AU, Awari A. Groundnut-husk ash concrete. Nigeria J Eng and Mater. 2001; 2 (1): 8-15.

14. Manasseh JA. Review of partial replacement of cement with some agro wastes, Nigeria J Technol. 2010; 29 (2): 12-20.

15. Cordeiro TS. Pozzolanic relativity of sugarcane bagasse ash. J. of Applied Sci. 2006; 7 (6): 204-217.

16. Singh NB, Singh VD. Hydration of baggasse ash blended portland cement, Cem. Concr. Res. 2000; 30: 1485-1488.

17. Raheem AA, Ibiwoye E O. A Study of Neem Seed Ash as Partial Replacement for Cement in Concrete. International Journal of Sustainable Construction, Engineering & Technology 2018; 9 (2): 55-65.

18. Shetty MS. Concrete Technology, (Revised Edition), Ram Nagar, New Delhi, S. Chand, 2005.

19. BS EN 196 - 2. Method of testing Cement – 2: Chemical analysis of Cement, British Standard Institute, 389 Chiswick high Road, London, W4 4AL, 1995.

20. ASTM C 618. Standard Specifications for Fly Ash and other pozzolans for use with lime ASTM C593 – 95.West Conshohocken P.A American Society for Testing and Materials, 1994.

21. Adesanya DA, Raheem AA. A study of the workability and compressive strength characteristics of corn cob ash blended cement concrete. Construction and Building Materials 2009b; 23 (1): 311 – 317.

22. Fetra, VR. Application of RHA'S Pozzolanic Properties in the making of CEB. International Journal of Sustainable Construction Engineering and Technology 2011; 2(2): 32 – 36.

23. BS EN 197-1. Composition, Specification and Conformity Critaria for Common Cements. British Standard Institute, 389 Chiswick high Road, London, W4 4AL, 2000.

24. ASTM C 150. Standard Specification for Portland Cement. Annual book of ASTM Standards (Revised Edition). 4(2), 1 – 7, 2002.

25. Taylor HF. The Chemistry of Cements. Vol. 1. Academic Press, New York, 2004.

26. Dabai MU, Muhammad C, Bagudo BU, Musa A. Study on the effect of Rice Husk Ash as cement Admixture. Nigeria Journals of Basic and Applied Sciences 2009; 17 (2): 252 - 256.

27. Singh M. Influence of Blended Gypsum on the properties of Portland cement and Portland Slag cement. Cement and Concrete Research 2000; 30: 1185 – 1188.

28. Neville AM. Concrete Technology, Delhi, India: Pearson Education (Singapore) Pte Ltd. Indian Branch, Patparganj, 2003.

29. NIS 439. Standard for Cement. Lagos Nigeria: Standard Organisation of Nigeria, 2000.

18th International Conference on
Non-Conventional Materials and Technologies
"Construction Materials & Technologies for Sustainability"
(18th NOCMAT 2019)
24th – 26th July 2019
Nairobi, Kenya

INNOVATIVE SANDWICH PANELS MADE OF WOOD BIO-CONCRETES AND SISAL REINFORCED CEMENT COMPOSITES

M'hamed Y. R. Da Gloria[1], Daniele O. J. dos Santos[1], Vanessa M. Andreola[1], Romildo D. Toledo Filho[1]

[1] Department of Civil Engineering, COPPE, University Federal of Rio de Janeiro, University City, RJ, Brazil;

ABSTRACT

In this study, sandwich panels were produced with two lightweight cores containing wood bioconcretes, and facings in laminate mortar reinforced with long sisal fibers. To produce the core, two cement-to-wood mass ratios of 0.5 and 2.5 were used to analyze the behavior of the panels, with high and low wood shavings content. The facings were produced with 6% of long sisal fibers reinforcing a cementitious mortar free of calcium hydroxide. The density and longitudinal compressive behavior of the sandwich panels were accessed. The results showed that the panels could be considered as lightweight for presenting densities lower than 1500 kg/m3. Besides, the panel with low wood shavings content was more resistant and presented a better bond between the core and the laminates. Globally, the bio-based sandwich panels demonstrated technical potentiality for multi-use in wall system, and offered new range of use of vegetables in cementitious matrices.

Keywords: Sandwich Panels, Cement, Wood Shavings, Long Sisal Fibers, Wood BioConcretes.

Introduction

The development of a new construction material requires the study of many factors such as mechanical strength, durability, energy consumption, applicability and interaction with the environment. These factors can directly influence on the total production and production cost of the buildings. In this context, the construction industry must upgrade its constructive practices for being the first consumer of natural resources and the highest wastes producer.

The, practices that aim to maximize the performance of new constructions as well as offer innovative materials that meet the new needs of users should be one of the starting points of the construction industry. In the group of innovative materials, we can find the lignocellulosic composites that are composed of aggregates/fibers of vegetable origin and bound with synthetic or mineral binder.

Some materials were already developed and marketed on a large scale such as hemp concrete in France, produced from the mixture of hemp particles and lime. Out of being ecological, this concrete presents low density (500 kg/m^3) and a low heat conduction capacity [1]. In Brazil, the wood residues are abundant (around 80 millions of m3/year) and can be a good alternative for having composite materials similar to hemp concrete.

Beyond the wood, previous works [2] [3] [4] [5] highlighted promising results of the sisal fiber as reinforcement of cementitious composites of high performance.

An interesting proposal for the use of these vegetal residues is the combination of wood and sisal composites in a sandwich structure, consisting of a core containing wood bio-concrete, and the facings in cementitious laminates reinforced with long fibers of sisal.

Materials and Methods

The proposed sandwich panels are composed by two facings of long sisal fiber-cement composites and a core of wood bioconcrete.

Long Sisal Fiber-Cement Composites (LSIFCC)

The LSIFCC were produced using a matrix made of free hydroxide calcium mortar (Figure 1) with compressive strength at 28 days of 26 MPa. The binders used were the Brazilian Portland Cement CPII E-32, Metakaolin and Fly Ash, in the proportion of 33%, 27% and 40%, respectively, as developed by [6]. Quartz river sand with nominal diameter lower than 840 μm was used, and the mix design in mass was 1: 0.5: 0.4 (binder: sand: water). In order to guarantee a good workability, a third generation superplasticizer Glenium 51 with 33% of solids was used in the proportion of 1.72% of the total binder weight. The mix design is summarized in Table 1.

Based on the results obtained by [4], 6% of long sisal fibers were used as unidirectional reinforcement of the matrix. The characteristics of the sisal fibers are listed in Table 2. Before being used in the matrix, the fibers were washed in hot water (80ºC) for one hour, dried and combed.

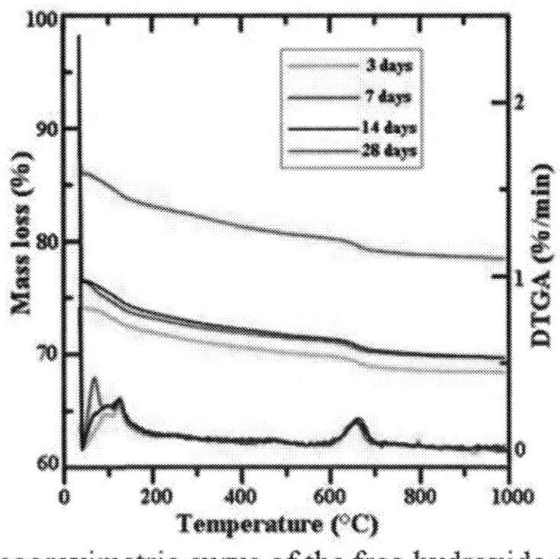

Figure 1: Termogravimetric curve of the free hydroxide calcium matrix

Table 1: Matrix mix design (kg/m^3)

Materials	Cement	Metakaolin	Fly Ash	Sand	Water	Glenium
Content	362	296	439	439	548.5	18.95

Table 2: Characteristics of the sisal fiber

Origin	Density	Tensile strength	Young Modulus
Bahia (Brazil)	900 kg/m^3	447 MPa	19 GPa

Wood Bio-Concretes

The WBC used in the current study were developed in a previous work [7]. From six WBC produced by the authors, two were used in the current study as cores of the sandwich panels: WBC 0.5 and WBC 2.5. Their mix design and characteristics are resumed in Table and Table , respectively [7].

The wood shavings (WS) were derived from woodworking activities of carpentry in Rio de Janeiro (Brazil). Four types were collected in random proportions: Manilkara salzmanni, Hymenolobium petraeum, Cedrela fissilis and Erisma uncinatum Warm. After a uniform mix of the WS, the particles passing through the 1.18 mm sieve were removed in order to avoid high water absorption. The general aspect and the granulometric curve of the WS are shown in Figure 2. The apparent density and the moisture content of the WS were 760 kg/m3 and 13.66%, respectively.

a)

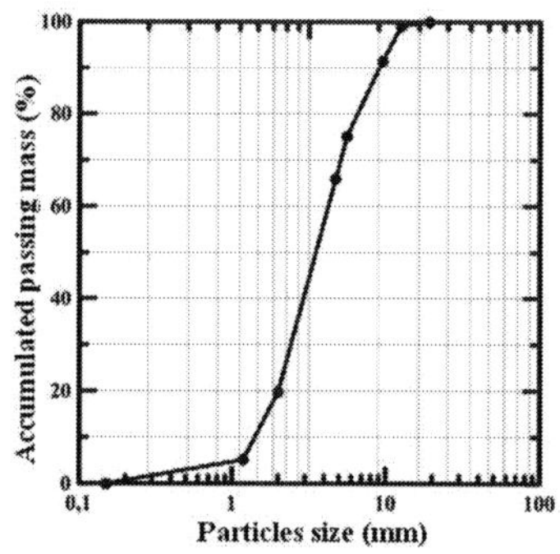

b)

Figure 2: WS a) General aspect b) Granulometric curve

The Brazilian Portland cement CP II E-32 was used as binder and the calcium chloride as setting accelerator. Two WBC were chosen as cores of the sandwich panels: WBC 0.5 and WBC 2.5. Their mix design and characteristics are resumed in Table 3 and Table 4, respectively [7].

Table 3: Mix design of the WBC based on the cement weight

	Cement	WS	Water	$CaCl_2$
WBC 0.5	1	2	0.45	0.03
WBC 2.5	1	0.40	0.40	0.03

Table 4: Density, thermal conductivity and compressive strength of the WBC

	Density (kg/m^3)	Thermal conductivity (W/mK)	fc_{28} (MPa)	Young modulus (GPa)
WBC 0.5	700	0.06	0.44	0.029
WBC 2.5	1100	0.46	9.45	2.47

Panels Manufacturing

The sandwich panels were produced with dimensions of 900 x 250 x 100 mm^3. The nominal thicknesses of each facing and the core were 10 mm and 80 mm, respectively. The casting was done through a hand layup technique, in the following sequence: inferior facing, core and superior facing.

The inferior facings were molded with 6% of long sisal fibers, divided in three layers interspersing four equal layers of mortar (Figure 3 a, b, c). A first layer of mortar was introduced in the molds, followed by one layer of fibers slightly pressed to ensure their good penetration into the mortar. Then, the second and third layers of mortar and fibers were introduced, and the cast ended with the fourth mortar layer.

To produce the WBC, the calcium chloride was dissolved first in water; next, the cement and the water were introduced into the mixer and mixed for 2 min. Then, the WS were added gradually in saturated-surface-dry conditions, and mixed during 4 min. The fresh WBC were molded over the inferior face, in three layers hand compacted.

The superior facings were cast over the core of WBC, following the same process of the inferior facings. At the end of the molding, the whole panels were vibrated on a vibratory table (68 Hz) during 30s. Then, the casted panels were kept in their molds, protected to the humidity and demolded after 24h. After demolding, they cured into a vapor chamber at 20 ± 3°C and 95 ±2% RH until they reached 28 days of old. The manufacturing steps are shown in Figure 3.

Figure 3: Panel Manufacturing Steps a) Laying of the long sisal fibers b) spreading of the mortar over the fibers c) Surface adjustment of the mortar d) hand compacting of the WBC e) surface adjustment of the WBC (0.5) f) WBC (2.5) g) Spreading of the first mortar layer over the core h) Spreading of the mortar layer on the long fibers i) ending of the molding

Longitudinal Compressive Strength Test

The longitudinal compressive tests were realized at 28 days of curing based on the ASTM C364M-07, using the universal testing machine Shimadzu-1000 kN at the speed of 0.3 mm/min. Five samples per sandwich panel (SP) were used with dimensions of 100 x 100 x 200 mm^3. During the test, the compressive force was applied in a direction parallel to the sandwich facing plane. Four linear variation displacement transducers (LVDTs) were positioned at mid-height of each lateral side of the samples, and measured the axial displacement over a gage length of 100 mm (Figure 4) of both cores and facings.

a)

b)

Figure 4: Longitudinal compressive test a) SP 0.5 b) SP 2.5.

RESULTS AND DISCUSSION

Densities

Table 5 presents the densities values of the SP. As expected, the panel SP0.5 was the lighter because its core also was lighter than SP2.5 core. The panel SP2.5 showed a density 30% higher than SP0.5, nevertheless, both SP could be classified as lightweight materials according to the RILEM Functional classification (1978).

Table 5: Densities of the SP.

	Density (kg/m^3)
SP 0.5	1020
SP 2.5	1450

Longitudinal Compressive Strength Test

The load *vs* strain curves of the facings and cores of SP 0.5 and SP 2.5 are illustrated in Figure 5 (a) and (b), respectively. The maximum load and equivalent compressive strength values are shown in Table 6.

It can be observed that the sandwich panels showed compressive strengths higher than their WBC cores. When compared with their cores, SP 0.5 and SP 2.5 presented strength 4 times and 1.34 times higher, respectively. This gain of strength can be attributed to the presence of the LSIFCC faces, stronger than the WBC cores.

The SP 2.5 was 6 times more resistant than the SP 0.5, illustrating that the strength did not depend only on the facings. During the test, the facings presented lower strains, more evident after the peak load. At the end of the test, PS 0.5 showed the higher difference of strains between facings and core, around 7 times.

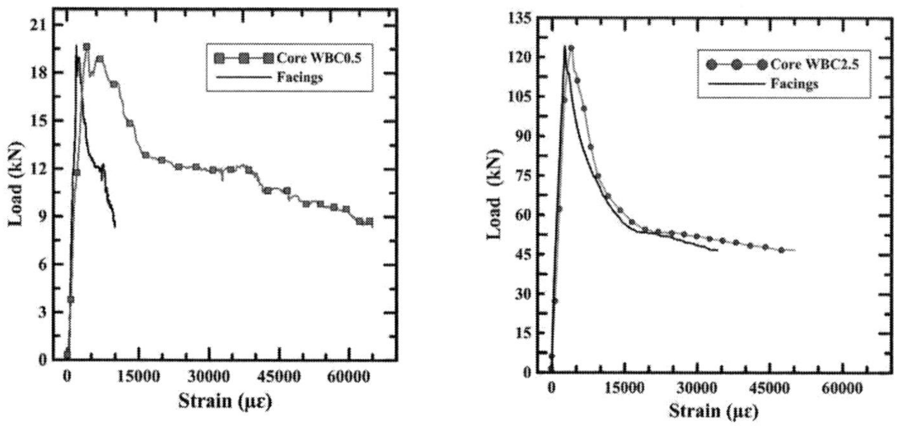

a) b)

Figure 5: Typical load - strain curves in edgewise compressive tests a) SP 0.5 b) SP 2.5.

Table 6: Compressive test results of the sandwich panels

	F_{max}(kN)	fc_{28} (MPa)
SP 0.5	19.67	2.01
SP 2.5	123.95	12.67

The samples after the test are showed in Figure 6. It can be observed that SP0.5 collapsed by cracking of the core. One sample also showed a delamination in the facing between the long fibers and the mortar layers. The SP 2.5 showed a collapse by shear and both cores and facings presented cracks. The facings of SP 0.5 showed less cracks than those of SP 2.5, for being submitted to a lower load during the test.

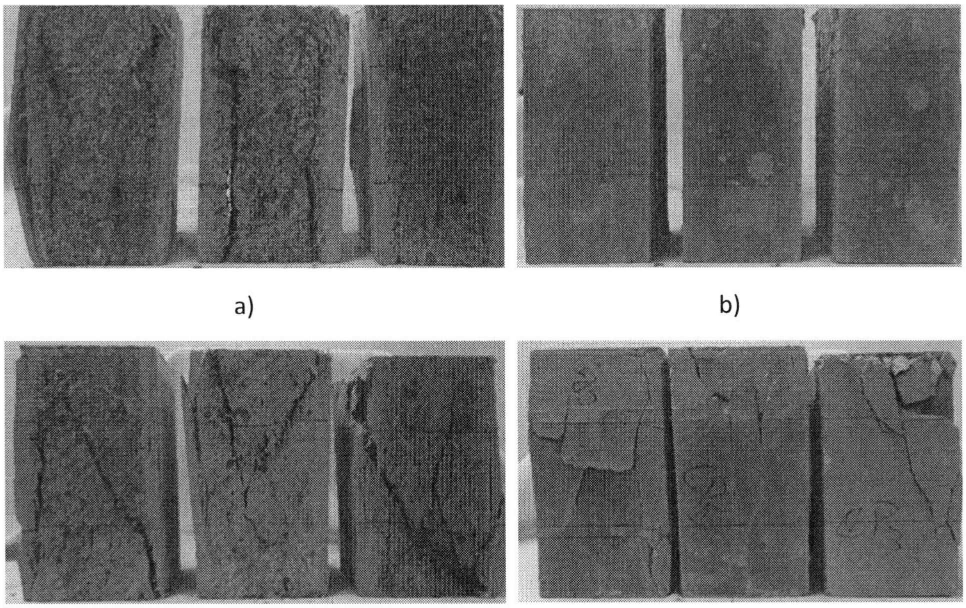

a) b) c) d)

Figure 6: The samples after the test a,b) SP 0.5 c,d) SP 2.5.

CONCLUSION

The experimental results of this work revealed that it was possible to produce sandwich panels using vegetables materials as aggregates and reinforcement for construction. The WBC 2.5 and WBC 0.5 allowed obtaining lightweight sandwich panels that can be used as structural and non-bearing element, respectively. The compressive test also showed a good bond between the faces and the cores of both sandwich panels. In fact, none of the panels presented a collapse by separation of the core and the faces. Further, other mechanical and physical test will be realized to have a complete characterization of the sandwich panels.

REFERENCES

[1] AMZIANE, S.; ARNAUD, L., Les bétons de granulats d'origine végétale (application au béton de chanvre). Hermes Science Publication, 2013.

[2] TOLEDO FILHO, R. D.; SCRIVENER, S.; ENGLAND, G. L.; GHAVAMI, K., Durability of alkali-sensitive sisal and conconut fibres in cement mortar composites. Cement & Concrete Composites, 22:127–143, 2000.

[3] LIMA, P. R. L.; Análise teórica e experimental de compósitos reforçados com fibras de sisal. PhD thesis, Universidade Federal do Rio de Janeiro, Rio de Janeiro, 2004.

[4] MELO FILHO, J.;. Desenvolvimento e caracterização de laminados cimentícios reforçados com fibras longas de sisal. Master's thesis, Universidade Federal do Rio de Janeiro, Rio de Janeiro, 2005.

[5] SILVA, F.A.; CHAWLA, N.; TOLEDO FILHO, R.D., Tensile behavior of high performance (sisal) fibers. Composites Science and Technology, 68:3438–3443, 2008.

[6] FERREIRA, S.R.; LIMA, P.R.L.; SILVA, F.A.; TOLEDO FILHO, R.D., Influência de ciclos molhagem-secagem em fibras de sisal sobre a aderência com matrizes de cimento Portland. Revista matéria pp.1024-1034, 2012.

[7] DA GLORIA M.Y.R.; TOLEDO FILHO, R.D. Influence of the wood shavings/cement ratio on the thermomechanical properties of lightweight wood shavings-cement. 6th Amazon & Pacific Green Materials Congress and Sustainable Construction Materials Lat-Rilem Conference; Cali, Colombia, 2016.

IC-NOCMAT 2019

18th International Conference on
Non-Conventional Materials and Technologies
"Construction Materials & Technologies for Sustainability"
(18th NOCMAT 2019)
24th – 26th July 2019
Nairobi, Kenya

ASSESSING THE MECHANICAL PROPERTIES OF BAMBOO CULTIVATED IN ITALY

Silvia Greco[1], Luisa Molari[1], Mirko Maraldi[2]

[1] DICAM, University of Bologna, Bologna, Italy;

[2] DISTAL, University of Bologna, Bologna, Italy;

ABSTRACT

The excellent mechanical performance of bamboo, coupled with its sustainability, have boosted the use of this plant as a structural material in the last decades. There are countries in which this material has been used in construction for millennia and represents an asset. On the other hand, there are countries where still there is not enough knowledge of the structural properties of locally-grown bamboo. The available studies and data in the literature refer to American or Asian bamboo, and constructions in other parts of the world are made with these imported materials. Expanding the knowledge of bamboo mechanical properties to other species can facilitate its diffusion as a structural material in parts of the world where bamboo is not traditional as such. In order to accomplish this, a very important step is the development of standardised testing procedures. In this paper, the results of an experimental campaign for the mechanical characterization of Phyllostachys Viridiglaucescens grown in Italy is presented. This species is one of the most common bamboo species cultivated in temperate climate. Compressive tests, tensile tests, shear tests and bending tests were performed. The methodology is mainly compliant with the International Standards; however, to cope with the particular physical properties of Italian bamboo, the testing methodology was modified, where needed, with respect to that prescribed by the International Standards.

Keywords: Phyllostachys Viridiglaucescens, Mechanical properties, European bamboo.

INTRODUCTION

We are facing a broad range of challenges that claim sustainable choices. There is growing recognition and awareness that nature can help provide viable solutions. For this reason, we are looking with increasing interest at the use of natural materials for construction (see for example Maraldi et al. [1] and Mazhoud et al. [2]) and, in particular, at the use of bamboo [3-5]. Bamboo is, in fact, able to combine a high mechanical resistance with a low specific weight and a very high growth speed. These characteristics can give bamboo a new crucial role among future sustainable materials for construction, even if bamboo has been used as a structural material for millennia.

The inner structure and the mechanical properties of some species of bamboo like Moso (Phyllostachys Edulis) and Guadua (Guadua Angustifolia Kunth) have been widely studied. In Dixon & Gibson [6] and Dixon et al. [7], an experimental investigation on two-years-old Moso bamboo was performed, and the shape of the culm of bamboo illustrated in great detail. In these studies, Young's modulus and density were taken as global mechanical indicators for tensile strength. In Archila-Santos et al. [8], the relation between the mechanical properties of bamboo culms and their microstructure was further investigated. In Awalluddin et al. [9], the flexural properties of Moso were compared with those of Guadua and Tre Gai bamboo. Other studies also referred to Guadua Angustifolia Kunth (Guadua) species commonly used for construction in Colombia [8], while several Undergraduate theses at the Universidad National de Colombia focused on the mechanical properties of Guadua [10-13]. In Mat et al. [14], the mechanical characterization of a variety of common Malaysian bamboo species (Bambusa Vulgaris, Dendrocalamus Asper, Schizostachyum Grande, and Gigantochloa Scortechinii) was reported. Recently, more tests have been conducted on Bambusa Vulgaris [15].

Despite to papers referred to, knowledge of the mechanical properties of bamboo species that can be cultivated in temperate climate is missing. Expanding knowledge to other bamboo species can greatly facilitate the diffusion of bamboo as a structural material in parts of the world where bamboo is not traditional as such.

This paper addresses the full mechanical characterization of Phyllostachys Viridiglaucescens, a bamboo species commonly grown in Italy.

MATERIALS AND METHODS

The mechanical characterization of Phyllostachys Viridiglaucescens was performed in agreement with the Standard ISO 22157-1:2017 [16] as much as possible. In particular, as suggested by the Standards, tests to evaluate the compressive strength parallel to the fibres (labelled as C), the tension strength parallel to the fibres (labelled as T), the tension strength perpendicular to the fibres (labelled as T⊥), the shear strength parallel to the fibres (labelled as S), and the bending strength perpendicular to the fibres (labelled as B) were performed. An overview of the test campaign is provided in Table 1.

Tests were performed using a Galdabini PMA10 Ultimate Testing Machine (UTM, Galdabini Cesare S.p.a., Cardano al Campo, Italy) at the Laboratory of Structural Engineering (LISG) of the University of Bologna. Test specimens were cut from bamboo culms grown in a bambusetum located in Selvaiana Capezzano Pianore (Camaiore, Lucca, Italy).

The dry-basis moisture content was determined for each specimen according to the Standards [16] at the end of each test. Samples were oven-dried at 103 °C and, after 24 h, weighed at regular intervals until a stable value of specimen weight was reached.

TABLE 1 – TESTS OVERVIEW.

Type of test	Label	Node	No. of specimens
Compressive test parallel to the fibres	C	with	6
		without	6
Tensile test parallel to the fibres	T	with	6
		without	6
Tensile test perpendicular to the fibres	T⊥	without	6
Shear test parallel to the fibres	S	without	6
Bending test perpendicular to the fibres	B	without	6

Compressive test parallel to the fibres

In accordance with ISO Standards [16], specimens dimensions were such that L < D or L < 10 t, being L the height of the specimen, D the external culm diameter and t the thickness of the culm. To determine specimens dimensions, 3 measurements were performed on the culms and values were averaged.

The testing machine was equipped with a spherical joint metal head that could oscillate about the horizontal axis to ensure uniform force distribution over the entire specimen load section (Fig. 1a). A layer of Teflon tissue of extremely low thickness was interposed between the specimen ends and the UTM load plates, in order to limit friction. Two orthogonal bidirectional strain gauges were applied to each specimen to measure the deformation along the direction of the compressive load (i.e. parallel to the fibres) and that in the circumferential direction (Fig. 1b).

Normal stress in the direction of the load (σ_z) was calculated as follows:

$$\sigma_z = \frac{F}{A}, \tag{1}$$

being F the force applied by the UTM and A the specimen cross section area.

(a) (b)

Fig. 1 – C-test: a) machine setup; b) strain gauges application.

Assuming a linear elastic orthotropic constitutive law and a thin-wall model for the specimen, the longitudinal Young's modulus under compression (E_c) and the ratio of the circumferential to the longitudinal deformations (r) could be determined [17]:

$$E_c = \frac{\sigma_z}{\varepsilon_z} \tag{2}$$

$$r = -\frac{\varepsilon_\theta}{\varepsilon_z}, \tag{3}$$

where ε_z is the longitudinal deformation and ε_θ is the circumferential deformation. As prescribed by the Standards, both E_c and r were calculated in the interval between 20% and 60% of the maximum attained stress:

$$E_c = \frac{\left(\sigma_{z,60} - \sigma_{z,20}\right)}{\left(\varepsilon_{z,60} - \varepsilon_{z,20}\right)} \tag{4}$$

$$r = -\frac{\left(\varepsilon_{t,60} - \varepsilon_{t,20}\right)}{\left(\varepsilon_{z,60} - \varepsilon_{z,20}\right)}, \tag{5}$$

where the subscripts 60 and 20 are used conveniently.

The value of the compressive strength ($\sigma_{c,ult}$) was calculated as prescribed by the Standards as well:

$$\sigma_{c,ult} = \frac{F_{c,ult}}{A}, \tag{6}$$

where $F_{c,\,ult}$ is the maximum load reached during each test.

Tensile test parallel to the fibres

For this test, specimens were longitudinally cut from the culms; the resulting cross section had one dimension equal to the culm wall thickness (t) and the other one (b, the specimen width) equal to 5 mm. Due to the small value of the culm thickness of Phyllostachys Viridiglaucescens, the length of the specimen and the specimen ends design were different from what prescribed by the Standards. Further details on this topic can be found in Greco & Molari [18]. Specimens dimensions were determined by averaging the values measured at 3 different locations of the specimen.

Furthermore, also the specimens clamping setup was different from the Standards prescription and tailored to the reduced culm thickness of Phyllostachys Viridiglaucescens species: the ends of the specimen were buried in resin and covered with two pieces of tube for anchorage to the UTM clamping devices (Fig. 2). Further details on the modified setup are reported in Greco & Molari [18]. As depicted in Fig. 2, a unidirectional strain gauge and a deformometer were placed in the longitudinal direction (parallel to the fibres) to measure longitudinal deformation ε_z.

(a) (b)

Fig. 2 – T-test: a) machine setup; b) detail of the specimen.

Assuming a linear elastic constitutive law, the longitudinal Young's modulus under traction (E_t) was calculated in the same fashion as for the compressive tests:

$$E_t = \frac{\left(\sigma_{z,60} - \sigma_{z,20}\right)}{\left(\varepsilon_{z,60} - \varepsilon_{z,20}\right)}. \tag{7}$$

Analogously, the ultimate tensile strength was evaluated as:

$$\sigma_{t,ult} = \frac{F_{t,ult}}{A}, \tag{8}$$

where $F_{t,\,ult}$ is the maximum tensile load reached during each test.

Tensile test perpendicular to the fibres

To measure the tensile strength of the culm along the direction perpendicular to the fibre, a test was performed according to the Standards [16]. Specimens were drilled (Fig. 3b) using a 20 mm diameter milling cutter; the hole allowed the insertion of a split cylinder for specimen loading (Fig. 3a). The length of the specimen (L) was equal to the sum of the diameter of the culm (D) and the diameter of the hole (d). As prescribed by the Standards, the ratio of d to D ranged between 0.25 and 0.5 for all the specimens.

Tests were then carried out at force-control loading; the loading speed was 6 N/s to ensure a duration of the test compatible with what prescribed by the Standards: in all the cases, thetotal duration of the test was not less than 3 minutes.

(a) (b)

Fig. 3 – T⊥-test: a) machine setup; b) the specimen.

The value of the tensile strength perpendicular to the fibres was then calculated as the force provided at failure by the UTM divided by the area of the two surfaces at which the specimen split:

$$\sigma_{\perp,ult} = \frac{F_{\perp,ult}}{2t(L-d)}. \tag{9}$$

Shear test parallel to the fibres

The shear strength along the direction of the fibres was determined as prescribed by the Standards, enforcing four shear planes by means of four alternating triangles interposed between the two specimen ends and the load plates of the UTM (Fig. 4).

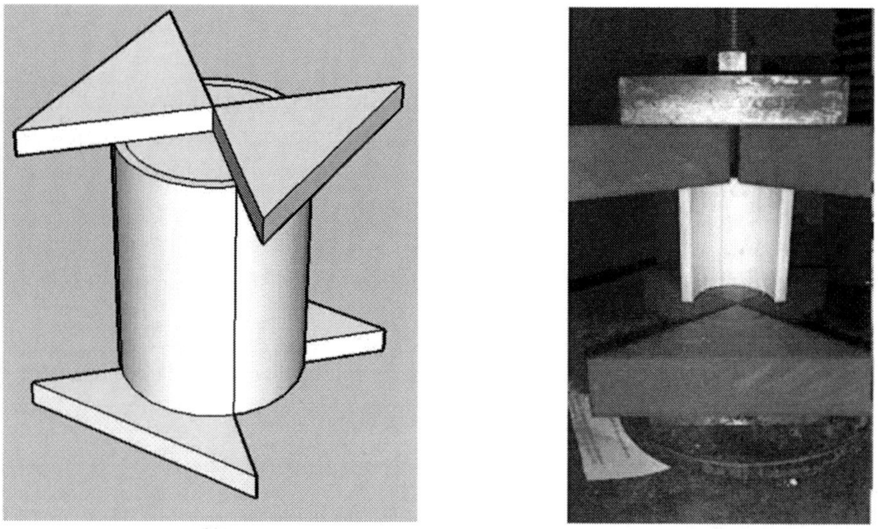

(a) (b)

Fig. 4 – S-test: a) sketch of the loading scheme; b) picture from the test.

The value of the ultimate tangential strength τ_{ult} was calculated according to the Standards:

$$\tau_{ult} = \frac{F_{s,ult}}{\sum_{i=1}^{4} t_i L_i}, \tag{10}$$

being $F_{s,ult}$ the UTM compressive force at specimen failure, and t_i and L_i , respectively, the thickness and length of the four enforced shear planes.

Bending test perpendicular to the fibres

In this test, the specimen was subjected to a compression in the direction orthogonal to the culm axis (Fig. 5). This load generated a stress state within the specimen mainly due to a transverse bending moment. The specimens had length (L) equal to the diameter (D) and were chosen with a circularity tolerance of 0.05 D, as suggested by the Standards [16]. For each specimen, two diameters were marked which divided the specimen in 4 quadrants. Four segments named N - E - S - W were traced (Fig. 5) and, in correspondence to these segments, measurements relative to the specimen outer diameter (D) and thickness (t) were taken.

Tests were carried out at force-control; the loading speed was 6 N/s to ensure a duration of the test compatible with what prescribed by the Standards: in all the cases, the total duration of the test was not less than 3 minutes. Between the loading plates and the specimen, a Neoprene layer of 1 cm thickness was interposed (Fig. 5).

Fig. 5 – B-test setup.

The normal stress in the E-W plane was calculated considering the specimen subjected to normal force and bending moment:

$$\sigma_{b,ult} = \frac{12 M_{ult,EW}}{L t^3}\left(\frac{t}{2} + h\right) - \frac{F_{ult}}{2Lt} \tag{11}$$

where:

$$M_{ult,EW} = \frac{F_{ult} D}{2\pi}\left(1 - \frac{t^2}{3D^2}\right) - \frac{F_{ult} D}{4} \quad (12)$$

$$h = \frac{D}{2} - \frac{t}{\ln\left(\frac{D+t}{D-t}\right)} \quad (13)$$

RESULTS AND DISCUSSION

Compressive test parallel to the fibres

The values of the ultimate compressive stress, of the Young's modulus under compression and of the ratio of the circumferential to the longitudinal deformations are collected in Table 2. Overall, the mean value of the ultimate strength was 71.3 MPa (standard deviation 5.4 MPa). For the specimens with node, the mean value of the ultimate compressive strength was 72.4 MPa (standard deviation 6.0 MPa), while for the specimen without node the mean value was 70.2 MPa (standard deviation 5.1 MPa). As it regards the Young's modulus, the mean value was 18.1 GPa overall (standard deviation 4.3 GPa); the mean value for the specimen with node was 20.6 GPa (standard deviation 4.9 GPa), while for the specimen without node was 15.5 GPa (standard deviation 1.2 GPa).

TABLE 2 – C-TEST DATA (mean value of moisture wet-basis: 12%).

specimen		D [mm]	t [mm]	L [mm]	$\sigma_{c,ult}$ [MPa]	E_c [GPa]	r [-]
with node	1	58.0	6.2	58.2	79.0	15.4	0.41
	2	61.2	9.3	62.5	65.2	25.6	0.57
	3	65.3	7.6	68.3	65.6	17.8	0.50
	4	63.4	6.4	59.8	75.6	27.6	0.52
	5	55.3	6.6	54.1	77.5	19.3	0.52
	6	66.3	7.4	67.2	71.7	17.9	0.20
without node	1	59.3	5.8	58.0	74.5	16.8	0.61
	2	61.5	9.3	57.8	64.5	14.0	0.77
	3	64.7	6.7	60.8	74.0	16.5	0.82
	4	64.1	7.0	63.8	63.2	14.1	0.70
	5	57.0	5.9	55.2	71.0	15.6	0.43
	6	62.5	7.1	61.0	74.2	16.3	0.42

The diagram of the longitudinal stress against the longitudinal strain and the circumferential strain is reported in Fig. 6a for the specimens with node and in Fig. 6b for the specimens without node. The trend of the stress versus the longitudinal strain was similar both for the specimens with node and for those without node. On the other hand, the trend of the stress versus the circumferential strain was different. In particular, in the case of specimens without node the curves show local instability (right-hand side of Fig. 6b), while in the case of specimens with node such trend does not appear (right-hand side of Fig. 6a). Thus, the node played a stabilizing role under compression, as confirmed by the failure morphology of the specimens (Fig. 7).

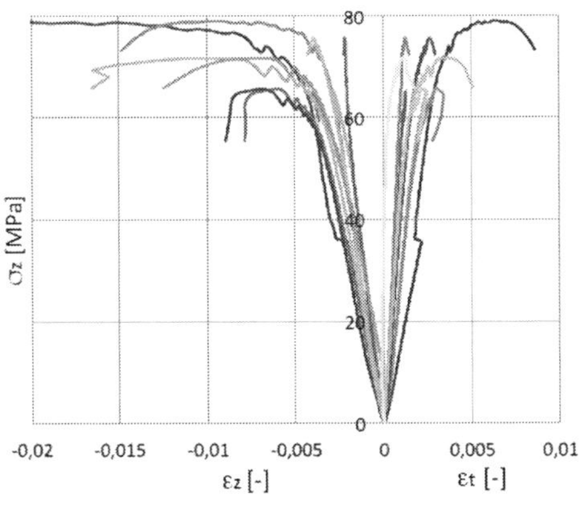

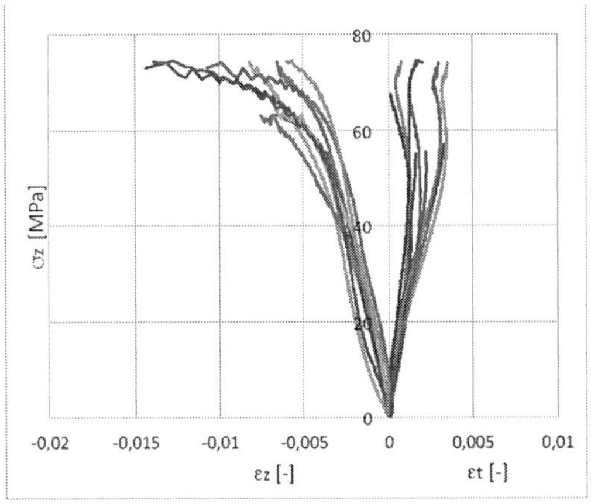

(a) (b)

Fig. 6 – Stress-strain diagrams of C-test: a) specimens with node; b) specimens without node.

(a) (b)

Fig. 7 – Specimens morphology after failure on C-test: a) specimen with node; b) specimen without node.

Tensile test parallel to the fibres

The ultimate tensile strength and the Young's modulus under traction are collected in Table 3. Overall, the mean value of the ultimate strength was 216.5 MPa (standard deviation 35.4 MPa). For the case of specimens with node, the mean value of the ultimate tensile strength was 200.3 MPa (standard deviation 26.6 MPa), while in the case of specimen without node the mean value was 234.6 MPa (standard deviation 35.6 MPa). As it regards the Young's modulus under tension, the mean value was 18.8 GPa overall (standard deviation 2.9 GPa); for the case of specimens with node, the mean value was 19.4 GPa (standard deviation 3.5 GPa), while for the case of specimens without node the mean value was 18.0 GPa (standard deviation 1.8 GPa). The ultimate tensile strength was found to be lower in the case of specimen with node, due to the discontinuity of some fibres (Fig. 8) which did not extendthrough the whole length of the specimen, but were interrupted at culm node.

TABLE 3 – T-TEST DATA (mean value of moisture wet-basis: 9.5%).

specimen		b [mm]	t [mm]	$\sigma_{t,ult}$ [MPa]	E_t [GPa]
with node	1	4.6	6.4	172.8	14.7
	2	4.5	6.4	222.9	20.5
	3	4.7	8.1	185.1	15.0
	4	4.8	5.9	170.8	22.7
	5	4.5	6.0	204.5	22.7
	6	4.6	6.9	234.5	20.0
without node	1	5.1	5.6	247.3	-
	2	6.7	8.0	165.1	16.5
	3	5.1	6.3	240.2	16.8
	4	5.8	6.0	265.7	19.9
	5	4.8	6.1	253.4	19.4
	6	4.8	6.8	206.0	-

In Fig. 9a and 9b, the diagrams of the longitudinal stress against the longitudinal strain are reported for the specimens with and without node, respectively. In both cases, the behavior can be considered elastic and the failure brittle.

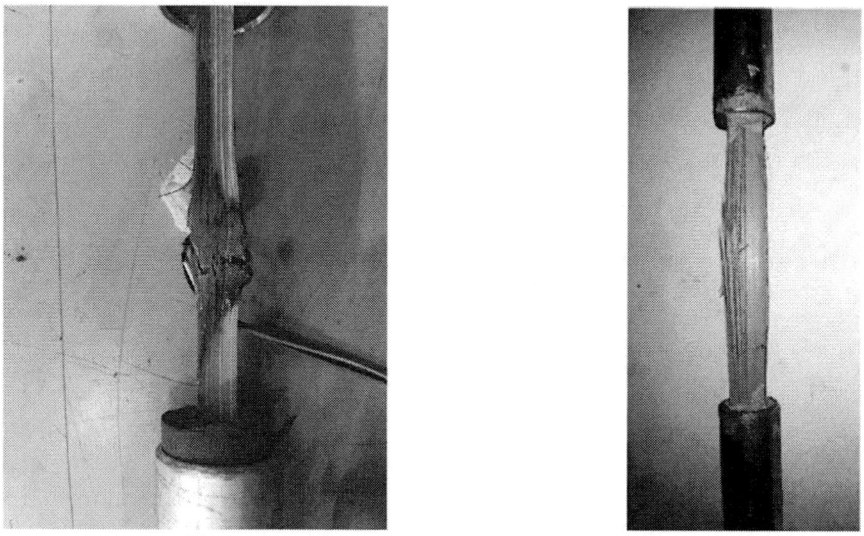

(a) (b)

Fig. 8 – Specimens morphology after failure on T-test: a) specimen with node; b) specimen without node.

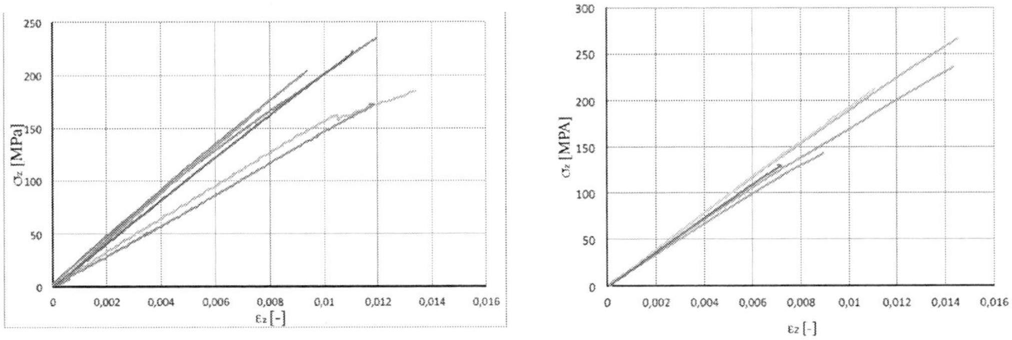

(a) (b)

Fig. 9 – Stress-strain diagrams of T-test: a) specimens with node; b) specimens without node.

Tensile test perpendicular to the fibres

In this test, once the maximum load was reached, the specimen split as shown in Fig. 10. The failure was essentially brittle. The values of the ultimate tensile strength are collected in Table 4. The mean value of the ultimate tensile strength was 2.0 MPa (standard deviation 0.62 MPa).

Fig. 10 – Specimen morphology after failure on T⊥-test.

TABLE 4 – T⊥-TEST DATA (mean value of moisture wet-basis: 10.4%).

specimen	D [mm]	t [mm]	L [mm]	$\sigma_{\perp,ult}$ [MPa]
1	64.0	6.1	84.1	1.99
2	64.1	5.9	84.2	2.31
3	55.6	5.6	75.8	1.84
4	68.1	5.2	88.3	0.85
5	55.2	6.0	75.2	2.32
6	54.6	5.6	74.5	2.62

Shear test parallel to the fibres

The values of the ultimate tangential strength and of the failure modes are reported in Table 5 for all the tested specimens. Data regarding specimens 1 and 2 were disregarded from subsequent computations, since the failure was not activated at the enforced shear planes, but elsewhere. The mean value of the ultimate tangential strength was 2.2 MPa (standard deviation 0.04 MPa).

TABLE 5 – S-TEST DATA (mean value of moisture wet-basis: 10.9%).

specimen	D [mm]	t [mm]	L [mm]	failure mode	τ_{ult} [MPa]
1	54.5	5.0	60.3	out of enforced planes	3.25
2	61.2	9.5	60.1	out of enforced planes	3.22
3	65.8	6.1	59.2	along enforced planes	2.19
4	65.2	5.2	63.2	along enforced planes	2.25
5	56.2	5.8	57.2	along enforced planes	2.21
6	61.3	7.5	65.0	along enforced planes	2.14

Bending test perpendicular to the fibres

For all the tests carried out, failure occurred at the E-W plane first, and only at a later stage at the N-S plane (Fig. 11). The values of the ultimate normal stress $\sigma_{b,ult}$ are collected in Table 6 for all the tested specimens. The mean value of the ultimate normal stress was 14.2 MPa (standard deviation 2.0 MPa).

TABLE 6 – B-TEST DATA (mean value of moisture wet-basis: 10.4%).

specimen	D [mm]	t [mm]	L [mm]	$\sigma_{\perp,ult}$ [MPa]
1	54.6	6.2	54.1	16.9
2	62.5	7.5	61.9	11.3
3	56.1	5.0	55.8	15.2
4	52.4	6.2	51.5	16.7
5	53.1	5.0	52.5	15.4
6	53.8	5.2	53.5	13.8

Fig. 11 – Specimen morphology at failure during a B-test.

CONCLUSIONS

The paper aimed at expanding the knowledge of the mechanical properties of bamboo species grown in temperate climate, in order to increase the awareness on this kind of material in areas where it is not considered a viable building material yet. To this end, a mechanical characterization of Phyllostachys Viridiglaucescens bamboo grown in Italy was performed through a series of tests included in the ISO 22157-1 Standard. In particular, the following tests were performed:

- compressive tests along the fibres direction;
- tensile tests along the fibres direction;
- tensile tests perpendicular to the fibres direction;
- shear tests parallel to the fibres direction;
- bending tests perpendicular to the fibres direction.

In the cases of compressive test and of tensile test along the fibres direction, tests were performed both on specimens including culm node and on specimens without culm nodes. The overall results for all the tests are reported in Table 7. For the compressive tests and for the tensile tests parallel to the fibres, results show that the presence of culm node does affect the strength of the material.

In addition, strength values for one of the most popular bamboo species, Guadua Angustifolia, are also reported in Table 7 as collected from the literature. In terms of specific strength values, it can be observed that Phyllostachys Viridiglaucescens shows mechanical characteristics comparable or even higher to those of Guadua Angustifolia. On the other hand, the fact that these species are grown in a climate which is not ideal, results in smaller culm and culm-wall dimensions.

TABLE 7 – PERFORMANCE COMPARISON BETWEEN PHYLLOSTACHYS VIRIDIGLAUCESCENS AND GUADUA ANGUSTIFOLIA KUNTH (GUADUA).

	Phyllostachys Viridiglaucescens	Guadua	[ref.]
$\sigma_{c,ult}$	71.3 MPa	56.21 MPa	[12]
$\sigma_{t,ult}$	216.5 MPa	221.5 MPa	[19]
$\sigma_{\perp,ult}$	2.0 MPa	0.22 MPa	[10]
τ_{ult}	2.2 MPa	2.3 MPa	[11]
$\sigma_{b,ult}$	14.2 MPa	-	-
E_t	18.8 GPa	18.9 GPa	[19]
E_c	18.1 GPa	17.9 GPa	[12]

Acknowledgements

The authors wish to thank Mr Roberto Carli, Mr Mario Marcolongo and Mr Roberto Bianchi for the technical support during the tests and for the helpful discussions.

REFERENCES

[1] Maraldi, M., Molari, L., Regazzi, N., Molari, G., Analysis of the parameters affecting the mechanical behaviour of straw bales under compression, Biosystems Engineering, Volume 160, 179-193, (2017).

[2] Mazhoud, B., Collet, F., Pretot, S., Lanos, C., Mechanical properties of hemp-clay and hemp stabilized clay composites, Construction and Building Materials, 165, 1126-1137, (2017).

[3] Kaminsky, S., Lawrence, A., Structural use of bamboo, Part 1: Introduction to bamboo, The Structural Engineer, 95(4), 40-43, (2016).

[4] Yu, W.K., Chung, K.F., Chan, S.L., Axial buckling of bamboo columns in bamboo scaffolds, Engineering Structures, 27, 61-73, (2005).

[5] Minke, G., Building with bamboo, Birkhauser, Basel, (2012).

[6] Dixon, P.G., Gibson, L.J., The structure and mechanics of Moso bamboo material. J R Soc Interface. 2014; 1, 20140321 (2014).

[7] Dixon, P.G., Ahvenainen, P., Aijazi, A.N., Chen, S.H., Lin, S., Augusciak, P.K., Borrega, M., Svedström, K., Gibson, L.J., Comparison of the structure and flexural properties of Moso, Guadua and Tre Gai bamboo, Construction and Building Materials 90, 11- 17, (2015).

[8] Archila-Santos, H. F., Ansell, M.P. and Walker, P., Elastic properties of thermo-hydromechanically modified bamboo (Guadua Angustifolia Kunth) measured in tension, Key Engineering Materials 600,111-120, (2014).

[9] Awalluddin, Mohd Azreen Mohd Ariffin, Mohd Hanim Osman, Mohd Warid Hussin, Mohamed A. Ismail, Han-Seung Lee, and Nor Hasanah Abdul Shukor Lim, Mechanical properties of different bamboo species Dinie, MATEC Web of Conferences 138, 01024 (2017).

[10] Pacheco Puentes, C.A., Resistencia a la tracción perpendicular a la fibra de la Guadua Angustifolia (Perpendicular-to-grain strength of Guadua), Undergraduate Dissertation, Bogota, Colombia, Facultad de Ingenieria, Universidad Nacional de Colombia, 2006.

[11] Clavijo, S., Trujillo, D., Evaluacion de uniones a traccion en guadua (Study into tension connections in Guadua). Undergraduate Dissertation, Bogota, Colombia, Facultad de Ingenieria, Universidad Nacional de Colombia, 2000.

[12] Gonzalez Quevedo, C. E., Resistencia a la compresion paralela a la fibra de la Guadua angustifolia y determinacion del modulo de elasticidad, dissertation, Bogota, Colombia, Facultad de Ingenieria, Universidad Nacional de Colombia, 2006.

[13] Alarcon Gutierrez, E. Olarte Florez, J. O., Esfuerzo maximo de tension paralela a la fibra y detrminacion del modulo de elasticidad de la guadua angustifolia del municipio de pitlito-Huila, dissertation, Bogota, Colombia, Facultad de Ingenieria, Universidad Nacional de Colombia, 2013.

[14] Mat Zain, M. Y., Ali, M. T. and Hussim, A.N.H., Bambusa vulgaris: determination of the mechanical strength as bio-composite material, Journal of Fundamental and Applied Science, 20189, (2018).

[15] Po-Hua Lee, Marty Odlin, Huiming Yin, Development of a hollow cylinder test for theelastic modulus distribution and the ultimate strength of bamboo, 51, 235-243, (2014).

[16] ISO 22157-1, Bamboo structures: Determination of physical and mechanical properties of bamboo culms, International Standard, (2017).

[17] Young W.C., Budynas R. G., Roark's Formulas for Stress and Strain, McGraw-Hill, (2002).

[18] Greco, S., Molari, L., New set up for tensile test performed on thin bamboo, Fifth International Conference on Sustainable Construction Materials and Technologies (SCMT5), July 14-19, 2019, Kingston University London.

[19] Laroque, P. Design of a low cost bamboo footbridge, Master thesis In Civil and Environmental Engineering, Massachusetts Institute of Technology, 2006.

18th International Conference on
Non-Conventional Materials and Technologies
"Construction Materials & Technologies for Sustainability"
(18th NOCMAT 2019)
24th – 26th July 2019
Nairobi, Kenya

CARBON FOOTPRINT AND WATER SCARCITY ASSESSMENT OF WOOD BIO-CONCRETES: EVALUATION OF DIFFERENT LIFE CYCLE IMPACT METHODS

Lucas R. Caldas[1], Ana. Paula N. Menezes[1], M'hamed Y. R. Da Gloria[1], Vanessa M. Andreola[1], Daniele J. Santos[1], Romildo D. Toledo Filho[1]

[1] Department of Civil Engineering, COPPE, Universidade Federal do Rio de Janeiro, Rio de Janeiro, RJ, Brazil;

ABSTRACT

Normally, when bio-based materials are used as aggregates, especially wood shavings, and mixed with cementitious materials, it may have setting problems due to the extractives that are present in these materials. One alternative to solve this problem is the process of washing the bio aggregates in hot water and sometimes setting accelerators can be used. On the other hand, these processes and products can increase the environmental impacts of the bio-concrete production, due to the water and energy consumption. Then, different alternative for the wood shavings (WS)treatment must be evaluated in terms of environmental impacts and mechanical performance. This paper aimed to evaluate environmental aspects and the relation with mechanical performance(using the compressive strength as indicator) of different mixes of a wood bio-concrete (WBC) that is being developed in Brazil. The greenhouse gases (GHG) emissions and Water Scarcity were evaluated. As a way to perform a sensitivity analysis of the study, we used different Life Cycle Impact Assessment (LCIA) methods. For the carbon footprint, the Dynamic LCA and IPCC at 100 years were employed. For the Water Scarcity, we applied four methods: (1) Pfister, (2) Boulay, (3) Hoekstra and (4) Berger. The modelling was performed in the SimaPro software using the Ecoinvent database adapted to the Brazilian context. When the mechanical performance indicator was used, the WBC produced with not washed WS and CaCl2 (NW_A) mixture presented the best results, for the carbon footprint and water scarcity assessment. Even with use of different LCIA methods, the results pointed out that the NW_A mixture was he best option. This study showed that the wood shavings treatment is an important stage in the wood bio-concrete production and deserve special attention, mainly in relation to the water consumption.

Keywords: Wood bio-concrete, Dynamic LCA, LCIA Methods.

INTRODUCTION

In Brazil the use of concrete in the building sector is very common due to its availability, low costs, good performance and durability. However, as pointed out by Meyer [1], the use of concrete is responsible for great amount of raw materials consumption, such as mineral resources, energy, water and greenhouse gases (GHG) emissions.

On the other hand, the use of bio-based materials has been encouraged in the last years as a way to develop a bio-economy in order to have alternatives for climate change mitigation and adaption [2]. Currently, the quest for environmental sustainability has become the goal of many building projects, including the United Nation (UN) Sustainable Development Goals (SDG) that has as one of its goal the provision of sustainable, dignified and resilient housing for all. It is expected that the main global construction demand will be centered in emerging economies, such as Asia, Africa and South America, including Brazil [3].

Brazil has adequate conditions for the development of a bio-based economy, especially in the building sector, which is expected to grow in the next years and to attend the national housing deficit. Since concrete is the main material used in the Brazilian building sector, an alternative, using wood shavings as aggregates, from wood waste process has been developed, as presented by Da Gloria and Toledo Filho [4]. When bio-based aggregates are mixed with cementitious materials, it is possible to produce lightweight concrete, with good thermal properties [5]. The wood shavings contain organic components named extractives, which disturb the concrete setting and hardening [6].

When the wood shavings are used as aggregates in concretes it is important to wash them to remove their extractives in order to improve the setting time of the mixture and the mechanical performance [4]. Nevertheless, during the washing it is normally used high amount of water elevated at 80° C, which will require energy consumption and consequently more GHG emissions. One way to improve the wood cement compatibility during the production of bio-concrete, setting accelerators, such as the calcium chloride (CaCl2) are used, but they also have environmental impacts.. Then, different treatment methods will lead to different environmental impacts and mechanical performance of the bio-concrete.

To measure these environmental impacts, the Life Cycle Assessment (LCA) can be a very useful methodology, since it is pointed as a scientific and robust way to quantify and evaluate the potential environmental impacts of products, process and services during their life cycle [7]. In the LCA there are different Life Cycle Impact Assessment (LCIA) methods, such as CML-IA, IPACT 2002+, IPCC (2013), and others [8]. These methods have different characterization factors and environmental impacts categories, for example: climate change, acidification, water depletion, depletion of fossil fuels, etc. In the literature of LCA applied to the building sector, as pointed out by Anand and Amor [9] most of studies are concerned with the Climate Change impact, since most of countries are worried with this environmental impact and most of them have committed themselves to achieving reduction GHG emissions in the Paris Agreement at Conference of the Parties - COP 21 in Paris, 2015. Looking specifically for the wood bio-concrete, the water consumption and scarcity becomes an important environmental indicator to be evaluated, since a large volume can be consumed in the treatment of the wood shavings.

In this context, some research questions emerged: what is the best way to treat the wood shavings for use in the bio-concrete, when environmental impacts and mechanical performance are considered? Which LCIA method is more adequate to evaluate the carbon footprint and water scarcity of the bio-concretes? In order to answer these questions the aim of this article was to evaluate environmental aspects and mechanical performance of different mixes of wood bio-concrete (WBC), the GHG emissions and water scarcity, during its production, using different LCIA methods.

METHODOLOGY

Characterization of Wood Bio-concrete

The WBC were manufactured using treated wood shavings and different cementitious materials, such as Brazilian Portland cement of high early strength (CPV ARI), metakaolin (MK) and fly ash (FA), mixed with CaC12 and water. The mixes were produced in laboratory, with the following materials proportions: cement 60%, MK 30% and FA 10% (of cementitious materials), wood volumetric fraction of 45%. In Table 1 the bio-concrete composition and results of the compressive strength test are presented. The following mixtures were evaluated in this study.

- WBC with not washed WS - without CaC12 – NW_NA
- WBC with not washed WS - with CaC12 – NW_A
- WBC with WS washed one time - without additive CaC12 – 1W_NA
- WBC with WS washed one time - with additive CaC12 – 1W_A
- WBC with WS washed two times - with additive CaC12 – 2W_A
- WBC with WS washed three times - with additive CaC12 – 3W_A

TABLE 1 – MIXTURES EVALUATED IN THE STUDY.

Mixtures	Cement (kg/m³)	Metakaolin (kg/m³)	Fly ash (kg/m³)	Wood shavings (kg/m³)	Calcium chloride (kg/m³)	Mixing water (kg/m³)	Compressive strength (MPa)
NW_NA	428	214	71	140	0	551	2.91
NW_A	428	214	71	140	21	551	4.87
1W_NA	428	214	71	140	0	551	4.15
1W_A	428	214	71	140	21	551	5.03
2W_A	428	214	71	140	21	551	6.03
3W_A	428	214	71	140	21	551	6.43

Definition of Goal, Scope and Functional Unit

For the scope definition, the standard EN 15804:2012 [10] was used to the LCA model, considering the following modules: raw materials supply (A1); raw materials transport (A2); bio-concrete manufacturing (A3).

We choose as functional unit the volume of wood shaving bio-concrete (in m³). In this study, we also used a mechanical performance indicator to verify the differences between the compressive strength of the wood shaving bio-concretes related to the GHG emissions and water scarcity, a similar approach was used by Van Den Heede et al. [11] and Celik et al. [12]. At the end, the carbon footprint (in kgCO-2eq) and water scarcity (in m^3 of water) impacts to produce $1m^3$ with 1MPa strength of the bio-concrete, the Carbon-performance indicator (in kgCO2-eq/m^3. MPa) and the Water-performance indicator (in m^3 of water/m^3. MPa) were calculated, according to the experimental data of compressive strength values presented in Tab. 1.

Life cycle Inventory

The life cycle inventory was performed based on data collected in laboratory, literature and Ecoinvent v. 3.3 database. The electricity consumption of original Ecoinvent data was adapted to Brazilian energy mix for year of 2014. GHG emissions, removals and blue water (from rivers and lakes), were accounted over all stages.

A high initial resistance cement, CPV, with 90% of clinker, 5% of crushed limestone and 5% of gypsum, was considered, as CPV normally is used in Brazil. The average thermal energy consumption and electricity used in Brazilian cement plants for clinker and cement production was used, according to WBCSD [13] and MME [14].

A kaolin extraction and calcination process using charcoal as energy source, based on Borges et al. [15], was considered for metakaolin in the study. For FA it was just considered the treatment normally performed to use as a SCM and data from Chen et al. [16] was adopted considering the Brazilian electric energy mix for year of 2014.

As wood shavings are generated as waste in commercial production of wooden products, without any economic value, it was considered that they do not have any environmental impacts in their acquisition, besides the treatments performed in laboratory, presented in Fig. 1, such as screening, homogenization and washing in 80° C. It was accounted, in the laboratory, 1.5 kWh of electricity and 10 L of water (for each washing process) per kg of wood shavings.

Brazil is a country with continental dimensions which impacts in the transportation distances of building materials. Thus, different distances for transportation of raw materials (A2) as well as bio-concrete presumed to be transported from factory to site (A4) were assumed. Based on map research we considered 50 km as minimum distance, 200 km as average and 500 km as maximum, similarly to Lecompte et al. [17]. The road modal with EURO 3 and 16-32 tones, was adopted for transportation from Ecoinvent database, considering that Brazilian trucks normally have these characteristics for building materials transportation.

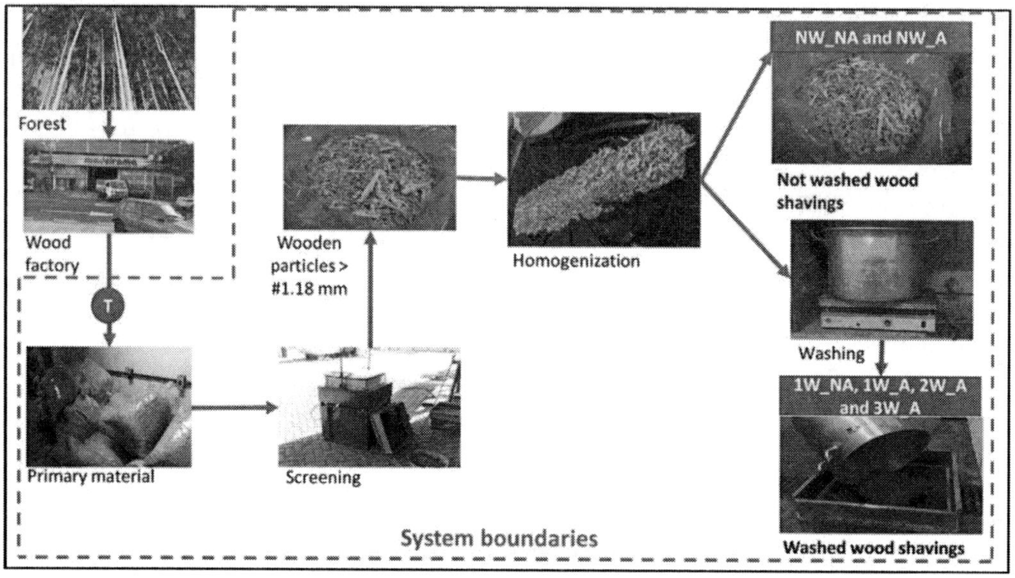

FIGURE| 1 – SYSTEM BOUNDARIES OF THE WOOD SHAVINGS TREATMENT.
Source: Authors (2018)

Life Cycle Impact Assessment (LCIA)

Carbon footprint

For the carbon footprint we used two methods: (1) The Dynamic LCA method proposed by Levasseur et al. [18] and the IPCC GWP100y [19] method.

Normally, in a conventional LCA, the impact of GWP of biogenic CO2 is not considered. The absence of timing during the biogenic CO2 flow can lead to significant differences in GWP calculation [20]. Levasseur et al. [18] proposed the Dynamic Life Cycle Assessment, with focus on GWP, which, in this paper, will be named the Dynamic Life Cycle Carbon Assessment (DLCCA). In the last years, different authors adopted the DLCCA for calculation of global warming impact (GWI), for example: Levasseur et al. [21], Fouquet et al. [20], Pittau et al. [22] and Peñaloza et al. [23].

The method proposed by Levasseur et al. [18] uses time-dependent characterization factors (DCF) for the calculation of the climate change category, in order to measure the cumulative radiative forcing generated by release of GHGs to the atmosphere. The instantaneous GWI (GWIinst) at the end are summed up in the defined time horizon for cumulative GWI (GWIcum). The GWIinst and GWIcum can be calculated using Eq. 1-3, proposed by Levasseur et al. [18].

$$DCF_{inst;GHG}(t) = \int_{t-1}^{1} a_{GHG} C(t)_{GHG} dt \quad (1)$$

$$GWI_{inst}(t) = \sum GHG \sum_{i=0}^{t} g_{GHG}(t_i).\, DCF_{inst,GHG}(t - t_i) \quad (2)$$

$$GWI_{cum}(t) = \sum_{i=0}^{t} GWI_{inst}(t_i) \quad (3)$$

To calculate the amount of atmospheric CO2 captured by wood plantation in the photosynthesis process, the approach suggested by Pittau et al. [22] was adopted. We considered the wood waste carbon content in the bio-concretes, using 53% of carbon in eucalyptus dry matter. This value was obtained in CHN elementary analysis, and used also by Caldas et al. [24].

We defined that the eucalyptus regrow started to occur after the bio-concrete manufacturing, considering a sustainable forest management. eucalyptus is considered a fast-growing bio-based material, with an economical rotation period of 10 years [25]. Then, we assumed that the eucalyptus is fully regenerated within 10 years of collecting, considering just the equivalent amount of wood (other waste such as leaves, branches and roots were not accounted). Based on data of Ecaliptus pellita growth by Cruz Rios [26] and the model presented by Pittau et al. [22], a model was developed. (A) eucalyptus biomass kept at steady state, (B) the aboveground eucalyptus is collected and the carbon is stored in the bio-concrete. At the same time the same biomass starts to grow in eucalyptus-plantation, capturing and storing CO2 and (C) the same quantity of the carbon that was collected is regenerated once again in the end of the eucalyptus plantation rotation.

Water Scarcity

We want to verify if the change in the LCIA method would change the results of the Water Scarcity impact. Therefore, we evaluated four different Water Scarcity LCIA methods: (1) Pfister et al. [27], (2) Boulay et al. [28], (3) Hoekstra et al. [29] and (4) Berger et al. [30]. All of them are already present in the SimaPro v. 8.5.2.0 software.

RESULTS AND DISCUSSION

Carbon footprint

For the carbon footprint, considering the two LCIA methods, we found the results presented in Figs 2 and 3. The bars indicates the carbon footprint (in kgCO2-eq for each m^3 of WBC) and the lines represents the carbon-performance indicator (in kgCO2-eq /m^3.MPa).

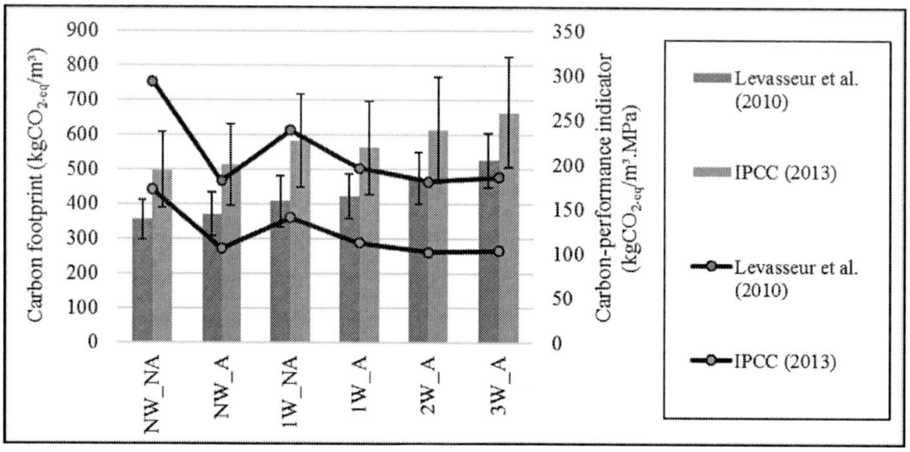

FIGURE 2 – CARBON FOOTPRINT AND CARBON-PERFORMANCE-INDICATOR

The error bars show the differences in the transport distances of the materials

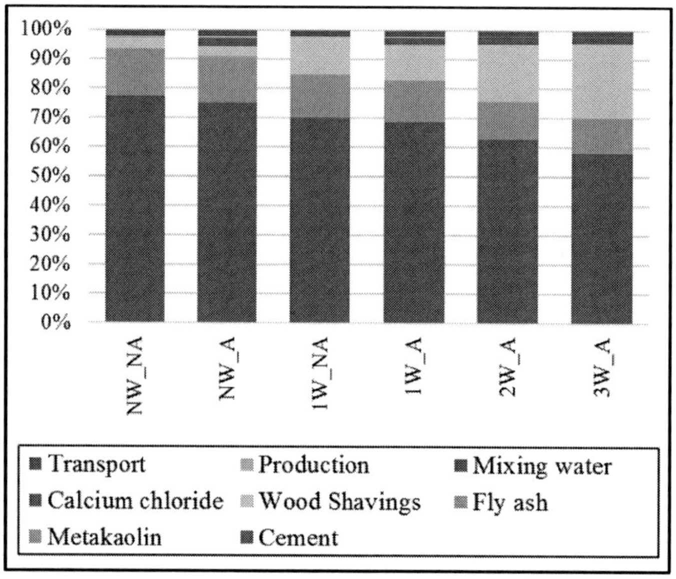

FIGURE 3 – CARBON FOOTPRINT PROFILE CONSIDERING THE SMALLEST TRANSPORTATION DISTANCES

There are considerable differences between the two methods for calculation of climate impacts. When the IPCC method was used, the CO2 captured during the eucalypt growth was not accounted for, which drastically increases the GWP impact of the bio-concretes. These finds were also verified for Pittau et al. [22]. The time of rotation period has an important influence on the captured CO2, according to the same authors, shorter rotation periods leads to less climate change impact. In this research we considered a forest with 10 years of rotation period, however, this can change, according to the tree species, climate and soil conditions. Then, the bio-concrete can have a less climate impact if we consider, for example, a 5 years rotation period forest, which in some cases and parts of Brazil is possible [31]. This uncertainty will be evaluated in future work.

The cement was the most impactful material due to the clinker production, by calcination of calcium carbonate and fossil fuels employed in its production. In second place there is the metakaolin that have its impacts from the kaolin extraction. Since we considered the use of charcoal for the calcination of kaolin, this choice has led to a decrease in the impact of climate change, because it is a biogenic source. If a fossil fuel was considered instead of charcoal, probably the impacts of metakaolin will be greater.

The impact of wood shavings grows with the increasing of the number of washing in 80 °C, because of the consumption of electricity. It is important to say that Brazil have an energy matrix with high number of renewable sources, and because of this the impacts of electricity for climate change in Brazil is smaller than it would be in other countries that use high amount of fossil fuels, such as China, Australia or Germany. The fly ash, CaC12, mixing water and the stage of production had a small share in the climate change impacts. The transportation stage can have a small or a large share in the carbon footprint depending on the transport distances of the materials.

Water Scarcity

For the waster scarcity and considering the four LCIA methods we found the results presented in Figs 4 and 5. The bars indicates the water scarcity (in m^3 of water for each m^3 of WBC) and the lines represents the water performance indicator (in m^3/m^3.MPa).

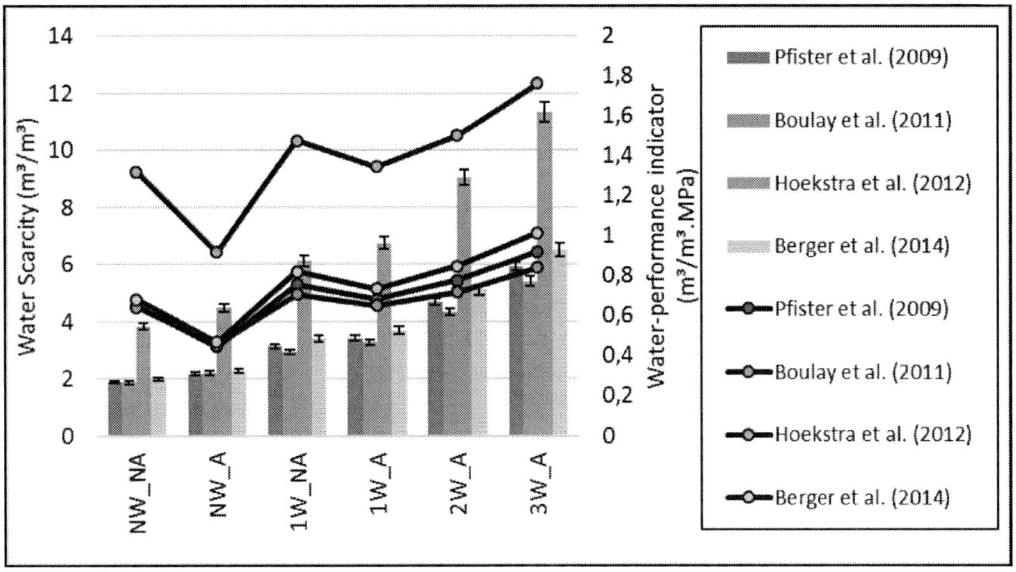

FIGURE 4 – WATER SCARCITY AND WATER-PERFORMANCE-INDICATOR

The error bars show the differences in the transport distances of the materials

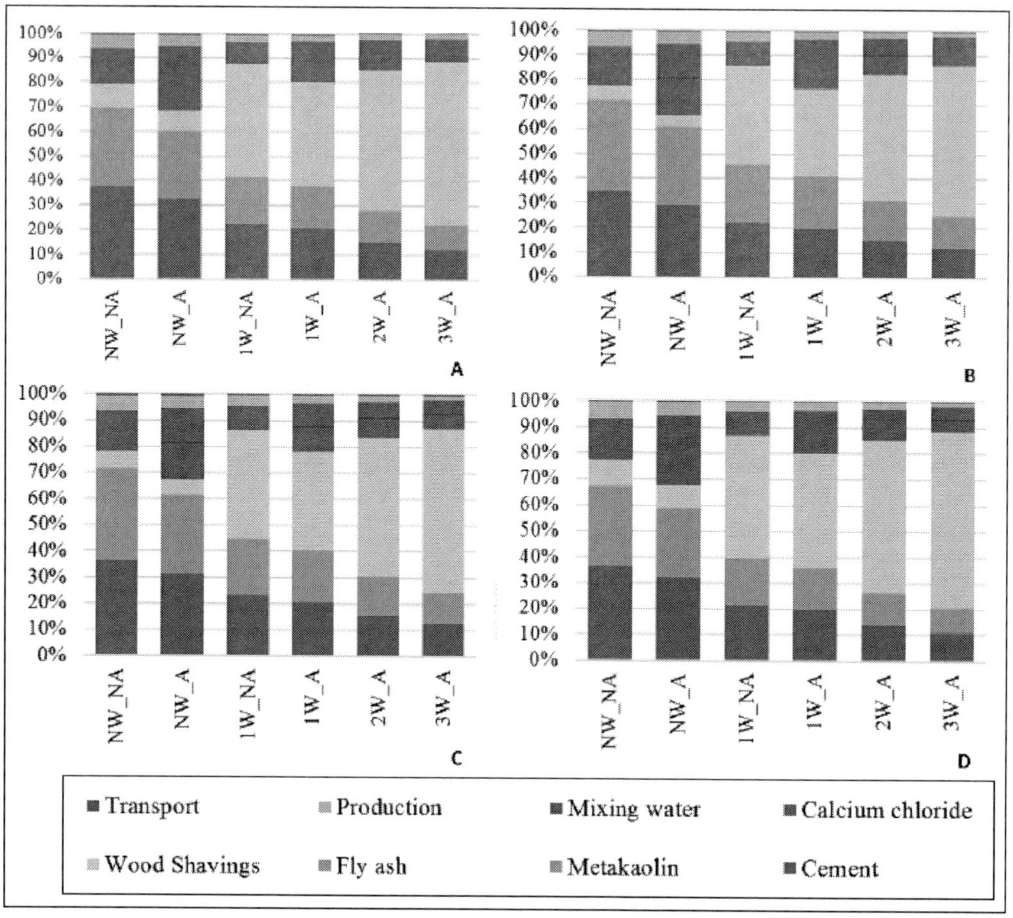

FIGURE 5 – WATER SCARCITY PROFILE FOR EACH METHOD CONSIDERING THE SMALLEST TRANSPORTATION DISTANCES. (A) PFISTER ET AL. (2009). (B) BOULAY ET AL. (2011). (C) HOEKSTRA ET AL. (2012). (D) BERGER ET AL. (2014).

When we evaluate the four Water Scarcity methods it is possible to see that the greatest difference between the mixtures occurred in the Hoekstra et al. [29], reaching 229% between 3W_A and NW_NA, while in the Boulay et al. [28] this difference was 188%. The water consumption for the wood shaving treatment was the main aspect that influenced the total water scarcity. The electricity consumption for the treatment impacts on the water scarcity because of the high share of hydraulic energy from Paraguay (the imported energy), that consumes high amount of water during the reservoir operation. The difference in the building materials transportation distances for the Water Scarcity was much lower than the Carbon Footprint, represented by the error bars in the graphs.

The metakaolin and cement have big shares for the NW_NA and NW_A. In these materials' life cycle the highest water consumption occurs during the charcoal, sodium hydroxide production and clinker production, respectively. The transportation stage does not have a great influence.

According to Kounina et al. [32], the Water Scarcity methods is not the same of Climate Change impact methods, since the first one is related with the regional and local water availability and the latter is a global impact. Then, in this study the target is not about having the number of water scarcity of the evaluated bio-concretes, but to see how the different consumptions of water for wood shavings washing increase this scarcity impact. In future works we have to adapt these methods for the Brazilian reality if we want to have more realistic values.

Gerbens-Leenes et al. [33] found that for the cement production it is necessary 0.21 m^3 of water/kg cement, while Mack-Vergara and John [34] verified that concrete production consumes 0.22 to 0.88 m^3 of water per each m^3 of concrete. However, these values are related to the inventory of water consumed and not with the water scarcity. When compared with the water consumed in the wood shavings washing process, 1.39 (one washing) to 4.17 m^3 (three washings) of water for each m^3 of bio-concrete, we can see that this process requires much more water than the water consumed in conventional concrete production.

From this finding, the next research that is being carried out will have the focus of knowing how this water can be treated and reused in the same process to decrease the consumption of fresh water. With these new findings, new water assessment will be evaluated in the future.

Main limitations of the study

There are some limitations that it is important to say. The first one is that the inventory of waste and energy consumption in the wood shavings treatment were measured in the laboratory scale. Then, if we think in an industrial scale, this consumption can be decreased to reach a more efficient process. Part of the inventory comes to the Ecoinvent database, which may not be represent the real process and technology, but this not invalidate the study, since it is a very common practice in LCA studies. Although wood shavings are waste, it is important to account how the impacts of allocation (using different methods, eg. mass and economic) can influence the environmental performance of WBC. Then, we hope that in future studies these limitations can be overcome.

CONCLUSIONS

In this study different mixtures of wood shaving bio-concretes were evaluated, with focus on the wood shaving treatment. When only the environmental performance is considered the WBC with not washed WS and without CaCl2 (NW_NA) presented the best results for carbon footprint, 355 to 498 kgCO2-eq/m^3 and water scarcity, 1.9 to 3.8 m^3 of water/m^3, while the three washed with CaCl2 (3W_A) presented the worst results, 526 to 664 kgCO2-eq/m^3 and 5.4 to 11.3 m^3 of water/m^3. However, when the mechanical performance is considered the mixture not washed and with CaCl2 (NW_A) was the most efficient with 76 to 105 kgCO2-eq/m³.MPa and 0.4 to 0.9 m³/m³.MPa. The treatment of wood shavings evaluated in this study showed higher influence in Water Scarcity indicator than the Climate Change impact. The use of the different LCIA methods allowed to confirm which mixtures are the best and the worst in terms of environmental impacts and mechanical performance.

REFERENCES

1. Meyer, C. The greening of concrete industry. Cement and Concrete Composites. [Online] 31 p. 601-605, 2009.

2. Federal Ministry of Education and Research (BMBF). Bioeconomy in Germany Opportunities for a bio-based and sustainable future. Alemanha, 2015.

3. UN Environment and International Energy Agency (2017): Towards a zero-emission, efficient, and resilient buildings and construction sector. Global Status Report 2017.

4. Da Gloria, M.Y.R.; Toledo, R.D.F. . Influence of the wood shavings/cement ratio on the thermo-mechanical properties of lightweight wood shavings-cement based composites. In: 6th AMAZON & PACIFIC GREEN MATERIALS CONGRESS and SUSTAINABLE CONSTRUCTION MATERIALS LAT-RILEM CONFERENCE, 2016, Cali. Proceedings of ….. April 27- 29th, 2016.. Cali: Track Comunicación, 2016. p. 365-374.

5. Nozahic, V. et al. Design of green concrete made of plant-derived aggregates and a pumice–lime binder. Cement and Concrete Composites, 2012, 34(2), (231-241).

6. Andreola, V.M.; Gloria, M. Y. R. ; Toledo Filho, R. D. . Behaviour Under Bending Loads Of Workable Bamboo Bio-Concrete. In: 17th International Conference - NOCMAT 2017 - Non-conventional Materials and Tecnhologie, 2017, Mérida. Behaviour Under Bending Loads Of Workable Bamboo Bio-Concrete, 2017. v. 1. p. 66-78.

7. ISO 14040, 2006. Environmental Management. Life Cycle Assessment. Principles and Framework. ISO, Geneva.

8. Bueno, C., Hauschild, M. Z., Rossingnolo, J. A., Ometto, A. R., Mendes, N. C., 2016. Sensitivity analysis of the use of Life Cycle Impact Assessment methods: a case study on building materials. J. Clean. Prod. 112 (20), 2208-2220.

9. Anand, Chirjiv K.; Amor, Ben. Recent developments, future challenges and new research directions in LCA of buildings: A critical review. Renewable and Sustainable Energy Reviews, January, v. 67, n. 1, p. 408-416, jan. 2017.

10. EN 15804. Sustainability of construction works – Environmental product declarations – Core rules for the product category of construction products. Brussels, 2012'.

11. Van Den Heede, P.; De Belie, N.; Environmental impact and life cycle assessment (LCA) of traditional and "green" concretes: Literature review and theoretical calculations. Cement and Concrete Composites, 2012, 34, 4, (431–442).

12. Celik, K.; Meral, C.; Gursel, A. P.; Mehta, P. K.; Horvath, A..; Monteiro, P. J. M.; Mechanical properties, durability, and life-cycle assessment of selfconsolidating concrete mixtures made with blended Portland cements containing fly ash and limestone powder. Cement and Concrete Composites, 2015, 56, (59–72).

13. World Business Council for Sustainability Devlopment (WBCSD). Global Cement Database on CO_2 and Energy Information. Availiable at: https://www.wbcsdcement.org/index.php/key-issues/climate-protection/gnr-database.

14. Ministério de Minas e Energia (MME). Balanço Energético Nacional. Relatório do Ano de 2017, Brasília, 2018.

15. Borges, P. H. R.; Lourenço, T. M. F.; Foureaux, A. F. S.; Pacheco, L. S. Estudo comparativo da análise de ciclo de vida de concretos geopoliméricos e de concretos à base de cimento Portland composto (CP II). Ambiente Construído, 2014, 14 (2), (153-168).

16. Chen, et al., 2010. LCA Allocation procedure used as an inciative method for waste recycling: an application to mineral additions in concrete. Res. Cons. Rec. 54, 12, 1231-1240.

17. Lecompte, T., Levasseur, Maxime, A. D. Lime and hemp concrete lca: a dynamic approach of ghg emissions and capture. In: S. Amziane, M. Sonebi. Proceedings of Second International Conference on Bio-based Building Materials, France: RILEM Publications.Levasseur, A.; Lesage, P. ; Margni, M. ; Deschênes, L. ; Samson, R. 2010. Considering Time in LCA: Dynamic LCA and Its Application to Global Warming Impact Assessments. Environmental Science and Technology, 2010, 44(8), (3169–3174).

18. Stocker, T.F.; Qin, G.-K.; Plattner; M.; Tignor, S.K. et al.; Climate Change 2013: The Physical Science Basis. Contribution of Working Group I to the Fifth Assessment Report of the Intergovernmental Panel on Climate Change. Cambridge University Press, Cambridge, 2013, United Kingdom and New York, NY, USA, 1535.

19. Fouquet, M. ; AnnieLevasseur, A.; Margni, M. ; Lebert, A.; Lasvaux, S. ; Souyri, B.; Buhém, C. ; Woloszyn, M. ; 2015. Methodological challenges and developments in LCA of low energy buildings: Application to biogenic carbon and global warming assessment. Building and Enviroment, 2015, 115, (759-776).

20. Levasseur, A.; Lesage, P. ; Margni, M. ; Brandão, M. ; Samson, R. 2012. Assessing temporary carbon sequestration and storage projects through land use, land-use change and forestry: comparison of dynamic life cycle assessment with ton-year approaches. Climatic Change, 2012, 115, (759-776).

21. Pittau, F.; Krause, F.; Lumia, G.; Habert, G.; Fast-growing bio-based materials as an opprtunity for storing carbon in exterior walls. Building and Environment, 2018, 129, (117-129).

23. Peñaloza, D.; Erlandsson, M.; Falk, A.; Exploring the climate impact effects of increased use of bio-based materials in buildings. Construction and Building Materials, 2016, 125, (219-226).

24. Caldas, L. ; da Gloria, M. ; Santos, D. ; Andreola, V ; Pepe, M.; Toledo Filho, R; Carbon Footprint of Bamboo Particles, Rice Husk and Wood Shavings-Cement Composites. In: S. Amziane, M. Sonebi. Proceedings of Second International Conference on Bio-based Building Materials, France: RILEM Publications.

25. Cruz Ríos, 2009. Bambú – Gradua – Bosques naturales y Plantaciones comerciales em México. 1 ed. Colombia: Colmex. 2009. 691 p.

26. Pfister, S.; Koehler, A.; Hellweg, S. (2009). Assessing the environmental impacts of freshwater consumption in LCA. Environmental Science and Technology, 43(11), 4098–4104; DOI: 10.1021/es802423e (download: http://pubs.acs.org/doi/full/10.1021/es802423e).

27. Boulay, A.-M.; Bulle, C.; Bayart, J.-B.; Deschenes, L.; Margni, M. Regional Characterization of Freshwater Use in LCA: Modeling Direct Impacts on Human Health Environmental Science & Technology 2011, 45, 8948-8957.

28. Hoekstra AY, Mekonnen MM, Chapagain AK, Mathews RE, Richter BD (2012) Global Monthly Water Scarcity: Blue Water Footprints versus Blue Water Availability. PLoS ONE 7(2): e32688. doi:10.1371/journal.pone.0032688.

29. Berger, M. et al. Ruud van der Ent, Stephanie Eisner, Vanessa Bach, and Matthias Finkbeiner. 2014. Water Accounting and Vulnerability Evaluation (WAVE): Considering Atmospheric Evaporation Recycling and the Risk of Freshwater Depletion in Water Footprinting. Environ. Sci. Technol., 2014, 48 (8), pp 4521–4528.

30. Silva, M. L., Ribeiro, C. A. A. S., 2006. Estabelecimento de rotação econômica para uma floresta regulada. Revista Árvore, Viçosa-MG, 2010, 30(1), (65-73).

31. Kounina, A. Review of methods addressing freshwater use in life cycle inventory and impact assessment. International Journal of Life Cycle Assessment, 18, 707–721.DOI 10.1007/s11367-012-0519-3

33. Garbens-Leenes, P. W., Hoekstra, A. Y., Bosman, R., 2018. The blue and grey water footprint of construction materials. Water Resources and Industry, 2018, 1-12.

34. Mack-Vergara, Y., John, V., 2017. Life cycle water inventory in concrete production – A review. Resources, Conservation and Recycling 122(2017) 227–250.

IC-NOCMAT 2019

18th International Conference on
Non-Conventional Materials and Technologies
"Construction Materials & Technologies for Sustainability"
(18th NOCMAT 2019)
24th – 26th July 2019
Nairobi, Kenya

STABILIZED COMPRESSED EARTH BLOCKS WITH TERNARY BINDER

Marvin Grover Alvarado Figueroa[1], Javier Quiñónez [2]

[1] Civil Engineers Co.;

[2] San Carlos University of Guatemala;

ABSTRACT

This document presents the experimental results in manufacture of compressed earth blocks with natural lightweight aggregates, using clay as main binder and a ternary material composed of high early strength hydraulic (HES) cement, calcium hydroxide and a local volcanic pozzolan, as a specific stabilizer of the main binder (clay). It also includes the evaluation of the physical characteristics and mechanical properties of blocks, according to the procedures and standards established by international regulations. The chemical and physical characterization tests of raw materials were made in specific laboratories of the Engineering Research Center at University of San Carlos of Guatemala. The blocks were manufactured with electric drive hydraulic equipment, available in a private company. Different dosages of natural volcanic aggregates, binder and stabilizing ternary material were used. The units were cured for 3 days using water at room temperature. The investigation records values of compressive strength for ages from 7 to 100 days; results of other tests required by international regulations are also included. The best compressive strength results were 5.1 MPa for 36 days edge. These results will allow the production of construction elements that reduce the environmental impact and provide a safe and affordable construction.

Keywords: blocks, earth, binder, stabilized, pozzolan.

INTRODUCTION

According to the Guatemalan National population Census in 2002, there were 2.2 million families living within Guatemala; of which, 410,097 didn´t own homes, and 611,495 lived in unsafe, unsanitary, or overcrowded housing. This data shows that there was a calculated housing deficit of at least 1,021,592 during that time in Guatemala. In other words, almost half of Guatemalan families and children have been living without adequate housing (1).

Expressing the quantitative data through percentages; 61% of the surveyed families where in a sub adequate housing one of the following reasons; 13% percent were living in temporal, unstable or unreliable housing, 29% had dirt floors, 22% had inadequate or mock walls, and 3% had inadequate roofing. Today, these numbers are only increasing due to the continued poverty and population growth that is currently affecting Guatemala.

The compressed earth block, also known as BTC, is a construction material made from mixing earth, stabilizing materials like; hydrated lime, hydraulic lime, and cement or clay. These materials are then mixed and compacted by a manual or mechanical hydraulic presses (2). This type of block could potentially replace the current, common, and highly pollutant brick that is used in many construction projects today.

A large part of Guatemala was formed through the immense volcanic activity that occurs all along the entire Ring of Fire. This left large reservoirs of volcanic materials throughout the territory. A specific group of these materials are called pyroclastic. Pyroclastic rocks are produced when highly heated magma is forced up to the Earth´s surface by pressure and accumulated gases and then undergoes a specific cooling process to create silicon aluminum compounds; as well as, iron compounds that uniquely function as binders. These highly pure pyroclastic compounds can readily mix with lime hydroxide and moisture. These compounds are usually referred to as naturally occurring Pozzolans (3).

Guatemala´s access to these natural resources allow it to substitute part of the need for Portland Pozzolan Cement in the fabrication of construction materials. Due to this fact, the use of these naturally occurring Pozzolans could help to reduce the environmental impact of construction, while simultaneously reducing the cost of production allowing for a greater possibility to reduce the housing deficit within Guatemala.

The materials used in the production of compressed earth blocks are; clay binders, ternary materials composed of highly resistant hydraulic cement, lime hydroxide, and volcanic Puzolan, as the principal binder. The production processes are completely detailed and the highlighted physical and chemical characteristic are also included within this study. Our production procedures, product, and testing methodology all presently comply with the International standards and requirements presently available. The investigation registered resistance values from day 7 through day 100 after the curing process, and also included the necessary testing for all international requirements. The best compression resistance values were collect on day 36 registering 5.1 mega Pascal pressure units.

Objectives

To evaluate the incidence of a natural binder and a ternary stabilizer on the physical characteristics and mechanical properties of blocks made with compressed earth.

Produce compressed earth blocks with natural lightweight aggregates using clay as a binder and a ternary material as clay stabilizer and evaluate their physical characteristics and mechanical properties according to procedures established by international regulations.

Methodology

The physical and chemical properties of the raw materials were tested in a specialized laboratory in the Engineering Research Center at the University of San Carlos, Guatemala. These tests were conducted based off of the requirements of the International Standars. After the construction processes and the proceeding tests were complete at a private company.

The materials used for the elaboration of compressed earth blocks were natural light weight aggregates, clay as main binder of the aggregates and a ternary binding material composed of high early strength hydraulic cement, calcium hydroxide and a local volcanic pozzolan, as a stabilizer of clay.

There were two categories of compressed block produced; those created with 10% and 15% clay binder. Within each group the blocks were stabilized with 25%, 50%, 75%, and 100% ternary materials, regarding the amount of clay

To evaluate the indicated uses for the ternary materials in the compressed clay blocks it was necessary to create a comparison or control group. For this reason a control group was created where only high early strength hydraulic cement was used in the production of the compressed blocks with all of the same percentages of ternary materials previously indicated.

Once the compressed blocks were fabricated they were cured for 3 days using water at room temperature.

The evaluation of the blocks was done based off of the norms established by CONGUANOR NGO 41 022 ¨Ladrillos de barro cocido¨; as a specific norm for the specification and evaluation of construction materials does not yet exist in Guatemala. The blocks were evaluated with compressive strength, wetting/drying, accelerated erosion SWINBURNE (SAET), and capillary absorption tests based on normal standard procedures to test construction materials in Spain known as UNE EN 41410:2008. "Bloques de tierra comprimida para muros y tabiques. Definiciones, especificaciones y métodos de ensayo." (5).

The necessary materials to complete the wetting/drying and accelerated erosion SWIMBURNE (SAET) testings were constructed based off of the norms established in UNE EN 41410:2008 "Bloques de tierra comprimida para muros y tabiques. Definiciones, especificaciones y métodos de ensayo."(5). The capillary water absorption of water testing norms were established in UNE-EN 772-11 "Métodos de ensayo de piezas para fábrica de albañilería. Parte II" (6).

The mix proportions that were used in the experimentation can be found detailed in the following tables. Tables 1 thru 4 correspond to group A and tables 5 thru 8 correspond to group B.

GROUP A

TABLE 1 – MIX PROPORTIONS 1 AND 5 GROUP A

Materials	Mixture 1		Mixture 5	
	kg	%	kg	%
Aggregates	79.55	87.81	79.55	88.89
Clay	8.84	9.75	8.84	9.87
HES Cement	1.1	1.22	1.1	1.23
Lime	0.33	0.37	0	0
Pozzolan	0.77	0.85	0	0
TOTAL	90.59	100	89.49	100

TABLE 2 – MIX PROPORTIONS 2 AND 6 GROUP A

Materials	Mixture 2		Mixture 6	
	kg	%	kg	%
Aggregates	79.55	85.72	79.55	87.81
Clay	8.84	9.52	8.84	9.75
HES Cement	2.21	2.38	2.21	2.44
Lime	0.66	0.71	0	0
Pozzolan	1.55	1.67	0	0
TOTAL	92.80	100	90.59	100

TABLE 3 – MIX PROPORTIONS 3 AND 7 GROUP A

Materials	Mixture 3		Mixture 7	
	kg	%	kg	%
Aggregates	79.55	83.72	79.55	86.75
Clay	8.84	9.3	8.84	9.64
HES Cement	3.31	3.49	3.31	3.61
Lime	1.00	1.05	0	0
Pozzolan	2.32	2.44	0	0
TOTAL	95.01	100	91.70	100

TABLE 4 – MIX PROPORTIONS 4 AND 8 GROUP A

Materials	Mixture 4		Mixture 8	
	kg	%	kg	%
Aggregates	79.55	81.82	79.55	85.72
Clay	8.84	9.09	8.84	9.52
HES Cement	4.42	4.55	4.42	4.76
Lime	1.33	1.36	0	0
Pozzolan	3.10	3.18	0	0
TOTAL	97.22	100	92.80	100

GRUPO B

TABLE 5 – MIX PROPORTIONS 1 Y 5 GROUP B

Materials	Mixture 1		Mixture 5	
	kg	%	kg	%
Aggregates	79.55	81.93	79.55	83.44
Clay	14.04	14.46	14.04	14.72
HES Cement	1.75	1.81	1.75	1.84
Lime	0.53	0.54	0	0
Pozzolan	1.23	1.27	0	0
TOTAL	97.09	100	95.34	100

TABLE 6 – MIX PROPORTIONS 2 Y 6 GROUP B

Materials	Mixture 2		Mixture 6	
	kg	%	kg	%
Aggregates	79.55	79.07	79.55	81.93
Clay	14.04	13.95	14.04	14.46
HES Cement	3.51	3.49	3.51	3.61
Lime	1.05	1.05	0	0
Pozzolan	2.45	2.44	0	0
TOTAL	100.60	100	97.09	100

TABLE 7 – MIX PROPORTIONS 3 Y 7 GROUP B

Materials	Mixture 3		Mixture 7	
	kg	%	kg	%
Aggregates	79.55	76.41	79.55	80.47
Clay	14.04	13.48	14.04	14.2
HES Cement	5.26	5.06	5.26	5.33
Lime	1.58	1.52	0	0
Pozzolan	3.69	3.54	0	0
TOTAL	104.11	100	98.85	100

TABLE 8 – MIX PROPORTIONS 4 Y 8 GROUP B

Materials	Mixture 4		Mixture 8	
	kg	%	kg	%
Aggregates	79.55	73.91	79.55	79.07
Clay	14.04	13.04	14.04	13.95
HES Cement	7.02	6.52	7.02	6.98
Lime	2.10	1.96	0	0
Pozzolan	4.91	4.57	0	0
TOTAL	107.62	100	100.60	100

RESULTS

Compressive strength test

To determine the compressive strength of the blocks, it was used an ELE International manual hydraulic press with a capacity of 250,000 pounds, figure 1.

The results obtained in the laboratory for both group A and B are shown in tables 9 and 10, respectively.

TABLE 9 – RESULTS COMPRESSIVE STRENGTH GROUP A

GROUP A - GROSS AREA EFFORT (MPa)					
Mixture	7 days	21 days	36 days	60 days	100 days
1	1.39	1.28	1.34	1.24	1.17
5	1.09	1.28	1.22	1.02	No data
2	2.07	2.61	2.22	2.94	2.88
6	1.64	1.44	1.77	2.07	1.57
3	2.37	2.99	3.31	3.90	3.80
7	2.30	2.88	3.26	3.78	3.31
4	2.65	3.61	3.43	4.82	4.39
8	2.92	3.56	3.61	4.59	4.38

TABLE 10 – RESULTS COMPRESSIVE STRENGTH GROUP B

GROUP B - GROSS AREA EFFORT (MPa)					
Mixture	7 days	21 days	36 days	60 days	100 days
1	1.56	2.09	2.03	2.15	2.14
5	1.34	1.66	1.73	2.05	2.02
2	2.03	3.07	2.92	4.22	3.65
6	2.58	3.29	3.42	3.55	3.46
3	2.37	3.84	3.41	5.09	4.67
7	3.05	3.85	4.35	4.74	4.71
4	2.65	4.07	4.42	5.82	6.11
8	3.71	4.20	5.06	5.80	5.78

The best results for both group A and B were the blocks made with mixtures 4 and 8. Figure 2 shows the behavior of the compressive strength with respect to the age of mixtures 4 and 8 of group B.

FIGURE 1 – ELE INTERNATIONAL MANUAL HYDRAULIC PRESS

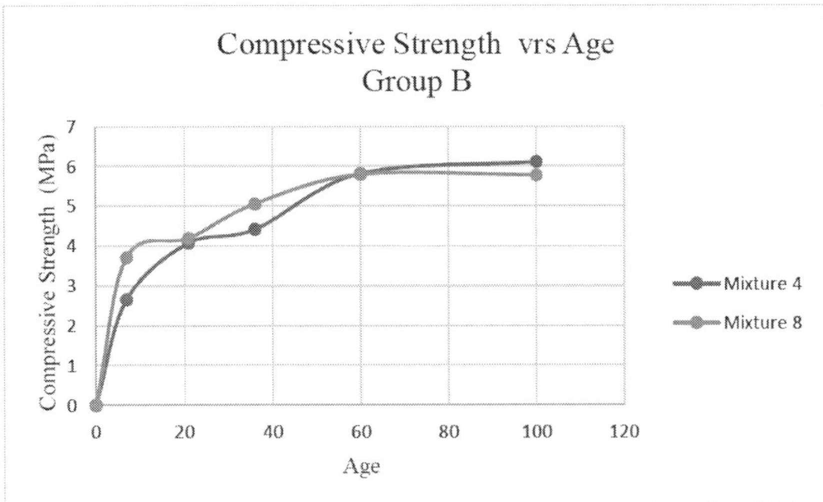

FIGURE 2 – COMPRESSIVE STRENGTH GROUP B MIXTURE 4 AND 8

Wetting/Drying test

The test procedure was based by the standar UNE EN 41410:2008. Water was added until the block was submerged 10 mm for 30 seconds, see Figure 3. Then, it was removed from the tray and allowed to air dry, until matching the color of the block that was used as a reference. The state of the block was observed, noting if certain rejection conditions appeared. The regulation requires that the immersion / drying / observation cycle be repeated 6 times. Tables 11 and 12 show the cycle number in which certain rejection conditions were appearing in the tested blocks.

TABLE 11 – RESULTS WETTING DRYING GROUP A

Mixtures	Cycle	Rejection conditions
1 y 5	2	Loss of soil layer and surface efflorescence
2 y 6	2	Loss of soil layer and surface efflorescence
3 y 7	1	Surface efflorescence
4 y 8	1	Surface efflorescence

TABLE 12 – RESULTS WETTING DRYING GROUP B

Mixtures	Cycle	Rejection conditions
1 y 5	2	Loss of soil layer and surface efflorescence
2 y 6	1	Surface efflorescence
3 y 7	1	Surface efflorescence
4 y 8	1	Surface efflorescence

Capillary absorption of water test

The test procedure, based on the standard UNE-EN 772-11, consisted in measuring the dimensions of the face of the block to be tested; then, the block was placed inside the tray on the supports, so that the face of the block did not touch the bottom of the tray and it was immersed in water to a depth of 5 mm for 10 minutes, see figure 3. After the time of immersion, the block was removed from the tray, the water was removed from the surface of the submerged face and weighed.

The values did not reflect and logical behavior and therefore have been omitted from the publication.

FIGURE 3 – EQUIPMENT USED DURING WETTING/DRYING AND CAPILLARY ABSORPTION OF WATER TESTS

Accelerated erosion test SWINBURNE (SAET)

Two compacted Earth blocks within each test group were tested in the accelerated erosion test. The test procedure is detailed in the norms establish in UNE-EN 41410:2008. To complete this test water was irrigated onto the blocks through a 5 mm pipe for a 10 minute period maintaining constant pressure and a 1.5 meter height difference between the block and the water tank. The Compressed block needs to be maintained at a 27 degree inclination respective to the horizon, like the figure represents in the photo below, figure 4. With a Vernier Caliper, the depth of the cavities, which were created by the impact of the water, were measured and the results can clearly be seen in table 13.

TABLE 13 – RESULTS ACCELERATED EROSION TEST SWINBURNE (SAET)

	GROUP A	GROUP B
Mixture	Depth of cavities (mm)	
1	2.45	0.95
5	5.7	1.15
2	2.15	0.8
6	1.65	0.6
3	1.95	0.9
7	0.65	0
4	0.5	0.4
8	0	0

FIGURE 4– ACCELERATED EROSION TEST SWINBURNE (SAET)

DISCUSSION

Compressive strength test

Group A

In mixtures 1 and 5, in general, there were no major differences. For ages between 21 and 36 days the blocks are classified as BTC 1. Taking into account that for ages greater than 36 days the resistance decreases. In mixtures 2 and 6, the results show an appreciable difference. The blocks made with the mixture 2 have twice the strength compared to the blocks made with the mixture 6. For the ages between 21 and 36 days the blocks are classified as BTC 1 for both mixtures. In mixtures 3 and 7, the behavior of the results is similar, they did not present major differences. For the ages between 21 and 36 days the blocks are classified as BTC 1. Emphasizing that for ages greater than and equal to 36 days the blocks made with both mixtures have an increase in resistance, which are classified as BTC 2. In the mixtures 4 and 8, the behavior of the results is similar, they did not present a significant difference. For the ages between 21 and 36 days the blocks are classified as BTC 2. Taking into account that for ages greater than 36 the blocks did not show a greater increase in resistance.

Group B

In mixtures 1 and 5, in general, there were no notable differences. For the ages between 21 and 36 days the blocks are classified as BTC 1, while taking into account that past day 36 the compressed block´s resistance maintained constant. In mixture 2 and 6 there was no notable difference. For the ages between 21 and 36 days the blocks are classified as BTC 2, and simultaneously showed no significant change to their resistance after day 36. In mixture 3 and 7 the behavior of blocks tended to show only minor differences. For the ages between 21 and 36 days the blocks are classified as BTC 2; however, after day 36 the blocks begin to develop a resistance that allowed them to be classified as BTC 3. For both mixes up through day 100 they did not show any diminishment in resistance. In mixtures 4 and 8 also showed very similar results to the previous study group. For the ages between 21 and 36 days the blocks are classified as BTC 2. However, mixture 8 had an increase in compression resistance after day 36 that permitted it to be classified as BTC 3. In mix number 4 the resistance increased after day 60 allowing it a similar classification of BTC 3.

Wet/ Dry cycling test

All of the blocks in both groups tested presented with superficial efflorescence.

CONCLUSIONS

Based on the experimental data collected during researche phases, both group A and B, independent of the differences noted, show a greater stability with the ternary material than with high early strength hydraulic cement alone.

The differences between the compressed blocks created with ternary material and the compressed blocks created with only cement did not present a considerable difference between both. In this case, the observation is made, that the natural binder (clay) used lacks the necessary plasticity to provide a considerable increase of resistance in the units, so it is concluded that, to elaborate blocks of compressed earth, it is only necessary to stabilize them with high early strength cement.

Taking into account the obtained results the compressed blocks made with mixes A and B, it can be concluded that due to the fact the that superficial efflorescence were observed the blocks are not apt for use in visible walls, and it will be necessary to apply a cover to protect the walls.

BIBLIOGRAPHIC REFERENCES

[1] Hábitat para la Humanidad y el Centro Latinoamericano para la Competitividad y el Desarrollo Sostenible (Clacds) del INCAE Business School. (2016). El estado de la vivienda en Centroamérica. 78-79.

[2]ALVARADO FIGUEROA, Marvin Grover. (2015). Bloques de Tierra Compactada estabilizados con aglomerante natural CP. 15.

[3]QUIÑÓNEZ DE LA CRUZ, Francisco Javier. (2013). Determinación y evaluación experimental del índice de reactividad puzolánica de diez bancos de materiales de la franja volcánica de Guatemala para la industria del cemento. 53.

[4]COGUANOR NGO 41 022. Ladrillos de barro cocido. Especificaciones. 1-7.

[5] UNE EN 41410:2008. Bloques de tierra comprimida para muros y tabiques. Definiciones, especificaciones y métodos de ensayo. 13-18.

[6]UNE-EN 772-11 "Métodos de ensayo de piezas para fábrica de albañilería. 7.

18th International Conference on
Non-Conventional Materials and Technologies
"Construction Materials & Technologies for Sustainability"
(18th NOCMAT 2019)
24th – 26th July 2019
Nairobi, Kenya

IC-NOCMAT 2019

MECHANICAL PERFORMANCE OF HOT MIX ASPHALT WITH ADDITION OF RECYCLED CONCRETE AGGREGATES AND GRANULAR PET

Ricardo Almeida de Meloa*, Vagner Costa Marquesb, Bismak Oliveira de Queirozb, Lêda Christiane de F. Lopes Lucena c, John Kennedy Guedes Rodriguesc

a Federal University of Paraiba (UFPB), Department of Civil and Environmental Engineering, Joao Pessoa, PB, Brazil

b UFPB, Program of PostGraduation of Civil and Envionmental Engineering

c Federal University of Campina Grande, Department of Civil Engineering, Campina Grande, PB, Brazil

ABSTRACT

The construction and demolition solid wastes (CDW) and plastic bottles can be recycled to modify asphaltic mixtures for using on surface course. The objective of this research was to analyze mechanical performance of modified asphalt mixtures with partial replacement of granite aggregates by up 25% to 75% of recycled concrete aggregates (RCA) and up 1% to 12% of granular PET (pellet). The mechanical performance tests were Marshall Stability, indirect tensile strength (ITS), resilient modulus (RM) and RM/ITS ratio. Although the modified asphaltic mixtures had had lower performance than conventional asphaltic mixtures, the majority met standards. Finally, the results could suggest maximum percentage of 25% for RCA and 2% for pellet in addition to modified asphaltic mixtures.

KEYWORDS: highway, asphalt mix, pavement layer, solid wastes, sustainability.

INTRODUCTION

The hot mix asphalt (HMA) is the main solution for paving surface courses on Brazilian highways, though stone matrix asphalt or asphalt modified with polymers also had been used. HMA surface course will be used on paving for a long time, due to lower costs, construction equipment's, laboratories and human staffs available in the Brazilian pavement construction companies.

Despite that, nowadays the pavement construction and maintenance is becoming costly because the distance to transport minerals aggregates is father away, so the transport costs become higher. However, use of solids wastes can be an alternative solution to substitute minerals aggregates and to reduce pavement costs. In accordance with this scenario, the paper references application of recycled aggregates from construction and demolition wastes, and PET wastes in asphaltic mixtures for paving surface courses on Brazilian highways.

Some studies investigated HMA made up 0% to 60% of recycled aggregates. The results showed that the recycled aggregates can provide high Marshall Stability, good resistance to permanent deformation and poor stripping behavior (PÉREZ et al. 2012; ZHU et. al. 2012; PASANDÍN and PÉREZ, 2013; PASANDÍN and PÉREZ, 2014).

Rafi et al. (2010) have replaced 25%, 50% and 75% of natural aggregates by recycled aggregates from construction and demolition wastes and reclaimed asphalt pavement (RAP) in asphalt mixtures, both in coarse and fine fractions. All the samples made by adding up to 50% of recycled aggregates met the specification requirements in terms of optimum asphalt contents, voids in mineral aggregates and stability/flow. Within the same percentages of RCA, Mills-Beale and You (2010) showed that asphaltic mixtures assessed by Superpave mix met specifications in terms rutting potential, Dynamic Modulus (E*), moisture susceptibility, resilient modulus and Construction Energy Index (CEI) for low volume roads. Besides that, the study suggests that percentage lower than 75% for asphaltic mixture met good performance.

In addition to that, the asphaltic mixtures within RCA had a lower performance than the asphaltic mixtures made with natural aggregates. By this way, Lee et al. (2012) used slag cement paste to pre-coat RCA for improving physical and mechanical proprieties in accord with crushing strength, indirect tensile strength moisture susceptibility and dynamic stability. Sojobi et al. (2016) evaluate effects of PET wastes by means of two procedures in HMA used for flexible pavement construction. The wastes were heated. After that, they were added according to dry process to pre-coat aggregates and by mixing with asphaltic mixture. The pre-coating aggregates process is more advantageous than mixing with asphaltic mixture, because it uses bigger volume of PET, so the recycling of wastes volume is higher. For both processes, the air voids, voids in mineral aggregate and Marshall Stability were higher than unmodified asphaltic mixture.

The PET added to asphaltic mixtures can decrease the aggregates porosity and absorption as well as improve soundness (VASUDEVAN et al., 2007; SOJOBI et al., 2016). These aspects can be improved by pretreatment of coarse recycled aggregate using liquid silicone resin in addition to strength, adhesion with asphalt and surface morphology (ZHU et al., 2012). In special, the recycled aggregates from solid wastes of civil construction and demolition can exhibit high absorption, so the optimum bitumen content is high. However, the PET wastes may be a solution for decreasing asphalt content and costs of HMA made with recycled aggregates. In our study, the principal objective was to analyze the mechanical performance of modified asphalt mixtures with partial replacement of granite aggregates by RCA and/or pellet.

MATERIALS AND EXPERIMENTAL METHODS

Selection of materials

The materials were natural aggregates (NA), recycled concrete aggregates (RCA), pellet (PET) and asphalt binder obtained from plants in João Pessoa, Brazil. The fractions of natural aggregates were crushed granite stone 19 mm, 12 mm and dust. The recycled aggregates were used in fractions of crushed stone 25 and 9 mm, sand and dust with percentages of 25%, 50%, 75% and 100%. However, the compositions of 50%, 75% and 100% RCA did not meet specifications in the DNIT ES 031/2006 Brazilian standard, and therefore the tests were limited to modified asphaltic mixture with 25% RCA in fraction of crushed dust. The addition of pellet in asphaltic mixtures aimed at diminishing absorption of recycled aggregates and consequently decreasing asphalt content. The pellet was in fraction passing 4.8 mm sieve and percentages of 2%, 4%, 8% and 12%. The specimen height with 8% and 12% of pellet did not meet specification in Marshall mix design, so they were discarded. The asphalt binder was 50/70 PEN.

Physical properties of aggregates and asphalt

The physical properties of natural, recycled aggregates and pellet were determined by the ASTM and Brazilian standards. Table 1 shows the results obtained by Los Angeles abrasion (LA), absorption and bulk specific gravity (BSG).

The crushed stone 19 mm and crushed stone 12 mm (NA) and the pellet have continuous particle size distribution, but with few fine material. The crusher dust (NA) and dust (RCA) had dense and well-graded particles size distributions. From the results of particles size distributions, it was determined the percentages of aggregates for asphalt mixtures in accord with limits of C particle size range at DNIT ES 031/2006 Brazilian standard. The C range is the most used particle size distribution for dense asphaltic mixtures (Marques et al. (2014) and Pinto et al. (2014)). According to Brazilian specifications, LA abrasion to aggregates for surface course is less than 50%. The results carried out lower than the required standards, so the aggregates must resist crushing, degradation and disintegration. The absorption of crushed stones were lower than 1%, which results little absorption of binder during the asphalt mixture production. The bulk densities of natural and recycled aggregates are in accordance with values specified.

TABLE 1. PHYSICAL PROPRIETIES OF NATURAL AND RECYCLED AGGREGATES

Test	Natural aggregates			Recycled aggregates
	Crushed stone 19 mm	Crushed stone 12 mm	Crushed dust	RCA dust
LA abrasion (%)	42.10	34.70	-	-
Absorption (%)	0.68	0.68	-	-
BSG (g/cm³)	2.61	2.63	2.66	2.44

The tests of physical characterization of 50/70 PEN asphalt binder were performed and certified by a Brazilian oil refinery. The results are shown in Table 2. According to it, the asphalt binder's characterization tests met the technical standards. Thus, it was concluded that the 50/70 PEN asphalt binder provides good characteristics for the production of asphalt mixtures.

TABLE 2. CHARACTERIZATION OF 50/70 PEN ASPHALT BINDER. SOURCE: LUBNOR OIL REFINERY, BRAZIL

Feature	Standard	Specification	Result	Unit
Penetration	D 5	up 50 to 70	59	10^{-1} mm
Soft point	D36	46 min	48.2	ºC
Brookfield Viscosity 135GC-SP21 20RPM	D 4402	274 min	468	cP
Brookfield Viscosity 150GC- SP21	D 4402	112 min	229	cP
Brookfield Viscosity A 177GC SP21	D 4402	57 to 285	83	cP
Flash Point	D 92	235 min	> 300	ºC
Thermal Susceptibility Index	X 018	- 1.5 to 0.7	- 1.3	-

Mechanical laboratory tests

Table 3 shows the labels for asphaltic mixtures produced with natural and recycled aggregates. The percentages of aggregates were defined in accord with the literature and based on the particles size curves of themselves as shown in Figure 1.

TABLE 3 – LABELS FOR CONVENTIONAL AND MODIFIED ASPHALT MIXTURES

Mixtures	Code	NA (%)	RCA (%)	Pellet (%)
Conventional	NA100	100	0	0
NA + RCA	NA75-RCA25	75	25	0
NA, RCA + pellet	NA75-RCA23-P2	75	23	2
NA, RCA + pellet	NA75-RCA21-P4	75	21	4
NA + pellet	NA98-P2	98	0	2
NA + pellet	NA96-P4	96	0	4

Legend - NA: natural aggregates; RCA: recycled concrete aggregates; P: pellet.

The asphaltic mixtures specimens were designed by Marshall Mix Design. First, the aggregates were dried at 100 °C for 24 hours. So, aggregates were separated and weighed according to the particle size for each mixture. After that, the asphalt binder and aggregates were heated at 160 and 170 °C, respectively. Then, they were mixed, compacted and molded in cylinders by dry process at 150 °C. The compaction energy was 75 blows per side of each specimen. So, some graphs were elaborated for the following parameters: Marshall Stability, flow, air voids (AV), voids in mineral aggregate (VMA), voids filled with asphalt (VFA) and bulk specific gravity (Gmb). After that, the optimum binder content (obc) was calculated by average of four binder contents corresponding to maximum values of Marshall Stability and Gmb, medium values AV and VFA. Finally, the obc was used to make asphaltic mixtures specimens for analyzing mechanical performance Indirect Tensile Stress (ITS) and Resilient Modulus (RM) tests.

Indirect Tensile Stress

The Indirect Tensile Stress (ITS) is the capacity of an asphaltic mixture to resist indirect tension. The test consists of applying a compression load to asphaltic mixtures along a diametric plane through two opposite loading strips. So, for each asphaltic mixture were molded three specimens in optimum binder content. The calculation of the ITS was made by equation 1.

$$ITS = \frac{2 \bullet F}{\pi \bullet D \bullet H} \quad (1)$$

In which:
F: tensile strength (Kgf);
D: specimen diameter (cm);
H: height of the specimen (cm).

Resilient Modulus

The Resilient Modulus (RM) test consists of repeated application of loads on a vertical diametric plane. During the test, stresses and recoverable and non-recoverable deformations are measured. The relation between stress and equivalent recoverable deformation represents the RM. It was molded three specimens of asphalt mixtures for each optimum binder content. After that, the specimens were left 2 hours at 25° C temperature before to realize resilient modulus test. The loads applied to specimens were estimated in stresses, resulting in 10% of ITS previously calculated for all asphaltic mixtures. That level of stress is not sufficient to break off the specimen during the RM test. The load application time was 0.1 s and unloading time was 0.9 s.

MECHANICAL PERFORMANCE AND DISCUSSION

This section show results tests in some tables and charts. Table 4 presents parameters carried out by Marshall Mix Design and specifications by DNIT ES 031/2006 standard.

TABLE 4. PARAMETERS OF MARSHAL MIX DESIGN FOR ASPHALTIC MIXTURES

Mixtures	Ob/c (%)	Gmb (g/cm³)	AV (%)	VFA (%)	Marshall Stability (N)	Flow (mm)	ITS (MPa)
NA100	5.7	2.36	3.3	79.9	10,522	6.2	0.84
NA75-RCA25	7.9	2.27	4.1	81.1	7,111	7.0	0.69
NA75-RCA23-P2	8.0	2.22	4.5	79.2	9,372	6.2	0.68
NA75-RCA21-P4	8.3	2.17	4.7	78.8	7,795	6.8	0.59
NA98-P2	6.5	2.29	3.5	80.4	10,468	6.4	0.53
NA96-P4	7.3	2.22	3.7	80.8	9,386	8.5	0.42
Standard	-	-	3 to 5	75 to 82	≥ 4,900	-	≥ 0.65

In accordance with Table 4, the asphalt contents were higher to modified asphaltic mixtures than conventional asphaltic mixture. In the mixtures added RCA crushed dust, this result can be explained by increasing of specific surface area of recycled aggregates after compaction and its high porosity. Thus, the absorption is higher and, as a result, it reduces effective asphalt content increasing air voids and decreasing voids filled with asphalt (MILLS-BEALE and YOU, 2010).

With addition of pellet, asphalt content increased because asphalt accumulates on the surface of the PET particles. This issue results in the reduction of the bitumen film thickness around the aggregate particles and reduces the aggregate–bitumen adhesion. Consequently, air voids and voids filled with asphalt are higher to modified asphaltic mixtures than conventional asphaltic mixture (MODARRES and HAMEDI, 2014a; SOJOBI et al., 2016).

Marshall Stability and flow

Marshall Stability is the maximum load at diametrical compression that a specimen of asphaltic mixture supports until maximum flow. Figure 1 shows the Marshall Stability tendency in function of percentages of RCA and pellet to modified asphaltic mixtures.

FIGURE 1. MARSHALL STABILITY AS A FUNCTION OF RCA AND PELLET

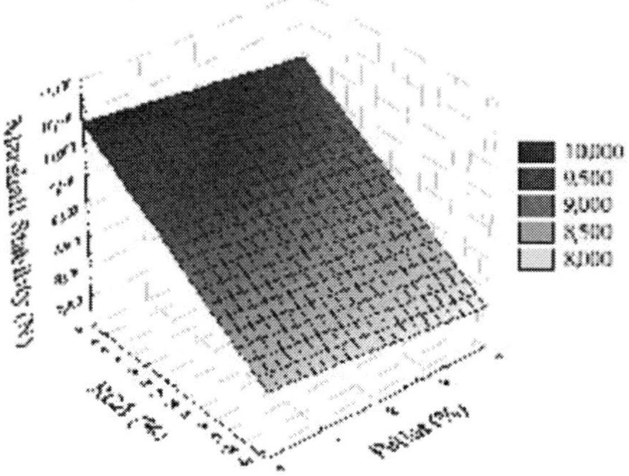

The Marshall Stabilities of modified mixtures were smaller than for conventional mixture, but all of them were higher than the minimum 4,900 N specified by DNIT ES 031/2006 standard. According to figure 1, the reduction of Marshall Stability was higher to mixtures made with RCA than with pellet. The diminishing of resistance for RCA asphaltic mixtures can be related to lower resistance and friction of recycled concrete aggregates (RAFI et al., 2010; ZHU et al., 2012). Taking into account the addition of pellet, Marshall Stability of the modified asphaltic mixtures also decreased. The diminishing of Stability with increment of PET content may be attributed to low friction between PET granules (HASSANI et al., 2005; MOGHADDAM and KARIM, 2012; MOGHADDAM et al., 2014). Besides that, the asphaltic mixtures modified by pellet in both situations pre-coating aggregates and modifying asphalt can provide Marshall Stability higher than conventional asphaltic mixtures (SOJOBI et al. 2016).

Flow is total strain of the asphaltic mixture specimen measured up null to maximum load during stability test. Figure 2 shows the flow tendency of asphaltic mixtures in relation to percentages of RCA and pellet.

FIGURE 2. FLOW AS A FUNCTION OF RCA AND PELLET

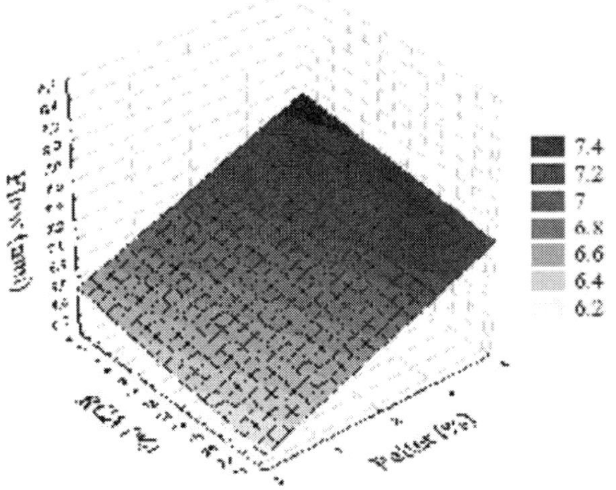

According to Figure 2, the increasing of flow is proportional to increment of percentages of pellet and RCA plus pellet. Besides that, the flow decreased when only percentages of RCA were taken into account. The asphaltic mixtures with RCA are more susceptible to bigger flow than conventional asphaltic mixtures (PÉREZ et al., 2012; PASANDÍN AND PÉREZ, 2013). In relation to modified asphaltic mixtures with pellet, the flow increases due to increment of pellet percentages. This flow increasing may be due to lower friction between PET granules exceeds good adhesion between the bitumen and PET granules (HASSANI et al., 2005; AHMADINIA et al., 2011; MOGHADDAM and KARIM, 2012).

Indirect Tensile Stress (ITS)

Figure 3 shows the tendency of ITS of asphaltic mixtures in relation to the RCA and pellet percentages. According to Figure 3, the ITS decreased to asphaltic mixtures modified by RCA and pellet. Besides that, the ITS values for NA100, NA75-RCA25, NA75-RCA23-P2 asphaltic mixtures were higher than ITS of 0.65 MPa specified at DNIT ES 031/2006 standard. In fact, the literature shows that ITS in RCA asphaltic mixtures is lower than NA asphaltic mixtures (ZHU et al. 2012; PASANDÍN AND PÉREZ, 2013). However, Lee et al. (2012) conclude that ITS in asphaltic mixtures with pre-coat RCA by slag cement paste was bigger than NA asphaltic mixtures. Moreover, increasing of ITS was more significant with pellet in asphaltic mixtures. In relation to the pellet added in asphaltic mixture, the asphalt content is bigger so it accumulates on the surface of pellet particles. As a result, the asphalt film thickness on aggregates is lower, thus the adhesion asphalt-aggregates is unsatisfactory, so ITS decreases (AHMADINIA et al., 2012; MODARRES and HAMEDI, 2014a; 2014b).

FIGURE 3. ITS AS A FUNCTION OF RCA AND PELLET

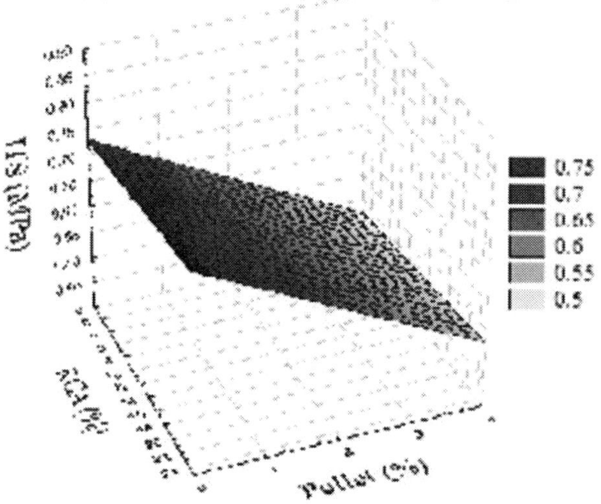

Resilient Modulus (RM)

The results show that resilient modulus diminished in relation to the increase of RCA and pellet in asphaltic mixtures. Figure 4 shows tendency of RM in relation to the percentages of RCA and pellet in asphaltic mixtures. According with Figure 4, the reduction of RM is proportional to reduction of RCA percentages on modified asphalt mixtures (MILLS-BEALE and YOU, 2010; QASRAWI and ASI, 2016). Besides that, the literature shows that RM performance is dependent on temperature and ageing time before the compaction. In this case, the absorption is higher and loss of volatile bitumen compounds carry out on hardening that affects stiffness of asphaltic mixtures. However, the RM for RCA modified asphaltic mixtures can be higher than the RM for conventional mixtures (PASANDÍN AND PÉREZ, 2013; 2014).

Taking into account pellet added to the asphaltic mixtures, the RM could decrease due to pellet percentages. To asphaltic mixtures modified by pellet percentages up 5% to 25%, with 2 mm diameter, the RM was inversely proportional to increasing of pellet (RAHMAN and WAHAB, 2013). Besides that, the results are divergent because lower percentages of pellet can increase the modulus resilient (AHMADINIA et al., 2012; MODARRES and HAMEDI, 2014a, 2014b), though RM can decrease for modified asphaltic mixtures with higher pellet contents.

FIGURE 4. RM AS A FUNCTION OF RCA AND PELLET

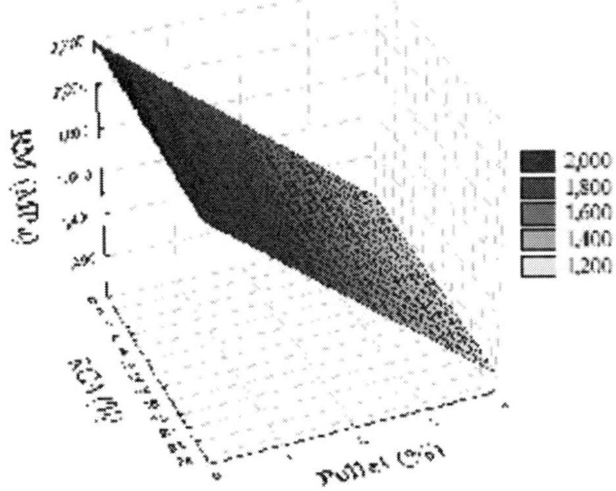

RM/ITS ratio

The resilient modulus and indirect tensile strength ratio (RM/ITS ratio) is an indicator of fatigue life of asphaltic mixtures. The lower RM/ITS ratio is better because it represents bigger strength and lower fatigue cracking on the surface course. Table 5 shows the results carried out.

According to the results in the Table 5, the DNIT ES 031/2006 standard specifies minimum ITS of 0.65 MPa, so NA100, NA75-RCA25, NA75-RCA23-P2 asphaltic mixtures can be analyzed. Thus, these asphaltic mixtures reached a MR/ITS ratio up 2,000 to 3,000 range, so the results are consistent with many conventional asphaltic mixtures made by some Brazilian asphalt plants.

TABLE 5. RESULTS OF RM, ITS TESTS AND RM/ITS RATIO

Mixtures	RM (MPa)	ITS (MPa)	RM/ITS
NA100	2,140	0.84	2.548
NA75-RCA25	2,049	0.69	2.970
NA75-RCA23-P2	1,575	0.68	2.316

NA75-RCA21-P4	1,176	0.59	1,993
NA98-P2	1,888	0.53	3,562
NA96-P4	1,040	0.42	2,476

CONCLUSIONS

RCA and pellet can modify asphaltic mixtures for using on surface course. However, in this research, the modified asphaltic mixtures had a lower mechanical performance than the conventional asphaltic mixtures. Besides that, the majority of modified asphaltic mixtures met specifications established in DNIT ES 031/2006 standard for surface course.

The optimum asphalt contents for modified asphaltic mixtures are higher than conventional asphaltic mixtures, so the production costs are also higher. However, the reutilization of wastes can diminish environmental and wastes disposal costs, so these ones can be a compensation of costs. The Marshall stability of modified asphaltic mixtures was lower, but the values were higher than minimum of 4,900 N established by Brazilian standard.

The modified asphaltic mixtures NA75-RCA25, NA75-RCA23-P2 met minimum ITS of 0.65 MPa specified at DNIT 031/2006 standard. In addition, both modified asphaltic mixtures carried out of RM and RM/ITS ratio values consistent with some asphaltic mixtures designed in Brazil for surface course. Thus, ITS, RM and RM/ITS ratio can suggest maximum percentage of 25% for RCA and 2% for pellet in addition to modify asphaltic mixtures.

Despite the lack of specific standards for alternative materials, this study concluded that RCA and pellet could be used to produce asphalt mixtures for paving of surface course, with advantages and limitations according to the technical standards analyzed herein.

REFERENCES

AHMADINIA, E.; ZARGAR, M.; KARIM, M. R.; ABDELAZIZ M.; SHAFIGH, P. Using wasteplastic bottles as additive for stone mastic asphalt. Materials and Design 32 (2011) 4844-4849. http://dx.doi.org/10.1016/j.matdes.2011.06.016.

AHMADINIA, E.; ZARGAR, M.; KARIM, M. R.; Abdelaziz M.; AHMADINIA, E. Performance evaluation of utilization of waste

Polyethylene Terephthalate (PET) in stone mastic asphalt. Construction and Building Materials 36 (2012) 984-989. http://dx.doi.org/10.1016/j.conbuildmat.2012.06.015.

DNIT ES 031/2006. Flexible Pavements – Asphaltic Concrete. Service Standard. National Department of Transport Infrastructure (DNIT). Rio de Janeiro. (In Portuguese).

HASSANI A.; Ganjidoust H.; Maghanaki, A. A. Use of plastic waste (polyethylene terephthalate) in asphalt concrete mixture as aggregate replacement. Waste Management & Research 23 (2005) 322–327. HTTP://DX.DOI.ORG/10.1177/0734242X05056739. Lee, CH.; JC. Dub; DH. Shen. Evaluation of pre-coated recycled concrete aggregate for hot mix asphalt. Construction and Building Materials 28 (2012) 66–71. doi:10.1016/j.conbuildmat.2011.08.025.

MARQUES, V. C.; Queiroz, B. O.; Lacerda, D. M.; Gouveia, A. M. A.; Melo, R. A. Mechanical Performance of Asphalt Mixtures with Natural Aggregates and Recycled Aggregates for Surface Course. Key Engineering Materials 600 (2014) 657-666. http://dx.doi.org/10.4028/www.scientific.net/KEM.600.657.

Mills-Beale J., You Z. The mechanical properties of asphalt mixtures with Recycled Concrete Aggregates. Construction and Building Materials 24 (2010) 230–235. http://dx.doi.org/10.1016/ j.conbuildmat.2009.08.046

MODARRES, A.; HAMEDI, H. Effect of waste plastic bottles on the stiffness and fatigue properties of modified asphalt mixes. Materials and Design 61, (2014a) 8-15. http://dx.doi.org/ 10.1016/j.matdes.2014.04.046

MODARRES, A.; HAMEDI, H. Developing laboratory fatigue and resilient modulus models for modified asphalt mixes with waste plastic bottles (PET). Construction and Building Materials 68 (2014b) 259–267. http://dx.doi.org/10.1016/j.conbuildmat.2014.06.054

MOGHADDAM, T. B.; Soltani, M.; KARIM, M. R. Experimental Characterization of rutting performance of Polyethylene Terephthalate modified asphalt mixtures under static and dynamic loads. Construction and Building Materials 65 (2014) 487-494. http://dx.doi.org/10.1016/j.conbuildmat.2012.02.054.

MOGHADDAM, T. B; KARIM, M. R. Properties of SMA Mixtures Containing Waste Polyethylene Terephthalate. International Journal of Chemical and Environmental Engineering 6, 2012.

Pasandín, A. R.; PÉREZ, I. Effect of ageing time on properties of hot-mix asphalt containing recycled concrete aggregates. Construction and Building Materials 52 (2014) 284-293. http://dx.-doi.org/10.1016/j.conbuildmat.2013.11.050.

Pasandín, A. R.; PÉREZ, I. Laboratory evaluation of hot-mix asphalt containing construction and demolition waste. Construction and Building Materials 43 (2013) 497–505. http://dx.doi.org/10.1016/j.conbuildmat.2013.02.052.

PÉREZ, I.; Pasandín, A. R.; Medina, L. Hot mix asphalt using C&D waste as coarse aggregates. Materials and Design 36 (2012) 840–846. http://dx.doi.org/10.1016/j.matdes.2010.12.058.

Pinto, H. K. R. F.; Assis, S. R. H.; Melo, R. A. Performance of Natural Aggregates and Recycled Aggregates for Unbound Granular Pavements Layers. Key Engineering Materials 600 (2014) 677-688. http://dx.doi.org/10.4028/www.scientific.net/KEM.600.677.

Qasrawi, H.; Asi, I. Effect of bitumen grade on hot asphalt mixes properties prepared using recycled coarse concrete aggregate. Construction and Building Materials 121 (2016) 18–24. http://dx.doi.org/10.1016/j.conbuildmat.2016.05.101.

Rafi, M. M.; Qadir, A. E; Siddiqui, S. H. Experimental testing of hot mix asphalt mixture made of recycled aggregates. Waste Management & Research, 1 – 12. HTTP://DX.DOI.ORG/10.1177/0734242X10370379.%202010.

RAHMAN, W. M. N. W. A.; WAHAB, A. F. A. Green Pavement Using Recycled Polyethylene Terephthalate (PET) as Partial Fine Aggregate Replacement in Modified Asphalt. Procedia Engineering 53 (2013) 124-128. http://dx.doi.org/ 10.1016/j.proeng.2013.02.018.

SOJOBI1, A. O., Nwobodo, S. E. and Aladegboye, O. J. Recycling of polyethylene terephthalate (PET) plastic bottle wastes in bituminous asphaltic concrete. Cogent Engineering, 3: 1133480, 1- 28. http://dx.doi.org/10.1080/23311916.2015.1133480. 2016.

VASUDEVAN, R.; S. K. Nigam, R. Velkennedy, A. R. C. Sekar, and B. Sundarakannan. Utilization of waste polymers for flexible pavement and easy disposal of waste polymers. In proceedings of International Conference on Sustainable Solid Waste Management, Chennai, India, pp. 105-111. 2007.

ZHU, J.; Wu, S.; Zhong, J.; Wang, D. Investigation of asphalt mixture containing demolition waste obtained from earthquake-damaged buildings. Construction and Building Materials 29 (2012) 466–475. http://dx.doi.org/10.1016/j.conbuildmat.2011.09.023.

STANDARDS, SPECIFICATIONS & INFORMATICS

IC-NOCMAT 2019

18th International Conference on
Non-Conventional Materials and Technologies
"Construction Materials & Technologies for Sustainability"
(18th NOCMAT 2019)
24th – 26th July 2019
Nairobi, Kenya

THE POTENTIAL UTILITY OF MATERIALS INFORMATICS IN DEVELOPING NON-CONVENTIONAL MATERIALS

Chuyuan Zheng[1], Yusuf Akintayo Akinbade[2], Ian Nettleship[1], Kent Harries[2]

[1] Department of Mechanical Engineering and Materials Science;

[2] Department of Civil and Environmental Engineering, Swanson School of Engineering, University of Pittsburgh, USA;

ABSTRACT

Materials informatics is widely acknowledged as a means of accelerating the rate of new materials discovery and material development. Such an approach will require widespread use of data driven methodologies and it is anticipated that downstream stakeholders, particularly industry, will be heavily involved in providing the required resources. This paper will propose workflows that illustrate how data driven methods could reasonably be applied to the processing-structure-properties relationships of locally sourced non-conventional materials. It is anticipated that simple material tests and imaging techniques will provide sufficient digital information for data driven approaches at relatively low cost. This utility will be illustrated using the example of bamboo and the levels of structure observable using widely available digital phone cameras. Finally, the possible role of the NOCMAT community in facilitating materials informatics will be considered.

Keywords: Non-Conventional, Materials, Informatics.

INTRODUCTION

New materials discovery and development has traditionally been a subject dominated by experimental approaches with relatively long development cycles of 20 years or more. The Materials Genome Initiative in the USA is a multidisciplinary, collaborative approach focused on shortening development cycles for materials by promoting major roles for computational modeling and simulation techniques of different types. Unfortunately, physics based modeling approaches – whether density functional theory for electron structure [1], thermodynamic modeling of phase stability [2], phase field modeling of microstructure or finite element analysis at the macroscopic scale – are often challenged by structural complexity. The multiple levels of structure in many materials have further limited the wider contributions of physics based modeling. Data driven modeling approaches are more suited to solving complex problems. Materials informatics [3] is a more recent application of data modeling approaches and while it does involve the core requirements of data capture, database management and data analysis it is also a basis for information based learning [4], through which new materials science could be discovered. Implicit in the practice of materials informatics is the participation of a committed community that has agreed to devote resources to the collection of trusted data and its consequent aggregation into curated databases that can be made available to those working in the community. This will take significant investment, especially if the data collection is expensive, requiring sophisticated, specialized equipment.

This paper will describe the potential of materials informatics to address the issues that contribute to the variability in the properties of locally sourced, non-conventional materials and also speculate on how such an approach might be implemented with suitable work flows. The specific example of full-culm bamboo will be considered. Finally, the expected challenges and potential benefits will be discussed.

MATERIALS INFORMATICS AND ITS APPLICATION TO NON-CONVENTIONAL MATERIALS

Briefly, materials informatics [3] is the application of information engineering to materials science and engineering. Data driven approaches such as machine learning and artificial intelligence, discover new knowledge through the generation, management and analysis of data. These techniques are able to efficiently deal with degrees of complexity that challenge physics based methods traditionally used in materials science and engineering.

The widely accepted materials science and engineering paradigm has within it three sources of data. Processing data derived from all the materials refining and manufacturing processes, materials structure data from all the levels of structure and finally all the different types of material property data. Data driven methods are used to predict relationships between designated inputs and outputs and thus it is necessary to decide which of the three types of data will be inputs and which will be outputs. Directly inferring relationships between materials processing and material properties is not desirable because it implicitly denies the central role of materials structure in developing new materials. It is therefore necessary to develop data driven models for both processing-structure and structure-properties relationships as shown in Fig.1. It is also necessary to assemble suitable information into accessible databases. Indeed, the availability of suitable data in appropriate formats will be of vital importance to the success of this type of research.

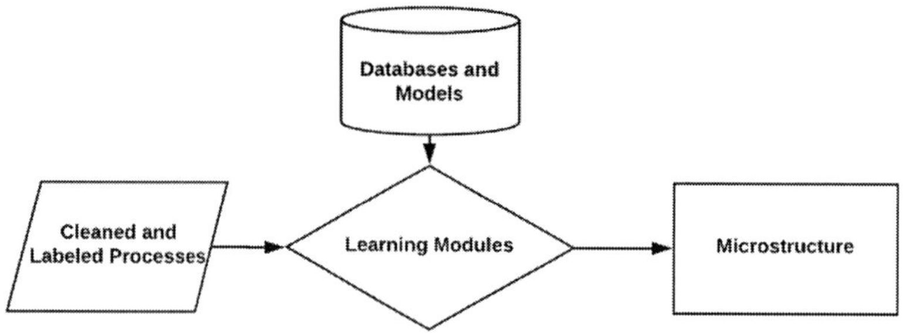

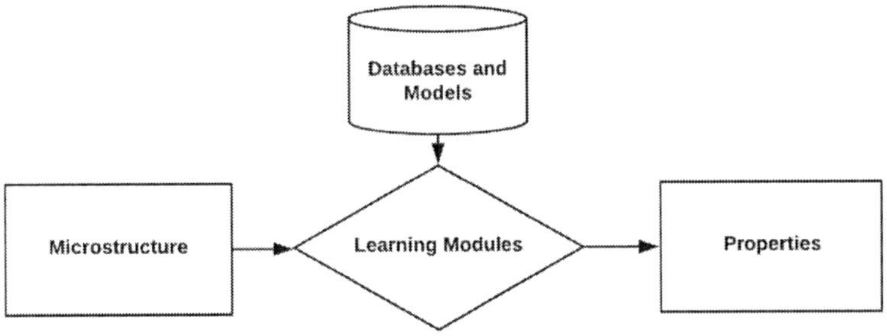

FIGURE 1: THE APPLICATION OF DATA DRIVEN METHODS TO THE MATERIALS PROCESSING-STRUCTURE-PROPERTIES PARADIGM.

In general, the databases needed for data driven materials discovery and development are not well developed across the materials processing-structure-property paradigm. Some traditional subjects such as basic crystallographic and thermodynamic information have been assembled into databases with some success while other topics suffer from a lack of coverage. The aggregation of data into searchable databases must accelerate if materials informatics it to fulfill its promise; several initiatives are being developed in this realm. In the US, for example, the National Institute of Standards and Technology (NIST) has taken a leading role in coordinating the necessary infrastructure and data resources in their Materials Genome Initiative [5], consistent with their traditional role in evaluation of data, data curation and data integration for materials. Specific materials data resources include the Materials Resource Registry [6], an inventory of materials databases, tools and repositories that can be accessed from anywhere in the world. In addition, the NIST Materials Data Repository [7] has the objective of improving the quality of materials data by establishing mechanisms for sharing data and best practices for management of materials databases.

Once databases of sufficient utility become available it will be necessary for materials science and engineering practitioners to become familiar with widely available data science techniques that could be used for establishing new processing-structure and structure-property relationships. At present even relatively simple methods such as principal component analysis, which correlates a set of recorded observations into a set of linearly uncorrelated outputs called principal components, is not well represented in materials science and engineering curricula at universities. More complex methods such as neural networks are capable of capturing non-linear relationships between each desired material output and the sum of all the material input observations [8]. The neural network "learns" by changing the weighting of the relationships between the input and output nodes called neurons as more data is added. While techniques such as neural networks have demonstrated great utility, they require proper model training and the user must have significant knowledge and understanding to correctly apply them.

At present, materials informatics has not been applied to non-conventional materials in any meaningful way. There are some databases that contain information on locally sourced materials such as wood. MatWeb [9], for example, contains property data for 306 different woods and wood based composites but such sources are rare and not sufficient to implement data analysis techniques for structure-property relationships. In particular there is relatively little archived information concerning the structure of non-conventional materials. This is somewhat inhibited by the wide range of important lengthscales in the structure of many non-conventional materials that, if they are to be fully quantified, would require characterization tools ranging from optical microscopes to transmission electron microscopes to micro tomography and x-ray diffraction. Many such tools are not easily accessed by practitioners working with non-conventional materials. Such data would also have to be compiled into standard formats that could be inputted into microstructure analysis kernels. These requirements raises real questions of social equity in materials informatics given that resources required for database development, curation and maintenance are, in theory, to come from downstream stakeholders

If data driven techniques are to be applied by the non-conventional materials community there is a need to come to agreement on their foreseen role, the nature of the information to be collected, data formats, accompanying metadata requirements and the organizations and institutions that would host such resources. Since non-conventional materials are locally sourced the processing-structure information should be collected in the field and so the required information must be constrained in ways that would make collection and submission of information feasible. Structure-property information is expected to be more limited but of higher fidelity if standard tests are used. Property information could also come from field tests if they were standardized but are more likely to come from laboratory testing. Fig.2 summarizes a proposed work flow by which materials processing-structure information collected in the field using a phone camera and information from laboratory tests could be combined into a materials informatics approach to non-conventional materials engineering. This will now be further explained in the context of bamboo.

THE EXAMPLE OF BAMBOO INFORMATICS

Bamboo, like wood, is derived from living tissue and therefore the levels of materials structure have evolved to function while the plant is alive and then degrade after the life cycle is complete. In contrast, for engineering applications, the plant is grown to a suitable maturity before harvesting, drying and treating. With the natural function of bamboo in mind, qualitative comparisons have been made between the levels of structure in bamboo and other natural materials and their biological function [10]. These levels of structure have been identified and imaged by a variety of techniques for bamboo [11-14]. The structure at some of these different scales is known to vary from species to species [15] and be sensitive to grow location and growth conditions. Once harvested, dried and treated, the properties of bamboo are dominated by fibers aligned with the axis of the culm and the fact that the culm is hollow. It is therefore reasonable to study the properties of bamboo in analogy with composites containing stiff aligned fibers. Hence, the axial properties, like elastic modulus Ec, are quantified using familiar concepts such as the rule of mixtures in Eq. 1.

$$E_c = (V_f E_f) + ((1\text{-}V_f)E_m) \qquad (1)$$

Where Ef is the elastic modulus of the fiber, Em is the elastic modulus of the matrix and Vf is the volume fraction of the fibers. This basic relationship has been adapted to take into account the variation in fiber volume fraction through the culm wall [16-19].

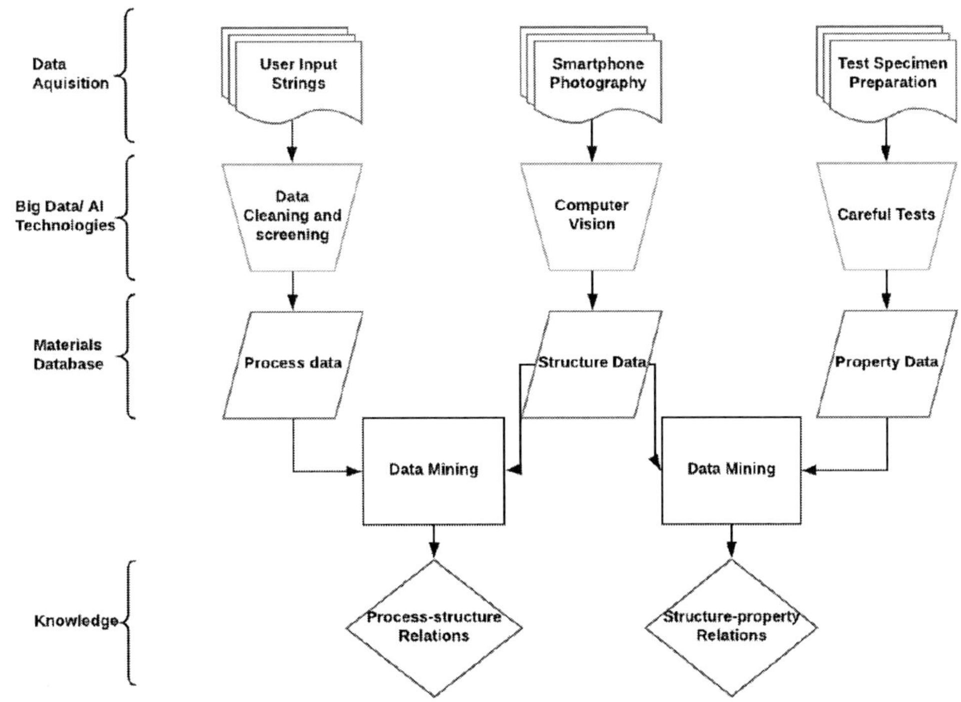

FIGURE 2: PROPOSED WORKFLOWS FOR PROCESSING-STRUCTURE AND STRUCTURE-PROPERTY RELATIONSHIP DETERMINATIONS FOR NON-CONVENTIONAL MATERIALS

Eq. 1 contains only one structural parameter, the volume fraction of the fibers and assumes two distinct and homogeneous phases. Examination of the structure at higher magnifications using electron microscopy and atomic force microscopy reveals a rich complexity that defies the notion of homogeneous reinforcing and matrix phases. While not representing all the levels of structure, Fig.3 shows a scanning electron micrograph that reveals the fibers are collected into fiber bundles that also contain hollow capillaries. The shape and internal structure of the fiber bundles varies from species to species and also through the thickness of the culm wall. Surrounding the fiber bundles is the parenchymal matrix made up of relatively large hollow cells.

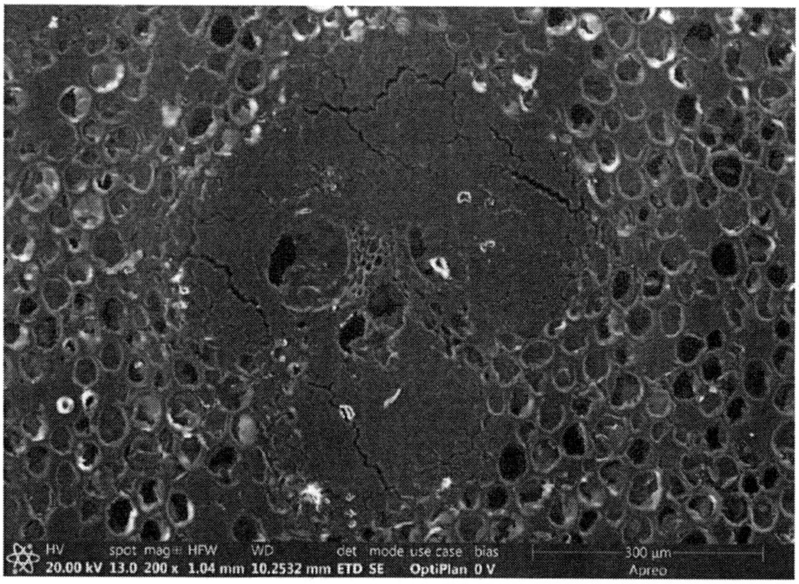

FIGURE 3: SEM MICROGRAPH OF A POLISHED SECTION OF BAMBOO

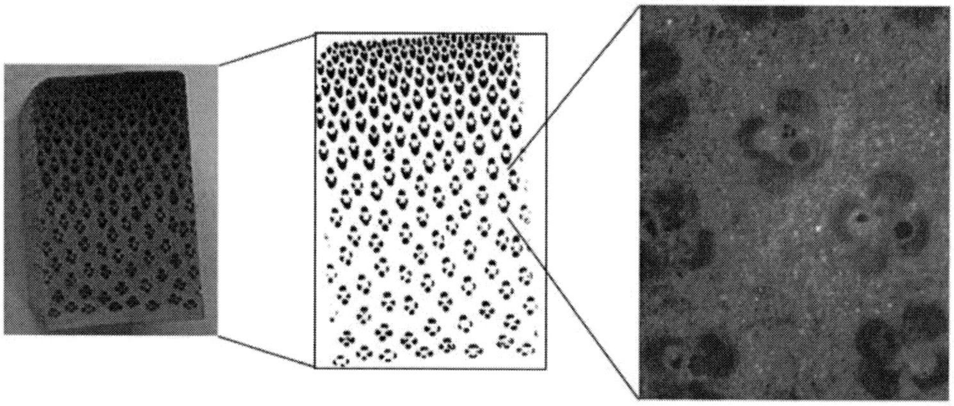

FIGURE 4: OPTICAL MICROGRAPH REPRESENTATIVE OF THAT TAKEN WITH A PHONE CAMERA AND THE CONSEQUENT PROCESSED IMAGE THAT ASSUMMES TWO PHASES. A HIGHER MAGNIFICATION OPTICAL MICROGRAPH OF THE BAMBOO STRUCTURE

If the rule of mixtures is a representative relationship, it is possible to make some interesting assumptions that would allow the structure of bamboo to be imaged and quantified using a simple phone camera suitable for use in the field. Fig. 4 shows such an optical image in which fiber bundles are distinctive due to their dark contrast. The accompanying processed image that assumes the material is two phases is also shown. The higher magnification optical micrograph shows that the vessels associated with the fiber bundles will also have the same dark contrast and thus be included in the measured fiber bundle area fraction. Thus it is clear that the practitioners working on bamboo will have to agree on how the contrast is to be interpreted during data processing by computer vision methods. While these assumptions do not take into account any variation in the internal structure of the fiber bundles, it does, under the constraints of the contrast conditions, allow the gradient in fiber volume fraction along the thickness of the culm wall to be estimated. It is thought that the variation in the gradient of fiber bundles volume fraction is responsible for much of the variation of axial properties. If this proves to be the case, one could envision that examination of the natural variation of the gradient of fiber bundles across the culm wall and the influence of species and growth locations and conditions would be able to explain the variation in axial properties of bamboo. Much more data will be needed if the application of data driven approaches are to develop a better understanding of the variability in the axial properties of bamboo.

The transverse properties of bamboo, on the other hand, are dominated by weak interfaces that commonly lead to splitting. In this circumstance, rule of mixtures for transverse loading is not expected to be particularly meaningful, especially when the variability in experimentally measured properties is large. Therefore, the correlation of the transverse properties of bamboo to structure observed using a simple phone camera will be more challenging. In particular, the transverse properties of bamboo appear to be dominated by the strength of the interfaces between the cell walls in the parenchyma [20] and cracking within the fiber bundles [21]. The phone camera might be able to determine if the transverse failure occurred through the fiber bundles or through the parenchyma cell walls but the interpretation will be more nuanced.

Although imaging of the volume fraction of fibers can be done in the field with a camera phone, this does not realize the database(s) required to generate, clean, store and maintain this information in a form that can be input into common data analytic tools. This must be done under the guidance of the NOCMAT community with an enduring commitment of its members to participate. It is important that the community agree on the standards used for capturing, transmitting, sharing and curating such information.

SUMMARY

Materials informatics is a data driven approach that has been proposed as an alternative means of discovering new materials and managing materials development. This is particularly attractive when the materials structure is complex with several levels that can influence desired properties. This complexity can lead to significant variability between experimental measurements on nominally the same material and confound physics based modeling approaches. While it is quite obvious that data driven approaches are proving their worth for complex problem solving in a variety of fields, they do require research communities to commit to developing standards and infrastructure for data acquisition, transmission, storage and sharing. In the case of materials informatics, there is also a need to divide the processing-structure-properties paradigm into processing-structure and structure-properties relationships to facilitate the application of data science techniques. These techniques could be as simple as principal component analysis or they might require more complicated models such as neural network and the attendant specialized knowledge. The infrastructure and standards for materials informatics have been slow to evolve since it was proposed over 10 years ago. However, recent developments such as the Materials Genome Initiative are beginning to gain influence in this respect.

The ability of materials informatics to address complexity is quite attractive in the case of natural, locally sourced non-conventional materials which are characterized by a continued lack of effective materials standards and the wide variability in properties. It is not very difficult to propose that materials information from simple field tests and observations could be collected at practically any location with a camera phone and shared with the community. This prospect does indeed play to the strengths of data driven approaches for materials development. However, the challenges associated with resources development and the collective commitment required to develop standards for data acquisition, transmission, storage and sharing should not to be underestimated. In the world of highly controlled materials the resources for database curation and management will probably come from the downstream stakeholders including multinational industries and governments. In the case of locally sourced non-conventional materials that serve relatively poor communities, the resource requirement raises significant issues of social equity. This manuscript has described how, in the example of full-culm bamboo, it may be possible to constrain the data acquisition to relatively inexpensive and widely available technologies that lower the barrier to the application of materials informatics to non-conventional materials. Since, this would be a collective endeavor it is critically important that the non-conventional materials community provide leadership in the development of standards and infrastructure so that participation would be broad and sustained with the benefits being widely shared.

ACKNOWLEDEMENTS

The authors would like to acknowledge the funding provided by the US National Science Foundation under award 1634739.

REFERENCES

1. Togo A. and I. Tanaka I., First principles phonon calculations and materials science, Scripta Materialia, 2015, 108, 1-5.

2. Liu Z.K., First-principles calculations and CALPHAD modeling of thermodynamics, Journal of Phase Equilibria and Diffusion, 2009, 30, 517.

3. Rajan K., Materials informatics, Materials Today, 2005, 8, 38-45.

4. Lecun Y., Bengio Y. and Hinton G., Deep learning, Nature, 2015, 521, 436-444.

5.The Materials Genome Initiative, US National Institutes of Science and Technology, https://www.nist.gov/mgi.

6.The Materials Resource Registry, US National Institutes of Science and Technology, https://www.nist.gov/programs-projects/nist-materials-resource-registry

7.Materials Data Repository, US National Institutes of Science and Technology, https://materialsdata.nist.gov/

8. Dengiz O., Smith A.E. and Nettleship I., Two-stage data mining for flaw identification in ceramic manufacture, International Journal of Production Research, 2006, 44 2839-2851.

9. MatWeb http://www.matweb.com/, 2018,

10. Wegst U.G.K., Bai H., Saiz E., Tomsia A.P. and Ritchie R.O., Bioinspired structural materials, Nature Materials, 2015, 14, 23-36.

11. Ray A.K., Das S.K., Modal S. and Ramachandrarao P., Microstructure characterization of bamboo, 2004 Journal of Materials Science, 39 1055-1060.

12. Palombini F.L., Kindlein Jr. W., de Oliveira B.F., Araujo Mariath J.E., Bionics and design: 3D microstructural characterization and numerical analysis of bamboo based on X-ray microtomography, Materials Characterization, 2016, 120, 357-368.

13 Krause J.Q., de Andrade Silva F., Ghavami K., da Fonseca Martins Gomes O., Dias Toledo Filho R., On the influence of Dendrocalamus Giganteus bamboo microstructure on its mechanical behavior, Construction and Building Materials, 2016, 127, 199- 209.

14. Zou L., Jin H., Lu W.Y. and Li X., Nanoscale structural and mechanical characterization of the cell wall of bamboo fibers," Materials Science and Engineering, 2009, C 29, 1375-1379.

15. Clark L.G., London X., and Ruiz-Sanchez E., Bamboo taxonomy and habitat, W. Liese, M. Kohl (eds.), Bamboo, Tropical Forestry 2015, 10, 1-30 Springer International Publishing Switzerland.

16. Amaada S., Munekata T., Ichikawa N.Y., Kirigai A. and Zhifei Y., 1996. The mechanical structures of bamboos in the view point of functionally gradient and composite materials, Journal of Composite Materials, 1996, 30, 800-819.

17. Nogata F. and Takahashi H., 1995. Intelligent functionally graded material: bamboo, Composites Engineering, 1995, 5, 743-751.

18. Li H. and Shen S., The mechanical properties of bamboo and vascular bundles, Journal of Materials Research, 2011, 26, 2749-2755.

19. Dixon P.G. and Gibson L.J., The structure and mechanics of Moso bamboo materials, Journal of the Royal Society, Interface, 2014, 11, 1-12.

20. Wang F., Shao Z., Wu Y., Mode II interlaminar fracture properties of Moso bamboo, Composite B, 44, 2013, 242-247.

21. Akinbade Y., Flower C.V., Harries K.A., Nettleship I., Papadopoulos C. and Shawn Platt S. Determining Through-Culm Wall Distribution of Mechanical Properties of Bamboo, submitted to Construction and Building Materials, 2019.

IC-NOCMAT 2019

18th International Conference on
Non-Conventional Materials and Technologies
"Construction Materials & Technologies for Sustainability"
(18th NOCMAT 2019)
24th – 26th July 2019
Nairobi, Kenya

NEGOTIATING THE STANDARDS MAZE TO ENABLE THE USE OF NOVEL CONSTRUCTION MATERIALS

Saad S. Yahya

University of Nairobi

ABSTRACT

One attribute of a cultured and organized sedentary community is the existence of rules and regulating on how buildings and supporting services should be constructed. And these rules get more complex as time passes, mainly because of changes in demographics, prosperity (or otherwise), lifestyle trends, technological advance and so on. Then there are new construction materials being developed and old materials finding new uses. And that is the essence of sustainability, that is making the future more livable and today's comforts and safety durable. It is for that reason that regulatory regimes and practice codes are essential. This paper reflects on the efforts of government and municipalities, especially Kenyan, to design and administer appropriate standards and building codes that will suit the diverse needs of citizens and industry, while not forgetting the staggering levels of social inequality. Much of the material comes from research work done in the period 2016-18 on first, standards for basic infrastructure in unplanned settlements in five cities; and secondly investigations to support the development of a housing policy for the City of Nairobi. Both initiatives were sponsored by the government of Kenya.

INTRODUCTION

One attribute of a cultured and organized sedentary community is the existence of rules and regulating on how buildings and supporting services should be constructed. And these rules get more complex as time passes, mainly because of changes in demographics, prosperity (or otherwise), lifestyle trends, technological advance and so on. Then there are new construction materials being developed and old materials finding new uses. And that is the essence of sustainability, that is making the future more livable and today's comforts and safety durable.

It is for that reason that regulatory regimes and practice codes are essential. This paper reflects on the efforts of government and municipalities, especially Kenyan, to design and administer appropriate standards and building codes that will suit the diverse needs of citizens and industry, while not forgetting the staggering levels of social inequality. Much of the material comes from research work done in the period 2016-18 on first, standards for basic infrastructure in unplanned settlements in five cities; and secondly investigations to support the development of a housing policy for the City of Nairobi. Both initiatives were sponsored by the government of Kenya.

DO STANDARDS MATTER?

Regulatory regimes: Why building codes are always work in progress: The early years of post-colonial Kenya saw a drastic review of the nation's laws. The Building Code 1968, passed as subsidiary legislation straddling both local legislation and public health legislation1, is to all intents and purposes still operational in spite of numerous attempts over the last four decades to review and update it. Its resilience lies in the coverage of the full range of building construction work in the formal sector, although its weakness when it comes to low cost and informal housing could be attributed to its somewhat rigid prescriptive approach rather than a benign, and probably more realistic, performance-driven stance.

Numerous studies have since been undertaken to address this issue, and one little known achievement is "Code 98" which was passed by the national Government as a set of adoptive rules under the Building Code in the section dealing with Scheduled Special Areas. The code aimed at enabling housing standards and procedures through reducing building costs, promoting innovative designs, encouraging use of local materials and putting emphasis on performance rather than specific construction materials was introduced. However this was adopted by only few Local Authorities. It allowed municipalities to adopt those rules if they wished, but only a handful (Nakuru being one) were to take advantage of this facility, although the then Housing Department in the central government, together with civic society activists went to great pains to promote the tool with training materials, workshops, prototype projects and so on.

All the same the 1968 code had some redeeming features, including Grade 2 Bylaws the Special Scheduled Areas device as envisaged in the Building Code and Code 98. Relaxed performance specifications can be used for development in such an area.

It is important that a more creative approach to building regulations is sought, rather than continuing to blame the colonial past for bequeathing the industry with unattainable standards. Code 98 was the first step, now the sector has to move farther forward. A big setback is the scarcity of building control staff who could effectively

- Administer the existing three-tier system and

- monitor the condition of buildings already built

- engage with designers and artisans on improving construction site practices.

Additional reforms done in the last decade are in the form of the Planning and Building Regulations 20092. This was an ambitious omnibus which came in five volumes but is still waiting for formal approval and use. For instance the document contains elaborate guidelines on non-water borne sanitary disposal, especially important as the emergence of low income multi-storeyed apartments and tenement blocks is posing new challenges, mainly on the following fronts:

- Congestion, typically in the form of high densities and exceedingly small rooms.

- Sanitation and hygiene issues, e.g. a toilet per floor, serving 30 or even 50 people; water scarcity; absence of adequate drainage and waste disposal facilities; limited food preparation space; polluted and smelly air.

- Structural instability putting residents in great danger. Building collapse is not uncommon, and lives are lost every so often; there is also an economic loss as precious materials go to waste.

Exposure to an unsafe environment, the risks including fire, burglary and violence after dark; women and children are overly exposed in this respect. Police surveillance in such neighborhoods is scant.

Skill shortfalls: More than half (54.5%) of all BEPs with graduate degrees and above operate in Nairobi (NHA 2012 p62). It is important that architects and engineers develop appropriate solutions that can address the tenement market in terms of safety, space standards, ventilation, safety and other attributes of a liveable home. There are new technologies that could make their work easier e.g. remote monitoring, through sensors, of different performance measures including how the premises are used, or particular structural elements are responding to stress or excessive use. Such monitoring and dispatch of relevant data could also be eased with the development of suitable smart phone applications.

The expansion of the building materials industry will help reduce costs, especially if the use of new types of materials was encouraged outside the traditional natural stone, cement and timber. A large percentage of materials, fittings and equipment are imported, mainly of low quality, and at low prices, making it difficult for local manufactures to innovate so that they can compete with imports from the near and far east. Existing facilities for testing construction materials need to be improved in terms of quality, precision and efficiency.

However better and less expensive materials will be of little benefit if the construction industry is not well organized. The majority of builders have had no formal training, while only a small percentage of contractors (20%) are registered or adequately equipped. Generally there is serious shortage of skilled artisans and technicians.

Built environment professionals enjoy statutory protection through registration boards, which is good for the purpose of maintaining professional and integrity standards, but more needs to be done to make professionals more responsive to the needs of the poor and how homes could be made better and more affordable through creative approaches to design tasks and project management.

GROWTH TRENDS

It is projected that by the year 2030, about 50 per cent of the Kenyan population will be urban residents. The rapid rate of urbanization continues to put more pressure on services to meet the needs of the growing population3.The demand for urban housing will continue to grow. In 2010 the demand for urban housing was estimated at around 80,000 units a year, projected to increase to nearly 300,000 units by 2050. By comparison, in 2013 only 15,000 housing construction permits were issued in Nairobi, and most of these were for high-income apartments.

Over 50 per cent of Kenyans live below the poverty line. The median income levels for employed people are between Ksh20,000 and Ksh25,0004. This means that apart from the population living below the poverty line, an average income earner cannot access housing from the market. Only 2 percent of formally constructed houses are targeted to the lower income segments of the market, which account for the largest share of demand. Furthermore, these estimates of housing demand are for new housing only and do not speak to the high qualitative housing deficit, as manifested in the high level of informality. Renting, mostly in the informal market, is more accessible and affordable.

Roughly 60 percent of Kenya's urban households live in housing that would be defined as a slum under the Millennium Development Goals5. Formal housing supply is not keeping pace with the growing urban population. On the other hand, investment by private sector players in low income housing has been minimal because returns are not as high as in the high income bracket. The private sector has tended to concentrate on the high end of the market. Therefore, Informal housing has become the only housing choice for most urban Kenyans.

A World Bank report on the status of African Cities hints that Eastern Africa has been identified as having the largest share of households in overcrowded conditions, the smallest share of households with permanent floors and access to toilets, and the largest percentage of households with electricity connections6. Although the capital investment shortfall that makes African cities crowded appears across all building types, it is most severe in housing. In Nairobi, commercial and industrial structures explain 55 percent of the total value of building stock — even though these structures occupy just 4 percent of the city's area. Residential development is urgently lacking.

Nairobi city and metropolis: East African cities are growing fat, and metropolitan Nairobi is a case in point. Although Nairobi City County has an estimated 1,813,000 formal and informal jobs, these jobs are not all held by city residents—an estimated 187,000 persons commute into Nairobi City County every day, and some 41,000 city residents commute to outlying counties7. That is for every one person who leaves the city for work every morning, four and a half come the other way.

The region has a population of about 6 million. Nairobi City, also known as Nairobi County, is the largest of the three cities in Kenya, followed by Mombasa and Kisumu. Of the 20 urban centers in the country, the largest is 28 times the size of the smallest.

Using the average inter-censal growth rate of 4.1%, the projected population of Nairobi City in 2017 was 4,327,716, to reach 7,296,561 by 2030; by then it is anticipated that Kenya will be predominantly urban with 61.5 per cent of its population living in urban areas. Not only are people becoming more numerous but they are living closer to one another, a trend which is continuing. Under the existing growth scenarios, it is anticipated that the Nairobi Metro Region will carry a majority of the national urban population.

The average household size in Nairobi is 3.2 members, and there are significant differences by location, poverty status, and gender of head of household. Male-headed households in informal areas are also significantly larger than female headed households, at 3 vs. 2.4 on average. Nearly half (44%) of Nairobi households have monthly expenditures below the poverty line. The proportion is lower among households with tenure i.e. 19% as opposed to 46% of those without.

Fixed capital specifically commercial and industrial structures in Nairobi account for 55 per cent of the total value of the building stock occupying only four percent of the city's area. By the end of 2017, it was envisaged that a further 130,000 square meters of modern retail space would complete development in the city.

But an updated estimate of population and housing need indicates that the average number of units required over the next ten years is approximately 123,599 units per annum, to gradually rise to a higher figure subsequently, although the population growth rate is likely to see a mild dip.

A critical component of urban expansion is building materials. It is estimate that over the next ten years the City of Nairobi (i.e. core city rather than the metropolis) will need Ksh1.1 trillion to be invested in building materials for housing alone8.

Materials supply chain and use: The high cost of approved building materials e.g. cement and steel is a disincentive. The building code prohibits use of alternative building materials and technologies. Prescribed materials in the building code include sand, cement and natural stones. As sand and natural stone are running out fast (not to mention the negative environmental degradation) there is need to explore substitutes. New materials and ABMT uptake has been slow. Most contractors and developers are conservative. There is need for a lot of campaigns, education and awareness raising. There is also the need to synthesize designs and plans to available approved building materials and components.

There are quite a number of new alternative materials and technologies in the construction industry. However, a few builders are venturing away from traditional methods either in the form of new building techniques or new materials. The City planning department has recorded a few instances where different types of alternative building materials and methods, such as expanded polystyrene (EPS) panels and waffle precast slabs have been used.

To control the use of substandard materials and building components, there is need for rigour in the testing and certifying of materials both local and imported by Kenya Bureau of standards. The roles of the Kenya Building Research Centre should be enhanced to actively research and support the local engineering and architectural talent to produce new and better materials and fittings locally. Indeed incentives to use available technical innovations and improve research capacity are needed. Industry/universities (of which Nairobi has many, and good ones too) / NCCG collaboration would go a long way.

There is a dire need for increased prefabrication to minimize wastage and improve on the quality of the finished product and delivery time. This is an area where the suppliers could capitalize on as the players are quite few at the moment.

The mushrooming of semi-permanent structures, made of galvanised iron sheets and timber framing, for rent in the peri-urban areas is a cause for worry. The occupants of these units are subjected to poor living standards, but since there is no regulation on non-permanent structures, it is difficult to monitor and control such housing ventures.

Building standards for industrial production require a limited number of possible sizes of building components. This allows for designing and adapting a common applicable system of dimensions e.g. as in a module.

Access to professional services is also restricted. There are many players in the construction industry. The National Construction Authority (NCA) plays a key role in registering the contractors, but more needs to be done - e.g. approval of buildings and inspection of ongoing buildings should also include monitoring the contractors and other professionals involved e.g. are they registered, qualified and so on. NCA should play a greater role in ensuring housing construction is undertaken by competent workers and artisans; and city inspectorate department needs to be allocated more human and other resources given the size of construction work ongoing. Related issues include the following:

- Skills and competency for each level should be in harmony with the county reality since currently there is a missing link. Data from NCA indicates that the construction industry in Kenya is dominated by 75% of semi-skilled and unskilled construction workers.
- The industry is experiencing skills shortages in key areas such as plumbing, electrical, masonry, paint works, welding and carpentry.
- 55% of contractors do not own equipment of their own and rely fully on leased and hired equipment.
- 78% of contractors rely on credit financing through banks and other sources.
- There are few women contractors with only 7% owning construction firms.
- Construction of houses sometimes takes long resulting in cost overruns and hence become unaffordable.

According to the NCA by 2013 only 20% of the contractors were licensed. This partly explains the cases where buildings are substandard despite a developer/investor fully financing construction in accordance with approved plans, bills of quantities and structural standards.

The situation also arises from risky cost cutting measures adopted by some developers to cut down on costs and time. Indeed, one cause for buildings collapsing is the speed at which they are built and use of say a weaker cement mix than that which is recommended. This raises the issue of building control and compliance. The barriers in the housing market delivery value chain will need to be addressed in a very realistic and coordinated manner.

SHACKS, TENEMENTS AND MEGA-RESIDENCES

Coping with diversity: It is important that policies and actions aimed at delivering adequate housing for an urban population addresses the broad range of expectations and aspirations. Large cities have to provide housing to cater for a wide range of accommodation, ranging from the most basic to the unbelievably luxuries apartments and villas costing two hundred times as much. Houses are a good indicator of inequality. This is reflected in the quality and availability of infrastructure services, which in turn is mirrored in regulatory regimes such as land use zoning and construction materials standards for building and services. When natural are scarce (e.g. timber, gravel, sand, thatch, lime, even water) questions arise as to whether appropriate controls are needed in the interest of fair distribution.

A recent phenomenon is the creation of tenements, a gross type of housing reminiscent of the industrial revolution in Europe.

Informal settlements: Taking preferences into account is crucial in order to ensure a close match between the types of housing that households want to live in and what the market offers. Affordability is even more critical for households in informal settlements in view of their low incomes and the lack of budgetary resources from the public sector to finance direct subsidies.

At the end of 2012, over a half (53%) of the households in informal settlements had incomes of less than Ksh13,000 per month or about Ksh17,000 in current prices. Most of the FGD participants largely depend on the informal sector (i.e. vegetable vending, mitumba,9 bauty salons, food kiosks and bars), or on casual labour in industries and construction sites with incomes averaging Ksh14,000 per month. This low income level points to the limited amount households in informal settlements would be able to pay for housing. Behavioural data for such households suggests that the propensity to pay for rental housing is not more than 20%, excluding water and electricity, and 27% if these services are included. This translates to a net rent not exceeding Ksh3,400 per month for a half of the population in Nairobi's informal settlements.

The vast majority of informal settlement households are renters and nearly a half of them (48%) pay rents below Ksh 2,000 in 2012. Focus groups conducted showed that the typical rent -- usually for a single room – ranges between Ksh1000 to Ksh3000 per month, and therefore an average of Ksh2000 per month. The cheapest room often has an earth floor and is made of timber walls and rusty iron sheets. They are often in areas that are most insecure or prone to environmental hazards. This finding, although not statistically robust because it is not based on household survey data, is more or less in line with the median rent obtained from survey data. The rental rooms available in the informal settlements are affordable by the majority of households but do not offer an acceptable housing solution because of their very poor standards: the building fabric is typically of low quality, and infrastructure services are either lacking or of an appallingly low standard. While some have illegal electricity connections others have no power and therefore residents rely on paraffin and candles for lighting. Numerous buildings in the City of Nairobi stand on sewer lines, roads reserves or on high voltage electricity lines. Green areas and public open spaces are rapidly dwindling. Sanitation is a major challenge as latrines and bathrooms are either non-existent or are too far to access especially at night or by children. Consequently disease epidemics, flooding and fires are common hazards. Yet the rates of return on informal housing are inordinately high, pointing to the extortionate behaviour of landlords in slum areas, probably driven by the perceived risks of investing in slum housing. Policy will need to address how slums can be phased out by improving security of tenure as well as raising the quality of infrastructure services and building structures.

Besides renting, it is important to consider owner occupation. Behavioural data shows that house buyers would be willing to pay more for housing than renters. Applying the typical yardstick of 30% of income,10 more than a half of households (53%) in Nairobi's informal settlements would not able to pay more than Ksh5,000 per month towards house purchase; and about a third cannot pay more than Ksh3,500 per month.11 However, formal dwellings with a monthly repayment in this range simply do not exist. Indeed, an estimate of household demand, based on current household income, clearly shows that in informal settlements there are practically no households that can afford to buy a house built by a formal developer. To illustrate, only about 10% of Nairobi's households could afford the cheapest newly built house by a formal developer in 2015, priced at around Ksh1.7 million12. The alternative for low-income and middle-income households who wish to own a home is to purchase a small residential plot and build incrementally over a number of years. Even this option would most likely be unviable for households in the first quintile (20%) on the income distribution in view of the prevailing high prices for serviced land in the city and its immediate peri-urban belt. However, prices decline sharply with distance from the core city,13 offering lower-income households the option of living in better housing further away from the inner city. This alternative comes at a cost because of higher commuting expenses over the long distances involved, compounded by heavy traffic jams and the opportunity cost in time.

Savings and credit cooperatives (SACCOs) and housing cooperatives have responded to the strong demand for small repeater loans for land purchase and incremental housing construction. These lenders provide the bulk of housing finance, accounting for more than 90% of housing credit.14 Most of this credit is unsecured and is taken out in the form of small loans of maturities that usually do not exceed 7 years.

Tenements: Nowhere are construction materials more abused than in the building of tenements. These are multistoried "apartments" in unplanned areas with high levels of congestion, say five or six families living in tiny rooms on ne floor and sharing a kitchen and toilet. Sanitary facilities are extremely poor. Liquid waste drains into open trenches which are also often used to irrigate urban agriculture.

Hence the need for a pragmatic approach, and recognition of the necessity to balance the needs of the poor against those of the more fortunate. As those needs change over time, policy adjustments may be necessary so as to keep it fit for purpose. More than half (54.5%) of all BEPs with graduate degrees and above operate in Nairobi (NHA 2012 p62). It is important that architects and engineers develop appropriate solutions that can address the tenement market in terms of safety, space standards, ventilation, safety and other attributes of a liveable home. There are new technologies that could make their work easier e.g. remote monitoring, through sensors, of different performance measures including how the premises are used, or particular structural elements are responding to stress or excessive use. Such monitoring and dispatch of relevant data could also be eased with the development of suitable smart phone applications.

Developers have to face inordinate hurdles, some of them avoidable, and some occasioned by integrity lapses. Ongoing attempts to make the building permit system more efficient and friendly should be designed to benefit all income groups, in order to facilitate rather than hinder mass production. But that is not to say that building control standards should be unduly relaxed, both in the interest of safety and hygiene. For example the irresponsible use of materials and labour as now seen in the tenements sub-sector, and the resulting risk of collapses, often fatal, far outweighs the benefits in increased housing stock. Designers, innovators and building control professionals ought to do more to protect lives in these apartment blocks. It is also unfortunate that there is a worrying scarcity of construction artisans at a time when technical colleges are almost empty for lack of eligible entrants.

Mega-residences: These are the homes of the rich and super-rich, typically corporate executives, diplomats, senior politicians and professionals. In the big cities, say Nairobi, Darussalam and so on) the price of a luxury apartment or house can be two hundred times that of a basic house occupied by say a cab driver or office clerk. Middle class household occupy the affordability zone in-between. Favorite locations for the super-rich are slopes and hilltops, usually west of the city centre, and water fronts. But refinement is not necessarily a quality they appreciate, since materials used for finishes and fittings are largely laminates and other synthetics, with hardly any timber or other natural materials. Sustainability issues relate to the frivolous use of materials, fittings, energy and other resources. Then there is infrastructure investments to take into account. However economic realities are such that such neighbourhoods are seen as reliable markers of urban prosperity; unfortunately addressing inequality is hardly viewed as the business of building codes and materials standards.

WHY DO STANDARDS FAIL?

In a study undertaken in 2006-08 some attention was paid to why and how building standards, especially infrastructure standards for informal settlements, fail; and the effects that failure has on not only prevailing pubic policies and technical approaches but also the reputation of sector players15. The questions addressed included:

- The extent, if any, to which slum improvement work is inhibited by the prevailing infrastructure standards regime
- What types of standards in the delivery process are really important?
- What are the performance criteria or indicators? What are the risks posed by standards failure or malfunction?
- What type of harm or damage should be expected, and who should bear the cost?
- What needs to be done to ensure that standards are fit for purpose and durable?

A suitable approach for analyzing and resolving these issues was developed in the form of a grading system represents in the form of a matrix and accompanying text. The focus was on infrastructure projects for informal settlements. The levels of risk associated with different categories of standards were ranked from Grade A (lowest) to Grade D (highest), which led to a set of preliminary interventions grouped into specific themes. These recommendations are subject to further refinement as work progresses. In short the top two risk categories are as follows:

Grad D threats: tenure/ destitution; participation; procurement lapses

Grade C Threats: maintenance/durability; cost and affordability; communications and operations.

Grade B threats: Planning /development control; inclusivity; sustainability; resilience; environment; workmanship.

Grade A threats: plot sizes (and by implication house size); materials; fittings.

For instance supply chain fraud – whether perpetrated by suppliers, subcontractors, employees, or a combination of those – could come in many forms, including:

- Falsified labour
- Inflated bills or expenses
- Bribery and unwarranted "gifts"
- Inflated vendor invoices
- Bid manipulation
- Counterfeit or cheap uncertified products
- Flouting specifications
- Unauthorized disbursements.

Technical issues like materials use, fittings, workmanship and resilience generally appeared lower down the scale. The point is that while technical standards are indispensable, for them to be effective there needs to be a strong foundation based on openness, inclusion and integrity.

INSTITUTIONAL DEVELOPMENT AND COOPERATON

When you have a complex web of policy makers it becomes difficult for administrators and professionals to develop new initiatives or even to reconcile the requirements relating to existing materials with those relevant to building methods. For example the following list agencies or groups was identified by planners when consulting stakeholders in Nairobi in 2018, as commanding an important say on how or where houses should be built.

Core policy makers constituted the state ministries16 involved in housing and related services including the Ministry of Transport, Infrastructure, Housing and Urban Development; Ministry of Devolution and Planning M&E component, the Disasters and Emergency Response Coordination department under the Ministry of Interior and Coordination of National Government, the Task force on Urban and Nairobi Metropolitan Development and the National Museums of Kenya. Others consulted include the National Lands Commission, the Kenya Informal Settlements Improvement programme (KISIP), National Environment Management Authority (NEMA), Kenya Urban Roads Authority (KURA), Nairobi Water Supply and Sanitation Company. Most of these were engaged through key informant interviews and in focus groups organized for the SWOT analysis of the housing sector and policy makers. These departments are crucial as they understand the current legal and legislative playing ground, the hurdles of housing development in the absence of an articulated policy.

But government on its own has not been able to improve industry capacity and responsibility. Some of the innovations mentioned above, like the slim two-page code known as Code 99 was the fruit or relentless efforts by civil society activists and built environment professionals in academia. Such reforms need unconventional approaches to the making of laws and administrative procedures. Developers and builders, I the interest of promoting their business interests, often provide solutions from which everybody could benefit. It is the professional societies which have somehow lagged behind when it comes to technical and managerial innovations, especially in the realm of shelter for the poor.

Built environment professionals enjoy statutory protection through registration boards, which is good for the purpose of maintaining professional and integrity standards, but more needs to be done to make professionals more responsive to the needs of the poor and how homes could be made better and more affordable through creative approaches to design tasks and project management. Green buildings is a fashionable phrase among designers but the realisation of that goal is not a priority although the amount of literature and good examples is rising. For instance International Finance Corporation's Ten Ways to Encourage Green Building Growth17 is a crisp prescription from which professionals and administrators could benefit.

CONCLUSION: Addressing Change

There is unrelenting pressure on prevailing rules and regulatory regimes to evolve and improve, consonant with changing times. But even on a day to day basis, it is important for practitioners to have the capacity to innovate and develop creative interpretations of existing rules so as to meet new challenges in an imaginative manner. For instance the following approaches are possible, and indeed one can find examples of a handful of imaginative solutions already in use on the ground, e.g. in the work of the Kenya Informal Settlements Improvement Programme.

1. Improvisation: Where the rules are silent, one should be able to conceive ad hoc rules to suit a particular situation, or create an exception to the rule. An example is the storm water sumps/collection basins built in Malindi town on Kenya's east coast, where conditions were right and the county/ municipality had to turn a blind eye. Under normal circumstances the rising water table would make pit latrines in the neighbourhood overflow. There is thus an element of experimentation in the approach, but it has been undertaken after careful studies and thought by the engineers and project manager.

2. Stretching or making standards more elastic: Prescription can be made more elastic or flexible in the spirit of performance standards; for example already available in the statutes book, in the context of housing, is Code 98, a simple set of adoptive rules which counties can adopt for development in Special Scheduled Areas. There is no similar code for infrastructure, and one is not likely to emerge soon owing to the rigid standards employed by utility companies. The performance standards approach (also known as the deemed-to-satisfy approach) is hugely amenable to technical innovations and advancement.

3. Broadening the band: The idea here is to introduce a min/maximum band of requirements, e.g. a plot shall be not less than 100m2 and not more than 400m2; or water consumption shall be not less than 50lit. per day and not more than 150lit. The upper limit is the desirable and the lower the absolute minimum.

4. Progression/ gentrification: Since life in some slums is getting better, either through upgrading or general economic growth in a particular city coupled with private regeneration efforts, the prevailing standards need to take into accountant market demand for higher standards of dwellings and services. Policies should support transformation of today's slums into tomorrow's middle income neighbourhoods. That process is sometimes accelerated by seed investment as briefly described below.

5. Catalytic initiatives: Seed investment in hardware needs to be leveraged with targets that elevate the aspirations of slum residents, e.g. investment in an iconic asset/facility like a local college or university18; a hospital; a social hall incorporating youth club, or theatre or art gallery or media and open data hub. A well designed shopping/high street (rather than a shopping mall) could work wonders, bringing safety and ambience while capturing the local spending power and creative talents and maturing the land and property market. Where is today's equivalent of Nairobi's Pumwani Social Hall which nurtured boxers, artists, dancers, dramatists for several generations in the last century? Planning and design standards for such soft infrastructure need to be developed.

The above approaches, though unconventional, can be pursued within the confines of existing laws. Building codes in general and materials standards in particular are perennial work in progress, but they are a mere tool in the larger task of building affordable, comfortable and durable homes for all.

18th International Conference on
Non-Conventional Materials and Technologies
"Construction Materials & Technologies for Sustainability"
(18th NOCMAT 2019)
24th – 26th July 2019
Nairobi, Kenya

PHYSICAL AND MECHANICAL BEHAVIOR OF POLYVINYL ALCOHOL IMPREGNATED PHYLLOSTACHYS PUBESCENS (MOSSO) BAMBOO

Lucas Muniz Valani[1], Fabrício de Campos Vitorino[1], Adriana Paiva de Souza Martins[1], Romildo Dias Toledo Filho[1]

[1] Sustainability Center for Research and Education on Environmental Friendly Materials and Technologies - NUMATS, Civil Engineering Program, Universidade Federal do Rio de Janeiro, Brazil;

ABSTRACT

Bamboo have outstanding carbon sequestration potential and excellent mechanical properties, amongst natural construction materials. However, due to its hygroscopic, humidity and temperature sensitivity, its outdoor use is limited. Such conditions lead to longitudinal cracks formation. Cracks occurs in bamboo's wall as a consequence of a radial moisture gradient that generate stress perpendicularly to natural fiber alignment. Moisture gradient is related to bamboo's microstructure, composed of vessels, fibers and parenchyma which densifies from inside to outside walls. Aiming to minimize aforesaid problems and to enhance mechanical properties, polyvinyl alcohol (PVOH) was used to impregnate strips of Phyllostachys Pubescens (Mosso) bamboo. Different impregnation temperatures (23, 60 and 100°C) were used to find the most effective polymer penetration. An optical active dye was mixed with PVOH to evidence the depth of impregnation in bamboo's wall thickness. Water absorption tests were carried out in plain and impregnated samples to evaluate the PVOH impregnation efficiency. Further, tensile mechanical tests were performed on most efficient impregnated specimen, to understand how PVOH affect this property. The results have shown that the optimal impregnation temperature was 60°C which led to a water absorption reduction of 33%. PVOH also affected the tensile mechanical behavior as compared to non-impregnated bamboo.

Keywords: Bamboo; Polyvinyl Alcohol; Impregnation; Physical Properties; Mechanical Behavior.

INTRODUCTION

Despite the many advantages of bamboo, some deficiencies of this material limit its use as building material, such as dimensional instability, hygroscopicity, low resistances to biological attacks and natural weathering.

The natural polymers present on cell walls and the presence of moisture make the material vulnerable to attack by microorganisms and insects, which feed on these compounds, extracting energy for survival. In addition to these biological agents, natural weathering agents such as ultraviolet radiation, oxygen, temperature and humidity variations, heat, air pollutants, and others also cause degradation of the material [1].

To minimize the problems of durability, many treatment methods have been developed. However, in general, methods are based on improving only one vulnerability. Thus, the use of preservatives improves biological behavior, but there is a need for complementary treatments to reduce water absorption and dimensional instability [2].

Bamboo, due to its porous structure, absorbs moisture or water when in contact with these agents, and retracts when it is dried [3].

Among the possible treatments to minimize water absorption and dimensional variations, polymer impregnation and heat treatment have been successfully adopted [4–9]. However, it should be noted that heat treatment causes a decline in mechanical properties [4, 9], as well as is an energy-intensive process.

Liese [10] states that the main route of penetration of the impregnant is the metaxylem vessels of the vascular bundles, and that the parenchymal cells that surround the vessels can be reached only by diffusion. Fiber protection also depends on diffusion.

Samples of Bambusa Tulda Roxb. presented higher dimensional stability, lower water absorption and higher mechanical strength after impregnation with thermosetting resins, as investigated by [5].

According to [5], the treatment of bamboo with water-soluble resins produces insoluble compounds that remain impregnated in the cell walls, guaranteeing the efficiency of the method and avoiding subsequent leaching of the impregnant.

Anwar et al. [7] verified the influence of immersion time and drying time (post-impregnation) on samples of Gigantochloa Scortechinii impregnated with phenol formaldehyde. The treatment was optimized for immersion time of 150 minutes and duration of drying at 60 ° C of nine hours.

The impregnation of wood (pine) with solutions of PVOH (3% by mass) reduced the ingress of water, and it was obtained a reduction of 80% in relation to untreated samples [11].

Yu et al. [12] studied the influence of impregnant content (phenol formaldehyde) on the physical and mechanical behavior of engineered bamboo (scrimber bamboo).They got aimpregnant content of 16% as optimum. The impregnations were performed at roomtemperature, with immersion time equals to 6 minutes.

The main objective of this work was to verify the influence of different temperatures (23° C, 60 ° C and 100 ° C) on the efficiency of the impregnation of bamboo (Phyllostachys Pubescens) with polyvinyl alcohol (PVOH), aiming to improve dimensional stability and mechanical properties. For the optimum temperature, the benefits were quantified through physical tests (water absorption) and mechanical tests (bending tests).

Microstructural tests (optical microscopy) were performed to qualitatively evaluate the penetration of the impregnant in the voids of the vascular system, in the lumens of the fibers and in the voids of the parenchyma matrix.

MATERIALS AND METHODS

Culms of Phyllostachys Pubescens (3 years old) were purchased from the company Takê Cortinas and Artefatos de Bambu Ltda, with a length of 2 meters and extracted from the median region of bamboo sticks. The preservative treatment used by the provider applied smoke in the mature bamboo. After receiving the culms, they were identified and stored in a dry and ventilated place. Strips of 300 x 20 x 5 mm (length x width x thickness) were produced from the unfolding of the culms.

The impregnant used was an aqueous solution of polyvinyl alcohol (PVOH). Polyvinyl alcohol is a crystalline polymer soluble in water and in some organic solvents with excellent properties of film forming, emulsification and adhesiveness. It has a molecular weight of 15,000 g / mol, pH equals to 4.8 (21 °C) and viscosity equals to 3793 cps (21 °C).

Firstly control tests were performed to obtain the time required for mass stabilization during impregnation with PVOH at temperatures of 23 °C, 60 °C and 100 °C. After impregnation, the samples were dried in an oven (103 °C) until mass constancy. Subsequently, the water immersion tests were carried out to identify the most effective impregnation temperature in the reduction of water absorption. For the optimum temperature (60 °C), 3-point bending tests and analysis of the microstructure (with the aid of an optical microscope) were performed.

The three-point flexural tests were run on a Shimadzu AGX model with a 100 kN load cell and displacement control at a rate of 1 mm / min. The distance between the end supports was 200 mm. The data acquisition system used the program Trapezium X. The displacements in the center of the span were obtained through an LVDT positioned in the central region. Four samples were tested in each condition (impregnated at 60 °C and not impregnated). The samples were previously dried in an oven (103 ºC), until constancy of mass. The flexural strength was obtained through Eq. (1), according to [13].

$$\sigma_{max} = \frac{3PL}{2bh^2} \tag{1}$$

where:

σ_{max}- maximum bending stress (MPa)
P- maximum load (N)
L - distance between end supports (mm)
b - prismatic sample width (mm)
h- sample thickness (mm)

A Zeiss Axioimager M2.m optical microscope coupled to an AxioCam MRc5 digital camera were used to capture reflected light microscopy images. The analysis was run in fluorescence contrast mode to illuminate the PVOH, which was mixed with an appropriate optical dye. For this purpose, prismatic samples (5 x 10 x 20 mm) of impregnated bamboo were cut, in the direction of the cross section of the fibers (as shown in Figure 1). Then the surface B was sanded and polished. A Zeiss Colibri 2 (445 nm) LED light source was used in combination with a Zeiss 38HE filter to ensure proper compatibility between fluorochrome excitation and the emission spectrum upon acquisition of the chamber. Similar procedure was also carried out by [14].

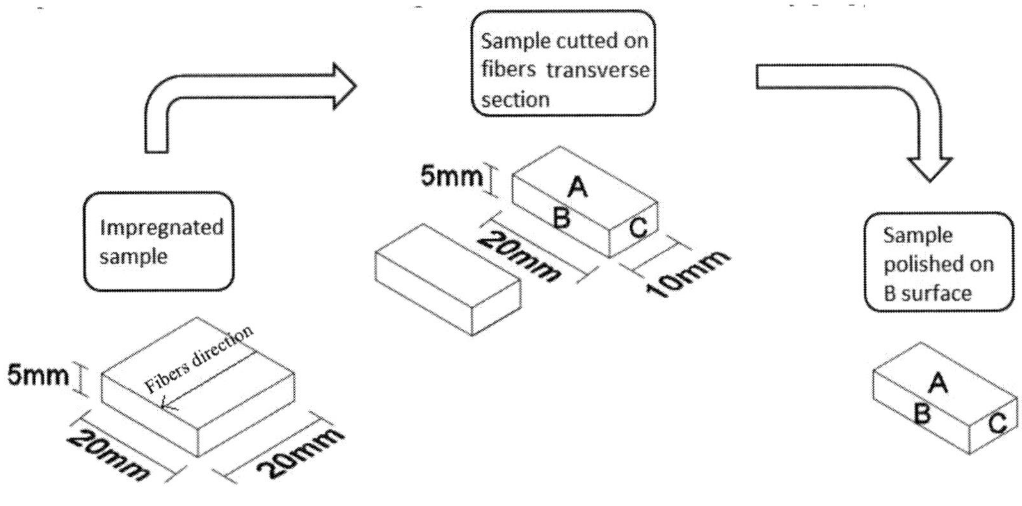

FIGURE 1 – SCHEME OF SAMPLE PREPARATION FOR OPTICAL MICROSCOPY OBSERVATIONS.

RESULTS

Impregnation of bamboo

Figure 2 shows the influence of time and temperature of PVOH polymer impregnation in bamboo. On Figure 2 is also possible to see the reference curve, which is the plain bamboo water absorption at 23°C. It is possible to see that there is a rapid polymer (water in reference case) mass increase at the first 24 hours of impregnations. After that, mass increases gradually until no further variation can be observed. At 23ºC, PVOH reach polymer mass constancy after 96 hours, while at 60 and 100ºC after 48 and 24 hours, respectively. Reference case absorbed water until 168 hours prior to mass constancy.

The rise in the temperature, up to 100ºC, altered the polymer impregnation capacity of bamboo. It can be seem that 60ºC is the optimal temperature for PVOH mass impregnation. Above 100ºC, the PVOH starts set and impregnation of the sample become impractical (see detail in Figure 2). The effective polymer mass gain, which can be seen after oven dry the impregnated bamboo sample until mass constancy at 100ºC for 24 hours, were 12%, 10% and 15% for 23, 60 and 100ºC, respectively. Notice that, plain bamboo is capable to practically regain 100% of water, after oven dry it lost whole absorbed water plus some extractives that is removed by water, which explains the negative mass value (-3% of mass).

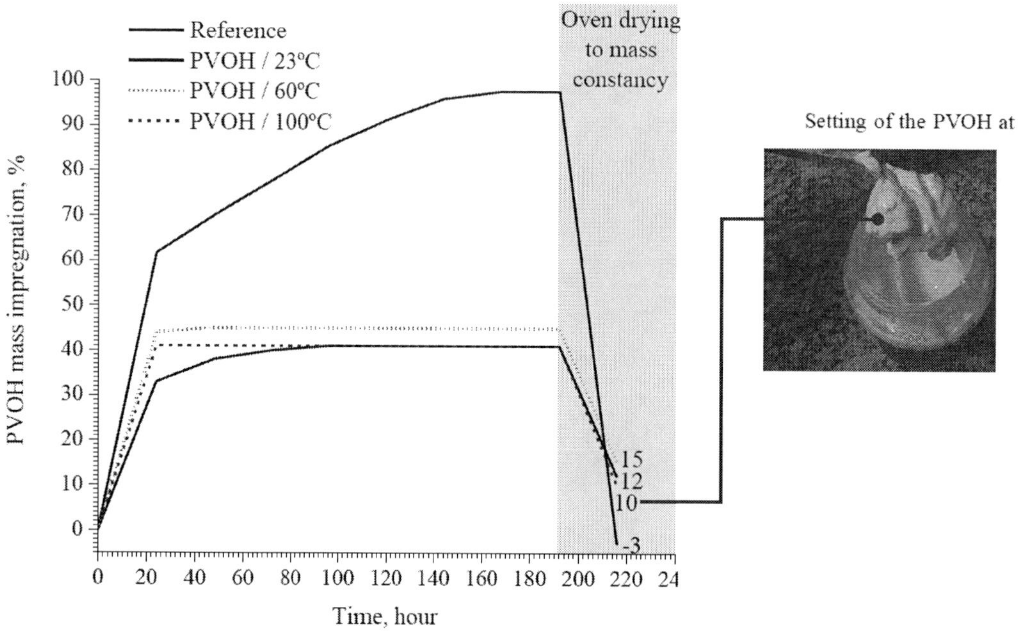

FIGURE 2 – IMPREGNATION OF BAMBOO WITH PVOH AT 23, 60 AND 100°C.

Water absorption of bamboo

The efficiency of the different temperature of PVOH impregnation of the bamboo against the water absorption can be observed on Figure 3. The water absorption of the plain bamboo can also be seen in the graph. The results show that the augment in the impregnation temperature lead to reduction in the water absorption of the bamboo. Compared to values found for the reference bamboo, the reductions observed on the water absorption of the impregnated bamboos were in the order of 27%, 32% and 29% for 23, 60 and 100°C impregnation temperatures. In this case, the 60°C impregnation was considered the optimal impregnation temperature.

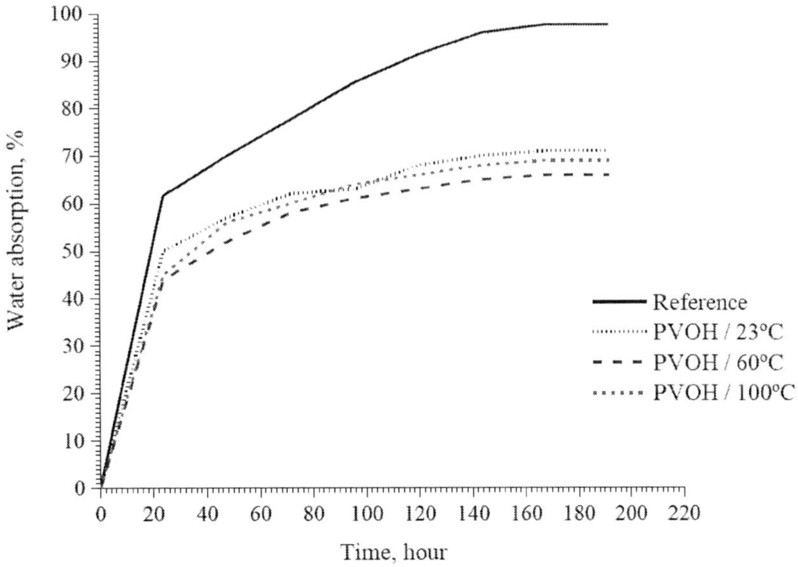

FIGURE 3 – WATER ABSORPTION OF BAMBOO AFTER 23, 60 AND 100°C PVOH IMPREGNATION.

Optical microscopy of 60°C impregnated bamboo

Figure 4A depicts the impregnation sample. It is possible to see, from the axial point of view, that the number of vessels increase from the outer wall of the bamboo to the inner wall, which also explain the increase in porosity in the same direction of bamboo.

Figure 4B shows the extent of 60°C PVOH impregnation of the bamboo on the transverse walls. The fluorescent green in the image is the PVOH optical activated by dye. One can see that, the PVOH penetrated through the bamboo walls forming a barrier-like in it. As can be seem in the scale of Figure 4 A, whole Pholem plus Xylem height (on inner wall) is approximately 0.5 mm.

In addition, it is possible to see that the PVOH also penetrate, axially, through some spots in the bamboo's Xylem and Parenchyma.

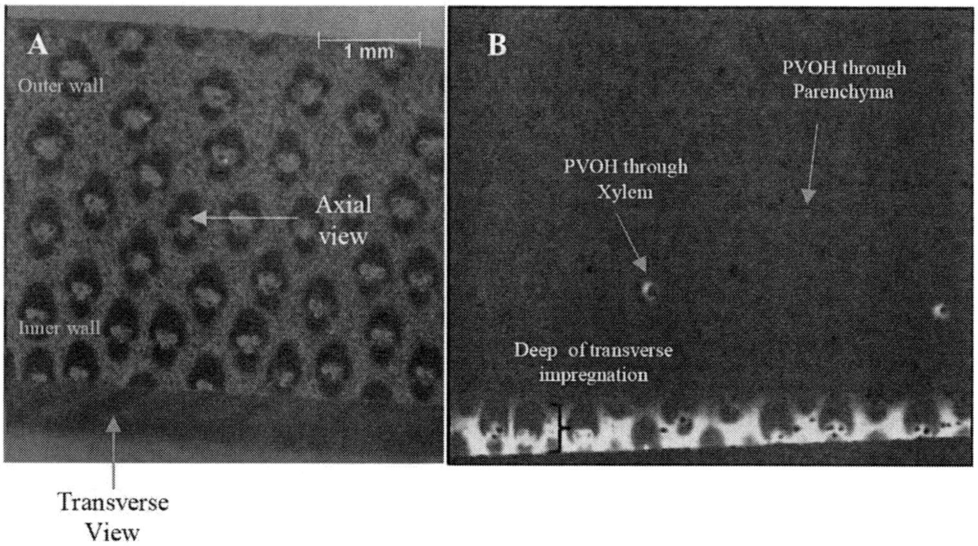

FIGURE 4 – A- VIEW OF THE IMPREGNATED SPECIMEN; B- OPTICAL MICROSCOPY SHOWING THE IMPREGNATION OF THE PVOH (AS GREEN) ON THE BAMBOO WALLS.

Flexure stress-deflection behavior

The typical stress-deflection curves of the Reference (plain bamboo) and the PVOH impregnated bamboo can be seem in figure 5. The tests were carried out on the 60ºC impregnated samples, once that was considered the optimal impregnation temperature. Standard deviation can also be seen in the figure 5 (green and grey areas).

From the typical curves on figure 5 is possible to conclude that the impregnation of the bamboo at 60ºC did not affected the peak stress and peak deflection significantly. As summarized on Table 1, small reductions on peak stress and peak deflection were found (16% and 26%, respectively). From the table 1 is also possible to conclude that the impregnation of the bamboo lead to decrease of the elastic modulus, which is plausible due to the polymeric nature.

However, as can be seem on standard deviation areas and on variation coefficients (VC on table 1), these differences were considered statistically similar. More tests will be carried out to reduce the standard deviation and confirm these tendencies.

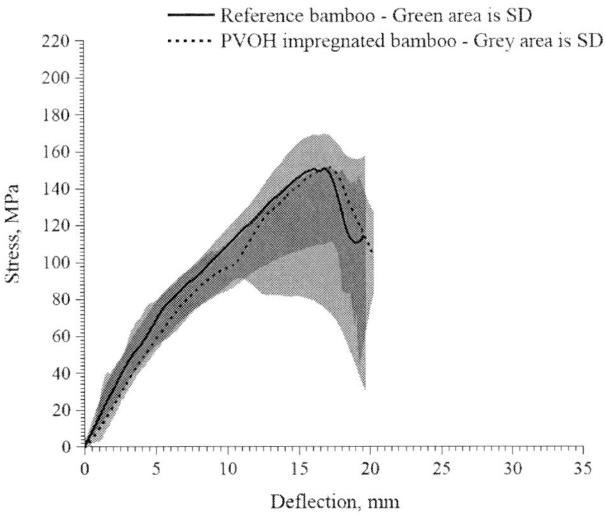

FIGURE 5 – FLEXURAL BEHAVIOR OF IMPREGNATED AND NON-IMPREGNATED BAMBOO.

TABLE 1 – PEAK STRESS AND DEFLECTION, AND ELASTIC MODULUS OF REFENCE AND IMPREGNATED BAMBOO.

	Reference bamboo		PVOH impregnated bamboo	
	Main	VC (%)	Main	VC (%)
Peak stress (MPa)	153.7	15.1	129.2	20.2
Peak deflection (mm)	16.7	12.4	12.3	58.2
Elastic modulus (GPa)	21.8	34.0	17.5	18.6

CONCLUSIONS

The present study evaluated the impregnation efficiency of PVOH polymer in a Mosso bamboo. The remarks of this study were:

- In the used methodology for polymer emulsion impregnation, the increase in temperature up to 60ºC showed to be the optimal condition;
- The impregnation above 100ºC was not feasible due to polymer setting;
- Water absorption significantly reduced after PVOH impregnations. In the optimal case, the reduction reached 24%;
- The use of optical active dye allowed to observe the formation of a barrier-like and small spots of PVOH in the bamboo sample;
- Mechanical tests showed that the impregnation method (up to 60ºC) did not affected negatively the stress-flexural behavior of the bamboo;
- There was a tendency of elastic modulus reduction with PVOH impregnation, however, more tests have to be done to confirm this tendency.

ACKNOWLEDGEMENTS

The authors wish to acknowledge the fund support of the Brazilian Council for Scientific and Technological Development – CNPq.

REFERENCES

1. SILVA JC (2007) Madeira preservada e seus conceitos. Rev da Madeira 103:9

2. Lee CH, Yang TH, Cheng YW, Lee CJ (2018) Effects of thermal modification on the surface and chemical properties of moso bamboo. Constr Build Mater 178:59–71. https://doi.org/10.1016/j.conbuildmat.2018.05.099

3. Azadeh A, Ghavami K (2018) The influence of heat on shrinkage and water absorption of Dendrocalamus giganteus bamboo as a functionally graded material. Constr Build Mater 186:145–154. https://doi.org/10.1016/j.conbuildmat.2018.07.011

4. Zhang YM, Yu YL, Yu WJ (2013) Effect of thermal treatment on the physical and mechanical properties of phyllostachys pubescen bamboo. Eur J Wood Wood Prod 71:61–67. https://doi.org/10.1007/s00107-012-0643-6

5. Deka M, Das P, Saikia CN (2003) Studies on dimensional stability, thermal degradation and termite resistant properties of bamboo (Bambusa tulda Roxb.) treated with thermosetting resins. J Bamboo Ratt 2:29–41. https://doi.org/10.1163/156915903321908549

6. PREECHATIWONG W, MALANIT P, KYOKONG B, KAMLANGDEE N (2011) Effects of Polymer Impregnation on Properties of Bamboo. Walailak J Sci Technol 3:79–91. https://doi.org/10.2004/wjst.v3i1.151

7. Anwar UMK, Paridah MT, Hamdan H, et al (2006) Impregnation of bamboo (Gigantochloa scortechinii) strips with low-molecular-weight phenol formaldehyde resin. J Bamboo Ratt 5:127–133

8. Anwar UMK, Paridah MT, Hamdan H, et al (2009) Effect of curing time on physical and mechanical properties of phenolic-treated bamboo strips. Ind Crops Prod 29:214–219. https://doi.org/10.1016/j.indcrop.2008.05.003

9. Yang TH, Lee CH, Lee CJ, Cheng YW (2016) Effects of different thermal modification media on physical and mechanical properties of moso bamboo. Constr Build Mater 119:251–259. https://doi.org/10.1016/j.conbuildmat.2016.04.156

10. Liese W (2004) A preservação do colmo de bambu com relação à sua estrutura. In: Simpósio Internacional Guadua. Pereira, Colombia

11. Shukla SR, Kamdem DP (2010) Swelling of polyvinyl alcohol,melamine and urethane treated southern pinewood. Eur J Wood Wood Prod 68:161–165. https://doi.org/10.1007/s00107-009-0361-x

12. Yu Y, Liu R, Huang Y, et al (2017) Preparation, physical, mechanical, and interfacial morphological properties of engineered bamboo scrimber. Constr Build Mater 157:1032–1039. https://doi.org/10.1016/j.conbuildmat.2017.09.185

13. American Society for Testing and Materials (2013) ASTM D1037 Evaluating Properties of Wood-Base Fiber and Particle. Annu B ASTM Stand 1–30. https://doi.org/10.1520/D1037-06A.1.2

14. Krause JQ, de Andrade Silva F, Ghavami K, et al (2016) On the influence of Dendrocalamus giganteus bamboo microstructure on its mechanical behavior. Constr Build Mater 127:199–209. https://doi.org/10.1016/j.conbuildmat.2016.09.104

IC-NOCMAT 2019

18th International Conference on
Non-Conventional Materials and Technologies
"Construction Materials & Technologies for Sustainability"
(18th NOCMAT 2019)
24th – 26th July 2019
Nairobi, Kenya

DEVELOPMENT OF PRECAST PLASTER ELEMENTS USING WOOD WASTE AND GLASS WASTE

Thibério César de Souza Vieira Lima[1], Eduardo da Cruz Teixeira[1], Marcos Antônio Padilha Júnior[1], Normando Perazzo Barbosa[2], André Luiz Santos Patriota[1]

[1] Instituto Federal do Sertão Pernambucano, Salgueiro – PE, Brazil;

[2] Universade Federal da Paraíba – UFPB – Brazil;

ABSTRACT

This paper aims at the development of precast elements of gypsum, wood residues produced by carpenters and glass residues were collected from the city of Salgueiro in northeastern Brazil. Were produced boards with the following dimensions (40x50, 33x33, 36x36 cm) were developed. After the collection of the residues, the tests of specific mass, grain size composition, pick time and apparent density were carried out. With the purpose of analyzing the mechanical resistance of the product resulting from the addition of gypsum with residues, 02 sample families were produced, the first consisting of gypsum and wood residue, and the second formed by the addition of gypsum and glass residues. Prototypes with 10% to 50% substitution of wood and vitreous silica residues were carried out, binder characterization tests were performed. with the reduction of gypsum and the addition of vitreous silica, thus contributing to a considerable reduction of environmental impacts. It was noted that the glass residues have good compactness. With the particle size analysis, it can also be concluded that the glass debris has well graded granulometry. The residue of the wood powder presents a grain size curve with continuous development. We performed substitutions of the gypsum mass for the corresponding volume of the wood residues, with the realization of the prototypes and possible to observe that the wood leaves a brownish aspect, already for the glass continues with the aspects of being only plaster. of time of handle, in which it can be observed that the binder has fast drying, making possible the production of precast elements.

INTRODUCTION

In the city of Salgueiro-PE there are commercial activities generating residues, among them glass debris, wood dust, and construction and demolition waste (RCD), these do not have adequate disposal, resulting in accumulations of debris in inappropriate places , burdening the environment. The wood is a heterogeneous material composed of cells produced by a living tree being chemically constituted by Carbon (C), Hydrogen (H), Oxygen (O) and Nitrogen (N) in the greatest quantities (KLOCK; ANDRADE, 2013). ´

The Glass is an amorphous ceramic material from organic materials at high temperatures (SMITH; HASHEMI, 2012). There are several studies that indicate the reuse of glass waste as a constituent of new materials, but the absence of public policies and government incentives results in discouragement for a more effective performance of organizations interested in generating income from the use of this residue.

Gesso is one of the ancient binders used by man, with vestiges of use before Christ. One of its main uses is in the production of precast elements for masonry and wall cladding, in addition to being widely used as decoration elements. For John and Cincoto (2010) in Brazil the application is directed to panels of linings and partitions and masonry linings.

Therefore, the development of research that analyzes the feasibility of using waste in new materials is of paramount importance. This research aimed to develop precast gypsum elements to be applied to masonry and wall coverings, using solid waste from the commercial activity of the city of Salgueiro.

MATERIAL AND METHODS

This study aimed to develop precast elements of gypsum plaster, initially were collected wood residues produced by carpenters, and glass residues. Wood waste from sawmills located in the city of Salgueiro in Pernambuco state were collected. The residues were identified and pre-classified as sawdust and wood shavings.

After the collection, the samples were analyzed for granulometric composition, unit mass in loose state and in compacted state, for the sawdust and woods shaving samples, which are fundamental for characterization of aggregates used in civil construction. A sieve test was carried out to obtain a granulometric range analysis. The samples were analyzed and classified according to their shape, and color.

The glass material was collected in glassworks stores in the commercial center of Salgueiro-PE. After the collection, were evaluated the specific mass, granulometric composition, setting time and apparent density. After the characterization of the aggregates, the characterization of the binding element (gypsum plaster) was carried out. The unit mass, specific mass and setting time were evaluated.

Gypsum plaster boards with the following dimensions (40 cm x 50 cm, 33 cm x 33 cm, 32 cm x 32 cm, 31 cm x 31 cm, 36 cm x 36 cm) were developed All samples have a thickness of approximately 2 cm. The images of the plaster boards are shown in figure 01.

FIGURE 01 - MODELS OF THE GYPSUM PLASTER BOARDS

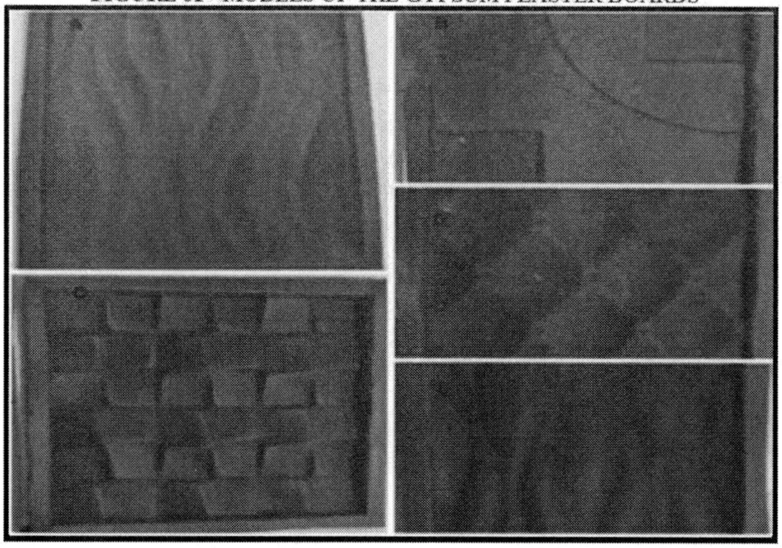

The plaster boards were produced in three families. Firstly, the reference samples consisting of only gypsum, followed by families formed of gypsum and glass waste, gypsum and wood sawdust and gypsum with wood shavings waste. The wastes were used in substitution of the volume of gypsum, in 10% up to 50%. See table 01.

TABLE 01- SUMMARY OF THE SUBSTITUTIONS IN VOLUME

GYPSUM % OF USE	GLASS % OF USE	WOOD SHAVINGS % OF USE	WOOD SAW DUST % OF USE
REFERENCE (100%)	0	0	0
90	10	10	10
80	20	20	20
70	30	30	30
60	40	40	40
50	50	50	50

RESULTS AND DISCUSSION

Glass characterization

Data obtained by the glass characterization tests were analyzed. Regarding the apparent density, The density of the glass in the loose state was 1531 kg/m³ and 1557 kg/ m³ when compacted.

Based on the sieving granulometry test, it was observed that the recycled glass has a continuous or well graded granulometry, as it presented a continuous format granulometric curve, the materials that went through the respective sieves, are presented in Table 04.

FIGURE 02. – IMAGE OF THE GLASS – EXPANDED 50 X.

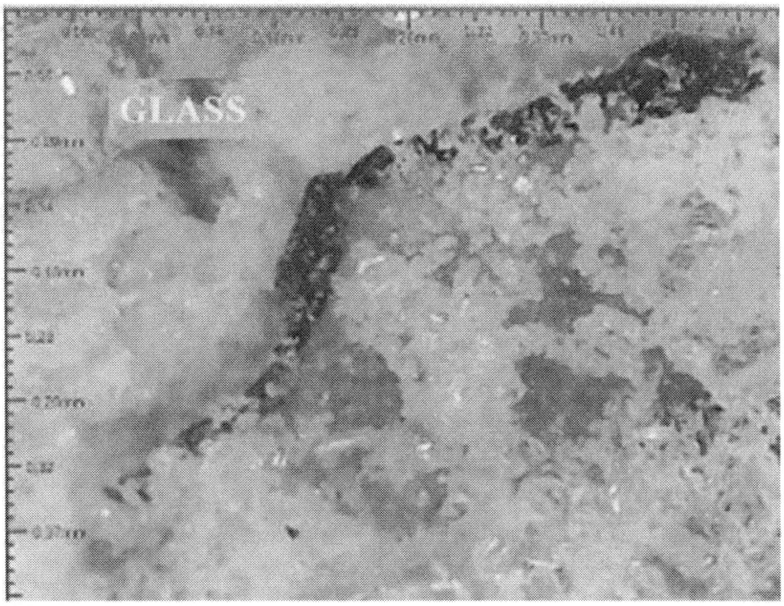

Characterization assay – Wood

FIGURE 03 – IMAGE OF THE WOOD SHAVINGS AND WOOD SAWDUST RETAINED IN # 600μm MM – EXPANDED 50 X.

The visual aspects of the wood shavings and wood sawdust are presented in figure 03. Figure 04 shows the grain size curve of wood sawdust and wood shavings, wood waste is thinner than glass.

FIGURE 04 – Particle size distribution of glass, saw dust and wood shavings

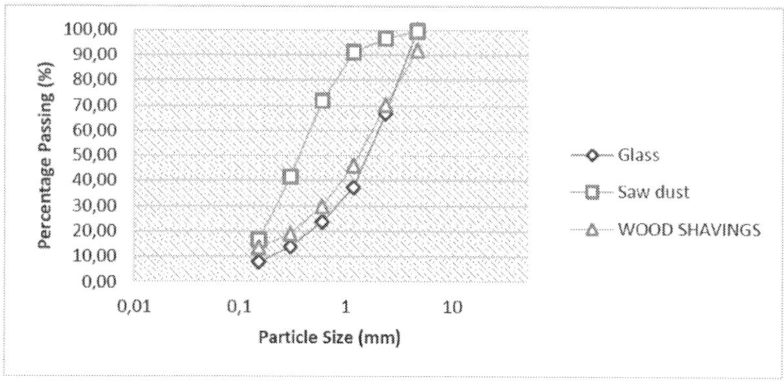

Table 05 shows the summary of the physical indexes evaluated. By analyzing the unit mass in loose state, the wood sawdust was of 159 kg / m³ while the wood shavings obtained was of 253 kg / m³.

The unit mass in the compacted state for the wood sawdust was of 359 kg / m³ and the wood shavings was of 326 kg / m³. It can be noticed that both materials have low apparent density and can be characterized as light materials.

Characterization test – Binder

Which indicates the time in which the binder is in the moldable state, it can be seen that the plaster only initiated the setting time after the first 10 minutes and finished at 25 minutes, it can be noticed that the hardening time of the plaster is very fast compared to the other binders. showing the unit mass in loose state of 858 kg / m³, and the specific mass assuming the mean value of 2,24 g/cm³.

Gypsum plaster boards

Table 02 shows the substitutions of the binder (gypsum plaster) by glass, wood sawdust and wood shavings waste. Masses of gypsum referring to 1 kg were used, in which substitutions were made in mass per volume for wood waste and mass per mass for the glass waste, the substitutions were performed from 10% to 50%.

TABLE 02- SUBSTITUTIONS BY THE WASTE MATERIALS - RESPECTIVELY

GYPSUM	WOOD SAW DUST (g)	WOOD SHAVINGS (g)	GLASS (g)
10% replacement			
1 Kg	18,56	29,50	100
20% replacement			
1 Kg	33,11	59	200
30% replacement			
1 Kg	55,67	88,50	300
40% replacement			
1 Kg	74,23	117,19	400
50% replacement			
1 Kg	92,78	147,49	500

Glass boards

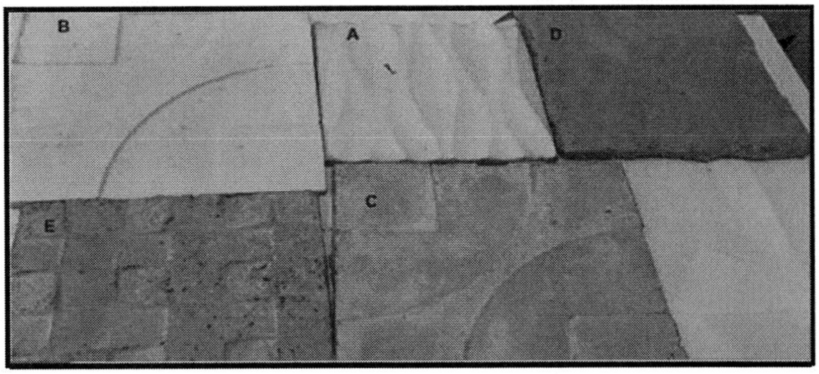

Prototypes using glass waste were developed as shown in Fig. 05, which illustrates gypsum plaster boards plus the addition of 10% to 50% of the gypsum residue respectively, the substitutions were performed by masses, visibly the plates did not change the color due to the colorless of the glass.

FIGURE 05- PROTOTYPES OF GYPSUM PLASTER WITH GLASS ADDITION

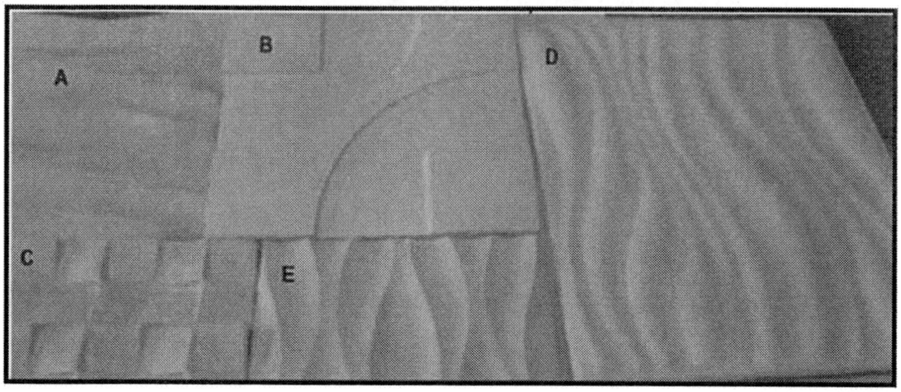

Wood shavings waste boards

The prototypes using wood shavings waste were developed, as shown in figure 06. The image shows boards made with gypsum plaster plus the addition of 10% to 50% of wood shavings in substitution of the gypsum volume. It can be observed that the boards "A and B" which corresponds to 10% and 20%, respectively, have brown staining at the edges, from the images "D to E" the boards had substitutions of 30%, 40% and 50% and presented a complete brown color due to the wood waste and thus, having higher amounts of the waste material.

Wood sawdust waste boards

Prototypes with substitution of wood sawdust were also made as can be seem in figure 07. With substitutions also made from 10% to 50%. It is possible to observe that the prototypes present brownish aspects like observed in the wood shavings boards, where the substitutions were made by mass per volume.

FIGURE 07- PROTOTYPES PRODUCED WITH WOOD SAWDUST.

Compressive Strength (MPa)

Figure 08 shows the compression ratio in relation to the reference sample. Samples with 50% glass substitution and 10% saw dust substitution showed the closest reference results. The wood to have higher porosity and fibers, figure 02, presented better performance among the others.

Figure 08 – strength of the samples

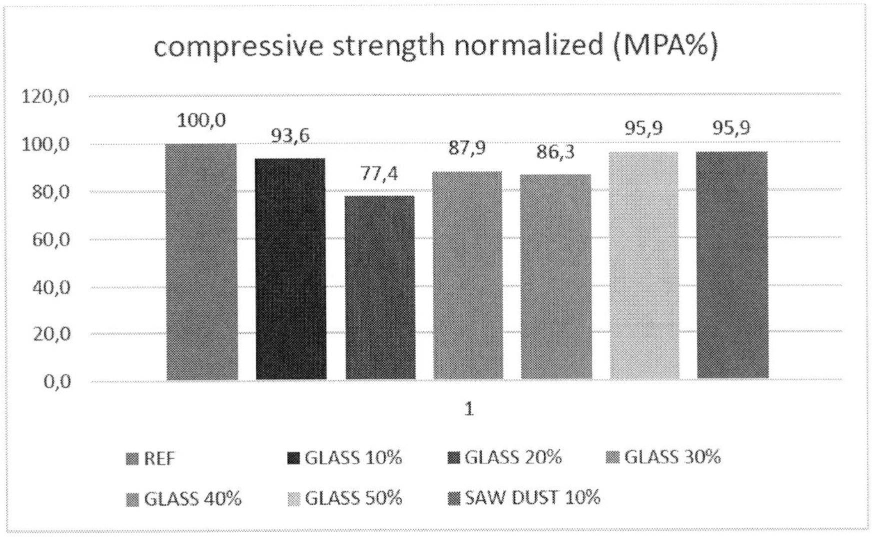

CONCLUSIONS

The glass presented higher bulk density than the other wood residues, so the samples were produced using a change of gypsum mass by the glass mass. The residues of the wood had lower apparent density than the plaster, being necessary to substitute materials in volume. Figure 4 shows that glass is the finest material.

Although the residues present similar morphological characteristics, Figure 02, the densities of the materials were quite different. The loose and compacted unit mass of the sawdust powder was 135% and 79% than the wood shavings, respectively.

The plates produced with gypsum presented a less rough surface and a more uniform coloration, whereas the surfaces produced with wood residues presented a rough surface and with a different coloration with the increase of the residual content.

The compressive strength of the samples with 10% wood dust residue and 50% glass showed higher mechanical strength. Probably the intrinsic fibers and porosity of the wood residue contributed to a greater aggregation with the gypsum and consequently greater mechanical resistance.

The studied residues present satisfactory results and the conclusions indicate potential for use in the production of new materials, aiming at the reduction of natural resources.

REFERENCES

ASSOCIAÇÃO BRASILEIRA DE NORMAS TÉCNICAS. NBR 12127: 1991 -Gesso para Construção - Determinação das Propriedades Físicas do Pó, Rio de Janeiro, 1991.

ASSOCIAÇÃO BRASILEIRA DE NORMAS TÉCNICAS. NBR 12128 (MB 3469) - Gesso para Construção: Determinação das propriedades físicas da pasta, Rio de Janeiro, 1991.

ASSOCIAÇÃO BRASILEIRA DE NORMAS TÉCNICAS. NBR 12129: 1991 (MB 3470) - Gesso Para Construção - Determinação Das Propriedades Mecânicas, Rio de Janeiro, 1991.

ASSOCIAÇÃO BRASILEIRA DE NORMAS TÉCNICAS. NBR 13207: 1994. Gesso para construção civil, Rio de Janeiro, 1994.

JOHN, Vanderley M.; CINCOTO, Maria Alba. Gesso para construção civil. In: ISAIA, Geraldo Cechella. Materiais de Construção Civil e Princípios de Ciência e Engenharia de Materiais. São Paulo: Ibracon, 2010. p. 727-759. Geraldo Cechella Isaia - Editor e Organizador. Editoração eletrônica e Diagramação - Arte interativa.

KLOCK, Umberto; ANDRADE, Alan Sulato de. QUÍMICA DA MADEIRA. 4. ed. Curitiba: Universidade Federal do ParanÁ, 2013. 85 p. UNIVERSIDADE FEDERAL DO PARANÁ SETOR DE CIÊNCIAS AGRÁRIAS DEPARTAMENTO DE ENGENHARIA E TECNOLOGIA FLORESTAL. Umberto Klock Eng. Florestal, Dr., Prof. Associado Alan Sulato de Andrade Eng. Ind. Madeireiro – Dr. Prof. Adjunto.

SMITH, Willian F.; HASHEMI, Javad. Vidros. In: SMITH, Willian F.; HASHEMI, Javad. Fundamentos de Engenharia e Ciência dos Materiais. 5. ed. Porto Alegre: Amgh, 2012. Cap. 11. p. 435-443. Tradução: Necesio Gomes Costa, Ricardo dia Martins de Carvalho, Mírian de Lourdes Noronha Motta Melo.

CHALLENGES AND OPPORTUNITIES FOR THE USE OF EXCAVATION MATERIAL IN THE BUILT ENVIRONMENT. A CASE-STUDY IN COLOMBIA

Edwin Zea Escamilla[1], Gnanli Landrou[4], Andrea Cifuentes Cuellar[2], Jesus Moreno[3], Guillaume Habert[4]

[1] Centre for Corporate Responsibility and Sustainability, University of Zürich, Switzerland;

[2] Catholic University of Colombia, Design Faculty, Colombia;

[3] University Santo Tomas, Architecture Faculty, Colombia;

[4] Chair of Sustainable Construction, Swiss Federal Institute of Technology (ETHZ), Switzerland;

ABSTRACT

To steer towards sustainable development we require buildings and infrastructure. As consequence to the production of conventional construction materials large amounts of CO2are emitted to the atmosphere. Thus, to achieve sustainable development it is necessary to find construction technologies able to supply the demand of buildings and have low CO2emission. In this study we explore the challenges posed by the regulation for compulsory use of Construction and Demolition waste from buildings in Colombia, especially in housing projects. Furthermore, we identify the opportunities related to this newly created material demand. Moreover, the quantities of potential material produced were calculated and related to the requirements of the Colombian Norm 0472 for use of C&D waste. These results showed that a significant amount of conventional materials can be replaced. To evaluate the environmental impact of this replacement a comparative life cycle assessment was carried out. The software OpenLCA was used and the database ecovinvet 3.1, and the environmental impact assessment used the evaluation method from the Intergovernmental Panel for Climate Change "IPCC2013". This study shows that the challenges can only be overcome when engineered solutions can bring to scale the use of excavation material in contemporary architecture in terms of volume and price. Finally, by reducing the amount of CO2 emission and presenting a potential economic savings, earth based construction materials can become a driver to achieve sustainable development.

Keywords: earth, excavation, materials, LCA, CO2

INTRODUCTION

One of the main prerequisites to achieve sustainable development goals is the presence of building and infrastructures to support the activities associated to them. As a consequence of the production of construction materials large amounts of CO2 are released into the atmosphere. Thus, we need to find a compromise between the provision of buildings for sustainable development and the reduction of CO2 emissions[1]. The main source of emissions can be associated to the energy used to produce cement and bricks and the transport of construction materials[2]. This shows that to achieve sustainable development we need strategies not only to produce construction materials with low levels of CO2 emissions but also to reduce the transport distances of construction materials.

On possible solution to this problem is to use Construction and Demolition Waste (C&DW) to produce construction materials close to the construction site. C&DW is one of the largest waste fractions in modern societies[3]. Even though this fraction is large, due to its intrinsic heterogeneity there are challenges for its use in the construction sector[4], mainly the requirement of special logistics associated to the storage of raw materials. These challenges render many potential solutions as impractical at the moment to apply them in an industrial set up. One of the most promising applications is the use of excavation materials for the production of non-structural masonry.

Excavation material is the main component of C&DW[5] accounting from more than 70% of the total weight. Moreover, excavation materials can be directly used in situ thus reducing transport distances of construction materials. Moreover, the masonry produced by stabilizing the excavation materials with geo-polymers[6] or cement can have lower CO2 emissions that those from fire clay bricks. Finally, excavation materials are mainly disposed in landfills which requires further transport of these materials and increases the pressure on land competition in urban and peri-urban areas.

Life Cycle Assessment (LCA) has been established as the main methodology used to assess the environmental impacts of products and services[7]. LCA proposes a relation input-output between human activities and the environment, which can be assessed through the production, use and disposal of products. LCA is described in the ISO norm 14040[8] and had been used for more than three decades in the building sector[9]. LCA can be used to model possible scenarios for the production of goods and provides information relevant is a decision making process.

In this paper we present the results of a case study in Colombia –South America. Colombia was selected for this case study due to the newly installed regulation from the Ministry of Environment that requires that buildings with areas superior to 1000m2 use C&DW in an increasing percentage starting with 2% in 2018 and up to 30% in 2030[10]. This regulation creates an ideal ground to test the potential of the use of C&DW in construction and specifically the use of excavated materials. For this case study, the main objective was to identify the potential CO2 reductions from the used of excavated materials in situ considering the level of use proposed by the Colombian Ministry of Environment Regulation 0472.

DATA AND METHODS

In order to calculate the environmental impact from construction materials and the Environmental Savings Potential (ESP) from the use of excavation in situ we carried out a comparative LCA. The comparison was don at the building level, having as functional unit an housing building of 1000m2. The amount of material required for the construction of the building are presented on Table 1. On this table we present the potential transport distances of construction materials in Colombia, these transport distances were estimated using the method developed by Zea & Habert[2]. The calculations were conducted considering the levels of use of excavation material, from 2 to 15%, proposed by the regulation 0472(REF). The models propose that an amount of fire clay bricks is replaced by soil stabilized bricks to fulfill the requirements of the regulation 0472.

The LCA models were developed using the software OpenLCA v1.7.4[11] and the LCA database Ecoinvent 3.1[12]. The datasets were adapted to the Colombia context following the work of Balzarini [13] and Zea[2]. The calculation of the environmental impacts were carried out suing the evaluation method from the Intergovernmental Panel for Climate Change IPCC2013[14]. The results are presented in ton of CO2 equivalents (tCO2eq) and represent the emissions associated to the production and transport of construction materials for the construction of one building. The uncertainty analysis was carried out using the Monte Carlo Simulations, using 9000 iterations per model.

The Environmental Savings Potential (ESP) is defined as the percentual difference between the total environmental impact of a benchmark model and the total environmental impact of the model being investigated. For this research, the base model were the excavation was not used was considered as benchmark.

Table 1 List of inputs w & w/o use of excavation materials

			Use of Excavation-based Materials															
		Unit	2016	2017	2018	2019	2020	2021	2022	2023	2024	2025	2026	2027	2028	2029	2030	Unit
		No Use	2%	4%	6%	8%	10%	12%	14%	16%	18%	20%	22%	24%	26%	28%	30%	
Excavation	Excavation Material	120 t	120	120	120	120	120	120	120	120	120	120	120	120	120	120	120	t
	Excavation hydraulic digger	49.2 **min**	49.2	49.2	49.2	49.2	49.2	49.2	49.2	49.2	49.2	49.2	49.2	49.2	49.2	49.2	49.2	**min**
	Machine operation Diesel	49.2 **min**	49.2	49.2	49.2	49.2	49.2	49.2	49.2	49.2	49.2	49.2	49.2	49.2	49.2	49.2	49.2	**min**
	Transport of excavation materials to disposal	10800 t * **km**	9390	6144	2897	349	0	0	0	0	0	0	0	0	0	0	0	t * km
Construction	Brick	948 t	905	861	818	774	731	687	644	600	557	513	470	427	383	340	296	t
	Concrete	980 t	980	980	980	980	980	980	980	980	980	980	980	980	980	980	980	t
	Mortar Cement	241.4 t	241.4	241.4	241.4	241.4	241.4	241.4	241.4	241.4	241.4	241.4	241.4	241.4	241.4	241.4	241.4	t
	Reinforcing Steel	3.4 t	3.4	3.4	3.4	3.4	3.4	3.4	3.4	3.4	3.4	3.4	3.4	3.4	3.4	3.4	3.4	t
	Transport Bricks	42660.0 t * **km**	40704	38749	36793	34838	32882	30927	28971	27016	25060	23105	21149	19194	17238	15282	13327	t * km
	Transport Concrete	44102.0 t * **km**	44102	44102	44102	44102	44102	44102	44102	44102	44102	44102	44102	44102	44102	44102	44102	t * km
	Transport Mortar Cement	10863.7 t * **km**	10864	10864	10864	10864	10864	10864	10864	10864	10864	10864	10864	10864	10864	10864	10864	t * km
	Transport Reinforced Steel	151.4 t * **km**	151.4	151.4	151.4	151.4	151.4	151.4	151.4	151.4	151.4	151.4	151.4	151.4	151.4	151.4	151.4	t * km
Production of excavation-based materials	Cement for stabilization	0.0 t	2	4	5	7	9	11	13	14	16	18	20	22	23	25	27	t
	Clay	0 t	36	72	108	144	180	216	252	289	325	361	397	433	469	505	541	t
	Clay infrastructure	0 **item**	2.9E-08	5.8E-08	8.7E-08	1.2E-07	1.5E-07	1.7E-07	2.0E-07	2.3E-07	2.6E-07	2.9E-07	3.2E-07	3.5E-07	3.8E-07	4.1E-07	4.4E-07	item
	Tap water	0 t	3	6	8	11	14	17	19	22	25	28	31	33	36	39	42	t
	Transport Cement	0 t * **km**	81	162	243	325	406	487	568	649	730	812	893	974	1055	1136	1217	t * km
	Transport Clay to site	0 t * **km**	0	0	0	174	1797	3421	5044	6667	8290	9913	11536	13159	14782	16405	18028	t * km

RESULTS AND DISCUSSION

Environmental Impact

Table 2 present the amount of materials required by the regulation 0472. At 2% level the regulation requires that 43.5t of C&DW based materials are used at the 30% level, in 2030, the amount increases to 651t per building. This presents on itself a challenge because it is estimated that a building of this size will have close to 120t of excavation materials. Thus, the regulation can only be fulfilled till the 8% level. Beyond this point, it would be necessary to bring additional raw materials to the construction site. To represent this process an additional amount of excavation material and its associated transport were added to the LCA models. The results for the change is amount of construction materials transported the unit ton times kilometer (t * km) was used, these results are presented on Figure 1. From this figure we can see that when using 30% of excavation-based materials the transport of Bricks is reduced significantly. Moreover, the total amount of transported materials, in t*km, is reduced in 20%.

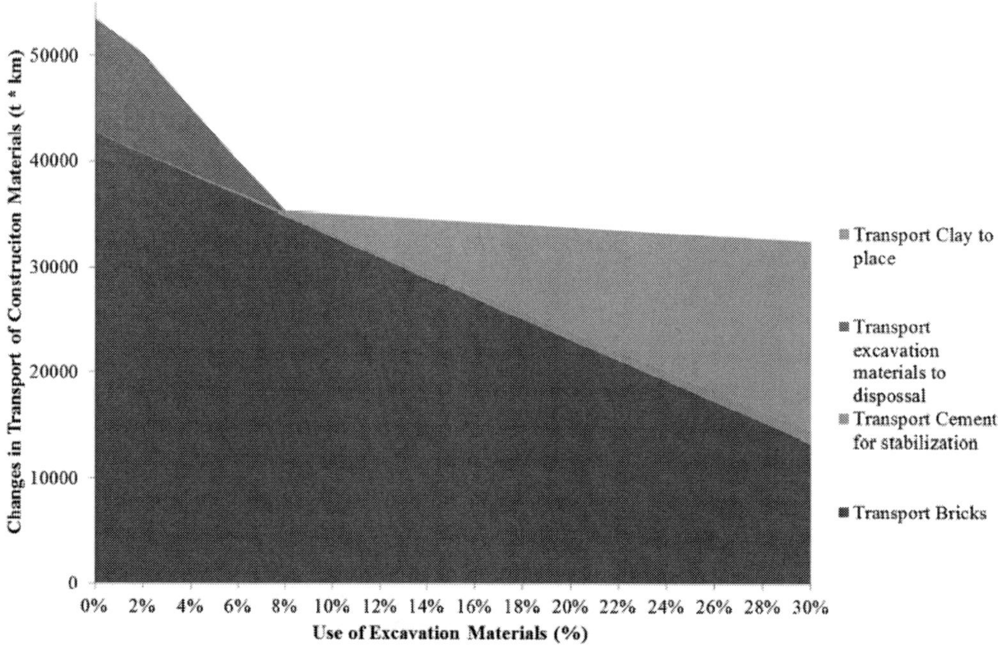

Figure 1 Transport distances of selected construction materials

Figure 2 presents the results for the LCA in terms of tCO2eq and Environmental Savings Potential. From this figure we can observe that by using a percentage, from 2 to 30%, of excavation-based materials the total CO2 emissions are reduced. This reduction is close to 2% for each 2% of excavation-based materials used. As a result when using 30% of excavation-based materials an Environmental Savings Potential of 27% can be achieved when comparing to the model with no replacement. It is important to note that these savings can be expected even when additional transport are calculated for the additional raw materials required to fulfill the 30% level of material replacement. Furthermore, the models represent the reduction in the use of fired clay bricks and the associated transport of these materials to the construction site. To better understand these process we carried out a contribution analysis for each model.

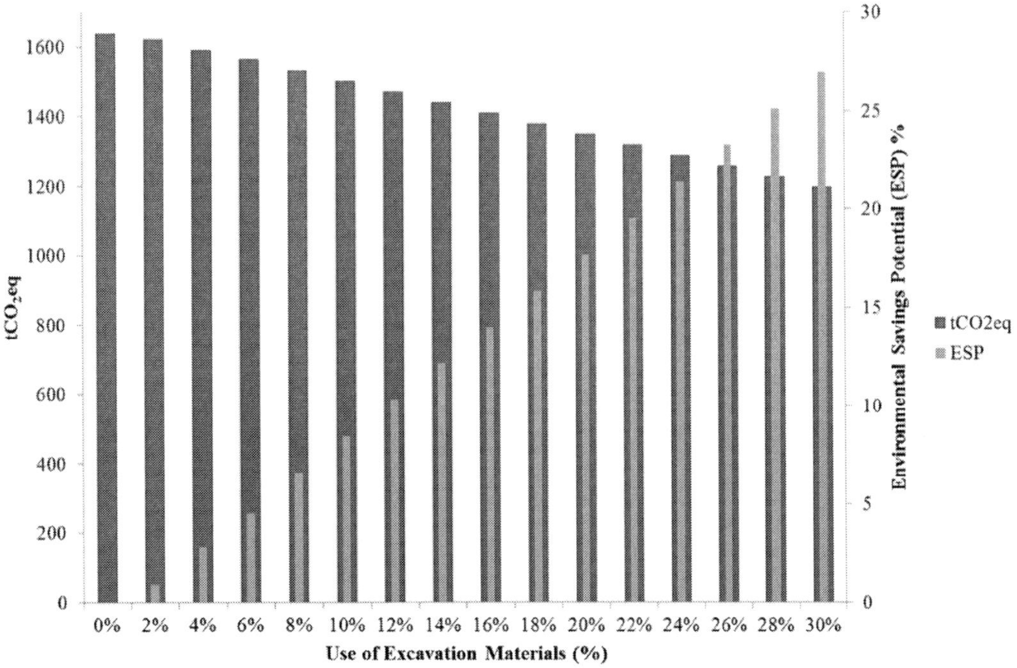

Figure 3 presents the results for the contribution analysis for three models: (i) benchmark, no replacement, (ii) use of 2% of excavation-based materials, and (iii) use of 30% of excavation-based materials. The contribution analysis shows how different process and products contribute to the total environmental impact. From this figure we can observe that Concrete is the main contributor to the environmental impact, followed by bricks and transport. Moreover, we can observe that with the replacement the contribution from bricks is reduced and the contributions from concrete and specially transport are increased. Considering that a 30% replacement produces a ESP of 27% we can see that the replacement of conventional construction materials with excavation-based construction materials has a positive impact even when additional transportation and materials is required. Moreover, we can observe that at 30% replacement the contribution from concrete increases significantly in comparison to the benchmark model. This is can be directly associated to the reduction in the amount of brick used in the model and the consequent increase in the proportion of concrete in the models.

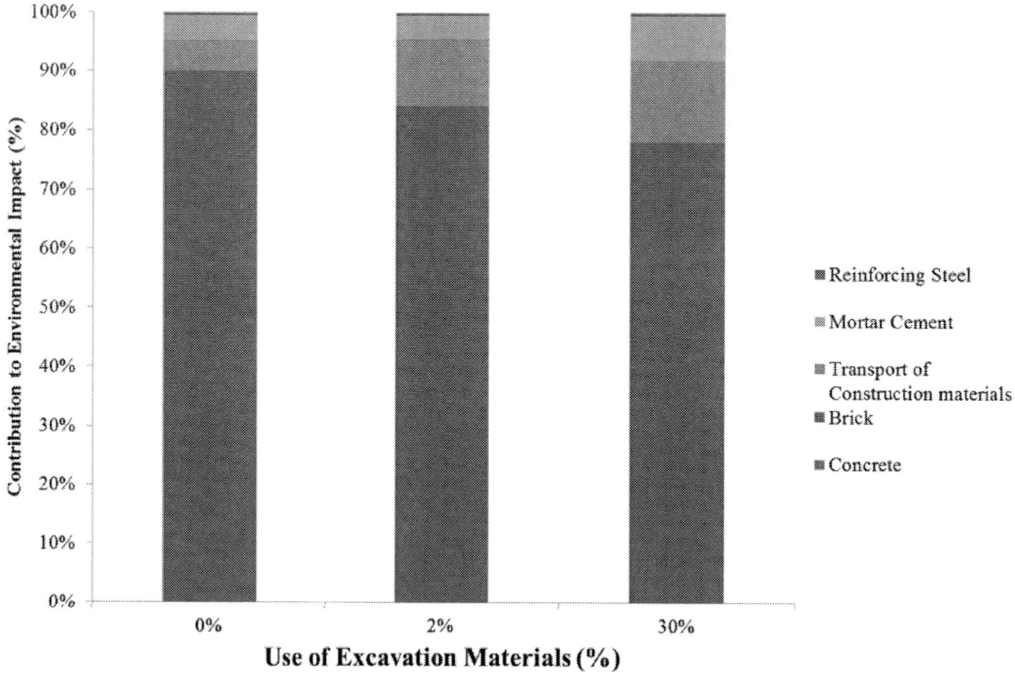

From these results we can observe that the optimal solution is achieved once 8% of replacement of conventional materials is reached. Beyond this level of use of excavation-based materials, the additional logistical challenges to acquire raw material might turn these solutions in practical. Nevertheless, an 8% replacement produces an ESP of 7% which becomes significant once we calculate this reduction in emissions in relation to the expected amount of new buildings being constructed in Colombia in the future. Figure 4 presents the results for the ESP at a national level considering the projections for the construction of housing building of +1000m2. From Figure 4 we can observe that nine million tons of CO2 can be avoided yearly if the requirements of the regulations are fulfilled. Moreover, these levels of replacement will require around eleven million tons of excavation-based construction materials. In return this will create a new market for this kind of solutions and a supply that needs cost effective solutions to fulfill this ever increasing demand.

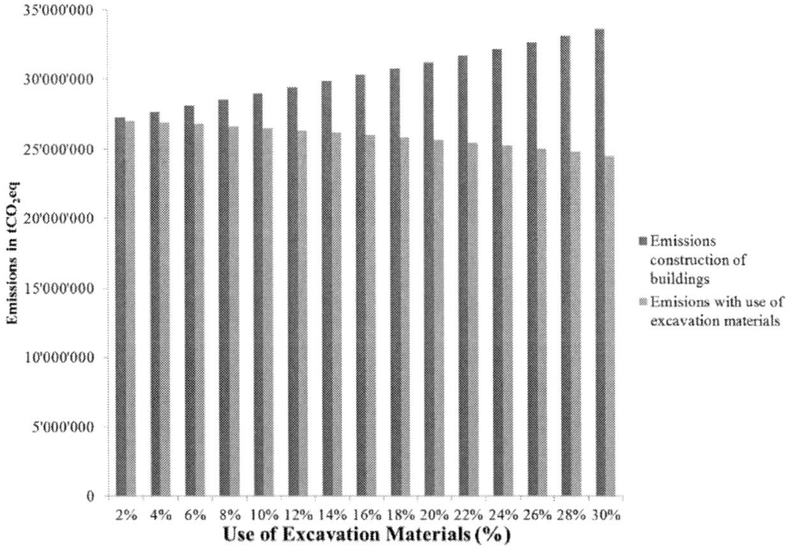

CONCLUSIONS

In this paper we explore the potential environmental benefits associated to the fulfillment of the regulation 0472 from the Colombian Ministry of environment. We carried out a comparative LCA and showed that the replacement of conventional construction materials, like fired clay bricks, with excavation-based materials can produce a significant reduction in the CO2 emissions. This reduction is directly related to the amount of materials being replaced and the transport of construction materials required to produce the excavation-based construction materials in situ. We showed that once an 8% replacement level, the excavation materials from the construction are completely used. Furthermore, in order to increase these levels of replacement additional raw materials needs to be brought to a construction site. This presents a significant challenge for constructors in the future, when the regulation will require them to replace 30% of conventional construction materials. We can conclude that the use of excavation materials have great potential not only reduce the CO2 emissions associated to the construction of buildings but also to generate new economic a social dynamics within the construction sector.

Acknowledgements

The authors would like to thank Swisscontact and the Project Sustainable Construction & Circular Economy from the Bucaramanga Chamber of Commerce for their support in the establishment of the interinstitutional relation that allowed the development of this article.

REFERENCES

1. UNFCCC. United Nations - PARIS AGREEMENT. 2018 [cited 2017 12.12.2017]; Available from: http://unfccc.int/2860.php.

2. Zea Escamilla, E. and G. Habert, Method and application of characterisation of life cycle impact data of construction materials using geographic information systems. The International Journal of Life Cycle Assessment, 2016: p. 1-10.

3. Akhtar, A. and A.K. Sarmah, Construction and demolition waste generation and properties of recycled aggregate concrete: A global perspective. Journal of Cleaner Production, 2018. 186: p. 262-281.

4. Iacoboaea, C., M. Aldea, and F. Petrescu, CONSTRUCTION AND DEMOLITION WASTE-A CHALLENGE FOR THE EUROPEAN UNION? Theoretical and Empirical Researches in Urban Management, 2019. 14(1): p. 30-52.

5. Hilgert, F., Excavation material as the new building resource for Zurich, in D-BAUG. 2018, ETH Zürich: Zürich.

6. Habert, G. and C. Ouellet-Plamondon, Recent update on the environmental impact of geopolymers. RILEM technical Letters, 2016. 1: p. 17-23.

7. Hellweg, S. and L. Mila i Canals, Emerging approaches, challenges and opportunities in life cycle assessment. Science, 2014. 344(6188): p. 1109-13.

8. ISO, ISO 14040: environmental management- life cycle assessment- principles and framework, ed. ISO. 2007, Geneva, Switzerland: International Organization for Standardization.

9. Cabeza, L.F., et al., Life cycle assessment (LCA) and life cycle energy analysis (LCEA) of buildings and the building sector: A review. Renewable and Sustainable Energy Reviews, 2014. 29: p. 394-416.

10. MoE&SD, Regulation 0472 for the integral management of construction and demolition waste, in 0472, C.M.o.e.a.s. development, Editor. 2017, Colombian Ministry of environment and sustainable development: Bogota, Colombia.

11. GD. OpenLCA The open source life cycle and sustainability assessment Software. 2017; Available from: http://www.openlca.org/.

12. SCLCI. EcoInvent Database. Swiss Centre for Life Cycle Inventories 2017; Swiss Centre for Life Cycle Inventories]. Available from: http://www.ecoinvent.org.

13. Balzarini, A., Environmental impact of brick production outside Europe, in Department of Civil, Environmental and Geomatic Engineering. 2013, Swiss Federal Institute of Technology ETH Zürich: Zürich.

14. Edenhofer, O., et al., IPCC, 2014: Climate Change 2014: Mitigation of Climate Change. Contribution of Working Group III to the Fifth Assessment Report of the Intergovernmental Panel on Climate Change. 2014: Cambridge, United Kingdom & New York, NY, USA.

18th International Conference on
Non-Conventional Materials and Technologies
"Construction Materials & Technologies for Sustainability"
(18th NOCMAT 2019)
24th – 26th July 2019
Nairobi, Kenya

DEVELOPING A PANELIZED BUILDING SYSTEM FOR LOW-COST HOUSING USING WASTE CARDBOARD AND REPURPOSED WOOD

Julio Diarte[1,2], Marcus Shaffer[1], Esther Obonyo[1]

[1] Penn State University, University Park, PA, USA;

[2] National University of Asuncion, San Lorenzo, Paraguay;

ABSTRACT

This paper presents research dedicated to prototyping a panelized building system developed in an academic setting, but not yet deployed in a "real world" context. The "building" in this system is fabricated using waste cardboard in combination with common repurposed wood. The system is conceptualized for use within low-income urban populations living in developing countries who cannot afford to build using conventional masonry or concrete-based approaches; and specifically, for populations who have easy access to waste cardboard and other discarded materials. The research presented here includes the design concept and production workflow for a panelized waste cardboard-based building system. The workflow includes the material collection and documentation, building parts design, building parts fabrication, and building parts assembly. The main component of this building system is a load-bearing wall panel that employs rolled cardboard tubes and plywood framing. In presenting this work, we articulate and illustrate the design criteria for the building components and share the results of construction experimentation using sheets of waste cardboard collected from our university waste stream. The experimentation concludes with the assembly of a full-scale sectional house prototype.

Keywords: Upcycling Waste Cardboard, Low-Cost Housing System, Alternative Construction Materials, Panelized Building System, Cardboard in Architecture.

INTRODUCTION

This paper presents research dedicated to prototyping a panelized building system developed in an academic setting, but not yet deployed in a "real world" context. The building in this system is fabricated using waste cardboard in combination with common repurposed wood. The system is conceptualized for use in low-income populations living in urban areas in developing countries who cannot afford to build using conventional masonry or concrete-based approaches; and specifically, for populations who have easy access to waste cardboard and other refused materials through informal "work" as self-employed collectors of recyclable materials.

In developing countries, waste cardboard is an underutilized and downgraded material that, if not used for recycling and production of more cardboard, ends up in landfills or open-air dumpsites. This material, however, presents several advantages that have been analyzed by academic research centers with the goal of extending its use from packaging to building construction [1]. Among all cardboard products, honeycomb boards, corrugated cardboard, and paper tubes have received special attention and there are several studies that propose these materials as multifunctional composite walls or building structures. The most remarkable studies on cardboard in architecture and construction have been published just in the last ten years [2–4]. However, these investigations have focused on brand-new cardboard products, while waste cardboard has received very little attention.

This paper presents the design concept and production workflow for a panelized waste cardboard-based building system. The workflow includes the material collection and documentation, building parts design, building part fabrication, and building parts assembly. The main component of this system is a load-bearing wall panel that employs rolled waste cardboard tubes and plywood framing. We describe and illustrate the design criteria for the building parts and show the results of a construction experiment performed using sheets of waste cardboard collected from the university waste stream. The experiment concluded with the assembly of a full-scale sectional house prototype.

The preliminary results of this work show that the proposed housing system can be prefabricated in a small workshop and assembled using unskilled labor, low-tech machinery, and conventional hardware frequently used for construction and woodworking. The machinery includes a custom-made parallel rolling device for making rolled waste cardboard tubes. The hardware includes wood screws, nails, threaded rods, nuts, bolts, and standard power tools, such as electric drills and screwdrivers.

The essential deliverables from this research include i) a methodology for empowering low-in-come populations to self-build adequate and safe structures using OCC; and ii) the environmental benefits of upcycling a very low-grade waste material into higher quality building components for housing for low-income communities. In subsequent work, a constructability assessment will be performed in Asuncion, Paraguay, to test the transferability of the tools and system. Asuncion has a fast-growing urban low-income population that needs housing, and there is an informal waste/recycling collection structure that includes waste cardboard. The focus here will be identifying any existing context-specific constraints. This information will be used to inform the design of the second iteration of the system.

ADVANCES IN CARDBOARD IN ARCHITECTURE

To date, cardboard in architecture and construction has been studied from different approaches. On the one hand, we have studies that focus on an architectural approach aspiring to design affordable housing systems that can be mass-produced. On the other hand, we have studies that focus on the engineering of the material; investigating the environmental and mechanical properties of the cardboard.

The Cardboard House experiment of Buckminster Fuller in the 1950s is a remarkable example of the first approach. This also represents one of the earliest experiments using cardboard in architecture; a relatively new product at that time [5]. But it was not until the 1990s when the interest in paper and cardboard in architecture accelerated. The main reason for this new-found enthusiasm for the material was the work of the Japanese architect Shigeru Ban, who used engineered paper tubes as a structural component in several of his buildings [6]. Ban's main inspiration for using paper tubes was related to the green factors inherent in the material: recyclable, strong, and easy to waterproof and fireproof in its industrial composition [7].

These factors have been, precisely, the ones that have been most investigated during the past fifteen years: recyclability, structural performance, and resistance to humidity and fire. The most comprehensive studies that touched on these topics come, for instance, from the Cardboard Technical Research and Developments group at Delft University of Technology [1], the Cardboard in Architectural Technology and Structural Engineering research group at ETH [2, 3], and recently from the "BAMP! – Building with Paper" research project at Technological University in Darmstadt in Germany [8]. These academic centers have published extensive research based on quantitative approaches applying experimental, statistical, and analytical methodologies.

Other no less important studies have also contributed with extensive reviews on the topic [9], investigated innovative ways of fabricating cardboard-based panels for temporary shelters [10, 11], and the development of environmentally-friendly additives for improving the water resistance of cardboard products [12]. In general, the most academically analyzed cardboard products are paper tubes and honeycomb boards, and the most common application are in space frames or linear systems (arches, shells, columns, and trusses) and in panelized systems (loadbearing and non-bearing walls). In terms of application in buildings, the trend is to use these materials for low-rise residential and commercial buildings, and some of these systems are already commercially available [13].

BUILDING SYSTEM DESIGN CONCEPT

Our goal is to implement a waste corrugated-based system in a recycling environment in an urban area in a developing country, working with people with little economic resources and construction skills but with easy access to waste cardboard. Consequently, the design concept was based on the following criteria: a) off-site prefabrication processes of building parts using waste cardboard as a main material; b) modular building parts to allow users to adapt the system to different locations; c) incremental construction to enable building by stages according to available resources, and d) low-tech construction considering possible economic limitations to buy expensive tools or machinery.

Following this idea, we designed a panelized building system comprised of three sub-systems: floor, load-bearing walls, and roof. This paper is focused on the sub-system of load-bearing walls. As a proof of concept, we designed a 9 sq. m. prototypical building unit that can serve a multipurpose use except for wet spaces (toilets, bathrooms, and/or kitchens), fabricated several of the panels needed, and assembled one part of the building. This unit can be extended or combined with other units to create larger spaces or a sequence of adjacent spaces. The maximum span proposed for the interior space is 3 meters and the minimum height is 2.4 meters. However, depending of the type of floor and roof structure, the span could be larger (figure 1).

The sub-systems floor and roof are not included in this paper. However, for the floor we can consider three options depending of the available resources or existing conditions: building a wooden platform, masonry, or concrete slab. The roof can be built using a wooden or metal structure covered with metal roof panels, synthetic textile, or canvas. The door and window frames – existing or custom made – can be plugged into the structure of the building without disrupting the system. To finalize the enclosure of the building, the panels can be finished with a facade made of cement plastering, drywall, plywood, metal sheeting, aluminum plates, glass fiber, reinforced plastic, or any other kind of siding that can protect the cardboard tubes from the natural elements.

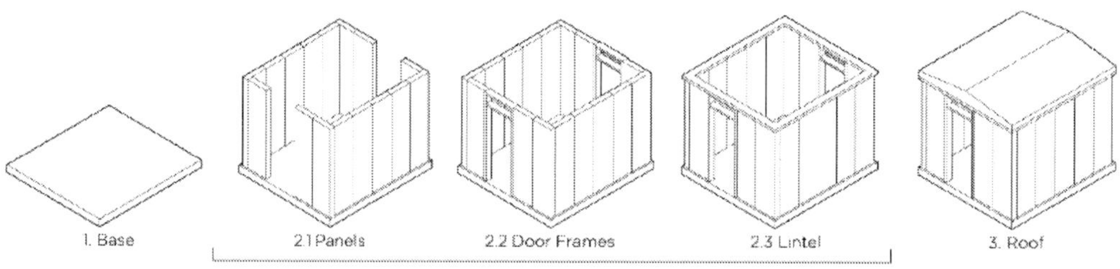

Figure 1: Sub-Systems

TESTING THE BUILDABILITY OF THE SYSTEM

The goal of the construction experiment was to assess the buildability of the design concept and identify the strengths and constrains of the building system. To do so, the experiment was held in the wood shop and building yard of the university working in a controlled environment. The production workflow of the system was divided in four phases. The material collection and documentation describe how the main material – waste cardboard – was obtained and what information was recorded, the building parts design explains the configuration of the panels, and the last two illustrate the fabrication of each part and the assembly of the prototype.

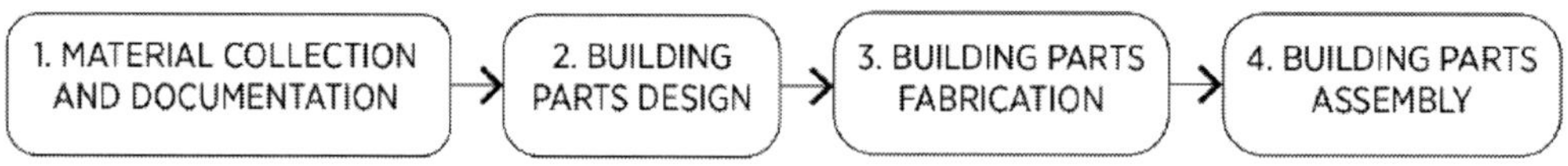

Figure 2: Building Parts Production Workflow

To assess the experiment, we paid specific attention to different variables. For example, the documentation of the material – amount and dimensions - and the tools needed to build the panels, how easy or hard was to produce the elements made with OCC, the fabrication time of the panel and each of its components, the number of laborers needed, the physical aspect of the finished panel, panel weight, ease of transportation, joints, and the actual physical space needed to perform the work. Variables that we did not measure in this experiment were compressive and flexural (bending) strength tests, fire, and moisture resistance.

Material Collection and Documentation

The sheets of waste cardboard were collected from university waste bins and only sheets that were not wet, contaminated with food, or ripped were used. The sheets were taken to the shop and sorted by size and thickness. We recorded the dimensions and flute direction of each sheet. The flute direction is very important because it has to be placed parallel to direction of the load to increase the load-bearing capability of the panel. A total amount of 69 sheets of two different dimensions but same thickness were collected. Although the information of the sheets was recorded manually, this can be done with the aid of smartphones using a free tape-measure application – smartphones are widely available in developing countries. In this sense, the corresponding author studied the possibility of incorporating digital tools to support the production workflow of building parts using sheets of OCC [20]. Table 1 summarize the material documentation.

TABLE 1 - Material Documentation

	Dimension	Flute Direction	Thickness	Quantity	Total Weight
Sheet A	2810 x 730 mm	730 mm	5 mm (single wall)	30	8 kg
Sheet B	2100 x 830 mm	830 mm	5 mm (single wall)	39	8.83 kg

Other supplies used for the fabrication of wall panels include plywood strips - mostly leftovers from the wood shop –, 1-1/2" x 1-1/2" pine wood sticks, starch-based adhesive – eco-friendly, economical, and easy to clean –, 5/16" threaded rods with nuts, packing tape, and wood screws. All these are standard supplies and easy to find in the target context of the project.

Building Parts Design

The question that arose when designing building parts with waste cardboard was what is the most efficient way of using this material to support a load? To answer this question, we looked at how cardboard is commonly used in packaging. Corrugated cardboard is a sandwich structure composed of two facings (liners) and a core (fluting). Because of the nature of its components – cardboard is an orthotropic material – and manufacturing process, the fibers that constitute the cardboard plates are aligned in two directions: Machine Direction (MD) and Cross-Direction (CD). Generally, the MD has better tensile and compressive strength than the CD and it is aligned vertically to the direction of horizontal loads. The CD, on the contrary, is frequently aligned with the direction of the application of vertical loads [21]. If a more resistant box is needed, the number of walls is increased by double or triple.

Following these criteria, we decided to align the CD of the sheets to the direction of the application of vertical loads – in this case the load of the roof. To increment the strength of the component we had two options: to increment the number of walls or plates by rolling or folding the sheets. The difference is that for the first, we don't need to score the sheets and we can roll the sheets continuously increasing the number of overlapping plates. For the second, although we can also increase the numbers of plates when folding, we will need to mark and score the sheets first. For this paper, we decided to use the first option creating "tubes" with the sheets of waste cardboard. The diameter of the tube/building part determines the length of the sheet needed. If the collected sheets are not long enough, they can be overlapped to reach the desired diameter. Figure 3 illustrates the criteria for designing the building parts.

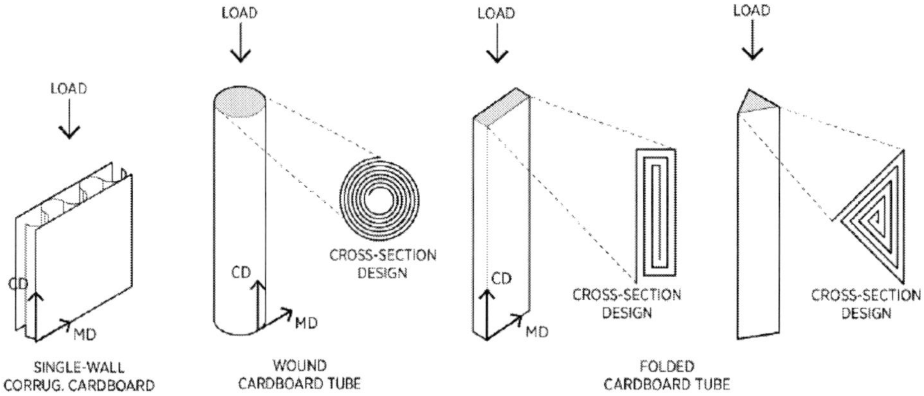

Figure 3: Building Parts Design Criteria

Building Parts Fabrication

The next step consisted on making the tubes using the collected sheets of waste cardboard. We used a parallel rolling/winding tool that consisted of a workbench with a shaft and a handle that was fabricated in the metal shop of the university. The design of the tool considered the methods and materials available in the target context as well as user-friendliness. The tool, also part of this research, was presented in a previous paper in a conference [22].

The procedure for making the tubes had three steps: First, the operator placed the sheets in the shaft and applied the adhesive on the interior face of the sheet. The direction of the flute had to be parallel to the shaft, otherwise, the winding tool will break the flutes and the tube will lose strength. Second, another operator wound up the sheets and controlled the tension created in the sheet. This tension, according to previous research, activates the micro fibers of the sheet and to improves the strength of the sheet [23]. Third, once the winding was over, both operators removed the shaft from the workbench and pulled the tube out.

Overall, the procedure took just a couple of minutes for each tube and the operation was simple – everything is mechanical and hand powered. To finish the tube, the operators added a conventional stain paint to improve moisture resistance. The tube was left to dry for 60 minutes and then was ready to use. The procedure was held in a building yard – concrete flooring and conventional exterior lights – and did not require any kind of special infrastructure. Figure 4 illustrates the main parts of the cardboard tube's fabrication process.

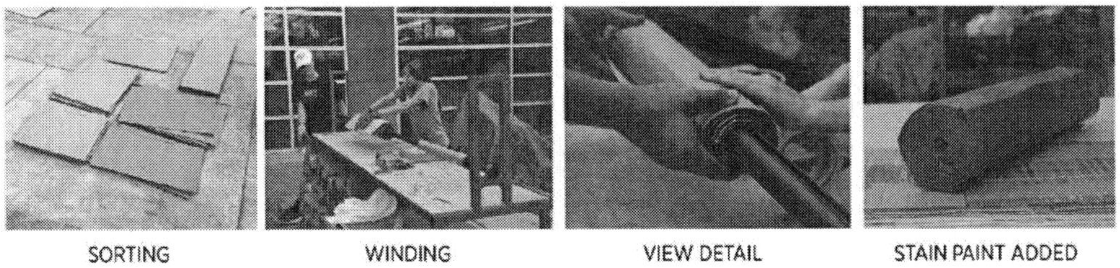

Figure 4: Production of Waste Cardboard Tubes

The design of the wall panels combined three elements: frame, infill, and facing. The frame was made of 9.52 mm (3/8") thick discarded plywood – but it can be made of any kind of flat board or studs. The infill was made of waste cardboard tubes, and the facing can be made of one the materials mentioned previously. The idea of using the waste cardboard tubes as infill comes from the good compressive strength of the sheets for loads parallel to the flute. Consequently, we placed the tubes vertically to make them receive the load of the roof together with the sides of the frame. The length of the tubes depends of the size of the sheets. Figure 5 illustrates the assembly process.

The first step was to prepare the frame. For this paper, the height of the panel was determined by the length of the plywood (2400 mm). However, if the ceiling needs to be higher, the sides of the frame can be extended by adding a connector between strips of plywood. In other panelized systems, the width varies between 500 to 1200 mm depending of the type of material used. For this case, we started with a width of 600 mm for four tubes of 150 mm diameter placed side by side.

In the center of figure 5 we show a diagram of the wall panel designed for this experiment. We used tubes of two lengths: 730 mm and 830 mm. To reach the length of the frame, we connect vertically three tubes using a pine wood stick of 1-1/2" x 1-1/2" inserted in the core of the tubes. Then, we place four of these tubes side by side to reach the width of 600 mm. To connect the tubes horizontally with the sides of the frame, we used a 5/16" threaded rod with nuts. The rods crossed the tubes and frame in four parts to compress the panel evenly. After adjusting the threaded rods, the final width of the panel was 570 mm.

We fabricated a total amount of 6 wall panels of 570 mm by 2400 mm. The overall weight of the finished panel was around 18 kg (cardboard + wood stick + frame + threaded rods) and it was relatively easy to lift and transport with two people. Additionally, we fabricated a small panel of 800 mm by 300 mm using 6 tubes of 300 mm long and 150 mm diameter. This panel was used as a lintel above the door frame. In the bottom left of figure 5 we show the custom-made wooden template that was prepared for perforating the tubes. In the right side of the figure, we show a finished panel without siding.

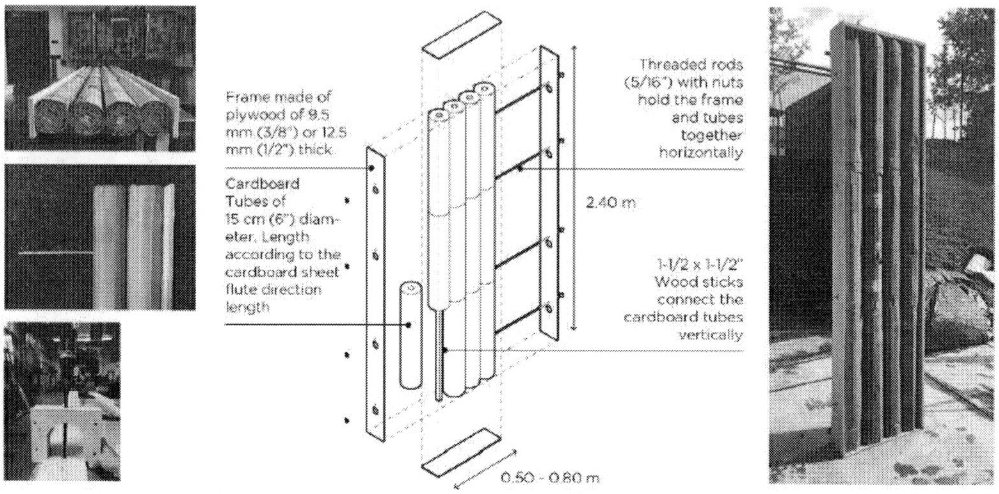

Figure 5: Wall Panel Assembly. Process, Diagram, and Finished Panel

Building Parts Assembly

We used the seven panels that were fabricated to assemble a section of the 9 sq. m. building unit. This full-scale section of the building offered information about how to erect the panels, how to connect them to the base platform, how to connect them to each other side by side, corner connections, and how to solve the situation when we have a door frame. Additionally, the section was also useful for having a physical idea of the stiffness of the assembly system. We divided the assembly in four main steps: a) building a base platform, b) raising the panels one by one, c) placing the concrete lintel, and d) building the roof. One single person performed the assembly and worked under a tent during two straight days to complete the first two steps. Figure 6 illustrates each step of the process.

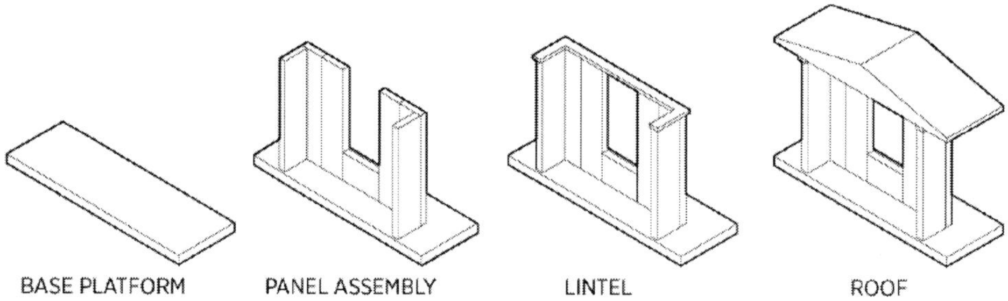

Figure 6: Building Unit Section Assembly

The base platform consisted on a conventional wooden platform placed on top of an existing concrete floor. The base platform can also be made of masonry or a concrete slab depending of the available resources. Next, the builder traced the position of each panel on the platform and raised the panels one by one connecting it to the platform and between them using wood screws. The connection to the base was effortless, but it was necessary to design a male-female joint to connect the panels between them and avoid cutting the threaded rod heads on the side of the panels. For connecting panels in a corner, the builder added a plywood strip to one of the panels to screw the next panel on it. Using electric hand tools like a drill, a screwdriver, a small circular saw, and other conventional tools to trace and level, the overall process performed smoothly and quickly. Figure 7 shows some of the issues identified both in the building parts and in the assembly process.

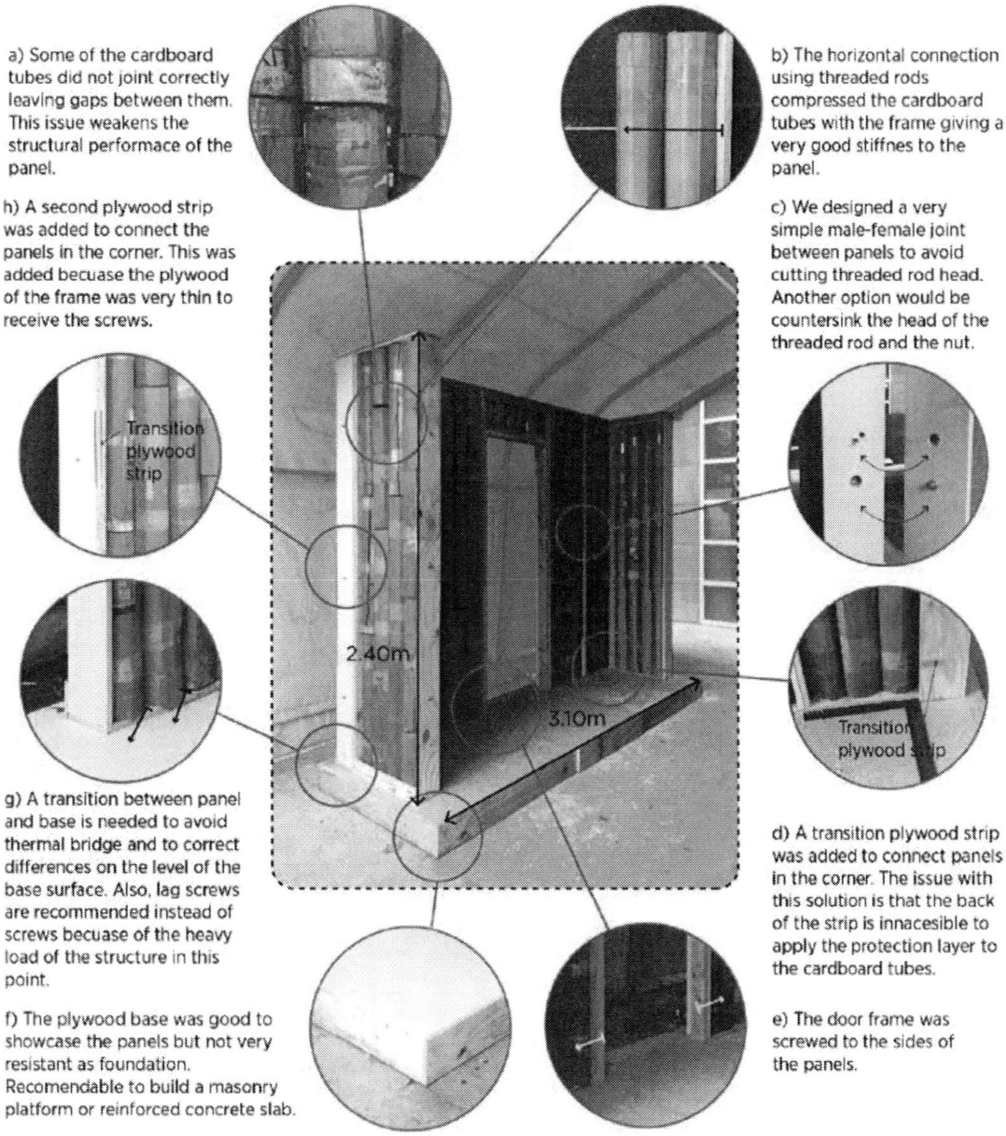

Figure 7: Preliminary Evaluation of a Full-scale Prototype Section

CONCLUSIONS AND FUTURE WORK

The goal of this research is to develop a building system, based in environments of recycling, that are applicable to low-income populations to build safe structures using waste cardboard. In this paper, we presented the design concept and production workflow for a panelized waste cardboard-based building system. We showed the development of two of the three phases of the production workflow. The presented experiment suggests the potential of an off-site prefabrication system based on modular building with incremental grow and assembled using low-tech tools. However, in concluding, we want to highlight some aspects that need further improvement before testing the system in a real-context situation.

• Variability of the material. Although the material is easy to get, transport, and manipulate, it is not uncommon for collectors to obtain sheets of dis-similar size and thicknesses, because the material comes from the urban waste stream. This variability of the found material required adapting the procedure for making same building parts with different materials. In this experiment we had only two types of sheets so the variability was insignificant. However, if we have several types of sheets, we would have to substantially increase the fabricating times of building parts.

- Precision of the fabrication of building parts. We had issues joining the cardboard tubes in both situations; vertical connections of tubes-to-tubes, and horizontal connections of tubes-to-frame. The gaps between cardboard tubes could have a critical impact in the overall strength of the panel (figure 7-a). To solve the first situation, we could wrap a final sheet of OCC to cover the gaps and to connect all the tubes vertically. For the second, we could fill the gaps with a cement-based plaster. This cement-based plaster will be useful for protecting the tubes from the elements and fire.
- Quality and strength of the connections. Screws are practical, but can be expensive and fragile. Snap-fit joints made of plywood for panel-to-panel connections could be a feasible alternative to screws. Snap-fit joins have demonstrated been a very practical solution for low-cost building systems and the studies that tested this type of mechanical connection for structural wood panels is positively increasing [18, 24].
- Facing application. We mentioned several options for facings: cement-based plastering, drywall, plywood, metal sheeting, aluminum plates, glass fiber, reinforced plastic, etc. However, each one of these should be tested and its use weighted in terms of cost and availability in the target context and application method. So far, cementbased plastering could be a good solution in cost, availability, and value-added durability. According to previous research, the application of a cement-based plaster could not only enhance the strength of the rolled cardboard tubes, but also protect the material from the elements [2]. Further experiments will focus on this area of work.

AKNOWLEDGMENTS

The first author wants to express his gratitude to:

- Mrs. Suzanne Pohland Paterno for the support to this research through the Alma Heinz and August Louis Pohland Graduate Fellowship in the School of Architecture and Landscape Architecture.
- The Stuckeman School Wood Shop staff: Allan Sutley and specially to Steve White.

REFERENCES

1. Eekhout M, Verheijen F, Visser R (2008) Cardboard in Architecture. IOS Press, Amsterdam

2. Pohl A (2009) Strengthened corrugated paper honeycomb for application in structural elements. ETH Zurich

3. Ayan Ö (2009) Cardboard in Architectural Technology and Structural Engineering. ETH Zurich

4. Salado G de C (2011) Painel de vedação vertical de tubos de papelão: Estudo, proposta e análise de desempenho. Universidade de Sao Paulo

5. Fuller B (1953) The Cardboard House. Perspecta - Yale Archit J 2:28–35

6. Ban S (2017) Shigeru Ban. Material, Structure, and Space, 1st ed. Toto Publishing Ltd., Tokyo

7. TED Shigeru Ban: Emergency shelters made from paper | Video on TED.com

8. Lübke J, Wettlaufer M, Kiziltoprak N, et al (2018) Honeycomb-Paperboard Glass Composite Beams. CE/Papers Spec Issue Eng Ttransparency Glas Archit Struct Eng 2:57–69. https://doi.org/10.1002/cepa.910

9. Bank LC, Gerhardt TD (2016) Paperboard Tubes in Architecture and Structural Engineering: A Review. 10

10. Asdrubali F, Pisello AL, D'Alessandro F, et al (2015) Innovative cardboard based panels with recycled materials from the packaging industry: Thermal and acoustic performance analysis. Energy Procedia 78:321–326. https://doi.org/10.1016/j.egypro.2015.11.652

11. Imperadori M, Salvalai G, Pusceddu C (2014) Air Shelter House Technology and its Application to Shelter Units: the Case of Scaffold House and Cardboard Shelter Installations. Procedia Econ Financ 18:552–559. https://doi.org/10.1016/S2212-5671(14)00975-7

12. Buckley HL, Touchberry CH, McKinley JP, et al (2017) Renewable Additives that Improve Water Resistance of Cellulose Composite Materials. J Renew Mater 5:1–13. https://doi.org/10.7569/JRM.2016.634109

13. Fiction Factory Wikkelhouse. https://wikkelhouse.com/. Accessed 17 Feb 2019

14. Otti L (2014) Home Poor Home - Strategies for Housing the New Poor: Typological and Technological Innovation for Contemporary Housing Habits. Arch e-journal Dissem Dr Res Archit 1:54–65

15. Richardson A, Udeaja C, Ekundayo D, et al (2013) Reuse of Materials and Byproducts in Construction. Waste Minimization and Recycling. Springer International Publishing, London

16. Raut SP, Ralegaonkar R V, Mandavgane SA (2011) Development of sustainable construction material using industrial and agricultural solid waste: A review of wastecreate bricks. Constr Build Mater 25:4037–4042. https://doi.org/10.1016/j.conbuildmat.2011.04.038

17. Buratti C, Belloni E, Lascaro E, et al (2016) Sustainable Panels with Recycled Materials for Building Applications: Environmental and Acoustic Characterization. Energy Procedia 101:972–979. https://doi.org/10.1016/j.egypro.2016.11.123

18. Elsayed K, Fioravanti A, Squasi F (2017) Low-Cost Housing: Testing snap-fit joints in agricultural residue panels. Proc 35th Int Conf Educ Res Comput Aided Archit Des Eur 2:167–174

19. Navaranjan N, Dickson A, Paltakari J, Ilmonen K (2013) Humidity effect on compressive deformation and failure of recycled and virgin layered corrugated paperboard structures. Compos Part B Eng 45:965–971. https://doi.org/10.1016/j.compositesb.2012.05.037

20. Diarte J, Vazquez E, Shaffer M (2019) A Digital and Analog Workflow for Up-cycling Old Corrugated Cardboard (OCC) as a Building Material . Tooling Cardboard for Smart Reuse :

21. Hahn EK, de Ruvo A, Westerlind BS, Carlsson LA (1992) Compressive strength of edgeloaded corrugated board panels. Exp Mech 32:259–265. https://doi.org/10.1007/BF02319364

22. Diarte J, Shaffer M (2018) Urban Recycling Intervention: Devising Low-Cost Tools for Transforming Old Corrugated Cardboard into Building Elements and Construction Materials. In: Lahtinen P, Raasakka V (eds) 10th International Conference on the Environmental and Technical Implications of Construction with Alternative Materials. Tampere, p 362

23. Sekulic B (2013) Structural cardboard: Feasibility study of cardboard as a long-term structural material in architecture. Polytechnic University of Catalonia

24. Robeller C, Mayencourt P, Weinand Y (2014) Snap-Fit Joints. CNC Fabricated, Integrated Mechanical Attachment for Struactural Wood Panels. In: Gerber D, Huang A, Sanchez J (eds) Proceedings of the 34rd Annual Conference of the Association for Computer Aided Design in Architecture. USC School of Architecture, Los Angeles, pp 189–198

18th International Conference on
Non-Conventional Materials and Technologies
"Construction Materials & Technologies for Sustainability"
(18th NOCMAT 2019)
24th – 26th July 2019
Nairobi, Kenya

ANALYSIS OF SCIENTIFIC PRODUCTION IN THE SCOPE OF THE BRAZILIAN NATIONAL NETWORK FOR RESEARCH AND DEVELOPMENT OF BAMBU - REDEBAMBU/BR: BIBLIOMETRY AND SOCIAL NETWORKS

Ohayon, Pierre [1]; Ghavami, Khosrow[2]; Siqueira, Cristiana[3]; Lopes, Sara[3]

[1] Departamento de Contabilidade/ Faculdade de Administração e Ciências Contábeis/Universidade Federal do Rio de Janeiro - FACC/UFRJ, Brazil

[2] Departamento de Engenharia Civil/ Pontifícia Universidade Católica do Rio de Janeiro (CIV/PUC-Rio), Brazil

[3] Departamento de Biblioteconomia e Gestão de Unidades de Informação / Faculdade de Administração e Ciências Contábeis/Universidade Federal do Rio de Janeiro - FACC/UFRJ, Brazil

ABSTRACT

Through the Call for Projects MCTIC/TRANSVERSAL ACTION/CNPq n°. 66/2013, the Brazilian Ministry of Science, Technology, Innovation and Communications (MCTIC) created Redebambu/BR within the PNMCB - National Policy for Incentive to the Management and Planting of Bamboo. The study aims to verify, among the projects granted, those related to bamboo´s theme, their scientific production in the last five years of its six projects coordinators, through bibliometric analysis and social networks. This exploratory study is based on specialized literature and documents retrieved from the *Lattes Curricula* of the six sponsored project coordinators. It presents the current panorama of the scientific production of the network, composed of a total of 88 researchers and shows the fields of application of bibliometrics within Redebambu/BR. Irregular distribution of researchers and scientific production is observed in the country. Finally, it is recommended: good use of bibliometrics; greater focus on users′ information needs for projects and programs; establishment and maintenance of a verifiable and relevant public database; closer integration of planning and evaluation to decision making.

Keywords: Project, RedeBambu/BR, Scientific production, Bibliometric indicators, Social networks.

CONTEXTUALIZATION

In 1972, the United Nations Conference on the Human Environment was held in Stockholm, the first major international meeting focused on environmental issues. The consequences of pollution on the quality of life of the population generated dissatisfaction in several society's sectors, especially in industrialized countries where the denunciations and warnings about damages to the environment came from non-governmental organizations and the scientific community. During the Stockholm Conference some concepts and principles were introduced which later became the basis for environmental policies [1].

In 1987 the Norway's Prime Minister, Gro Harlem Brundtland, head of the World Commission on Environment and Development, led the drafting of a document entitled Our Commom Future. This document defines sustainable development as "...the one that meets the needs of the present without compromising the ability of future generations to meet their own needs." [2].

The concept of sustainable development was ratified in 1992, during the United Nations Conference on Environment and Development (Rio's Conference), and also acknowledged that developed countries needed to provide financial and technological support to developing countries in order to enable that they could progress sustainably [1].

This movement also had an impact on civil construction, one of the oldest and most important activities in the history of civilizations. The construction industry consumes natural resources and energy resources, uses unskilled labor and produces waste, thus constituting an economic activity with a great negative impact. Then came the concept of sustainable construction presented by Professor Charles Kibert at the First World Conference on Sustainable Construction (1994), referring to the application of sustainability for construction activities [3].

The need to replace materials used in the construction industry, highly polluting and with high energy consumption in its production, led scientists and researchers to begin researching local low-cost materials to apply to buildings, especially for social constructions. Emerged a modern society concerned with producing in a sustainable manner.

In 1979, began researches on Non-Conventional Materials and Technologies (NOCMAT), proving their superiority when compared with industrialized material and their advantages for contributing to the search for sustainable development, already identified as a serious global problem. Despite this evidence, materials from non-conventional technologies have not been used on a large scale.

In 2015, the 21st Conference of the Parties (COP21) took place in Paris, as an important cooperation event between rich and non-rich countries. A historic agreement was signed where all representatives agreed to take steps for reducing the effects of climate change from 2020. Brazil has committed to restore and reforest 12 million hectares of forest until 2030. This goal not only contributes to the reduction of greenhouse gases, but also generates economic opportunities. But all this requires research and development. At the national level, the Law No. 12,484 / 2011 of the National Policy for Incentives to Sustainable Management and Bamboo's Cultivation (PNMCB) was created in 2011, aiming at the development of bamboo's culture in Brazil, promoting sustainable management of native species and cultivation of bamboo focused on the production and improvement of the activity as an instrument to promote socioeconomic development.

REDEBAMBU/BR

One of the applications of the Law No. 12,484 / 2011 is to encourage research and technological development oriented to the sustainable management and cultivation of bamboo. Thus, on September 23, 2013, the National Council for Scientific and Technological Development (CNPq) in partnership with the Ministry of Science, Technology and Innovation (MCTI) launched the so-called MCTI / Transversal Action / CNPq No. 66/2013, in order to select projects aimed at structuring the National Network for Research and Development of Bamboo (RedeBambu/BR).

RedeBambu/BR constitutes a national network of research and technological development, involving researchers, technicians and professionals from Universities and Research Institutes in an interdisciplinary and transversal work system, whose objectives are to insert bamboo in the socioeconomic development of Brazil. The result of call No. 66/2013 for structuring the Network was announced on November 23, 2013, comprising six research leaders, as shown in Table 1.

Table 1: Transversal Call Result N°. 66/2013 - MCTI/CNPq

Candidate's Name:	Institution's Name:
Khosrow Ghavami	Faculdade de Zootecnia e Engenharia de Alimentos da USP
Selma Elaine Mazzetto	Universidade Federal do Ceará
Jorge Luiz Colodette	Universidade Federal de Viçosa
Miguel Pedro Guerra	Universidade Federal de Santa Catarina
Rogério de Araújo Almeida	Universidade Federal de Goiás
Jonny Everson Scherwinski Pereira	Empresa Brasileira de Pesquisa Agropecuária

Source: adapted from CNPq´s site [4]

SOCIAL NETWORKS

Informal networking is a way of human organization in both business and academic environments. The concept of networks in this study is used as a way to verify the interactions that take place for the scientific production by the members of researchers' community forming the RedeBambu/BR. Information transfer happens when people (actors) connect themselves, share ideas, exchange opinions, and engage themselves in joint activities. Thus, "[...] the ties formed by information transfers between actors build networks" [5]. "Furthermore actors do not belong to just one network. They can be members of a number of different social networks, each based on different types of relationships. Co-workers may form a network based on the exchange of information relating to getting their job done" [6].

According to Marteleto [7] a social network "[...] represents a set of autonomous participants, integrating ideas and resources around shared values and interests". In this way, the network indicates either the space where the communication experience takes place, or the "changes and continuances" in the shared communication between the network. The author says that "[...] the term network, in the singular or plural, is associated to the adjective "social" for specifying the field, but without delimiting a specific discipline [...]" [8], enabling the participants' interdisciplinarity who come together in collaboration aiming at a common goal, in synthesis, makes possible the connection of knowledge. Social networks are a set of social units and relationships that these social units maintain with each other, directly or indirectly, through a chain of variable extensions. They can be formed by individuals, informal groups or more formal structures such as organizations, associations or companies. "In scientific context, [subjects] can play different roles, functions and responsibilities within the dynamics of knowledge production" [9]. Its dynamics are focused on perpetuating, consolidating and developing the activities of its members. The study of social networks consists of studying the relational links between social actors. Researchers working in a given area form a network of co-authorship, and in this way, it can be said that each of the six research leaders included in the Call No. 66/2013 for the structuring of RedeBambu/BR forms a network of co-authorship.

STUDY MOTIVATION

There are few initiatives to implement evaluation methodologies and the construction of ST&I indicators " [...] even acknowledging the need to establish new evaluation criteria for publicly funded programs and projects" [10]. It is considered as an evaluation within Redebambu/BR a "[...] systematic collection of information about activities, characteristics and outcomes of actions with the purpose of reducing uncertainties, improving effectiveness and making decisions, regarding what these actions are doing and influencing" [11]. Thus, there is a need to map the activities of projects' coordinators, as well as their collaborators, so that it can be demonstrated the progress made since the dissemination of the Call No. 66/2013.

GOALS

The study aims, as a general objective, to bring about in the context of the scientific production developed by the proponents sponsored by the Call No. 66/2013, themes related to bamboo. The specific objectives include (a) verify the proponents' scientific production, possible advances and / or possible barriers within the network; (b) indicate the contributions of librarianship through bibliometric and social networks analysis; (c) analyze the main aspects observed in the last five years, from the proponents of the Call.

METHODOLOGY

From the result of Call No. 66/2013, the Lattes Platform was used for data gathering, the basis for recording the past and current activities of researchers in Brazil, where the résumé of the six names included in the Call were sponsored. In the bibliographic production of the researchers' Lattes curriculum, a search was made through the Control + L (or Control + F) option to locate the term "bamb", truncated so that both bambu (Portuguese) and bamboo (English) could be retrieved. Considering that the result of the Call was announced in November 2013, the time cut adopted for the recovery of bibliographic production was between 2014 and 2018. The extracted data were inserted in a spreadsheet in which the title and authors of the bibliographic production were signaled, the institutions of which the authors and their respective Brazilian regions are included (South, Southeast, North, Northeast and Central West). The publications were also separated according to the type of document retrieved, namely congress proceedings, paper submission, articles published in periodicals, book chapters and a last category for any other publications that did not fit the previous alternatives. In addition, to perform the analysis and visualization of the social networks of each of the six researchers included in the Call, Gephi software, an open source software package, was selected, this one developed to analyze and allow the visualization of a network.

RESULTS

The study retrieved 37 documents (Figure 1), which were divided according to the methodology, showing that most of the publications are between articles published in periodicals and annals of congress (both with 35%); followed by book chapters (22%); other publications (5%) and work presentations (3%).

Figure 1 - Distribution by document type

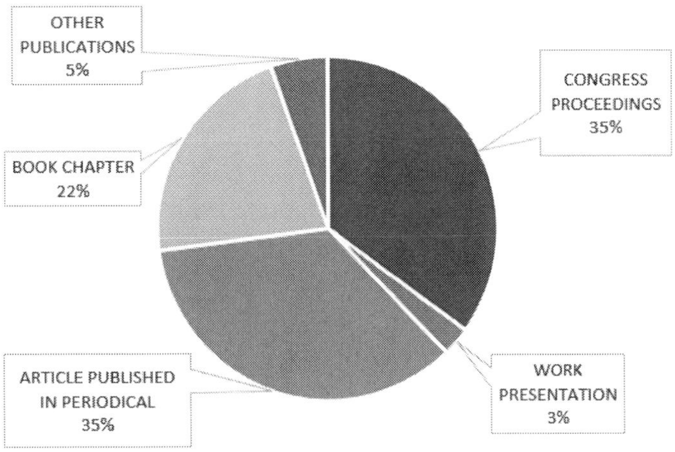

A total of 88 researchers were retrieved from the 37 documents, of which 6 are those referred by the Call N°. 66/2013 and 8 were not located on the Lattes Platform. Starting from the division of these 88 authors per institution, the graph of Figure 2 shows that approximately 23% of the researchers belong to the Pontifical Catholic University of Rio de Janeiro (PUC-Rio); followed by the Federal University of Ceará (UFC) with 11% of the authors. It should be noted that researchers not found represent 9% of the graph.

Figure 2 - Distribution of authors by institution

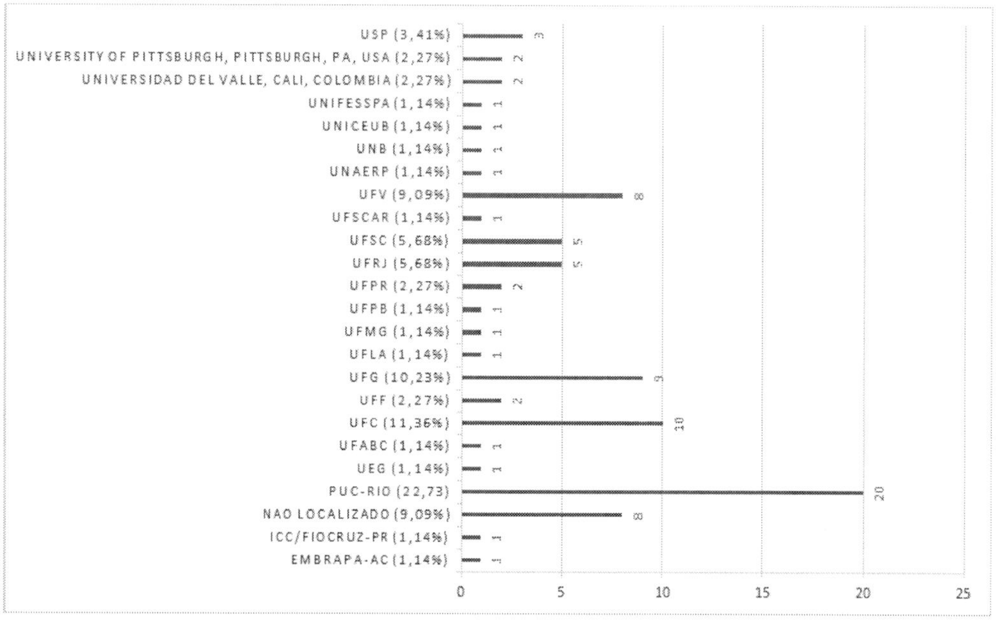

The division by institutions made it possible to analyze also which Brazilian regions have more and fewer publications.

As shown in Figure 3, there is a high concentration of research in the Southeast Region, corresponding to 48% of publications; while the North Region has the lowest index of research, only 3% of publications.

Figure 3 - Distribution of authors by region

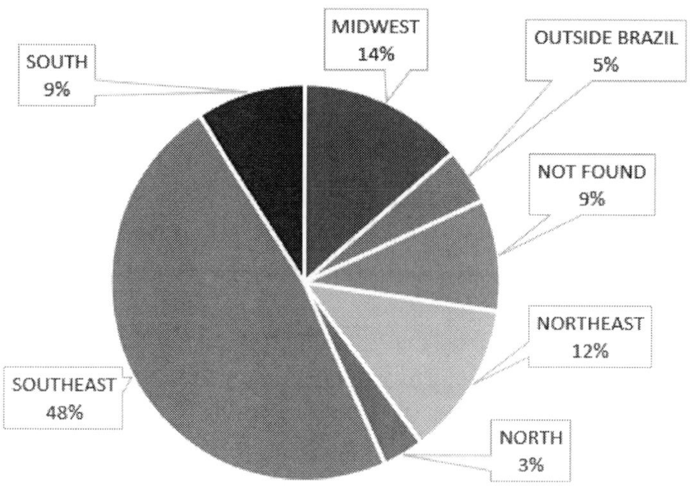

The Gephi software was used to illustrate the social network of each researcher sponsored by the Call N°. 66/2013.

The application allows to visualize with which authors the researcher writes more frequently, by means of highlights in their names, namely, in both the size of the font and the size of the representation symbol (sphere, in the case of this study).

In addition, it allows to visualize with which regions of the country the researcher usually relates, by means of assigning colors to each one.

The first researcher of the call, Khosrow Ghavami, has its network represented in Figure 5. The production of Ghavami covers 16 documents among the 37 recovered, in which there are the collaboration of another 38 researchers.

The purple color, predominant in Figure 4, represents the Southeast Region of Brazil and is composed by a network of 30 authors; the light green color illustrates the 4 authors who relate to the researcher and are from other countries.

The light blue color represents 2 authors of the network whose curricula were not found; already the colors dark green and orange represent, respectively, 1 author of the Center-West region and 1 author of the Northeast Region.

Figure 4 - Representation of the network of the Khosrow Ghavami Researcher

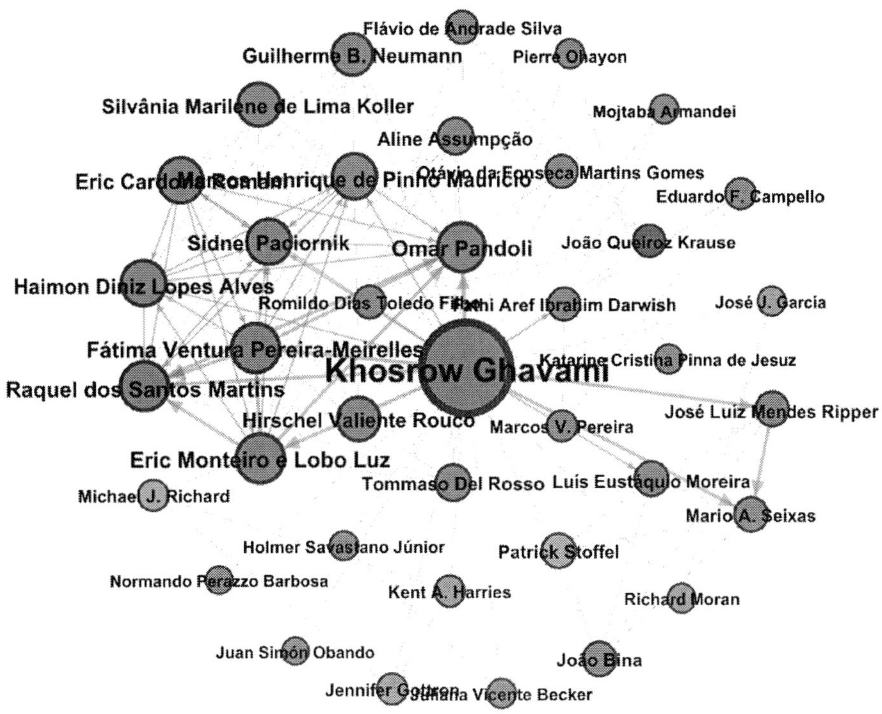

The network of the second Call's researcher, Selma Elaine Mazzetto, is composed of 12 authors, with 4 of the 37 documents recovered. Table 2 shows the most frequent Regions of the country in the network.

Table 2 - Regions in the network of Researcher Selma Elaine Mazzetto

South	Southeast	North	Northeast	Midwest	Authors not found
None	1 author (8,33 %)	None	10 authors (83,33 %)	none	1 author (8,33 %)

The network of researcher Jonny Everson Scherwinski Pereira is composed by 8 authors, with 6 of the 37 documents recovered. Table 3 shows the most frequent Regions of the country in the network.

Table 3 - Regions in the network of the Researcher Jonny E. S. Pereira

South	Southeast	North	Northeast	Midwest	Authors not found
None	1 author (12,5 %)	3 authors (37,5 %)	None	3 authors (37,5 %)	1 author (12,5 %)

The researcher Jorge Luiz Colodette has a network composed by 7 authors, with 5 of the 37 documents recovered. Table 4 shows that all the authors with whom the researcher related belong to the Southeast Region.

Table 4 - Regions in the network of the Researcher Jorge Luiz Colodette

South	Southeast	North	Northeast	Midwest	Authors not found
None	7 authors (100 %)	None	None	None	None

The next network analyzed is from the researcher Miguel Pedro Guerra, who represents 1 of the 37 documents recovered. The paper has 10 authors, whose 80% represent the South Region and 20% represent the Southeast Region according to Table 4.

Table 5 - Regions in the network of the Researcher Miguel Pedro Guerra

South	Southeast	North	Northeast	Midwest	Authors not found
8 authors (80 %)	2 authors (20 %)	None	None	None	None

The last researcher, Rogério de Araújo Almeida, has the network composed by 13 researchers, with 5 of the 37 documents recovered. Table 6 indicates the most frequent Regions of the country in the network.

Table 6 - Regions in the network of the researcher Rogério de A. Almeida

South	Southeast	North	Northeast	Midwest	Authors not found
None	1 author (7,69 %)	None	None	8 authors (61,54 %)	4 authors (30,77 %)

FINAL REMARKS

It has been verified that 43.24% of the total of the documents recovered indicates collaboration of the researcher Khosrow Ghavami, Emeritus Professor at PUC-Rio, founder of ABMTENC – Brazilian Association of the Science of Non-Conventional Materials and Technologies and active in the field for more than 40 years.

The authors of this study recognize the limitation in the retrieval of documents by choosing the use the Lattes Platform as a reference′ source, since it is the author who feeds the Platform. Future study on databases such as Scielo or Web of Science could broaden the research findings.

Based on the results, it can be seen that only few documents have been published since the result of the Call N°. 66/2013. This may be due to the non-release of financial resources, until the date of completion of this study. The lack of resource release makes difficult the full progress of scientific and technical research. Considering the findings related to the low adherence of authors from the North region producing articles on the topic, it can be said that there is a lack in the region requiring investments in order to encourage more researches on Non-Conventional Materials and Technologies.

ACKNOWLEDGEMENTS

The authors would like to thank ABMTENC Association which made available all the NOCMAT Conferences data and the *18th International Conference on Non-Conventional Materials and Technologies – Construction Materials & Technologies for Sustainability* (18th NOCMAT 2019), 24th – 26th, July 2019, Nairobi/Kenya organizers for their valuable effort and proceedings for the discussions, which form the basis of the meeting. Also, financial support given by FAPERJ and FACC/UFRJ are appreciated.

REFERENCES

1. Lago, A. A. C. do. Estocolmo, Rio, Joanesburgo: o Brasil e as três conferências ambientais das Nações Unidas. Brasília: FUNAG, 2006.

2. Siqueira, C. Práticas informacionais em comunidades de compostagem. 2017. 71 f. Trabalho de conclusão de curso (Biblioteconomia e Gestão de Unidades de Informação). Universidade Federal do Rio de Janeiro, Brazil, 2017.

3. Pereira, P. I. Construção sustentável: o desafio. 2009. 122 f. Monografia. (Engenharia Civil). Universidade Fernando Pessoa, Porto, 2009.

4. Brasil. Portal CNPq. Chamadas públicas: resultados. Retrieved from: http://memoria.cnpq.br/chamadas-publicas;jsessionid=13A063C80F0140D890DAEC877B9FADCD?p_p_id=resultadosportlet_WAR_resultadoscnpqportlet_INSTANCE_0ZaM&idDivulgacao=3981&filtro=resultados&detalha=chamadaDetalhada&exibe=exibe&id=209-121-2084&idResultado=209-121-2084. Access in: 25 abril 2019.

5. Haythornthwaite, C. (2009). Social network and information transfer. In: The encyclopedia of library and information science. Marcia Bates, & Mary Maack (Eds.). NY: Taylor & Francis.

6. Haythornthwaite, C. (1996) Social network analysis: An approach and set of techniques for the study of information exchange. Library and Information Science Research, 1996, 18(4), 323-342.

7. Marteleto, R. M. Análise de redes sociais: aplicação nos estudos de transferência da informação. Ciência da Informação, Brasília, DF, v. 30, n. 1, p. 71-81, jan./abr. 2001.

8. Marteleto, R. Redes sociais, mediação e apropriação de informações: situando campos, objetos e conceitos na pesquisa em ciência da informação. Pesquisa Brasileira em Ciência da Informação, Brasília, v.3, n.1, p.27-46, jan./dez., 2010.

9. Zattar, M.. Prática informacional em redes no domínio da governança da água: um estudo sobre o processo de produção do conhecimento. 2017. 159 f. Teses (Doutorado em Ciência da Informação)- Universidade Federal do Rio de Janeiro, Instituto Brasileiro de Informação em Ciência e Tecnologia, Rio de Janeiro, 2017.

10. Viotti, E. B.; Macedo, M. de M. (Orgs.) Indicadores de Ciência, Tecnologia e Inovação no Brasil. Campinas/SP: Editora da Unicamp, 2003.

11. Lima, N. P. C. Avaliação das ações de Ciência, Tecnologia e Inovação (C,T&I): reflexões sobre métodos e práticas. Brasília: Centro de Gestão e Estudos Estratégicos/CGEE, 2004.

IC-NOCMAT 2019

18th International Conference on
Non-Conventional Materials and Technologies
"Construction Materials & Technologies for Sustainability"
(18th NOCMAT 2019)
24th – 26th July 2019
Nairobi, Kenya

BIBLIOMETRIC ANALYSIS OF SCIENTIFIC AND TECHNICAL PAPERS WITHIN NOCMAT 1984-2017 INTERNATIONAL CONFERENCES

Ohayon, Pierre[1]; Sharafi Rad, Ali[2]; Ghavami, Khosrow[3]; Siqueira, Cristiana[4]; Lopes, Sara[4]

[1] Departamento de Contabilidade/ Faculdade de Administração e Ciências Contábeis/Universidade Federal do Rio de Janeiro - FACC/UFRJ, Brazil;

[2] Departamento de Engenharia Industrial/ Pontifícia Universidade Católica do Rio de Janeiro (DEI/PUC-Rio), Brazil;

[3]Departamento de Engenharia Civil/ Pontifícia Universidade Católica do Rio de Janeiro (CIV/PUC-Rio), Brazil;

[4]Departamento de Biblioteconomia e Gestão de Unidades de Informação / Faculdade de Administração e Ciências Contábeis/Universidade Federal do Rio de Janeiro - FACC/UFRJ, Brazil;

ABSTRACT

The last International Conference on Non-Conventional Materials and Technologies (NOCMAT) held in Mexico in 2017 presented more than 80 new papers among different NOCMAT themes, showing that there is a great number of scientific and technical studies being developed. 1,485 articles have been presented along the 16 International Conferences, since the first one held in Brazil, in 1984. The series of NOCMAT Conferences have proven to be a leading forum where scholars, governmental and non-governmental agencies, practitioners exchange innovations of low energy cement technologies, new ecological materials and systems such as bamboo and natural fibers. To show the reliability and durability of the newly developed materials and technologies, in addition to the results obtained in the laboratories, large scale constructions should be built and permanently monitored, requiring higher and continued investments from sponsoring agencies and private organizations. The general objective of this study is to analyze the betweenness of researches presented within all the thematic areas of the NOCMAT Conferences. It presents a bibliometric analysis of the network created, mapping the actors (nodes) and its relations (edges) by participating authors, involved Institutions; participating Countries and Year, and NOCMAT Themes, along the last 33 years.

Keywords: NOCMAT Thematic Trends, International Conferences, Papers, Bibliometric Analysis.

INTRODUCTION – NOCMAT RESEARCH CONTEXT

The understanding of sustainability in building construction has undergone changes over the years. First attention of specialists was directed towards the topic of limited resources, especially energy, and its impact on the natural environment. Now emphasis is placed on technical issues such as materials, building components, construction technologies and energy-related design concepts as well as on non-technical issues such as economic and social sustainability.

The pursuit of sustainable development as defined in the Gro Harlem Brundtland Report [1] as "development that meets the needs of the present without compromising the ability of future generations to meet their own needs" has become a major issue when trying to meet the challenge of providing proper housing for an increasing world population. Recently, Pierre Papon emphasized the necessity to take time because progress is not as fast as we think and also developments are becoming heavier [2].

To increase understanding of sustainable materials, also known as Non-Conventional Materials and Technologies (NOCMAT) using organic materials, which are used either alone or as reinforcement in different types of matrices such as soil, cement and polymers, many research programs have been carried out all over the globe. In recent years, there has been an increase in research into the use of natural and non-conventional construction materials, which have suitable physical and mechanical properties for structural applications, and are desirable as ecological alternatives to more commonly used industrialized materials.

The study of NOCMAT for construction began in the 1970's. This field of research has continued to grow in the decades since. The increased use of locally available natural and waste materials in construction can promote environmental sustainability and aid in the eradication of extreme poverty.

Many of the world's most poverty-stricken people live in remote locations, typically rich in natural resources. In many of these areas, the use of steel and reinforced concrete in construction has become a symbol of economic status. These materials are imported for use when local natural materials can also be used to create structures that can successfully meet the intended need at a lower cost and environmental impact (considering precautions are taken to prevent deforestation or depletion of other resources).

Organizations like ABMTENC - Brazilian Association of Non-Conventional Materials and Technologies [3] at the Pontifical Catholic University of Rio de Janeiro and the INBAR - International Network for Bamboo and Rattan strive to achieve sustainable development and poverty alleviation by promoting the use of these two natural resources Bamboo and Rattan [4].

Since the development of industrialized materials, including steel and concrete, natural materials have been abandoned for their structural use. While the mechanical properties of natural materials are less predictable than those of the industrial materials, the natural materials are readily available in many rural locations and can be used with significantly less processing. The availability of steel and cement, on the other hand, depend on the presence of factories and an abundant feedstock supply to create these materials, as well as adequate infrastructure to permit transportation to the locations for their use.

Now, with energy consumption and sustainability becoming increasingly important issues in the construction industry, industrialized materials are not always optimal when there are many natural materials that have excellent structural properties, are renewable, and require minimal energy input to be construction-ready. Therefore, as the progress of the civil engineering industry relies on the continued development of different materials in construction, NOCMAT should be considered in the search for the low-energy structural materials of the future.

ABMTENC´S CONTRIBUTIONS TO NOCMAT INTERNATIONAL CONFERENCES

The Brazilian Association of Non-Conventional Materials and Technologies – ABMTENC was founded on 29th of March 1996, at the Pontifical Catholic University of Rio de Janeiro – PUC-Rio, by professor Ghavami and the rector of the University, the Vice Rector Jesus Hortal Sanches S.J., the previous Rector - Father Laércio Moura S.J., and the deans and scientists, engineers, architects and many other professionals, who have been interested to the study and development of the science of non-conventional materials and technologies.

Once professionals, organizations and institutions in Brazil have been brought together to share their common goals of assisting science and developing non-conventional materials and technologies, ABMTENC intends to develop research and the expand the knowledge of this field, as well as encouraging communication, between scientist and universities, institutions and non-governmental organizations throughout the inside of country and abroad, also. ABMTENC intent to share knowledge of these sciences through the publication of books and magazines, as well as promoting meetings, congresses, conferences, courses, and techno-scientific reunions.

ABMTENC's success depends fundamentally on the interest and participation of its members, individuals and enterprises that want to share advances in research into non-conventional technologies and materials in many different fields of application such as: construction, housing, bamboo, recycling, environment, sanitation, bau-biologie, composites materials with vegetal fibers, earth architecture [3].

NOCMAT INTERNATIONAL CONFERENCES

The former International Conference on Non-Conventional Materials and Technologies (IC-NOCMAT) has been realized in Mérida/Mexico in November 2017, coinciding with the 33rd anniversary of NOCMAT.

This IC-NOCMAT follows previous successful events in Winnipeg/Canada (2015), Pirassununga, State of São Paulo/Brazil (2014), João Pessoa/Paraíba-Brazil (2013), Changsha, Hunan/China (2011), Cairo/ Egypt (2010), Bath/England (2009),

Cali/Colombia (2008), Maceió/Alagoas-Brazil (2007), Salvador/Bahia-Brazil (2006), Rio de Janeiro/RJ-Brazil (2005), Pirassununga/ São Paulo-Brazil (2004), João Pessoa/Paraíba-Brazil (2003), Hanoi/Vietnam (2002), Bhubanewar/India (1997) after the first NOCMAT 1984 held in Rio de Janeiro/Brazil.

Study Objetive and Methodology

The aim of this study is to map and analyze thematic trends of scientific and technical production within the NOCMAT conferences systematizing 1,485 papers published in them. Since its first edition in 1984 to his last in 2017 are highlighted: (1) the production of published articles and authorship; (2) the chronology of the thematic areas; (3) the geographical distribution by country; (4) the institutional origin; (5) axes and thematic trends. The technical themes of knowledge application considering the thematic areas identified in each NOCMAT conference were pointed out for trend analysis.

It is included in the field of studies on *State of the Art*[1] [5] that aims to understand the evolution and trends of scientific and technological knowledge produced on certain topics or areas of expertise. Articles published in conferences and magazines coming from dissertations, doctoral theses or even activities of research and development (R&D) are one of the main forms of scientific communication for knowledge in construction. This field permits to examine national policies, programs and strategies that aim to foster notably open access (OA) and discuss how OA policies are monitored and enforced [6, 7]. This knowledge capital oriented to practical applications has been studied by many intellectuals in order to assess high quality processes concerning research and technological development such as accountability, productivity and measurement, absorptive capacity, intellectual property, among many others [8, 9, 10, 11, 12, 13].

Bibliometric Tools for Scientific and Technical Paper's Analysis

Bibliometrics is a branch of science that quantitatively studies the processes of scientific communication, *"[...] with the purpose of explaining the dynamics of science and technology from indicators of outputs and inputs."* [14]. Initially known as statistical bibliography, it was popularized in 1969 from an article written by Pritchard [15], although originally created by Paul Otlet in 1934 [16].

Since its origins, it has had two major concerns: scientific production and search for immediate practical benefits for libraries, initially developed for measuring the number of editions and copies of books, the number of words contained in them, the space occupied on bookshelves and usage statistics. Over time, it has been oriented for the study of other formats of bibliographic production (periodical papers and other types of documents) pointing out authors' productivity, scientific research directions and networks, social collaborations, and citations analysis.

---

[1] "The term '*state of the art*' refers to the highest level of general development, as of a device, technique, or scientific field achieved at a particular time. It also refers to such a level of development reached at any particular time as a result of the common methodologies employed at the time" [5].

The analysis of scientific networks brings about important indicators such as network density (calculated by the number of relations existing by the number of possible relations in a network), degree of centrality (indicates the number of connections an actor has), index of centralization when an actor occupies a central position in the network and the others have to pass through it), degree of intermediation (possibility that an actor has to facilitate the communication between pairs of actors) and degree of proximity (ability of an actor to connect to all the other actors in the network) [17].

Different computer programs were developed for constructing and viewing bibliometric maps. More specifically, for the authors [18]:

"*VOSviewer pays special attention to the graphical representation of bibliometric maps. The functionality of VOSviewer is especially useful for displaying large bibliometric maps in an easy-to-interpret way. VOSviewer's ability to handle large maps is demonstrated by using the program to construct and display a co-citation map of viewing bibliometric 5,000 major scientific journals*".

Bibliometric data generate indicators that allow a global view of research development, providing relevant information in different scales, from the broadest (country) to the most specific (institution). In this way, it is possible to estimate a tendency of growth or decrease in many areas of research, providing a better decision making. The collectivization and internationalization of research is a growing trend verified through bibliometric mappings.

DATA AND ANALYSIS

In this topic are presented the level of production covering all the 14 NOCMAT conferences realized in different countries since 1984 to 2017, this one being its last edition. All the data were extracted from CD-ROMs and Books of abstracts/full papers and made available by ABMTENC Association.

Mapping papers by the participating authors

NOCMAT 1984 - the first of the series held in Rio de Janeiro/Brazil, began with a single overall theme entitled "Low-Cost and Energy Saving Construction Materials". Since the second NOCMAT held in Bhubaneswar/India in 1997 until now, emerged the term "*Non-Conventional*" in the Conference's title.

Table 1, Graphs 1 and 2 present the distribution of all the papers by year (NOCMAT Conference series), country of hosting NOCMAT Conferences, and number of authors per paper (1 to more than 6).

Table 1: Number of Papers per Participating Authors

				Participating Authors							
NOCMAT	**Year**	**City**	**Country**	**1**	**2**	**3**	**4**	**5**	**6**	**6<**	**Total p.**
1	1984	Rio de Janeiro	Brazil	31	16	11	4	1	-	1	**64**
2	1997	Bhubaneswar	India	8	5	7	3	-	1	0	**24**
3	2002	Hanoi	Vietnam	8	22	16	10	4	2	1	**63**
4	2003	João Pessoa	Brazil	3	13	24	11	6	1	1	**59**
5	2004	Pirassununga	Brazil	3	13	18	13	3	5	2	**57**
6	2005	Rio de Janeiro	Brazil	6	21	17	18	6	4	1	**73**
8	2006	Salvador	Brazil	5	23	34	7	12	3	2	**86**
9	2007	Maceió	Brazil	4	23	18	16	12	4	1	**78**
10	2008	Cali	Colombia	9	35	24	16	19	9	1	**113**
11	2009	Bath	UK	19	30	23	27	13	6	0	**118**
12	2010	Cairo	Egypt	16	26	34	23	14	5	1	**119**
13	2011	Hunan	China	6	28	44	39	13	6	3	**139**
14	2013	João Pessoa	Brazil	18	27	43	20	10	13	7	**138**
15	2014	Pirassununga	Brazil	2	30	33	26	18	18	9	**136**
16	2015	Winnipeg	Canada	10	28	35	21	24	8	6	**132**
17	2017	Mérida	Mexico	1	19	24	21	10	6	5	**86**
				149	359	405	275	165	91	41	**1,485**
				10%	24%	27%	19%	11%	6%	3%	100%

Graph 1: Number of Papers per Participating Authors

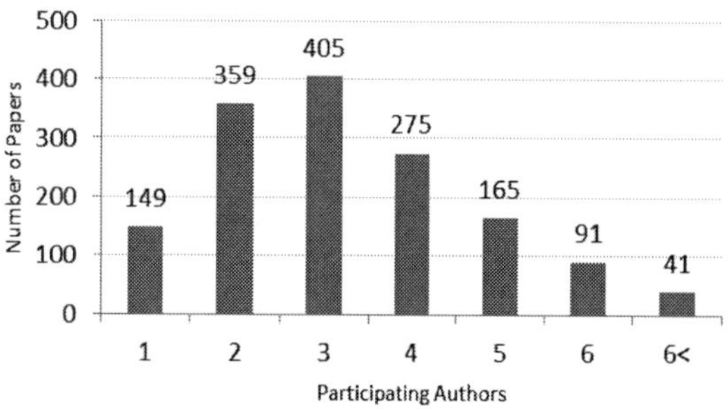

Graph 2: Number of Papers per Year

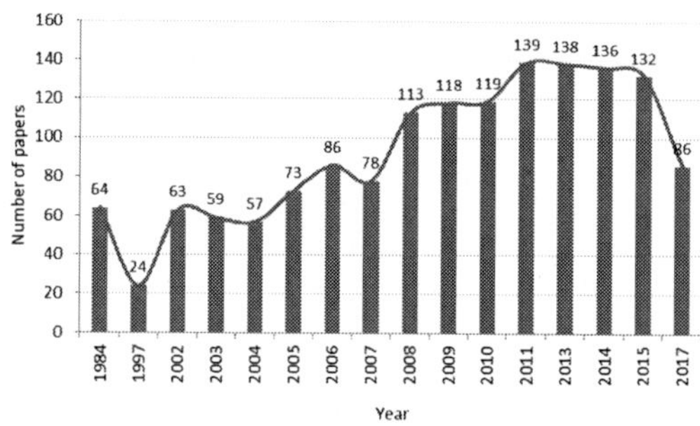

The data collected totalize 1,485 papers for all the NOCMAT Conference series. Table 1 shows that 51% of the papers (764 papers) are published by 2 or 3 authors and, 90% of the papers (1,336 papers) are published by more than one author. Graph 2 indicates a significant average growth of the number of papers presented from 1984 to 2017.

Mapping papers by the involved centers of NOCMAT knowledge (Institutions)

Tables 2 and Graph 3 present the distribution of all the papers by involved Institution in all NOCMAT Conference series from 1984 to 2017.

Table 2: Number of Papers per Participating Institutions

				Participating Institutions						
NOCMAT	Year	City	Country	1	2	3	4	5	6	Total p.
1	1984	Rio de Janeiro	Brazil	49	11	3	1	-	-	**64**
2	1997	Bhubaneswar	India	17	7	-	-	-	-	**24**
3	2002	Hanoi	Vietnam	41	17	5	-	-	-	**63**
4	2003	João Pessoa	Brazil	36	18	2	2	1	-	**59**
5	2004	Pirassununga	Brazil	36	20	1	-	-	-	**57**
6	2005	Rio de Janeiro	Brazil	46	21	4	1	1	-	**73**
8	2006	Salvador	Brazil	62	21	2	-	1	-	**86**
9	2007	Maceió	Brazil	46	26	5	-	1	-	**78**
10	2008	Cali	Colombia	61	43	8	1	-	-	**113**
11	2009	Bath	UK	64	38	14	2	-	-	**118**
12	2010	Cairo	Egypt	70	28	20	1	-	-	**119**
13	2011	Hunan Changsha	China	88	40	10	1	-	-	**139**
14	2013	João Pessoa	Brazil	84	33	16	5	-	-	**138**
15	2014	Pirassununga	Brazil	78	39	12	6	-	1	**136**
16	2015	Winnipeg	Canada	83	37	8	4	-	-	**132**
17	2017	Mérida	Mexico	51	12	14	6	3	-	**86**
Total papers				912	411	124	30	7	1	**1,485**
Percentage				61.4%	27.7%	8.4%	2.0%	0.5%	0.1%	100.0%

The data from Table 2 and Graph 3 show that about 36.0% of the papers (535 papers among a total of 1,485) are published by 2 and 3 institutions and about 38.6% of the papers (573 papers) are published by more than one institution in all NOCMAT Conference series. The level of multiple authorship increased indicating a greater integration of human resources and institutions.

Graph 3: Number of Papers per Participating Institutions

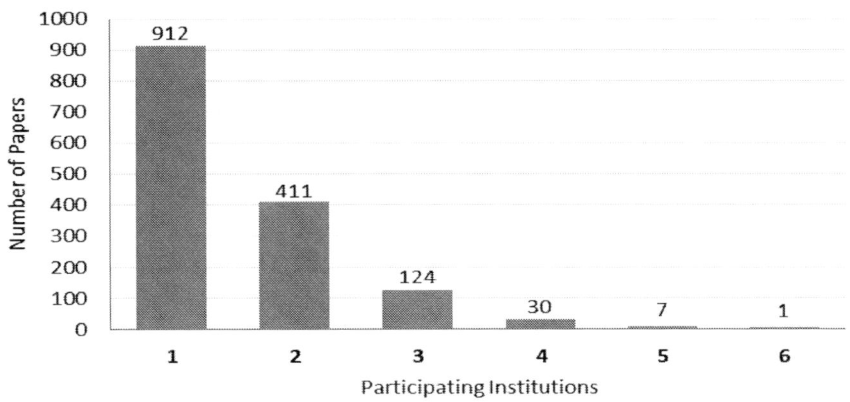

Table 3 and Graph 4 show the number of Institutions which participated in all NOCMAT Conference series.

Table 3: Number of Institutions per Year and Higher Number of Papers Presented by an Institution

Year	Total Ints.	Institution with highest nº of papers		nº of Papers	Country
1984	67	LEHIGH	Lehigh University	4	USA
1997	21	PUC-RIO	Pontifícia Universidade Católica do Rio de Janeiro	5	Brazil
2002	53	HANYANG	Hanyang University	6	Korea
2003	48	PUC-RIO	Pontifícia Universidade Católica do Rio de Janeiro	9	Brazil
2004	38	USP	Universidade de São Paulo	8	Brazil
2005	58	CEFET-PR	Centro Federal de Educação Tecnológica do Paraná	8	Brazil
2006	53	UFPB	Universidade Federal da Paraíba	14	Brazil
2007	55	UFAL	UFAUniv Federal de Alagoas	8	Brazil
2008	82	UNIVALLE	Universidad del Valle	16	Colombia
2009	122	BATH	University of Bath	19	UK
2010	68	HBRC	Housing and Building National Research Center	16	Egypt
2011	88	HNU	Hunan University	17	China
2013	108	UFPB	Universidade Federal da Paraíba	14	Brazil
2014	81	USP	Universidade de São Paulo	**22**	Brazil
2015	81	PUC-RIO	Pontifícia Universidade Católica do Rio de Janeiro	**22**	Brazil
2017	35	PUC-RIO & UFRJ	Pontifícia Universidade Católica do Rio de Janeiro & Universidade Federal do Rio de Janeiro	16	Brazil

Table 3 and Graph 4 show that the number of institutions participating in NOCMAT Conference series increased, in a very significant average growth, from 21, its lowest level (in 1997/Brazil) to 122, its highest level (in 2009/UK).

Graph 4: Number of Institutions per Year

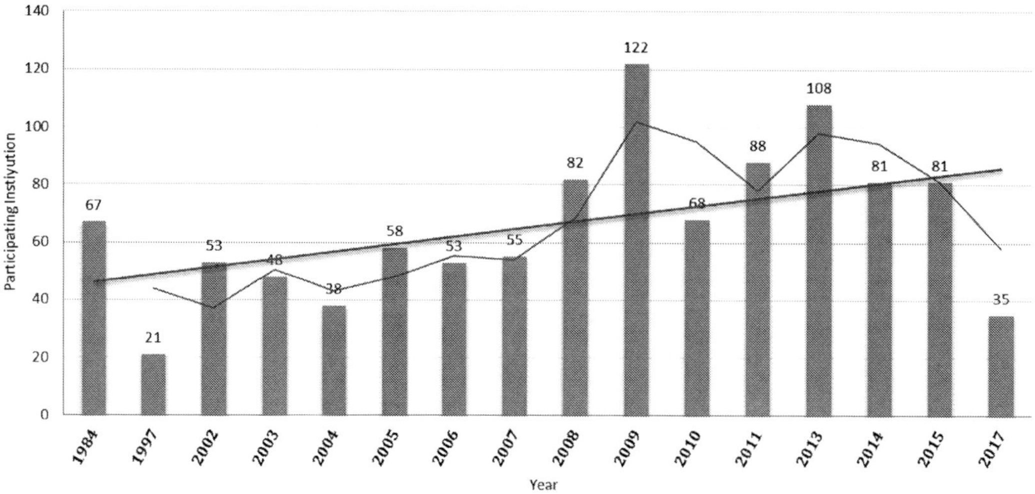

Table 3 shows that the highest number of papers by institutions is increasing from 1984 (4 papers from LEHIGH) to 2015 (22 papers from PUC-RIO). The data collected reveals that more than 641 Institutions have participated in all NOCMAT Conference series. In average, 66 Institutions have participated in each of them.

Table 4: Number of Institutions Participating in all NOCMAT Conference Series by each Country

Rank	Countries	N° of Institutions	%
1	Brazil	148	23.1%
2	UK	56	8.7%
3	China, USA (46 ins. each)	92	7.2%
4	Colombia	34	5.3%
5	India	28	4.4%
6	Mexico	22	3.4%
7	Argentina	20	3.1%
8	Canada, Egypt (17 ins. each)	34	2.7%
9	France	16	2.5%
10	Japan, Spain (15 ins. each)	30	2.3%
11	Italy	14	2.2%
12	Algeria, Iran (11 ins. each)	22	1.7%
13	Australia, Portugal (9 ins. each)	18	1.4%
14	Germany	8	1.2%
15	Romania, Vietnam (6 ins. each)	12	0.9%
16	Korea, Peru (5 ins. each)	10	0.8%
17	Cuba, New Zealand, Nigeria, Poland, South Africa, Thailand (4 ins. each)	24	0.6%
18	Belgium, Chile, Ghana, Libya, Malaysia, Venezuela (3 ins. each)	18	0.5%
19	Ecuador, Greece, Ireland, Nepal, Pakistan, Sweden, Switzerland, The Netherlands, Turkey (2 ins. each)	18	0.3%
20	Bolivia, Botswana, Cypress, Denmark, Ethiopia, Guatemala, Jordan, Kuwait, Mozambique, Paraguay, Philippines, Qatar, Russia, Senegal, Singapore, Taiwan, Tanzania (1 ins. each)	17	0.2%
		641	

Table 4 shows that among all institutions that participated in all the NOCMAT Conference series, Brazil is in first position with 148 participating institutions. One of its important impacts is the increasing number of calls for R&D projects sponsored by the Brazilian Ministry of Science, Technology and Innovation [19].

4.3 Mapping papers within NOCMAT Conferences by participating Countries and Year

Table 5, Table 6, Graphs 5 and 6 present the distribution of all the papers by participating country and year of NOCMAT Conference series.

The data from Table 5 show that 157 papers are published by authors from 2 countries and 180 papers among 1,485 are published by more than one country in all NOCMAT Conference series.

61[2] countries have participated in all NOCMAT Conference series.

Graph 5 shows significant average growth of the number of papers engaging authors of two countries or more. For NOCMAT Conferences held in Brazil and India, respectively in 1984 and 1997, just 3 papers were presented by authors from 2 countries. In 2017, the number increased to 12. Researchers of different countries are being involved in common projects and engaging resources from multiple sponsor agencies, increasing engagements at the international level.

Table 6 (which indicates more specifically the location of each NOCMAT Conference) and Graph 6 show that NOCMAT Conference 2009 held in United Kingdom (University of Bath) integrated 30 countries, the highest registered in all NOCMAT Conference series. In average, 18 countries have participated in all NOCMAT Conference series.

Table 5: Number of Papers per Participating Countries

			Participating countries					
Year	**City**	**Country**	**1**	**2**	**3**	**2&3**	**Total p.**	**Percentage**
1984	Rio de Janeiro	Brazil	61	3	-	3	**64**	4.3%
1997	Bhubaneswar	India	21	3	-	3	**24**	1.6%
2002	Hanoi	Vietnam	54	7	2	9	**63**	4.2%
2003	João Pessoa	Brazil	52	6	1	7	**59**	4.0%
2004	Pirassununga	Brazil	52	5	-	5	**57**	3.8%
2005	Rio de Janeiro	Brazil	66	6	1	7	**73**	4.9%
2006	Salvador	Brazil	81	5	-	5	**86**	5.8%
2007	Maceió	Brazil	64	12	2	14	**78**	5.3%
2008	Cali	Colombia	96	12	5	17	**113**	7.6%
2009	Bath	UK	101	15	2	17	**118**	7.9%
2010	Cairo	Egypt	104	14	1	15	**119**	8.0%
2011	Hunan Changsha	China	126	12	1	13	**139**	9.4%
2013	João Pessoa	Brazil	123	14	1	15	**138**	9.3%
2014	Pirassununga	Brazil	118	16	2	18	**136**	9.2%
2015	Winnipeg	Canada	114	15	3	18	**132**	8.9%
2017	Mérida	Mexico	72	12	2	14	**86**	5.8%
Total Papers			1,305	157	23	180	**1,485**	100%

[2] (1) Algeria, (2) Argentina, (3) Australia, (4) Belgium, (5) Bolivia, (6) Botswana, (7) Brazil, (8) Canada, (9) Chile, (10) China, (11) Colombia, (12) Cuba, (13) Cypress, (14) Denmark, (15) Ecuador, (16) Egypt, (17) Ethiopia, (18) France, (19) Germany, (20) Ghana, (21) Greece, (22) Guatemala, (23) India, (24) Iran, (25) Ireland, (26) Italy, (27) Japan, (28) Jordan, (29) Korea, (30) Kuwait, (31) Libya, (32) Malaysia, (33) Mexico, (34) Mozambique, (35) Nepal, (36) New Zealand, (37) Nigeria, (38) Pakistan, (39) Paraguay, (40) Peru, (41) Philippines, (42) Poland, (43) Portugal, (44) Qatar, (45) Romania, (46) Russia, (47) Senegal, (48) Singapore, (49) South Africa, (50) Spain, (51) Sweden, (52) Switzerland, (53) Taiwan, (54) Tanzania, (55) Thailand, (56) The Netherlands, (57) Turkey, (58) UK, (59) USA, (60) Venezuela, (61) Vietnam.

Graph 5: Number of Papers for two Participating Countries per Year

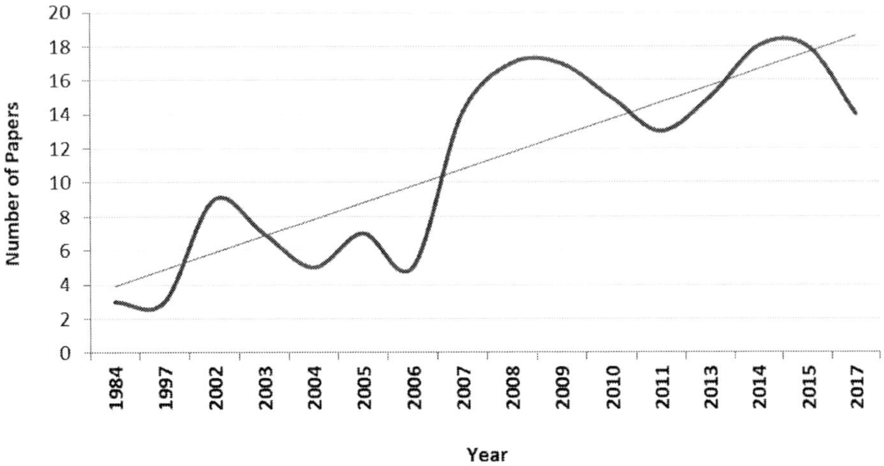

Table 6: Number of Countries per Year

Year	City	Country	Participating Countries
1984	Rio de Janeiro	Brazil	17
1997	Bhubaneswar	India	9
2002	Hanoi	Vietnam	21
2003	João Pessoa	Brazil	17
2004	Pirassununga	Brazil	11
2005	Rio de Janeiro	Brazil	17
2006	Salvador	Brazil	12
2007	Maceió	Brazil	16
2008	Cali	Colombia	17
2009	Bath	UK	30
2010	Cairo	Egypt	15
2011	Hunan Changsha	China	16
2013	João Pessoa	Brazil	27
2014	Pirassununga	Brazil	19
2015	Winnipeg	Canada	21
2017	Mérida	Mexico	20

Graph 6: Number of Countries per Year

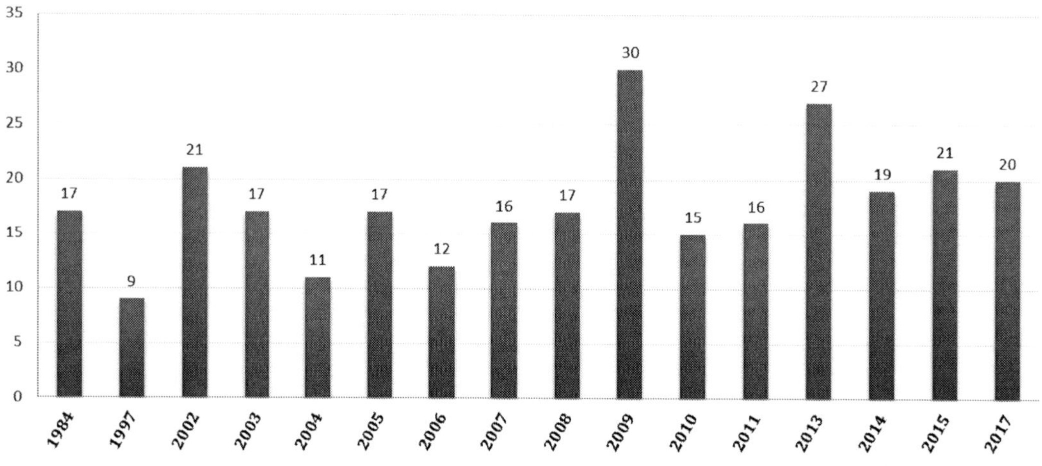

Mapping papers within NOCMAT Conferences by Themes

Tables 7, 8, 9 and 10, Graphs 7, 8 and 9 present the distribution of all the papers included in different Themes.

Table 7 shows the number of papers presented by each NOCMAT Conference and by different themes. It can be observed in this Table that NOCMAT Conferences of 1997, 2003, 2004, 2005 and 2006 were not included. The reason is that the proceedings of these one didn´t associated papers to themes.

Table 8 shows the number of Themes increased from 1 (in 1984/Rio de Janeiro-Brazil) to 14 (in 2014/Cairo-Egypt) and average is 8 Themes for all NOCMAT Conference series. It is observed that the numbers of knowledge areas are increasing and being included in new Themes along the NOCMATs.

Regarding the number of papers presented in each Theme, Tables 7 and 9 show those being most focused by the authors in each NOCMATs. For instance "Inorganic Matrix Composites" with 44 papers is the most focused Theme in NOCMAT Conference 2008/Cali-Colombia.

Table 7: Themes per NOCMAT Conferences Series

Themes List	Nº of Papers
1984	64
- Low Cost and Energy Saving Construction Materials	64
1997	24
- Not reviewed	24
2002	63
- Bamboo, Soil and Waste Materials	30
- Concrete and Composite Materials	33
2003	59
- Not reviewed	59
2004	57
- Not reviewed	57
2005	73
- Not reviewed	73
2006	86
- Not reviewed	86
2007	78
- Bamboo as construction material	10
- Composites materials	21

- Durability aspects of non-conventional materials	3
- Environmentally friendly, energy efficient construction	3
- Non-conventional materials and technologies management	2
- Other topics of interest	3
- Technology applied for low costing houses	7
- Vegetable fibers and soil construction	7
- Waste materials in building construction	22
2008	**113**
- Bamboo, Guadua and Others	14
- Inorganic Matrix Composites	44
- Organic Matrix Composites	15
- Residues and Subproducts Valorization	8
- Sustainable Technologies	16
- Techniques and Characterization	16
2009	**118**
- Bamboo	8
- Case studies	18
- Cement, lime and concrete materials	20
- Earth building	21
- Innovative masonry	11
- Natural fibers and fiber composites	11
- Recycled materials	6
- Straw bale	7
- Timber	16
2010	**119**
- Advances in research methodologies and materials testing	15
- Applications in improving building safety through materials and constructional development	7
- Construction materials and technologies to deliver affordable housing	13
- Construction materials and technologies to reduce climate change	2
- Durability and performance of construction materials	12
- Environmentally friendly, energy efficient construction	16
- Infrastructures systems and materials	6
- Innovations in development of low carbon materials and technologies	8
- Life cycle assessment of materials	3
- Low cost housing concepts, prototypes and applications	5
- Miscellaneous	1
- Recycling of industrial, agricultural and urban waste stream materials	21
- Seismic engineering resistance, flood mitigation and disaster prevention	5
- Standards and guidelines	5
2011	**139**
- Bamboo materials	17
- Bamboo structures	19
- Case study of non-conventional buildings	7
- Cements and concretes	16
- Low carbon, energy saving design and practice	39
- Natural fibers and materials	8
- Non conventional mortars, renders and earth building	4
- Recycled materials	21
- Timber and masonry	8
2013	**138**
- Advances in research methodologies and material testing	8
- Bamboo and wood	22
- Cement, mortar and concrete	31
- Construction materials and technologies to deliver affordable housing and low energy	15
- Earthen Materials and Constructions	19
- Life-cycle assessment of materials and durability	3
- Natural fibers and materials	20
- Recycling of industrial, agricultural, and urban waste stream materials	13
- Standards, guidelines and policy issues	7
2014	**136**
- Advances in research methodologies and material testing	15
- Applications in improving welfare and comfort	8
- Construction materials and technologies for sustainability, energy, efficiency, affordable housing	16
- Durability, life cycles and performance of building materials	9

- Earth architecture and construction	11
- Natural fibers and plant-based materials (bamboo, straw, plant fiber)	22
- Non conventional and innovating research and technology	16
- Non-conventional structures	6
- Recycling of industrial materials, agricultural and urban waste	33
2015	**132**
- Academic reviews, Regulations & Feasibility Studies	6
- Bamboo	27
- Concrete / Earth	1
- Concrete / Earth	6
- Concrete / Mortar	26
- Concrete Aggregates	9
- Earth / Soil	14
- Fiber characterization, Panelized Straw Bale construction	7
- Material properties and behaviours	14
- Microscopic binder studies + Mycelium insulation	14
- Roofs, Rainwater, Cooling	8
2017	**86**
- Advances in research methodologies and materials testing	11
- Applications in improving building safety through materials and constructional development	5
- Construction materials and technologies to deliver affordable housing	15
- Construction materials and technologies to reduce climate change	4
- Durability and performance of construction materials	24
- Environmentally friendly, energy efficient construction	6
- Infrastructure systems and materials	2
- Low cost housing concepts, prototypes and applications	2
- Recycling of industrial, agricultural and urban waste stream materials	15
- Standards and guidelines	2
Total number of papers	**1,485**

Table 8: Number of Themes per Year

Year	1984	2002	2007	2008	2009	2010	2011	2013	2014	2015	2017
Nº of themes	1	2	9	6	9	14	9	9	9	11	10

Table 9: Themes with most focused in each NOCMAT Conference

Year	Theme
1984	Low Cost and Energy Saving Construction Materials
2002	Concrete and Composite Materials
2007	Waste materials in building construction
2008	Inorganic Matrix Composites
2009	Earth building
2010	Recycling of industrial, agricultural and urban waste stream materials
2011	Low carbon, energy saving design and practice
2013	Cement, mortar and concrete
2014	Recycling of industrial materials, agricultural and urban waste
2015	Bamboo
2017	Durability and performance of construction materials

Table 10: Number of Papers per Main and Sub-Groups of Themes

Main and sub-groups	Number of Papers
Bamboo and Natural fibers	**220**
Bamboo	139
Building	1
Natural fibers	71
Vegetable fibers	9
Cement, Composite and Concrete Materials	**217**
Cement, Composite and Concrete Materials	217
Construction Material	**362**
Building	31
Climate	2
Earth	59
Housing	165
Masonry	17
Straw bale	9
Timber	25
Technology	21
Durability	33
Energy	**24**
Energy	24
General	**61**
Case Study	21
Miscellaneous	6

Residues	12
Seismic	5
Technology	17
Non-conventional materials	**75**
Building	11
Case Study	9
Housing	1
Technology	22
Structures	17
Durability	9
Management	6
Recycling	**113**
Recycling	113
Safety	**7**
Safety	7
Standards	**14**
Standards	14
Material and Testing	**93**
Infrastructures	8
Life cycle	6
Low carbon material	13
Testing	66
Others2	**299**
Others2	299
Total number of papers	**1,485**

Table 10 and Graphs 7, 8 and 9 were created by the authors of this paper. Data collected were categorized in Main and Sub-groups of Themes.

Table 10 and Graph 7 show a strong presence of papers related to the Sub-themes "*Construction Materials*" with 362 papers, "*Bamboo and Natural fibers*" with 220 papers, and "*Cement, Composite and Concrete Materials*" with 217 papers among a total of 1100 papers.

Graph 7: Number of Papers per Main Group of Themes

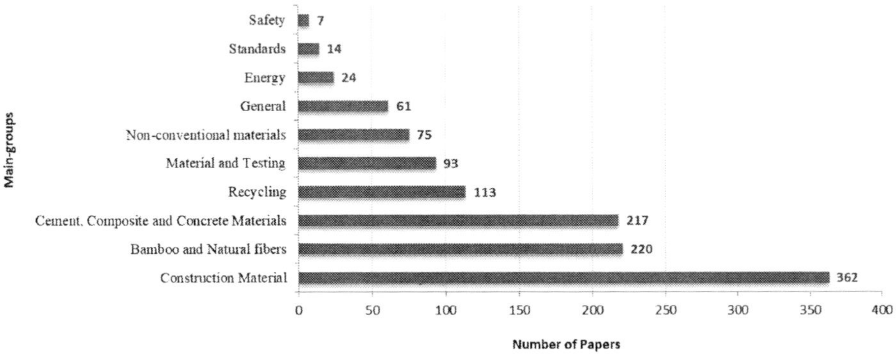

Graph 8: Number of Papers for 3 main groups per NOCMAT Conference Series

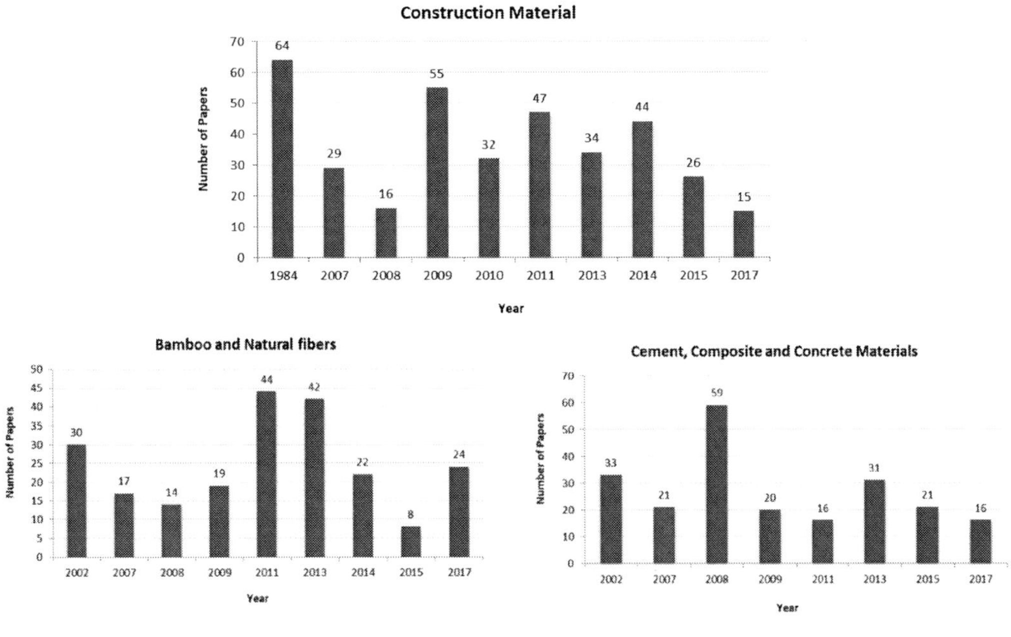

Graph 8 presents the concentration of papers in the 3 main groups of Themes by NOCMAT Conference series. For "*Construction Material*" Theme, the highest number of papers has been presented in 1984/Brazil (64 papers). For "*Bamboo and Natural Fibers*" Theme, the highest number of papers has been presented in 2011/China (44 papers). For "*Cement, Composite and Concrete Materials*" Theme, the highest number of papers has been presented in 2008/Colombia (59 papers).

Graph 9: Number of Papers per Sub-groups of Themes

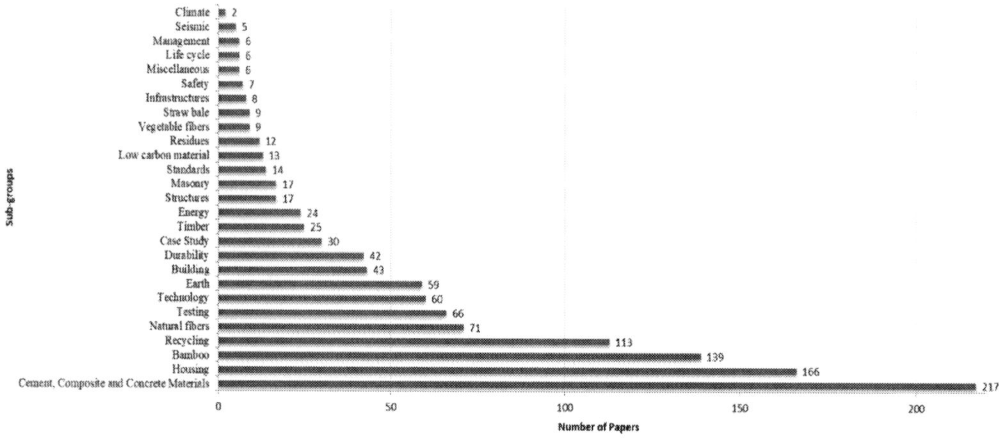

Graph 9 presents all the papers in each Sub-groups of Themes. Among 27 sub-groups of Themes, the five more expressive cover 706 papers (i.e., 60% of a total papers). Sub-group "*Climate*" presents one of the lowest numbers of papers.

FINAL REMARKS

This study mapped 1,485 papers along 33 years and 16 NOCMAT Conferences showing regularity with increasing production in a wide range of NOCMAT sub-groups of Themes. This varied interest is of great importance, because since the 1st NOCMAT Conference in 1984, they become a core subject in scientific and technological policies aiming at a more environmentally friendly and sustainable construction, and also allowing a new "NOCMAT culture" within an increasing number of countries and public and private institutions notably of education and research. Universities are creating multidisciplinary courses integrating Engineering with Management, Economy, Biology, Agriculture Sciences among others traditional knowledge areas. R&D projects integrate more and more researchers specialized in these areas.

These transformations sponsored especially by research funding agencies with increasing grants and scholarships for NOCMAT Themes in civil engineering, led to a greater appreciation of studies and research results presented in different editions of these NOCMAT Conferences. Thus, guidelines for studies that integrate different areas of knowledge, such as nanotechnology for civil engineering [20] or non-conventional materials in civil construction [21], with significant implications for academic backgrounds to be created, gain increased importance.

In turn, thematic application areas identified in this study, negligible as compared to others, permit to identify thematic gaps that require more attention from now on, notably a greater and deep sense of accountability.

For the future NOCMAT Conferences some improvements are suggested. Among existing data structure, it is recommended a database and integrated information system mainly on authors, academic level and area, countries and institutions, and themes. The database related to each submitted paper should consider one or more different defined themes. It also should be considered a defined application area regarding the papers content by the Organizer Committee for each submitted paper into the database.

Suggestions for future studies in the field of studies on *State of Art* should map and analyze authors' network regarding the institutions and countries, and also, themes centrality and closeness. This mapping should be continued to recognize the new themes and areas trends, their increasing or stagnation.

A Science and Technology Observatory for "NOCMAT" not yet existing should be created. This one should have the function of "observing", "tracking", "anticipating" and "monitoring" the development of knowledge areas with a "keen" eye on critical issues, notably those presented in NOCMAT Conferences [22]. As a central issue for a transition to create the NOCMAT Observatory, institutions concerned with these materials and technologies should be initially involved. Their role and success for the creation of a pertinent NOCMAT Observatory depends fundamentally on the interest and participation of its members, individuals and enterprises, and more specifically, in an initial step, the IC-NOCMAT (International Committee). This paper shows that 641 institutions engaged, 61 countries involved in NOCMAT Conference series along 33 years of experience justify such an integrative initiative.

Finally, for this engagement new fundamental values for NOCMAT Conference series should be considered, those presented by ABMTENC: "*The rapidly changing world is facing many threats and serious challenges. The glory of modern industrialization has been in doubt with ample facts of environmental deterioration problems. Yet, Development and survival are still the central issues facing a great portion of mankind. The world needs new role models of development and industrialization that are characterized by environmentally friendliness and sustainability. Construction industry is in the forefront facing today's challenges. Providing durable, efficient and effective infrastructure system, affordable housing, clean water, energy, communication and sanitation is fundamental to economic prosperity, social justice, political stability and quality of Life*".

Acknowledgements

The authors would like to thank ABMTENC Association which made available all the NOCMAT Conferences data and the *18th International Conference on Non-Conventional Materials and Technologies – Construction Materials & Technologies for Sustainability* (18th NOCMAT 2019), 24th – 26th, July 2019, Nairobi/Kenya organizers for their valuable effort and proceedings for the discussions, which form the basis of the meeting. Also, financial support given by FAPERJ and FACC/UFRJ are appreciated.

REFERENCES

[1] UNITED NATIONS. *Our Common Future*. Report of the World Commission on Environment and Development. Oslo, March 20, 1987. Available at: <http://www.un-documents.net/our-common-future.pdf>. Access in: April 10, 2017.

[2] PAPON, P. *FUTUR*: les avancées technologiques 2025, 2050, 2100. Paris: Le Monde, 2013. (Special Edition, February-April, 2013).

[3] ASSOCIATION OF NON-CONVENTIONAL MATERIALS AND TECHNOLOGIES. Available at: <http://www.abmtenc.civ.puc-rio.br/17nocmat>. Access in: April 10, 2017.

[4] INTERNATIONAL NETWORK FOR BAMBOO AND RATTAN. Available at: <http://www.inbar.int/#1>. Access in: April 10, 2017.

[5] WIKIPEDIA.ORG State of the art. Available at:<https://en.wikipedia.org/wiki/State_of_the_art>. Access in: May 15, 2017.

[6] ARCHAMBAULT, E.; CARUSO, J.; NICOL, A (2014) *State-of-art analysis of OA strategies to peer-review publications*. Science Metrix, October 22, 2014. (Document: RTD-B6-PP-2011-2: Study to develop a set of indicators to measure open access Deliverable D.2.1, 2014 Update, Version 5b. Available at: <http://science-metrix.com/sites/default/files/science-metrix/publications/d_2.1_sm_ec_dg-rtd_oa_policies_in_the_era_update_v05p_0.pdf>. Access in: May 15, 2017.

[7] Côté, G.; Roberge, G.; Archambault, E. *Bibliometrics and Patent Indicators for the Science and Engineering Indicators 2016 — Comparison of 2016 Bibliometric Indicators to 2014 Indicators*. Science Metrix, January 14, 2016. Available at: <http://science-metrix.com/sites/default/files/science-metrix/publications/science-metrix_comparison_of_2016_bibliometric_indicators_to_2014_indicators.pdf>. Access in: May 15, 2017.

[8] Weber, K. M.; Hemmelskamp, J. *Towards Environmental Innovation Systems*. Heidelberg: Springer, 2005.

[9] Kibert, Ch. J. *Sustainable Construction* – Green Building Design and Delivery. 4th Ed. Hoboken: Wiley, 2016.

[10] Moraes, L. A. F. de; Ohayon, P.; Ghavami, K. Application of Non-Conventional Materials: Evaluation Criteria for Environmental Conservation in Brazil. *Key Engineering Materials*, Trans Tech Publications, Switzerland, v. 517, p. 20-26, 2012.

[11] Harries, K. A.; Sharma, B. *Nonconventional and Vernacular Construction Materials* – Caracterisation, Properties and Applications. Cambridge: Woodhead Publishing/Elsevier, 2016.

[12] Huang, A. L.; Chapman, R. E.; Butry, D. *Metrics and Tools for Measuring Construction Productivity*: Technical and Emprirical Considerations. Gaithersburg/ Maryland: NIST – National Institute for Standards and Technology, September 2009.

[13] *A Policy Study on the Sustainable Use of Construction Materials*. OECD: Paris, September 11, 2015. (Working Party on Resource Productivity and Waste, ENV/EPOC/WPRPW(2014)4/FINAL).

[14] Canchumani, R. M. L. *Scientific Domains in UFRJ*: mapping of fields of knowledge. Thesis (PhD in Information Science) – Federal University of Rio de Janeiro, School of Communication, Brazilian Institute for Information in Science and technology. Graduate Program in Information Science, Rio de Janeiro, 2015, 185p.

[15] Pritchard, A. A Statistical Bibliography or Bibliometrics? *Journal of Documentation*, Bingley/UK, v. 25, n. 4, p. 348-349, January 1969.

[16] Otlet, P. M. G. *Traité de Documentation* – le livre sur le livre - théorie et pratique. Bruxelles: Editiones Mundoneum – Palais Mondial, 1934.

[17] Vasudevan, R. K.; Ziatdinov, M.; Chen, C.; Kalinin, S. V. Analysis citation networks as a new tool for scientific research. *Materials Research Society*, Warrendale/PA, v. 41, p. 1009-1015, December 2016.

[18] Eck, N. J. van; Waltman, L. Software survey: VOSviewer, a computer program for bibliometric mapping. *Scientometrics*, Budapest, v. 84, n. 2, p. 523-538, August 2010.

[19] Ohayon, P.; Ghavami, K.; Jesuz, K. *Specifications of Building Environmental Evaluation Methods for within Redebambu Network in Brazil.* In: NOCMAT 2015 - 16th Nonconventional Building Materials and Technologies International Conference 2015 – Construction for Sustainability - Green Materials & Technologies. August 10-13, 2015, Winniped/Canada.

[20] Pandoli, O.; Martins, R. S.; Barbosa, B. A.; Toni, K. L. G. de; Paciornik,S.; Maurício, M. H. P.; Lima, R. M. C.; Padilha, N. B.; Letichevsky, S.; Avillez, R. R.; Ghavami, K. Biocomposite Bambbo: A Regioselective Impregnation with Silver Nanofillers for Antifungal Application. *Materials Research Forum*, Millersville/PA-USA, v. 7, p. 415-426, 2018. ISSN 2474-395X (Online). http://dx.doi.org/10.21741/9781945291838-39.

[21] Carvalho, M. C. R.; Estevão, A. C.; Fialho, P. B.; Alves, L.; Cordeiro, L. E. C.; Cardoso, E. P.; Figueiredo, T. A. Non-Conventional Materials in Civil Construction: A Case Study in Curvelo – MG. *Materials Research Forum*, Millersville/PA-USA, v. 7, p. 750-755, 2018. ISSN 2474-395X (Online). http://dx.doi.org/10.21741/9781945291838-73.

[22] Ohayon, Pierre; Barreiros, D. S.; Ghavami, Khosrow. Science and Technology Observatory for "NOCMAT" in Brazil: Role and Proposed Framework. *Key Engineering Materials*, v. 600, p. 399-412, 2014. Trans Tech Publications, Switzerland, doi: http://dx.doi.org/10.4028/www.scientific.net/KEM.600.399.

ANALYZING FIVE EVALUATION APPROACHES FOR R&D PROJECTS WITHIN THE BRAZILIAN NATIONAL BAMBOO´S NETWORK – REDEBAMBU/BR

Pierre Ohayon[1], Khosrow Ghavami[2]

[1] Universidade Federal do Rio de Janeiro – UFRJ, Brazil;

[2] Pontifícia Universidade Católica do Rio de Janeiro – PUC-Rio, Brazil;

ABSTRACT

The Brazilian Ministry of Science, Technology, Innovations and Communications (MCTIC) and its National Council for Scientific and Technological Development (CNPq) launched the call for research and development projects nº 66/2013 to select proposals for financial support able to contribute significantly for the scientific and technological development and innovation in the country, specifically for structuring the National Network for Research and Development of Bamboo namely called Redebambu/BR. The Redebambu/BR network supports the implementation of PNMCB - National Policy to Encourage Handling, Sustainable Planting and Preservation of Bamboo which has been created by the Law 12.484/2011. In the context of sustainable development, there is a need for a general framework of analysis from impacts to economic and social improvements to understand innovation process and for a greater coordination of research efforts. Systematic evaluation issues are becoming of increased concern notably for state support agencies for sponsored R&D programs and projects. This qualitative study presents and analyzes five important evaluation approaches responding to policy needs for Redebambu/BR network and its NOCMAT – *non-conventional materials and technologies* sponsored projects, specifically: (1) Impact evaluation; (2) Implementation evaluation; (3) Goalfree evaluation; (4) Process evaluation; and (5) Pluralist evaluation. A number of strengths and weaknesses of each approach is highlighted. Finally, questions regarding the adequate evaluation objectives, the appropriate report's content and the effective way for evaluation results meet real users needs are considered and discussed.

Keywords: NOCMAT Projects, Evaluation approaches, Network, Redebambu/Br

THE BRAZILIAN REDEBAMBU/BR NETWORK AND ITS R&D SPONSORED PROJECTS

The Ministry of Science, Technology, Innovations and Communications (MCTIC) of Brazil and its National Council for Scientific and Technological Development (CNPq) launched the call for R&D projects Nº 66/2013 to select proposals for financial support able to contribute significantly for the scientific and technological development and innovation in the country, specifically for structuring the National Network for Research and Development of Bamboo namely called Redebambu/BR.

The Redebambu/BR network supports the implementation of PNMCB - National Policy to Encourage Handling, Sustainable Planting and Preservation of Bamboo which has been created by the Law 12.484/2011. The Redebambu/BR network has representation in five Brazilian regions through the Regional Centers. These are "diffusion technology centers for strengthening the culture of bamboo in the country, offering technology solutions to all levels of the bamboo´s production chain, from the family farmer to the processing industry" [1].

The activities of Redebambu/BR network are managed by a Steering Committee composed by the heads managers of the six Regional Centers, a representative of the MCTIC and a representative of the industrial sector.

The sponsored projects cover one or more research areas within the main Thematic Area entitled "Chain development bottlenecks in production of bamboo" which applies to large knowledge areas such as Engineering, Biology, Agriculture or subareas such as Management of Research and Technology, among others. The motivation to develop a methodological framework, for scientific and technological activities involving this main Thematic Area covering 17 sub-themes[3] comes essentially from the new priorities of the brazilian Government in meeting the technological demands concerning the implementation of PNMCB.

It is therefore suggested a model for evaluating the activities of scientific and technological development within the Redebambu/BR network. Proposed model(s) for the evaluation system within the Redebambu/BR network should be able to still achieve the following specific objectives:

(i) Measurement of the degree of implementation of the objectives and strategic goals for Redebambu/BR network under the national policy PNMCB and its projects adherence to international trends in scientific and technological research which were submitted and supported by the CNPq/MCTIC sponsor agency;

(ii) Sorting and analysis of relevant results from the point of view of the scientific and technological development policy, based on the goals of PNMCB, S&T sponsor agencies and Redebambu/BR network which constitutes an "operator" of the supported projects;

(iii) Registration of all possible divergences between expected results and actual results, analyzing its causes and implications;

---

[3] (1) Identification of native and exotic species; (2) propagation of bamboo´s species; (3) bamboo´s propagation and cultivation for temperate climate; (4) Production of reinforced concrete, activated coal and glued laminated bamboo ("plyboo"); (5) Planting for large-scale biomass and cellulose; (6) Production and manufacturing of bamboo buds; (7) Cultivation of tropical bamboo and use in landscaping; (8) Structural application of bamboo; (9) Sustainable handling of *Guadua sp* forest; (10) Identification of species with greater economic potential; (11) Technologies Demonstration for energetic use of bamboo; (12) Application of bamboo for restoration of degraded lands; (13) Strategy for introducing the culture of bamboo in family farming; (14) Technology transfer for production and planting of seedlings; (15) Introduction of new species of bamboo in Brazil; (16) Use of bamboo for high performance composite materials; (17) Production of raw materials for cosmetics and food industry.

(iv) Balance of projects performance, those supported under the main Thematic Area and its 17 sub-themes;

(v) Evaluation of the direct impacts of research activities within the Redebambu/BR network (those relative to their more immediate goals) and indirect impacts of supported projects regarding policies and involved institutions;

(vi) Production of a summary of statistical data covering all aspects of the supported projects and their implementation;

(vii) Construction of a system of indicators to analyze and monitor the Redebambu/BR´s projects.

Six research projects were selected by the CNPq/MCTIC from the following Brazilian research centers/Universities:

(1) Faculty of Animal Science and Food Engineering/University of São Paulo (State of São Paulo) and Department of Civil Engineering/Pontifical Catholic University of Rio de Janeiro (State of Rio de Janeiro);

(2) Department of Plant Science-Agricultural Sciences Center/Federal University of Santa Catarina (State of Santa Catarina);

(3) Department of Organic and Inorganic Chemistry/Federal University of Ceará (State of Ceará);

(4) Thematic Area of Biotechnology-EMBRAPA Genetic Resources/EMBRAPA - Brazilian Agricultural Research Enterprise (Brasília/ Federal District);

(5) Department of Forestry Engineering-Center for Agricultural Sciences/Federal University of Viçosa (State of Minas Gerais);

(6) Department of Rural Engineering-School of Agronomy/Federal University of Goiás (State of Goiás).

Each project should be developed originally in 24 months beginning in January 2014. The amount of financial resources approved by the CNPq/MCTIC varied between 800,000.00 and 1,300,00.00 Reais *per* project (ie between 200 thousand and 325 thousand U$ Dollars). The total amount of estimated resources in the Call for R&D sponsored Projects has been 6 million Reais (ie about 1,5 million U$ Dollars, at the actual rate of 1 U$D = 4 Reais). The actual economic crisis in Brazil impacted on grants and the project's completion are delayed.

SELECTION AND EVALUATION FUNCTIONS

Science and technology have contributed largely in the last three decades to the economic development without considering adequately different social classes. The intensive R&D activities in the rapidly growing areas of ST&I (Science, Technology and Innovation) such as new high resistance cements, steel, petrochemical derived materials, among others have not given the opportunity to less developed nations to cut the vicious circle which maintained them technologically dependent on industrialized countries [2]. Since two decades, the Brazilian MCTIC brings about challenges notably related to low cost energy materials and technologies which are ecologically acceptable. It indicates that one of the main "bottle-necks" in terms of information, which restricts seriously the proper

ST&I planning and decision making process, is the development of adequate evaluation models and its pertinent indicators [3].

In order to overcome these difficulties, six interrelated key functions of technological resources management [4], characterizing the *what*, *why*, *when*, *where*, *how* and *who* for strategic and operational applications, are traditionally considered [5]. But they have not been systematically considered for the assessment of innovative projects related to the use of locally available materials in abundance and appropriate technologies, in developing countries [6, 7, 8]. These six strategic functions are: *to carry out an inventory* of technological resources (available technologies, expertise and skills); *to select and evaluate* technological resources, their strengths and weaknesses, and their economic potential; *to optimize* (make the best use of technological resources); *to enrich* technological resources through investigation, acquisition, alliance, research, development, improvement, innovation, renewal and replacement, as well as to further develop human technological expertise and skills by recruiting, training and team building; *to watch* developments in the scientific, technological and competitive environment employing an appropriate *technological vigilance and intelligence system*; *to protect* technological resources by safeguarding intellectual property, and by preserving human expertise and skills [4].

The accelerated rhythm in which the results of the research on NOCMAT – *non-conventional materials and technologies* are being introduced into a society, principally used to conventional materials and technologies imported from industrialized countries and not sufficiently prepared to receive them, create new economic, financial, administrative, organizational and human resources problems. Specifically NOCMAT projects which benefited from an unconditional enthusiasm by researchers are seen by the community as suspicious not because of their "few" results but of their "any" results obtained. To show the reliability and durability of the newly developed materials and technologies, in addition to the results obtained in the laboratories, large scale constructions should be built and permanently monitored, requiring higher and continued investments from sponsoring agencies and private organizations. Therefore, interest to establish rational framework integrating scientific institutions and sponsoring agencies for research programs and evaluation, which are directed to social, economic and technological advancement is increasing [9, 10, 11].

R&D of NOCMAT Projects Selection and Evaluation Capacity

Project is understood as a set of actions, performed in a coordinated way by a temporary organization, in which necessary inputs are allocated for, in a given period, achieving a specific goal [12]. There are over the last decades numerous conceptualizations for the management and evaluation for R&D - *research and development* projects and programs [13, 14, 15].

Several relevant aspects are considered in the management of R&D projects, namely: (i) project team; (ii) project life cycle; (iii) organizational climate project and environmental conditions.

Selection and Evaluation in most developing countries is becoming as an important tool for the management of technology, as a necessary link between R&D and society needs. A key feature of a successful S&T organization is the ability to learn from past experience and react to market or client responses. Selection and Evaluation capacities can play an important role in influencing policy analysis and formulation; improving resource allocation and budgetary process; improving investment programs and projects, examining fundamental missions. However, in these countries the adequate use of feedback in formulating projects, programs and policies and allocating resources is only incipient. Sensitivity to public criticism and the fear of political fallout from selection and evaluation findings are inhibiting factors. Many social appropriate technologies still lack the essential requirements of effective selection and evaluation. The quality of information and access to it is often insufficient, mechanisms for feedback into the decision making process are weak and a culture of accountability by using pertinent indicators is not firmly applied.

The barriers in the selection and evaluation of NOCMAT are mainly high cost of their procedures and lack of interest and commitment to the selection and evaluation functions at the political level; feedback mechanisms for applying selection and evaluation findings; more attention given to preparing and appraising programs and projects than to evaluating their performance on completion; involvement of institutional and national staff in selecting and evaluating externally financed programs and projects; attention to the quality of information; objectivity and independence in conducting selection and evaluation; access to the research result on low-cost energy materials and technologies; trained staff [16]. In addition, most experts receive their education in industrialized countries and are not necessarily aware of local conditions and local solutions for a sustainable program. These experts could even damage or hinder the development of the project. In state and federal sponsoring agencies, selection and evaluation does not have high priority for major reasons such as: little effective methods for selection and evaluation; lack of incentive for future productivity and limited freedom of action.

Traditional Dimensions for Selection and Evaluation Frameworks

For the elaboration of a NOCMAT project, Selection and Evaluation take traditionally into account five important dimensions, reflected into the NOCMAT project framework: efficiency, effectiveness, impact, relevance and sustainability.

Efficiency considers how well inputs are converted into outputs for example: the percentage of non-conventional materials included in products/objects and economic efficiency measured by energy and costs reductions, safety and life-cycle improvements.

Effectiveness defines the extent to which the innovative project is likely to achieve its main objectives. It should reveal the effective changes observed by government action and relative advantages of the innovation to the users, improvements in warranty, reduction in complaints, new norms created or adopted for constructions, reduction of construction deficit.

I*mpact* is related to the longer-term effects on the problem situation or need which relates to the Development/Overall Objective. It should present the economic and social advantages pointing out changes in local culture, increased applications for constructions, pollution reduction, waste reduction, number of inhabitants benefited by the project implementation, increased Human Development Index by NOCMAT utilization, improved dissemination of technical information.

Relevant projects are those, which are necessary to solve the problems/needs, and their outputs are necessary to achieve the objective through direct and verifiable cause-effect relationships. *Relevance* should indicate: the degree of urgency and rate of growth of private and public funds allocated for housing programs using non-conventional materials and technologies; improved state of art in selected scientific disciplines and contributions to solve technical problems; professional learning; adaptability of NOCMAT application for construction in specific environmental and socioeconomic conditions.

Sustainability refers to the extent to which the improved situation (as resulted from the achieved objective) can be maintained by users in their own way. It is linked to the local availability of funds for continued operation, maintenance of equipment and re-training of staff, and to the institutional and managerial capabilities. It should stand out the scientific and technological prestige using appropriate indicators, local/regional and international cooperation, financial saving and new contracts with state sponsoring agencies.

Selected Indicators for R&D NOCMAT Projects

Indicators are traditionally applied consistently throughout the selection and evaluation processes, such as: proposal application and selection stage; ongoing evaluation and monitoring stage; final evaluation stage. The last stage is probably the most important element of the overall continuous evaluation and monitoring process in terms of assessing the achievement of objectives of NOCMAT projects and thus encompasses aspects of networking, dissemination and outputs. The final set of measurement indicators include: (i) *additionality* (improvement of R&D to main objective); (ii) outcomes and achievements; (iii) quality; (iv) benefits; (v) continuity and; (vi) gaps (areas of research competence not covered or not represented by representatives) [17, 18].

Strategic objectives which express the specific ones with NOCMAT refer to: reduction of pollution shown by the increased use of NOCMAT; biodiversity protection through the number of families adopting NOCMAT; improvement of the quality of life of the population given by the number of dwellings using NOCMAT and the increased *per capita* of local revenue; organization/community and institutional participation reflected by the number of Non Governmental Organizations, NGO, and public institutions and by the number of NOCMAT courses promoted [19].

Sustainable Development is aimed at meeting the material needs of the present generations without compromising the ability of the future generations to meet their needs [20]. At the upstream stage of the sustainable technological development, indicators of *prevention* are pointed out such as: less energy inputs for raw materials measured by: (i) reduce use of natural resources not renewable; (ii) reduce ecological disturbances; (iii) reduce noise and vibration and emissions. At the process stage, *containment* is considered, still pollution is being generated and has to be controlled. At the downstream stage, *utilization* of NOCMAT is focused, assuring less wastes generation, energy savings and financial resources savings for needy populations [21].

EVALUATION APPROACHES FOR THE REDEBAMBU/BR NETWORK

Strategic choices within the Redebambu/BR Network need adapted elements for conceptualizing an evaluation. These one present numerous divergent or opposed problems, the question being *what is the usefulness of an evaluation, its expected objectives?* The methodological controversy reveals, regarding the usefulness of an evaluation, that the real issue is not methodological but essentially based at the level of the credibility of the evaluation near the social concerned actors. Evaluation is never a neutral practice outside of power relations. It is an integral part of the political context of ST&I programs [22].

Many projects submitted, applying for grants, are frequently rejected due to very limited and specific aspects of an evaluation, related just on the declared objectives and expected results.

In this way, this qualitative study presents and analyzes five important evaluation approaches responding to policy needs for Redebambu/BR network and its NOCMAT – *non-conventional materials and technologies* sponsored projects, specifically: (1) Impact evaluation; (2) Implementation evaluation; (3) Goalfree evaluation; (4) Process evaluation; and (5) Pluralist evaluation. A number of strengths and weaknesses of each approach is highlighted.

Impact Evaluation Approach

Traditional evaluation design involves the following question: "*To what extent has the program, the project produced the expected effects*? *Did the observable effects are in accordance with the pre-established objectives*?"

This approach – called *impact evaluation* compare initial goals to results. It presents many problems such as: ambiguous, inconsistent, even contradictory goals; difficulty to translate in operational terms real impacts (costs/benefits); most of the evaluations do not even include an analysis of resources' implementation initially planned (Figure 1).

Figure 1 – Impact Evaluation Approach

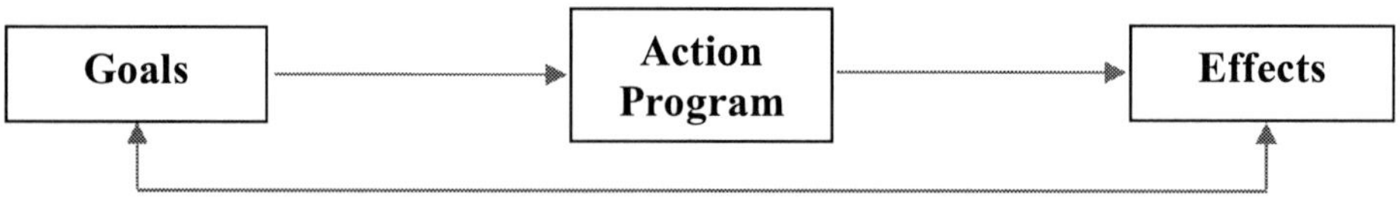

Source: [22, p. 102].

Many R&D goals change during the project implementation due for instance to the reduced amount of financial resources obtained. Frequently, impact evaluations do not consider the description of the real resources implemented within the program.

Implementation Evaluation Approach

In this approach called *implementation evaluation*, the "program" is not a homogeneous entity but it is considered as a set of implemented resources (financial, technical, data processing resources) and the question is "*did the choice and implementation of the resources enable the results to be achieved? the subsidiary question being: "were the resources effectively implemented in accordance to the program and the initial goals?"* [22] (Figure 2).

Figure 2 - Implementation Evaluation Approach

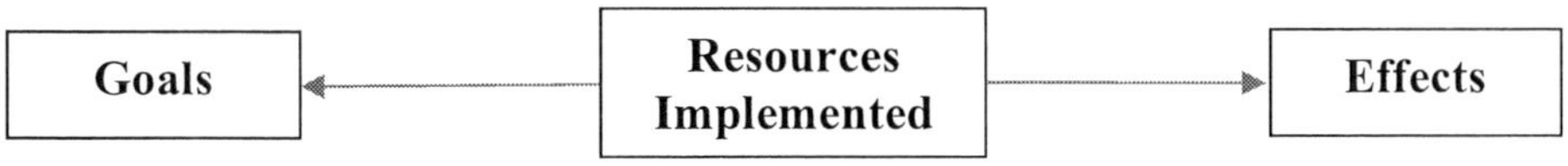

Source: [22, p. 103].

This typically operational assessments aim at overall effectiveness, the maximization of resources implemented in relation to the goals set, bringing about the description of the facts or events that occurred during the course of the program, but remaining insufficient to explain them [22].

Goalfree Evaluation Approach

This approach ignores the official objectives and the question is "*which of the effects of a program are those that meet basic needs*?" Many projects fail in relation to the initial goals but have positive spin-offs in different areas and are therefore recommendable [22] (Figure 3).

Figure 3 - Goalfree Evaluation Approach

Source: [22, p. 104].

Had this approach been effectively adopted, many projects submitted could have been selected able to produce boomerang effects. A number of technical norms suggested by scientists and engineers for to the use of bamboo in civil construction were not yet approved, mainly due to more economic and political reasons or even to the influence of very powerful industries that lead with conventional materials and technologies [23, 24].

Process Evaluation Approach

This approach aims at explaining the phenomena. It analizes the interactions between the program and its environment and the question is "*which are the processes that cause the observable effects*?" This type of process evaluation aims at identifying and comprehensively understanding the facts related to undertaken actions within programs [22] (Figure 4).

Figure 4 - Process Evaluation Approach

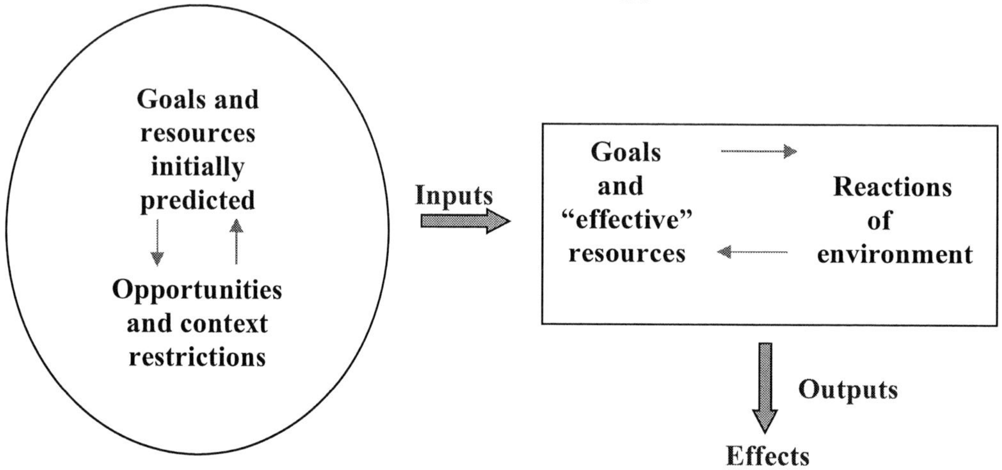

Source: Adapted from [22, p. 105].

Goals can be replaced by new operational ones; its hierarchy can be modified; its allocated resources may vary. This type of evaluation is proposed in response to the new requirements of decision makers and in this way the intentions of programs managers are privileged. But different barriers should be considered in this approach: abundance of independent or explanatory variables; costly model; conflicting articulations of different actors neglected; and diversity of values of evaluators left out [22].

Pluralist Evaluation Approach

The starting point of this approach is the recognition of a plurality of value systems that coexist within society. The program is understood as a never-ending process of collective learning and pluralistic research of solutions to social problems and the question is "*Which information does the social system need to improve its understanding of the effects or events that have arisen during the Program*? The first task of this type of evaluation is to identify the decision makers and users of the analysis' results. The second task is to clarify the value systems that will guide the evaluation goals and their own characteristics (method, criteria). Social actors, within this approach, are not supposed to seek an "optimal" solution but a "satisfactory" solution considering the context. The more the environment is uncertain, the less the objectives are stable and concrete. In other words, less the future is predictable and more the assessment is useful as it tries to contribute for the reduction of decision makers' uncertainty regarding the ecosystem. As a barrier that should be considered in this approach, is the complex system of values and references of each actor, whose expectations evolving according to their individual or collective experiences, their satisfaction or failure of these expectations [22] (Figure 5).

Figure 5 – Pluralist Evaluation Approach

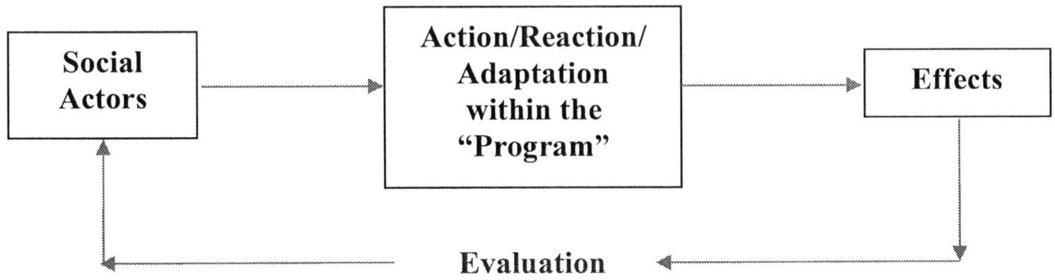

Source: Adapted from [22, p. 106].

Many researchers have tried to include in their projects proposal the elaboration of academic disciplines related to their thematic study, without any success. In this way, it is important to consider that education and research should not be considered as a *cost* by evaluators but as an *investment* for future, considering urgent social needs [25, 26, 27, 28].

FINAL REMARKS

The five evaluation approaches suggested for R&D sponsored projects should interact at all levels of the programmatic action of the Redebambu/Br Network within the Brazilian MCTIC (annual and multi-annual operational plans). They should also influence the behavior of all final actors involved (executor institutions ie, institutional operators of projects executed by different teams engaged) and which produce impacts on the external environment (notably the productive and educational sectors) within their respective domains.

In its turn, the external environment (economic, political, and social) influences both the sponsored project's development and the Redebambu/Br Network. These one do not compete with each other. On the contrary, they should all be adopted avoiding penalizing projects that would merit financial support.

For each evaluation approach, it is required to relativize, whenever possible, the criteria and indicators related the R&D sponsored projects, not only with what has been *planned*, but also with those of its larger *context*. This would permit to know how important is the project in thematic areas where it operates in relation to other government (and possibly private) actions in these same areas, taking into account both the results achieved by the research projects submitted/sponsored.

In this paper, it is shown that the implementation of NOCMAT could be well succeeded by applying the selection and evaluation framework for R&D projects within the Redebambu/Br Network, which should be seen by all concerned parties as a way to learn and improve their integrated action in a systematic form. This is possible when the efforts are made to build social, political, economic, environmental, cultural and technical indicators for innovative projects, which are designed to serve the real needs of decision makers at local, state and federal level.

Acknowledgements

The authors would like to thank ABMTENC Association which made available all the NOCMAT Conferences data and the *18th International Conference on Non-Conventional Materials and Technologies – Construction Materials & Technologies for Sustainability* (18th NOCMAT 2019), 24th – 26th, July 2019, Nairobi/Kenya organizers for their valuable effort and proceedings for the discussions, which form the basis of the meeting. Also, financial support given by FAPERJ and FACC/UFRJ are appreciated.

REFERENCES

[1] Brasil. Conselho Nacional de Desenvolvimento Científico e Tecnológico. *Chamada MCTIC/Ação Transversal/CNPq Nº 66/2013*. Available at: <file:///C:/Users/Pierre/Downloads/Chamada+66-2013.pdf>. Access on: Oct. 15, 2014, p. 15.

[2] Albuquerque, M. E. E.; Bonacelli, M. B. M.; Weigel, P. A questão ambiental e a contribuição dos institutos de pesquisa à geração de tecnologias ambientalmente sustentáveis. In: *Parcerias Estratégicas*, Brasília/DF, v. 15, n. 30, p. 9-24, January 2010.

[3] Brasil. Ministério da Ciência e Tecnologia *Livro Verde* – O debate necessário: Ciência, Tecnologia, Inovação – Desafios para a Sociedade Brasileira. Brasília: MCT/Academia Brasileira de Ciências, jul. 2001.

[4] Morin, J.; Rafferty, P. J. *The six key functions of technological resources management*. Miami/Florida-USA: Institute of Industrial Engineering. Proc. of the Second International Conference on Management of Technology, Feb. 28–Mar. 2, 1990, p. 621-627.

[5] Kuhlman, S. *Evaluation as a source of 'strategic intelligence*. In: Shapira, P.; Kuhlman, S. Learning from Science and Technology Policy Evaluation – Experiences from the United States and Europe. Northampton/MA: Edward Elgar, 2003. (Chapter 18, p. 352-379).

[6] Bellen, H. M. Van. *Indicadores de Sustentabilidade* – uma análise comparativa. 2ª Ed., São Paulo: FGV Editora, 2006.

[7] Wickremasinghe, S. I.; Gupta, V. K. *Science & Technology Policy and Indicators for Development* – Perspectives from Developing Countries. Dehli: Daya Publishing House, 2008.

[8] Soltman, C.; Stucki, T.; Woerter, M. *The performance effect of environmental innovations*. KOF Working Papers, n. 330. Zurich: Swiss Federal Institute of Technology, February 2013.

[9] Edler, J.; Georghiou, K.; Blind, E.; Uyarra, E. Evaluating the demand side: New challenges for evaluation. *Research Evaluation*, Oxford, v. 21, n. 1, p. 33-47, March 2012.

[10] Clark, W.; Crutzen, P.; Schellnhuber, H. Science for Global Sustainability: Toward a New Paradigm. *CID Working Paper* n. 120. Science, Environment and Development Group, Center for International Development Cambridge: Harvard University, 2005.

[11] Hong, H. D.; Boden, M. R&D programme Evaluation: theory and practice – a Comparative Analysis of Large Scale R&D Programme Evaluation. Surrey/UK: Ashgate Publishing, 2003.

[12] Barcaui, A. (Org.). PMO – Escritório de Projetos, Programas e Portfólio na Prática. Rio de Janeiro: Brasport, 2012.

[13] Ellis, l. Introduction to Evaluating R&D Process Management. In: ________. *Evaluation of R&D Processes*: Effectiveness through Measurements. Norwood/MA: Artech House, 1997.

[14] Geisler, E. *The Metrics of Science and Technology*. Westport/CT: Quorum Books, 2000.

[15] Dale, R. Evaluating Development Programmes and projects. 2nd Ed., London: Sage Publications, 2004.

[16] The World Bank Building Evaluation Capacity. Washington: The World Bank/Operations Evaluation Department. *Lessons & Practices*, n. 4, p. 1-11, 1994.

[17] Ministério da Ciência, Tecnologia e Inovação - MCTIC. *Plano Anual de Monitoramento e Avaliação*. Comissão Permanente de Monitoramento e Avaliação. Brasília/DF: ASCAV - Assessoria de Acompanhamento e Avaliação das Atividades Finalísticas/Secretaria Executiva, December 2015.

[18] TECHNOPOLIS GROUP & MIOIR. *Evaluation of Innovation Activities*: Guidance on Methods and Practices. Study funded by the European Commission. Brussels: Directorate for Regional Policy/European Union, 2012.

[19] Brasil. Ministério do Meio Ambiente. *Relatório sobre a Aplicação Preliminar dos Indicadores do Projeto AMA para Monitoramento do PPG7*. Brasília: MMA/ Secretaria de Coordenação da Amazônia. Projeto Piloto para Proteção das Florestas Tropicais do Brasil – PPG7. Projeto: Apoio ao Monitoramento e Análise – AMA. March 2001.

[20] UNEP Bergen Ministerial Declaration of Sustainable Development in the ECE Region. Bergen/Norway, 14-15 May, 1990, *Industry and environment*, v. 13, n. 2, p. 54-56, April/June 1990.

[21] Mullick, A. K. *Role of Cement and Concrete in Sustainable Societal Development*. 1st International Conference on Concrete & Development, Tehran/Iran, April 30–May 2, 2001, p. 573-582.

[22] Monnier, E. *Les Stratégies et les Dispositifs d'Évaluation*. In: Évaluations de l'action des pouvoirs publics. 2nd Ed. Paris: CPE/Economica, 1992. (Chapter V, p. 101-126).

[23] Moraes, L. A. F. de; Ohayon, P.; Ghavami, K. Application of Non-Conventional Materials: Evaluation Criteria for Environmental Conservation in Brazil. *Key Engineering Materials*, Trans Tech Publications, Switzerland, v. 517, p. 20-26, 2012.

[24] Ohayon, P.; Ghavami, K.; Jesuz, K. *Specifications of Building Environmental Evaluation Methods within Redebambu Network in Brazil*. In: Proceedings of the NOCMAT 2015 - 16th Non Conventional Building Materials and Technologies International Conference 2015 – Construction for Sustainability - Green Materials & Technologies. August 10-13, 2015, Winnipeg/Canadá.

[25] Geisler, E. *Science and Technology, The Economy, and Society*. In: Creating Value with Science and Technology. Westport: Quorum Books, 2001. (Part IV, p. 167-315).

[26] ASSOCIATION OF NON-CONVENTIONAL MATERIALS AND TECHNOLOGIES. Available at: <http://www.abmtenc.civ.puc-rio.br/17nocmat>. Access in: April 10, 2017.

[27] Ohayon, Pierre; Barreiros, D. S.; Ghavami, Khosrow. Science and Technology Observatory for "NOCMAT" in Brazil: Role and Proposed Framework. *Key Engineering Materials*, v. 600, p. 399-412, 2014. Trans Tech Publications, Switzerland, doi: http://dx.doi.org/10.4028/www.scientific.net/KEM.600.399.

[28] Ghavami, K.; Franco, P. J. H. (Org.). *Non-Conventional Materials and Technologies* – NOCMAT for the XXI Century. Millersville/PA: Materials Research Forum LCC, 2018. (Preface, Materials Research Proceedings, Volume 7).

DESIGNING & BUILDING WITH BAMBOO

18th International Conference on
Non-Conventional Materials and Technologies
"Construction Materials & Technologies for Sustainability"
(18th NOCMAT 2019)
24th – 26th July 2019
Nairobi, Kenya

EXPERIMENTS WITH PARTS OF RINGS TO DETERMINE THE INNER AND OUTER CIRCUNFERENTIAL STRENGTH OF BAMBOO

Richard Morán[1], Silvia Greco[2], Cliff Konrath[2], Luisa Molari[2], Jose J. Garcia[1]

[1] Escuela de Ingeniería Civil y Geomática, Universidad del Valle, Cali, Colombia;

[2] DICAM, Alma Mater Studiorum, University of Bologna, Bologna, Italy;

ABSTRACT

Bamboo is a natural sustainable material with a high potential to substitute high contaminating construction materials in various applications. It is a heterogeneous material axially reinforced with cellulose fibers that confers high axial mechanical properties. However, it is prone to longitudinal fissures, mainly due to tensile and shear stress components. Previous studies have been focused on determining the circumferential strength of bamboo using the edge bearing tests. To complement these efforts, two tests with parts of rings are proposed to determine the inner and outer circumferential tensile strength of bamboo. In the first set-up, a semi-ring is radially loaded at the middle section and supported on rollers at both ends such that the inner surface of the specimen is under tensile stress. In the second protocol, similar to the edge bearing test for a complete ring, a specimen bigger than a semi-ring is submitted to compression loading to generate tensile circumferential stresses on the outer surface. Both tests are compression based. The ongoing experimental program includes the bamboo species Guadua angustifolia, Phyllostachys edulis, Bambusoides, Vivax, Iridescens and Violascens. The inner failure strain was significantly higher than the outer for all the species. A brittle failure was observed with longitudinal fissures initiated at the tensile surface. Stress determination through the wall thickness is highly dependent on the constitutive equation. Taking into account the variation of the circumferential modulus with a radial position, the outer strength for the bamboo species Guadua angustifolia was 10.0 MPa (COV=0.09, n = 15), which is about 50% higher than the inner strength of 6.4 MPa (COV=0.26, n =15). Higher failure stresses were obtained for the European species calculated with a homogeneous model. More experiments have to be conducted to analyze the variation with radius of the circumferential modulus of the European species for a more accurate determination of strength.

Keywords: Bamboo, circumferential strength, functional graded material, mechanical characterization.

INTRODUCTION

Bamboo is a plant of rapid growth that plays an important role to control water cycles, to reduce erosion, and to sequester carbon. Bamboo culms are tubes reinforced with transverse diaphragms located at variable distance along the length. Due to its excellent mechanical properties, bamboo has become a promising alternative to replace traditional and highly polluting construction materials, such as steel and concrete. Bamboo culms are functionally graded and axially reinforced with strong cellulose fibers. Therefore, it displays excellent mechanical properties along the axial direction of the culm but it is prone to longitudinal splitting on the fiber planes. This failure mode becomes critical near the stress concentrators as for example when holes are drilled through the culm to construct traditional joints. Thus, the structural bamboo elements usually fail by longitudinal splitting at axial stress levels one order of magnitude lower than the axial strength of the material. Longitudinal splitting has been attributed to the low tensile circumferential strength of the material (Moreira & Ghavami, 2015; Ghavami & Moreira, 1996). Various experimental set-ups (Lee, Odlin, & Yin, 2014; Sharma, Harries, & Ghavami, 2013; Torres, Ghavami, & García, 2007; Virgo, Moran, Harries, García, & Platt, 2017) have been proposed to determine the tensile circumferential strength of bamboo. The simple edge bearing test (Sharma, Harries, & Ghavami, 2013) can be performed to determine the circumferential strength of bamboo. In this test, the ring is under bending moment that changes with the position along the ring circumference. Hence, the ring fails at the inner or at the outer location of the wall thickness. This does not allow determining the inner and outer circumferential strength.

The objective of this ongoing study is to propose two simple tests with parts of rings that allow the determination of both, the inner and the outer circumferential strength of the material. This information may be used for a more reliable prediction of longitudinal splitting using theoretical finite element models.

MATERIALS AND METHODS

Material

This inter-laboratory study was accomplished with the European bamboo species Bambusoides (BAM), Edulis (EDU), Iridescens (IRI), Violascens (VIO) and Vivax (VIV) as well as with the bamboo species Guadua angustifolia (GA) grown in South America.

The specimens of the European bamboo were tested in the green condition, e.g. they were not dried or immunized. On the other hand, the specimens of Guadua angustifolia (GA) were borax treated and oven dried with practices well stablished (García, Rangel, & Ghavami, 2012). Moisture content was measured following the provisions of the NTC5525 (Instituto Colombiano de Normas Técnicas y Certificación , 2007). Average moisture in the tested culms was 10.6% (COV = 0.14) and the data about the dimensions of the specimens are summarized in Figure 1. The diameter and thickness of GA is about 1.5 and 2 times higher, than the respective diameter and thickness of the European species.

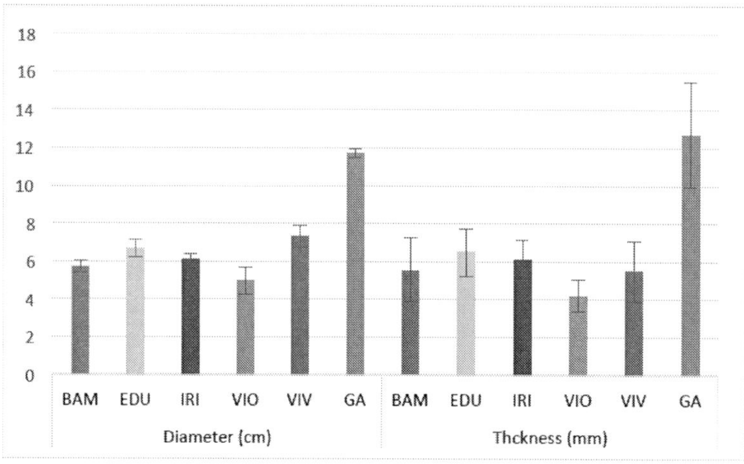

Figure 1. Summary of diameter (cm) and thickness (mm) of the specimens (n = 6 for the European species and n=15 for GA).

Experimental methods

The two experimental set-ups are shown in Figure 2. In the first assembly (Figure 2a), bamboo semi-rings were radially loaded at the middle section and supported at the ends using small steel plates resting on solid cylinders to avoid horizontal reactions. The moment at the center of the semi-ring can be calculated as the vertical reaction (half of the applied force) times the radius of the semi-ring. The load tries to open the semi-ring and the inner surface of the wall is exposed to circumferential tensile stress.

In the second set-up (Figure 2b), we used portions bigger than a semi-ring (with a shape similar to the letter C) in order to facilitate mounting the specimen. These samples were then subjected to two radially opposed compressive loads. In this case, the load closes the specimen and the circumferential tensile stresses occur on the outer surface of the wall.

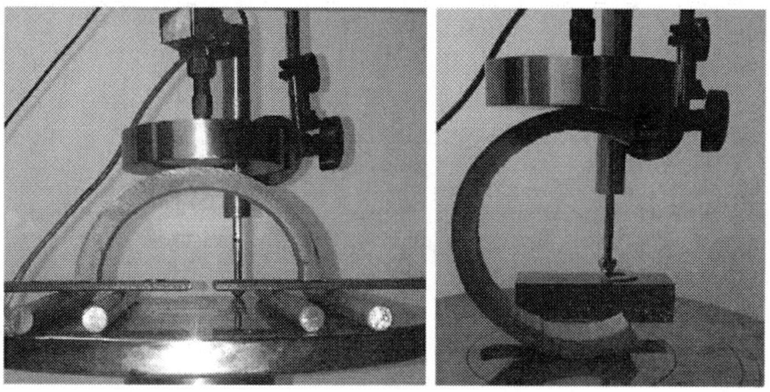

a. b.

Figure 2. Set-ups used to determine the circumferential tensile strength: a. On the inner surface using semi-rings, b. On the outer surface using C shaped specimens

The European species were tested in the Mechanics Laboratory at the University of Bologna, Italy. The load was applied using an universal hydraulic press, WOLPERT, while two STINGRAY cameras were used to perform digital image correlation (DIC) to measure the strain field at the zone of interest.

Twelve specimens were cut from three culms for each of the European species, and six specimens were tested under each of the two aforementioned testing protocols. For the DIC measurements, the zone of interest was carefully sanded with abrasive paper to ensure a smooth surface for the first layer of RAL 9010 matt white spray paint. Next, three to four layers of the white spray paint were successively applied to generate a consistent ground coat. Finally, to produce a random speckle pattern over the white varnish, a FOLIATEC spray film was applied, which was chosen after pilot tests with conventional varnish. The software VIC 3D 2012 (Correlated Solutions, Ltd.) was used to track the speckle pattern to calculate the strain contours. For a more consistent and precise measurement of maximum strains, an extensometer tool was circumferentially oriented along the inner arc of the middle section for the semi-ring test and along the outer arc of the middle section for the C test.

GA specimens were loaded with a Humbolt (Master load HM300) compression press. Load and displacements were respectively measured with load cell LC703-150 and an LVDT LD610-15 (Omega inc). Thirty specimens were tested for GA extracted from three culms. Fifteen were tested under each of the loading protocols.

Displacement control was used in all the cases at a speed of 1 mm/min for the first configuration and 2 mm/min for the second. A higher velocity for the second configuration was used since this set-up is more compliant than the first.

Calculations

Circumferential strength was calculated as the maximum tensile stress at failure. The maximum stress was calculated using the following formulas valid for homogeneous curve beams:

$$\sigma_{c,in} = \frac{P_{max}R}{2Lth}\frac{(R-R_i-h)}{R_i}, \qquad (1)$$

where Pmax is the maximum applied force, R is the mean radius, t is the thickness at the failure zone, L is the width, R! is the inner radius, and h is the distance between the centroid and the neutral axes, that can be calculated as:

$$h = R - \frac{t}{\ln\left[\frac{2R+t}{2R-t}\right]}. \qquad (2)$$

In the second configuration, the maximum stress occurs at the outer position of the middle horizontal section. The circumferential strength σ!,!"# was calculated as:

$$\sigma_{c,ext} = -\frac{P_{max}R}{Lth}\frac{(R-R_o-h)}{R_o} - \frac{P}{Lt}, \qquad (3)$$

where Ro is the outer radius and the other variables were defined before.

Complementary, results for the GA species were analyzed using finite element models including the variation of the circumferential modulus E! with radial position r. The following parabolic function was used in this analysis,

$$E_{\varphi}(r) = 925\left(\frac{r-R_i}{t}\right)^2 + 686 \quad \text{MPa.} \tag{4}$$

The finite element mesh was generated with eight-node hexahedral elements. A mesh size of 1 mm was prescribed along the wall thickness to accurately capture the variation of the circumferential modulus.

RESULTS

Typical load-displacement curves were approximately linear (Figure 3) and all the specimens exhibited brittle failures. The experimental strain field is similar to that obtained with the FE simulations (Figure 4). For all the species the inner failure strain was higher than the outer. In particular, for the European species, the measured inner failure strain (range 14048-32523 µε) was between 1.65 and 3.85 times larger than the outer (range 8330-16185 µε). For GA, results of finite element models yielded an inner failure strain of 9302 µε that is about 50% higher than the outer failure strain of 6211 µε.

The calculation of the circumferential strength (i.e. the maximum stress at the failure load) is strongly dependent on the constitutive equation. A homogeneous formulation yielded an inner strength higher than the outer for all the species. Relative high values of the circumferential stress were obtained for the European species (Figure 5). For GA the homogenous model predicted an inner hoop strength of 8.8 MPa (COV=0.25) that is significantly higher than the outer strength of 7.8 (COV=0.10). Opposed to this trend, the outer strength, using the graded model included in the FE analysis, of 10.0 MPa (COV=0.11) is significantly higher than the inner strength of 6.4 MPa (COV= 0.26).

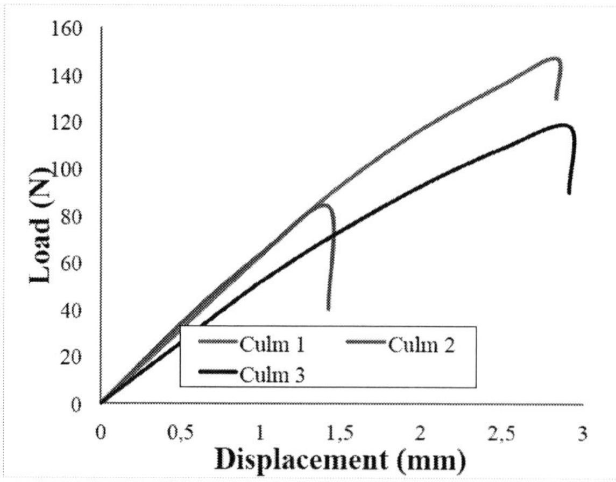

Figure 3. Typical load-displacement curves for the ring segments under inner tension in the test of semi-ring specimens

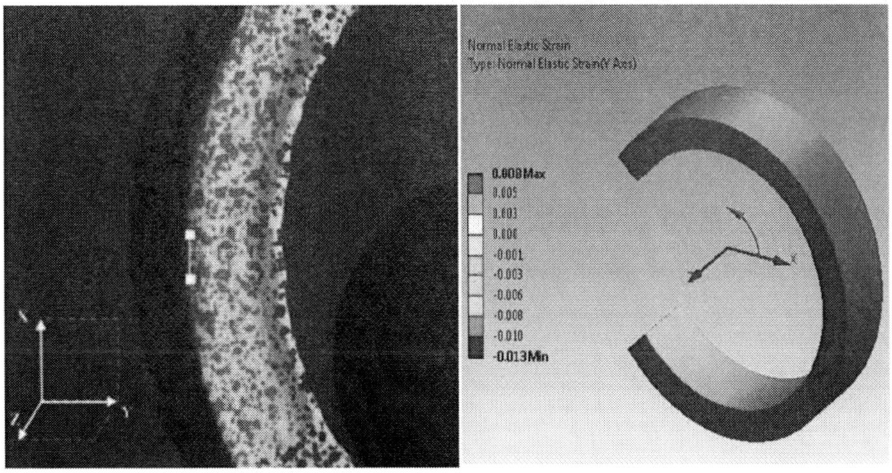

Figure 4. Typical stain field recored in the C test of European species (left) and results of FE simulations for GA

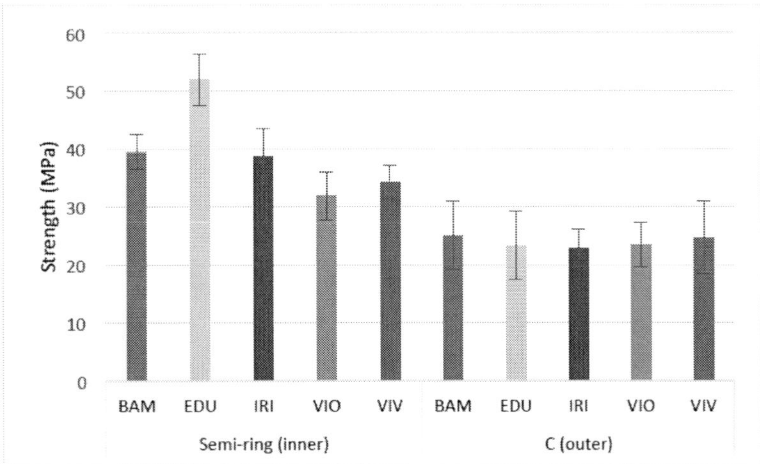

Figure 5. Summary of maximum circumferential stress calculated with a homogeneous formulation for the European species

The longitudinal splitting failures initiated on the inner surface for the first configuration and on the outer for the second, as depicted in Figure 6. All fissures were produced at the interface between the fibers and the matrix and rapidly propagated through the radius of the specimens to generate sudden collapse.

a b

Figure 6. Failure modes under a. Inner tension, and b. Outer tension

DISCUSSION

Two simple protocols have been proposed to determine the circumferential tensile strength of bamboo at the inner and outer culm wall. The tests are compression based and the specimens can be easily cut from bamboo rings. These tests complement previous information about the circumferential mechanical properties of bamboo, obtained with the edge bearing protocol. Results indicate that the outer failure strain is lower than the inner, which may be attributed to higher fiber density near the outer boundary. As the fibers are more closely packed near the outer boundary, the percentage of the flexible lignin matrix is lower, as well as the capacity to deform before the composite material reaches the maximum strength.

This study demonstrated the need to account for the heterogeneity of the material for a reliable calculation of the stresses. While an isotropic model predicts an inner strength higher than the outer, the functionally graded model predicts the opposite. Previous experimental studies using the edge bearing test (Sharma, Harries, & Ghavami, 2013) have shown that the neutral axis of the ring shifts toward the outer side, which is due to the higher outer circumferential Young's modulus, thus consistent with the adopted heterogeneous model adopted for GA to calculate the strength in this study.

The functionally graded nature of bamboo may explain in part the values of about two, three fold larger obtained for the circumferential strength of the European species compared to GA, since the stress for the European species was calculated with formulas valid for homogeneous materials. Studies with GA have shown that the outer circumferential modulus is about three times larger than the inner [9]. Thus, more studies have to be carried out with the European species to determine the variation of the hoop modulus with radial position for a more accurate calculation of the stress.

Differences in material treatment and size were also present in this study. First, the European specimens were in the green condition, as opposed to the GA specimens, which were oven-dried and immunized. In addition, the average diameter and thickness of

European species is smaller than those of GA. More analyses have to be accomplished in the future to fully understand the sources of this difference.

Acknowledgments

The first author is grateful to Administrative Department of Science, Technology and Innovation COLCIENCIAS for financial support through national doctoral grant No. 617. The other authors greatly appreciate the support given by Universidad del Valle to perform this study.

The authors form the University of Bologna wish to thank Roberto Carli and Mario Marcolongo for the technical support during the tests at LISG Lab of the DICAM Department.

References

1. Moreira E., Ghavami K. Limit State Design of Steel Pin Connections for Bamboo Truss Structures. In: 16th International Conference on Non-Conventional Materials and Technologies (16thNOCMAT 2015) , Winnipeg, 2015.

2. Ghavami K., Moreira L., Development of a new joint for bamboo space structures, Transactions on the Built Environment, pp. 3-12, 1996.

3. Lee P.H., Odlin M., Yin H., Development of a hollow cylinder test for the elastic modulus distribution and the ultimate strength of bamboo, Construction and Building Materials, pp. 235-243, 2014.

4. Sharma B., Harries K., Ghavami K. Methods of determining transverse mechanical properties of full-culm bamboo, Construction and Building Materials, pp. 627-637, 2013.

5. Torres L., Ghavami K., García J. A transversely isotropic law for the determination of the circumferential Young´s modulus of bamboo with diametric compression tests, Bahía Blanca: Latin American Applied Research, 2007.

6. Virgo J., Moran R., Harries K., García J., Platt S. Flat ring flexure ring test for full-culm bamboo. In: 17th International Conference on Non-Conventional Materials and Technologies (17thNOCMAT 2017), Merida, 2017.

7. García J., Rangel C., Ghavami K. Experiments with rings to determine the anisotropoc elastic constants of bamboo, Construction and building of materials, pp. 52-57, 2012.

8. Instituto Colombiano de Normas Técnicas y Certificación , NTC 5525 - Métodos de ensayo para determinar las propiedades físicas y mecánicas de la Guadua angustifolia Kunth, Bogotá: ICONTEC, 2007.

9. Moran R., Weeb K., Harries K., García J. Edge bearing tests to assess the influence of radial gradation on the transverse behavior of bamboo, Construction and Building Materials, pp. 574-584, 2017.

IC-NOCMAT 2019

18th International Conference on
Non-Conventional Materials and Technologies
"Construction Materials & Technologies for Sustainability"
(18th NOCMAT 2019)
24th – 26th July 2019
Nairobi, Kenya

ANALYSIS OF BOND-BEHAVIOR OF BAMBOO STRIP AND BAMBOO BIO-CONCRETE

Nathalia Andrade da Silva[1], Bruno Menezes da Cunha Gomes[1], Vanessa Maria Andreola[1] , M'hamed Yassin Rajiv da Gloria[1], Romildo Dias Toledo Filho[1]

[1] Department of Civil Engineering, COPPE, Universidade Federal do Rio de Janeiro, Rio de Janeiro, RJ, Brazil.

ABSTRACT

Bamboo presents great potential due to its mechanical properties, ease of planting and CO_2 capture. However, for its use as structural reinforcement in the form of strips, it is necessary to know its bond-behavior with the cementitious matrix. The present article aims to study the bond-behavior between bamboo strips and different bamboo bio-concretes. Bio-concretes containing volumetric fractions of 25, 35 and 45% of bamboo particles were used in cementitious matrices. The bamboo strips were submitted to three treatments: epoxy resin, epoxy with coarse sand, and epoxy resin with fine sand. The effect of these treatments on the bond-behavior was determined through pullout tests. The results indicated that the treatment of epoxy resin with the sands presented a more satisfactory result, requiring a force of 9 kN for the pulling out and resulting in a bond tensions of 4.5 MPa.

Keywords: bamboo strips, bio-concrete, bonding, pull-out.

INTRODUCTION

In a reinforced concrete structure it is understood that the concrete resists to compressive stresses and its reinforcement to tensile stresses. Due to the forces transfer that occurred between the concrete and the reinforcement, the system is efficient and responds adequately to the structure requests.

Since bamboo has mechanical characteristics that make it a good composite material for reinforcement in structures, it is important that its adhesion with concrete be more studied and improved. Out of being a renewable, low cost material and sustainable development, bamboo is emerging to replace wood and steel [1].

Over the last decades, many works have focused on the production of composite materials. Normally applied to a cement matrix, materials of natural and synthetic origin or even waste can be used in their production [2]. An interesting alternative is the combination of the bamboo with the bio-concrete in a structure.

There are some types of tests to determine bond strength and relative displacement between steel bars and concrete. These are: confined bar test, pull-out test and beam test [3].

The pull-out test is the most used in reinforced concrete and for bamboo reinforced concrete. According to Reis [4], they provide satisfactory qualitative and quantitative results that allow a good understanding of the parameters that influence bonding.

Salgado [5] analyzed and compared several types of treatments in bamboo samples in concrete, through pull-out test. Untreated bars were used, with tar and sand, wrapped in barbed wire and bars with fastening clips and grooves in the shell. The treatment that gave higher bond strength was that the tar with sand.

Lopes et al. [6] evaluated the bond strength and pull-out strength of bamboo reinforcing soilcement, through pull-out and push-in tests, using untreated and treated bamboo (species Dendrocalamus giganteus) with asphalt emulsion (Neutrol 45) impregnated with coarse sand. The value of the pull-out strength was higher in the specimens without treatment, showing the negative effect of the asphalt treatment on the adhesion between of bamboo and soil-cement matrix.

A relevant problem to be considered in the "bamboo-concrete" association, which directly interferes on the adherence, is the dimensional variation that occurs in bamboo due to the rapid absorption and release of water. For being a hygroscopic material, when in contact with the aqueous environment, bamboo varies in size, due to the cells' spacing. This fact results in a swelling of the bamboo before the concrete is cured. The water absorbed by the bamboo microstructure induces an expansion and promotes a local internal stress in the concrete that involves the bamboo, as shown in Figure 1 [7] [8].

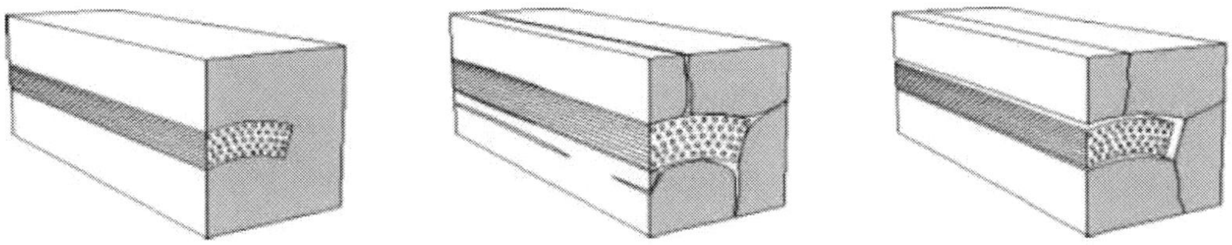

Figure 1 – Swelling and cracking associated with the use of raw bamboo in concrete [8].

Some waterproofing treatments have been used by researchers to minimize adhesion problems. Ghavami [9] [10] studied several products and methods of treatment of the bamboo surface, where the best results were observed with water-dispersed asphalt ink paint and structural adhesive epoxy resin-based, the two substances manufactured by Sika SA. The author observed that after being treated with the epoxy-based adhesive, the bamboo was immersed for 96 hours in water and showed an absorption rate of only 4%, and the epoxy resin coating increased the adhesion of the bamboo with concrete up to 5 times compared to untreated bamboo.

The use of sand with epoxy resin is a treatment option that has also been studied to improve the adhesion of bamboo to concrete. According to Azedeh et al [11], corrugation is an advantage in increasing the bond between bamboo and concrete.

This research aimed at studying the adhesion between bamboo strips coated with epoxy resin (type Sikadur) and sand, and bamboo bio-concrete with different volumetric fractions of biomass (25%, 35% and 45%). Pull-out test were performed in order to access the best treatment to improve adhesion of these materials.

MATERIALS AND METHODS

Preparation of bamboo strips

Bamboo culms of the species Dendrocalamus giganteus, aged over 3 years, treated with octoborate, were selected from Bambu Carbono Zero, located in Campo Belo, São Paulo, Brazil. The machinery used in the processing of the culms for the production of the strips was: circular saw, squaring machine, slicer and planer. The total of 30 strips with the following dimensions 6 x 12 x 150 mm (thickness x width x length) were produced, as shown in Figure 2. The strips were removed from the region closest to the culm wall, as it is a region with high fiber density, that provide a better mechanical response to tensile strength .The strip of bamboo without node was 163, 47 MPa and the density 0.65 g / cm^3.

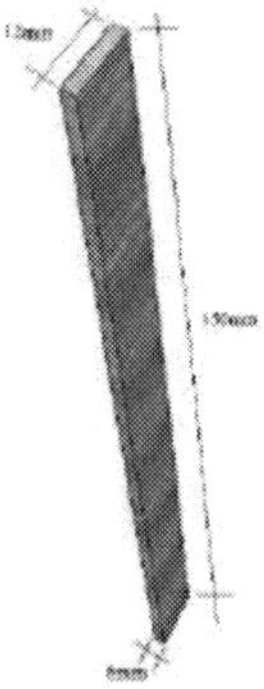

a)

b)

Figure 2- a) Dimensions of bamboo strips b)Strips used.

Bamboo Treatments

Four types of treatment were applied: REF (Reference), strips in natural condition without treatment; SIK, strips coated with epoxy resin Sikadur 32®; FSS (Fine Sand and Sikadur) which are the coated strips with epoxy resin and fine sand; and CSS (Coarse Sand and Sikadur), strips coated with epoxy resin and coarse sand. Figure 3 shows the strips with the treatments chosen for the study.

The coatings were applied 3 days before the bio-concretes casting in order to guarantee a complete cure of the epoxy resin.

Figure 3- Bamboo strips coated with sikadur/sand.

Bio-aggregates of Bamboo (BB)

The bio-aggregates of bamboo used in this study came from the State of Rio de Janeiro (Brazil). The bamboo species were: Dendrocalamus Asper, Phyllostachys aurea e Phyllostachys edulis. First, the stems were cut on table saw, in order to reduce their length from 120 cm to 40 cm. After the cut, the stems were processed in an industrial crusher to obtain the first particles with length between 5 and 19 mm. Then, in a knife mill, these particles were refined in a smaller scale to have the bio-aggregates with maximum diameter of 4.75 mm (Figure 4). The apparent density, the water absorption and moisture content were 590 kg/m³, 81% and 10.18%, respectively.

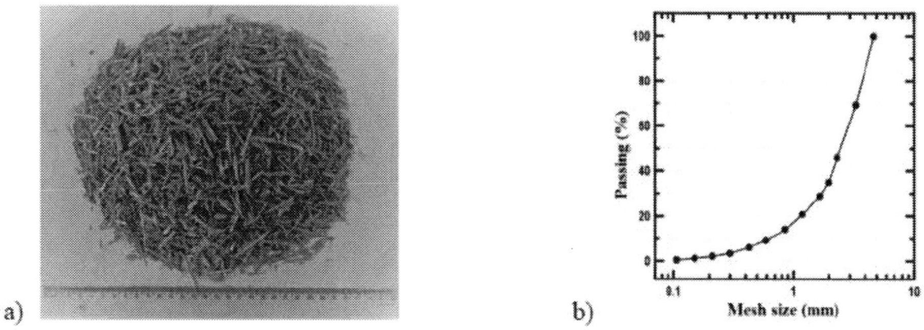

Figure 4 : (a) Bio-aggregates of bamboo (b) Granulometric curves

Bamboo bio-concrete production

Three mixtures of bamboo bio-concrete (BCB) were used, with cementitious matrix developed at the laboratory NUMATS of the Federal University of Rio de Janeiro. The matrix was composed of Brazilian Portland Cement (CPV-ARI), partially substituted in mass by 30% of commercial Metakolin and 10% of Fly Ash. Based on [12], the following parameters were defined: bamboo volumetric fractions 25%, 35% and 45%; water-to-binders ratio 0.4, additional water equal to the water absorption (81%).. Table 1 shows the proportion of the materials used in the prodution of 1m³ of BCB.

Table 1- Mixtures composition (kg) per cubic meter.

Materials	BCB 25%	BCB 35%	BCB 45%
Water	527.96	515.51	503.06
Cement	579.59	502.31	425.03
Fly ash	96.59	83.71	70.83
Metakaolin	289.79	251.15	212.51
Bamboo	132.50	185.50	238.50

The production of the bio-concretes began with a mix of the cementitious materials in a 2 liter mixer. Once the materials homogenized, the BB was added and the whole mixed for one minute before introducing slowly the water for 1 min. After completing 4 min of mixing, the mix process was stopped.

Before introducing the fresh BCB in the molds (50x50x50 mm³), the bamboo strips bases were slightly glued in the center of the bottom of the molds (Figure 5a) to ensure their verticality, and centrality during and after the filling of fresh BCB. Then, the fresh BCB were introduced in the molds until the top in order to have an embedment length of 50 mm of the strips. The samples were demolded after 24 hours (Figure b) and were cured in a 21° C room for 28 days. The average compression strength after 28 days of the BCB25%, BCB35% and BCB45% were 21.05, 14.10 and 12.30 MPa and their Young modulus 6.75, 4.38 and 3.03 GPa, respectively.

a)

b)

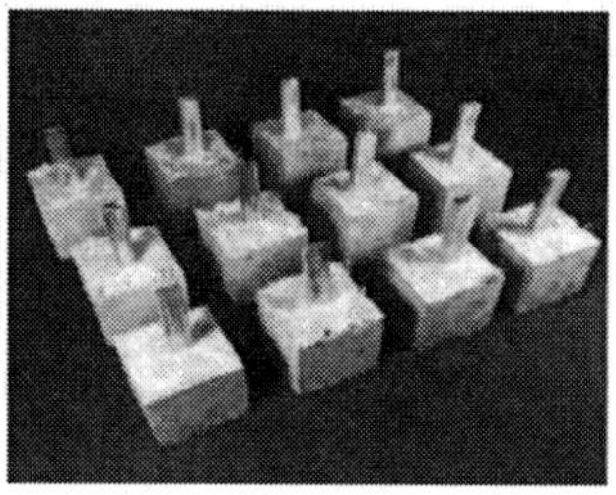

Figure 5- a) Bamboo strips glued in the molds b) specimens used in the tests

Pull out test

The pull out tests were performed using a Shimadzu AG-X machine (100kN), with speed rate of 1 mm/min. The test procedure consisted in applying tensile force on the strip in order to pull it out of the BCB, immobilized between two steel plates, according to Figure 6. First, the BCB 25% (lowest bamboo content) was chosen to evaluate the best bamboo strips treatment. Then, the specimens were produced with BCB 25% and treated bamboo strips (REF, SIK, CSS, FSS). Once the best treatment known, other pull out tests were performed with strips of that treatment, varying the bamboo fraction of the BCB (25%, 35%, 45%). That variation helped to understand the influence of the bio-aggregates volume on the bamboo strip and bio-concrete adhesion.

The bond strength (τ) was defined as the tensile force divided by the area of the embedded surface, as expressed in the equation 1.

Figure 6- Pull-out test set-up.

$$\tau = \frac{F}{(2e+2d).l_b}$$

(eq. 1)

Where,
τ = bond strength (MPa),
F= force (N),
e = thickness (mm),
d = width (mm),
l_b = embedment length (mm).

RESULTS

The results of the pull out test performed to determine the best bamboo strip treatment are presented in Table 2, while the typical curves of pull out are illustrated in Figure 7.

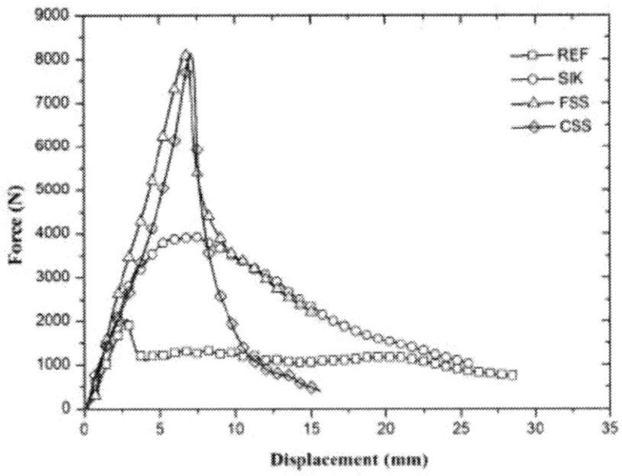

Figure 7- Typical pull out force – displacement curve.

According to the results, the greatest pull out forces were observed in the treatments with sand and Sikadur (CSS, FSS). CSS presented bond strength of 4.23 MPa, while FSS average adhesion tension was 4.14 MPa, a difference of 2.1% statistically insignificant.

Table 2- Pull out force and Bond Strength (25%).

Type of treatment	Pull out force (N)	Bond Strenght (MPa)
Reference (in natura)-REF	2129.27 (4.10)	1.07(4.10)
Sikadur 32 - SIK	3984.44 (4.5)	1.90(8.4)
Fine sand and Sikadur- FSS	8163.34 (8.19)	4.14 (0.75)
Coarse sand and Sikadur - CSS	8392.11 (5.6)	4.23 (5.6)

The untreated strips (REF) showed 56% of the bond strength of the strips coated with sikadur (SIK). The coating with sikadur created some dimensional variation along the embedded surface, which increased the adhesion. When compared with CSS and FSS, SIK presented 43% of their strength, proving the efficiency of using sand. The addition of sand on the sikadur created a roughness on the strip surface, which mobilized localized forces increasing significantly the bond. After the pull out test, it can be noted that the strips REF and SIK had a few amount of bioconcrete impregnated on their surfaces (Figure 8). No damage was observed in the holes created in the bio-concrete samples after the pulling out of the bamboo strip REF, while fracture was observed on the BCB surface after extraction of the SIK. Therefore, Sikadur promoted a friction on the surface of the BCB but not significant such as FSS and CSS (Figure 9).

a)  b) 

Figure 8- Samples after the test: a) Reference; b) Sikadur.

Once removed from the BCB, the bamboo strips FSS and CSS had relevant amount of bio-concrete adhered and that the cubic samples of bio-concrete suffered damage throughout the region around the strips (Figure 9).

a)  b) 

Figure 9- Samples after the test: a) Fine Sand and Sikadur; b) Coarse Sand and Sikadur.

As the CSS was the treatment which provided the highest bond strength, , it was used for the strips molded with the BCB35% and BCB 45%. The curves of Figure 10 shows the typical pull-out versus displacement curves obtained

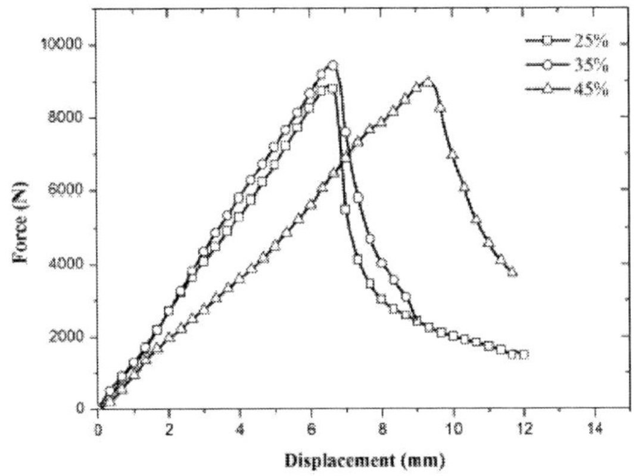

Figure 10- Typical pull out force – displacement for bioconcrete of bamboo.

Analyzing the bamboo strips treated with coarse sand and Sikadur in the mixtures of 25%, 35% and 45%, it is noticed that the cement and bio-aggregates contents did not significantly affect the pull out force and the bond-behavior. According to the results of Table 3, it is observed that when compared with the BCB bond strength of BCB 35%, which was the highest, BCB 25% and BCB 45% presented 93% and 98% of strength, respectively. The main difference observed in the curves is the higher displacement of BCB 45% before reaching the maximum strength.

Table 3- Pullout force and Bond Strength of different BCB

	Pull out force (N)	Bond Strength (MPa)
BCB 25%	8636.11 (3.55)	4.36 (3.55)
BCB 35%	9433.43 (0.13)	4.67 (3.99)
BCB 45%	9058.74 (1.77)	4.57(1.77)

CONCLUSION

The treatment of bamboo strips with Sand and Sikadur had a positive effect on their bond strength with bio-concrete matrix, increasing until four times the strength reached with the strips without treatment.

The combined effect of sikadur and coarse or fine sand in the coating of the strips revealed to be the best treatment to increase the adhesion.

The quantity of cement and bamboo bio-aggregates in the matrix did not interfere on the pull out forces and adhesion tensions.

It is necessary to coat the bamboo strips in order to improve the bond between the bamboo strips and the bio-concrete. The coating with epoxy resin provided a good bond behavior, but the combination of epoxy resin and sand gave the best results.

ACKNOWLEDGEMENTS

This research was supported by Núcleo de Ensino e Pesquisa em Materiais e Tecnologias de Baixo Impacto Ambiental na Construção Sustentável (NUMATS) and Coordenação de Aperfeiçoamento de Pessoal de Nível Superior (CAPES).

REFERENCES

1. Javadian, Alireza, et al. "Mechanical properties of bamboo through measurement of culm physical properties for composite fabrication of structural concrete reinforcement." Frontiers in Materials 6 (2019): 15.

2. Garcez, R. M; Santos, T.; Garcez, E.O; Gatto, D.A. Propriedades físicas de compósitos cimentomadeira com serragem de Pinus elliottii tratada. Revista Ciência da Madeira. (Brazilian Journal of Wood Science), v.7, n2, p.70-80, 2016.

3. Carvalho, E. P., Ferreira, E. G., Cunha, J. C., Rodrigues, C. S., Maia, N. S. Experimental Investigation of Steel-Concrete Bond for Thin Reinforcing Bars. Latin American Journal of Solids and Structures, 14(11), 1932-1951, 2017.

4. Reis, C. N. S. Influência da utilização de agregado miúdo de RCD na aderência aço-concreto reciclado. Bahia, 2009. Dissertação de Mestrado - Universidade Estadual de Feira de Santana, 2009.

5. Salgado, A.L. DE B. Resistência a la extracción de astillas de bambu enclavadas en concreto. In: Sustainable construction into the next millenium environmentally, friendly and innovative cement based materials. João Pessoa. João Pessoa: Federal University of Paraiba and The University of Sheffield. p.407-412,2000.

6. Lopes, W.G.R; Freire, W.J.; Ferreira, G.C.S; Ensaios de arrancamento e de empuxamento aplicados a taliscas de bambu encravadas em corpos-de-prova de solo-cimento. Revista Brasileira de Engenharia Agrícola e Ambiental, V.6, n.3, p.504-510,2002.

7.Ferreira, G. C. S. Vigas de concreto com armadura de bambu da espécie Dendrocalamus giganteus. Tese de Doutorado. UNICAMP- São Paulo,2007.

8. Javadian, Alireza, et al. "Bond-behavior study of newly developed bamboo-composite reinforcement in concrete." Construction and Building Materials 122 (2016): 110-117.

9. Ghavami, K. Ultimate load behaviour of bamboo-reinforced lightweight concrete beams. Cement and Concrete Conposites, London, v 17, n. 4, p. 259-351, 1995.

10. Ghavami, K. Bamboo as reinforcement in structural concrete elements. Cement and concrete composites, v. 27, n. 6, pp. 637–649, 2005.

18th International Conference on
Non-Conventional Materials and Technologies
"Construction Materials & Technologies for Sustainability"
(18th NOCMAT 2019)
24th – 26th July 2019
Nairobi, Kenya

FULL-CULM BAMBOO AS A FULL-FLEDGED ENGINEERING MATERIAL

Kent A. Harries[1], Ian Nettleship[1], Yusuf Akinbade[1], Christopher Papadopoulos[2], Felipe Acosta[2], Johanna Morales Albino[2], and Mairim Ariezaga Gonzalez[2]

[1] University of Pittsburgh, USA;

[2] University of Puerto Rico-Mayaguez, USA;

ABSTRACT

This paper provides an overview of the activities of a US National Science Foundation (NSF) funded project Full-Culm Bamboo as a Full-Fledged Engineering Material (Project Numbers NSF CMMI 1634739 and 1634828). The project, funded in 2017, is a collaboration between teams at the Universities of Pittsburgh and Puerto Rico Mayaguez.

Keywords: bamboo, geometric properties, material properties, modelling, uncertainty

INTRODUCTION

The primary objective of this project is to establish the framework and tools required to standardize the evaluation of the material and mechanical properties of full-culm bamboo and thereby place bamboo on a similar footing to timber as a conventional building material. This research consolidates and significantly extends the extant body of knowledge and establish materials- and mechanics-based constitutive models for the behaviour of full-culm bamboo as a functionally graded, fibre-reinforced material. Three representations of bamboo behaviour are being developed forming the framework and tools required to evaluate the material and mechanical properties of bamboo for engineered applications. First, detailed analytical modelling is key to understanding the engineering properties of bamboo as a functionally graded material. An analytical model is being developed informed by innovative experimental methods focused on establishing through-wall and along-culm variation of fundamental mechanical properties. Second, surrogate representation of difficult-to-obtain engineering properties suitable for field test methods is necessary if broad adoption of full-culm bamboo is to be reliable. The approach taken leverages the analytical platform in order to develop both empirical and mechanics-based representations of engineering design properties that may be obtained from practical field tests. Finally, a new framework for modelling uncertainty in bamboo material and mechanical properties, which can be highly variable, will enable reliable calibration of design equations. The approach includes stochastic generation of probability spatial distributions of mechanical properties implemented into the developed analytical model. The project advances the science of modelling both fibre reinforced and functionally graded materials. Indeed, a better understanding of the naturally evolved optimal design of the bamboo material and culm, including its nonhomogeneity and variability, will inspire optimization of other structural engineering functionally graded materials.

This paper reports work-in-progress and is intended to provide an overview of this NSF-funded project – the first of its kind funded by NSF – and to inspire others to develop similar data modelling techniques.

TRANSVERSE PHYSICAL AND MECHANICAL PROPERTIES OF BAMBOO

The circular cross section of bamboo is composed of unidirectional cellulosic fibres oriented parallel to the culm's longitudinal axis embedded in a parenchyma tissue matrix [1]. The parenchyma tissue matrix lignifies (hardens) as the culm matures leading to increased density and improved mechanical properties. As a functionally graded material, bamboo has evolved to resist its primary loading in nature: its own self-weight and the lateral effects of wind. The density of longitudinally oriented fibres increases from the inner culm wall to the outer culm wall [1]. The wall thickness is largest at the base of the culm and decrease with height up the culm. However, the size and quantity of vessels decrease with the height of the culm and are replaced with fibres. This addition of fibres compensates for loss in strength and stiffness due to reductions in diameter and wall thickness near the top of the culm, resulting in relatively uniform engineering properties along the entire culm height [2, 3].

Like any fibre-reinforced material, mechanical properties are highly correlated to the proportion and distribution of fibres in the cross section. Mechanical properties are influenced by density, which depends on fibre content, fibre diameter, and cell wall thickness [4]. The volume fraction of fibres ranges from approximately 60% at the exterior face of the culm wall to 10-15% near the interior face. The variation in density through the culm wall has been assumed by various researchers to be linear, quadratic, exponential, or a power function [5] and is known to be species-dependent [2, 6]. Due to the complex variation of the fibres and vascular bundles, the variation of material properties through the culm-wall thickness has been shown to be significant [7] and to have the effect of increasing gross culm stiffness by about 10% as compared to an assumed uniform distribution of the same volume of fibres [4]. A more refined analysis showed the effect of fibre gradient on gross culm stiffness to result in about a 5% increase for thin-walled culms ($D/t < 8$; P. edulis and Guadua angustifolia Kunth culms) and as much as 20% for thick-walled culms ($D/t > 8$; B. stenostachya culms) [3]. D is the culm outer diameter and t is the culm wall thickness.

In general, while highly variable, the longitudinal behaviour of bamboo is relatively well understood in a qualitative sense. From an engineering perspective, the longitudinal behaviour is most typically considered as a fibre-reinforced material in which longitudinal properties are obtained using a rule of mixtures approach. The dominant failure mode of bamboo however, is longitudinal splitting associated with bamboo carrying flexure, compression or tension loads. Janssen [8] describes the bending stresses in a culm as being characterised by the longitudinal compressive stress and transverse strain in the compression zone of the culm, with failure eventually occurring due to longitudinal splitting. This is ideally a Mode II longitudinal shear failure. However, in the presence of perpendicular stresses (as is the case where ever there is a non-zero shear-to-moment ratio), there is some Mode I component stress which significantly reduces the Mode II capacity. Richard et al. [9] demonstrate the effects of such mode mixity, reporting Mode I capacities equal to only 18% of the Mode II capacity. Beam tests having mixed mode behaviour exhibited shear capacities ranging from 40-70% of the Mode II capacity.

Both the Mode I and II behaviours described are primarily functions of the transverse properties of the fibre-reinforced culm which are believed to be dominated by matrix (parenchyma) properties. However, despite their importance in the dominant observed behaviour of full-culm bamboo, there are few studies of the transverse properties of the culm wall. In early work, Arce-Villalobos [10] concluded that there is no correlation between the density of bamboo and its transverse tensile strength. Janssen [4], based on flexural tests, reports that a transverse strain of 0.0013 results in transverse tensile failure of the culm wall. More recently, test methods have been proposed for obtaining transverse properties of bamboo culms [11-13] although these have not yet been widely adopted to obtain material properties over a range of species and conditions.

The objective of the present study is to investigate the transverse material property gradient through the culm wall and to connect the mechanical behaviour to physical characteristics, such as fibre density. In this study, a modification to the flat-ring flexure [13] test specimen, in which only portions of the culm wall cross-section are tested, is used to obtain a measure of the transverse tensile capacity of the bamboo. Microscopy analyses are used to qualitatively describe the culm wall architecture and to quantitatively assess the failure modes through the culm wall thickness. Throughout this study, all position-dependent data are presented with respect to normalised culm wall thickness ($x = 0$ is the inner wall and $x = 1$ is the outer wall).

Five different bamboo species were tested in this study, P. edulis, P. bambusoides, P. meyeri, P. nigra, and B. stenostachya. All are thin-walled (D/t generally greater than 8) except B. stenostachya which is a thick-walled species. The moisture content of all specimens at time of testing was between 13% and 17%.

Fibre Distribution

Digital images of full culm wall cross sections were taken at each of the four quadrants (N, S, E and W). These images were processed using a purpose-built MatLab script to obtain the distribution of fibres through the wall cross section. An example of a collected image and resulting MatLab pixel map is shown in Figure 1. The image is divided in the through-thickness direction into ten equal regions of thickness t/10 (Figure 1b) and the fibre volume ratio obtained for each (Figure 1c). From this analysis, the average fibre distribution (expressed as third-order polynomial best-fit curves, each having a coefficient of determination $R2 = 0.99$) is obtained as shown in Figure 1d and reported in Table 1. It is seen in Figure 1d that the fibre distribution among the four thin-wall Phyllostachys species is very similar. Indeed, a single relationship could be given for all four species having $R2 = 0.96$. A marked difference in fibre distribution is observed in the thick-walled B. stenostachya. Thus, fibre distribution differs by genera but perhaps less so among species in the same genera.

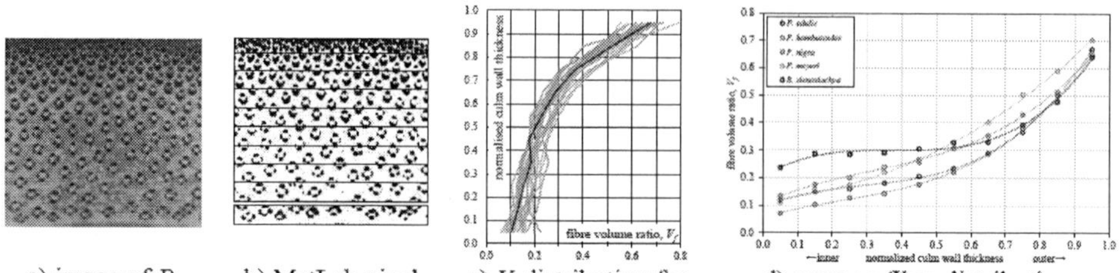

a) image of *P. edulis* culm wall | b) MatLab pixel map of fibres divided into ten layers | c) V_f distribution for 28 *P edulis* samples; black line is image in a) and b) | d) average fibre distributions obtained from image analysis

FIGURE 1 EXAMPLE OF DIGITAL IMAGE ANALYSIS OF CULM WALL [5].

Flat-ring Flexure Test

The flat ring flexure test [13] assesses the tendency of bamboo to fail via longitudinal cracks using a full cross-sectional specimen that is L ≈ 0.2D in length subjected to 4-point flexure as shown in Figure 2a. Due to the specimen geometry, the test gives the apparent modulus of rupture of the specimen (fr), as if the specimen acts as a beam in flexure with its extreme points along the diameter parallel to the loading pins (labelled N and S in Figure 2b); this value therefore, corresponds to the transverse tension capacity of the bamboo. Prior to this study, the flat-ring flexure test has been used on only full culm cross sections. Such full-culm cross section specimens are referred to in this work as the control specimens and the modulus of rupture thus obtained is denoted frC.

Modulus of rupture, frC, determined for the full-culm control specimens is reported in Table 1. Within the genus Phyllostachys, these values are similar and notably greater than that observed for B. stenostachya. Observed variation of test results is typical of bamboo and similar to that reported previously [13].

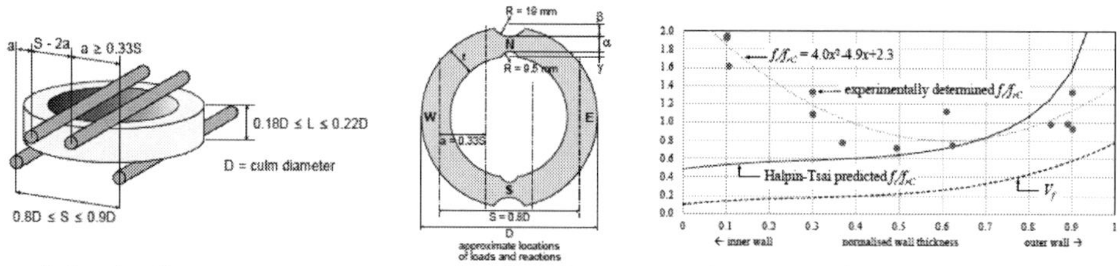

a) flat ring flexure test set-up | b) clipped specimen | c) representative results (*P. edulis*)

FIGURE 2 FLAT RING FLEXURE TEST, SPECIMEN AND REPRESENTATIVE RESULTS [5].

Clipped Flat-ring Test Specimens

This study adopts a modification to the flat-ring test specimen, in which only portions of the cross-section are tested in order to determine the effect of the material property gradient through the culm wall. 'Clipped' specimens (Figure 2b) had test regions machined using an end mill. Specimens were machined such that $\alpha \approx 0.20t$ or $0.25t$ and β and γ are in increments of approximately 0.20t or 0.25t such that the sum $\alpha + \beta + \gamma =$

t, the culm wall thickness. This approach divides the culm wall into segments for which the modulus of rupture is determined from each segment and is assumed to represent the average value for that segment; the value is then assigned to the centroid of the segment.

An example of the experimentally determined values of normalised fr/frC determined from the clipped tests for P. edulis are shown in Figure 2c and the corresponding best fit second-order polynomial relationships for all species are reported in Table 1. These all illustrate a similar trend although P. nigra specimens exhibit relatively little variation through the culm wall compared to the other species.

TABLE 1 SUMMARY OF CULM WALL FIBER DISTRIBUTION AND FLAT RING FLEXURE TEST RESULTS.

species	V_f	f_{rC} Mpa	f_r/f_{rC}	$\int \frac{f_r}{f_{rC}} dx$	f_{rE}/f_{rC}	δ_E/δ_C	$\frac{f_{rE}/\delta_E}{f_{rC}/\delta_C}$
P. edulis	$1.41x^3 - 1.23x^2 + 0.50x + 0.10$	17.3	$4.0x^2 - 4.9x + 2.3$	1.18	-	-	
P. bambusoides	$0.96x^3 - 0.91x^2 + 0.57x + 0.10$	15.7	$2.6x^2 - 2.5x + 1.5$	1.12	1.09	1.22	0.85
P. nigra	$0.94x^3 - 0.63x^2 + 0.36x + 0.06$	15.6	$0.7x^2 - 0.8x + 1.4$	1.23	0.98	0.98	1.00
P. meyeri	$0.15x^3 + 0.34x^2 + 0.17x + 0.11$	20	$2.4x^2 - 1.8x + 1.1$	1.00	0.93	1.27	0.70
B. stenostachya	$1.75x^3 - 1.98x^2 + 0.75x + 0.20$	9.4	$4.1x^2 - 3.6x + 1.6$	1.17	1.01	1.20	0.82

Effect of Outer Layer of Bamboo Culm

Integrating the fr/frC best-fit curves (Table 1) from x = 0 to x = 1 averages the local relative moduli across the section, and might be expected to represent the gross modulus across the section, i.e., the integral should equal unity. However, as shown in Table 1, with the exception of P. meyeri, the gross modulus obtained by integrating the clipped data exceeds unity by as much as 20%. A possible explanation for this behaviour – one in which the sum of the parts exceeds the capacity of the whole – is that failure of the full wall section control specimens is being initiated by a 'weak link'. A brittle failure of the outer layer of the culm wall initiating failure would explain this observation.

The extreme outer layer of a bamboo culm consists of a silica-rich outer skin (epidermis) and a thin region of densely packed fibres (this can be seen at the top of Figure 1a). It is believed that this layer will be more brittle than the rest of the culm wall and may help to initiate failures in specimens in which the outer wall is included. Therefore in the clipped specimen testing a question arises: is the outer layer contributing disproportionately to the observed behaviour, especially to the control and β = 0 tests? To investigate this effect, additional specimens were tested having β ≈ 0.05t and α ≈ 0.95t; essentially, these are full-culm sections with only the outer layer 'shaved' away. Flat ring flexure tests of this specimens having no epidermal layer were tested and the modulus of rupture, frE, and displacement of the loading points, δE determined. The ratios of these values to those obtained for full-culm control specimens are shown in Table 1. It is seen that the modulus of rupture, fr, is essentially unaffected by the removal of the outer layer.

With the exception of P. nigra, the stiffness of the specimen is observed to decrease on the order of 15 to 30% upon the removal of the outer layer (Table 1). This decrease is greater than can be attributed to the loss of 5% of the moment of inertia of the cross section (resulting from shaving the specimen) alone.

Discussion of Observed Transverse Behaviour

The data shown in Table 1 clearly indicate a parabolic distribution of modulus of rupture through the culm wall thickness with higher values at both the inner and outer walls and a minimum near the middle of the wall thickness. The fibre volume distributions, also shown in Table 1, indicate a typically observed distribution having few fibres at the inner wall and a greater number at the outer wall. Based on these fibre distributions, the predicted distribution of modulus of rupture using the Halpin-Tsai correction to the rule of mixtures [14] does not appear to capture the experimentally observed behaviour, particularly in the inner half of the culm wall where fibre volumes are lowest (Figure 2c). Therefore, bamboo does not appear to behave as a classic fibre-reinforced composite material in the direction transverse to the fibres.

The observed behaviour requires further study and may represent a material variation or morphological variation through the bamboo culm wall thickness. To investigate this further, the fracture surfaces of full-culm wall thickness in control specimens (Figure 3a) were investigated using a scanning electron microscope (SEM). Each failure plane was divided into a grid (Figure 3b) and images of each obtained, allowing the entire failure plane to be imaged. The images shown in Figures 3c and 3d are typical of images obtained near the inner and outer walls, respectively, of a P.edulis specimen. These were obtained along grid section C (Figure 3b), slightly above the neutral axis of the section in flexure, in this case. Image features did not vary considerably based on their vertical location in the specimen (A to F in Figure 3b).

In Figures 3c and 3d the failure plane can be seen to both follow the edge of the fibre bundles and also to go through the bundles themselves. The fracture surface can be seen to expose the vessels (voids) surrounded by the bundle (near top of images in Figures 3c and 3d). Where it is seen, the interface between the parenchyma and the fibre bundle appears intact in most cases, indicating that the interface between fibres in the bundle is weaker than the interface between the fibres and parenchyma. Additionally, the parenchyma shown in Figures 3c and 3d, appear to be behaving differently. Near the outer culm wall (3d), the failure appears to follow the interfaces between parenchyma cells; however near the inner culm wall (3c), the failure plane often appears to pass through the parenchyma cells (seen as non-intact cell walls in the image). This observation is typical of all images obtained in this study. Indeed, near the outer culm wall, the parenchyma is typically observed to fail in 'sheets' of intact cells as shown in Figure 3e. In other images (Figure 3f) the intact parenchyma close to the inner culm wall appear 'desiccated': the intact cell wall appears to be 'caving in' or concave rather than being slightly 'inflated' or convex nearer the outer culm wall (Figure 3e). This observation may suggest a gradient in moisture content through the culm wall or a residual effect of moisture gradient during the drying process. Such a gradient should be expected. The bamboo culm epidermis is relatively impermeable and resistant to wetting whereas the inner culm wall is permeable [1, 15].

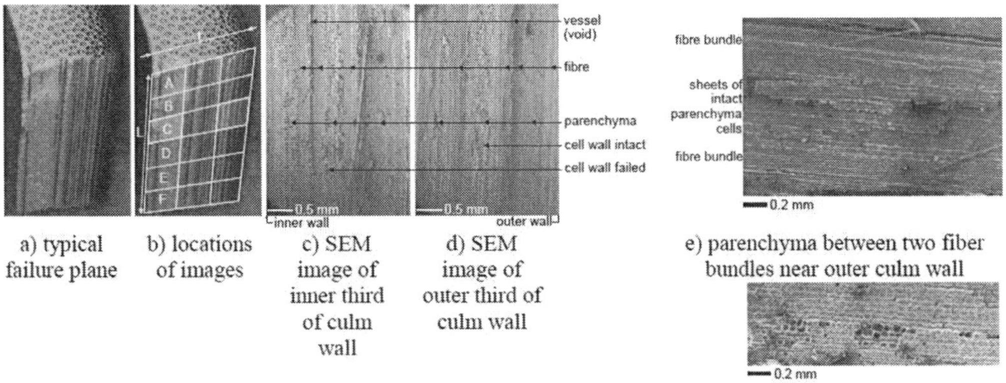

a) typical failure plane
b) locations of images
c) SEM image of inner third of culm wall
d) SEM image of outer third of culm wall
e) parenchyma between two fiber bundles near outer culm wall
f) parenchyma near inner culm wall

FIGURE 3 DETAIL OF CULM WALL IMAGES OF *P. EDULIS* SPECIMEN

Figure 4 shows an SEM image of a typical P. edulis vascular bundle (near the outer culm wall) showing the fibre bundles surrounding the vessel and the parenchyma into which the bundle is embedded. In Figure 4, the fibre bundle can be seen to be penetrated by inter-fibre cracks whereas the interfaces between fibres and parenchyma appear quite intact. It is noted that the parenchyma cell walls are relatively thick indicating a relatively old culm age [16]. The cracking of the fibres may therefore be a function of culm age (observed although not described by [16]) or have formed as a result of shrinkage associated with drying the bamboo. A recent study [17] identified significantly varying morphology of parenchyma cell walls through the culm wall thickness of P. edulis samples. Near the outer culm wall, parenchyma cell walls were tightly packed laminar structures with little annular space at interstices. Nearer the inner culm wall, the laminar structure of the cell wall was separating and a larger annular space is present at parenchyma cell interstices. It is unclear how these differences impact behaviour illustrated in Figure 3 but it is evident that parenchyma is not homogeneous through the cross section. Neither [16] nor [1] provide insight on the source of this inhomogeneity and the present authors can only speculate on its cause, although it does appear to affect the through thickness mechanical behaviour of the culm wall.

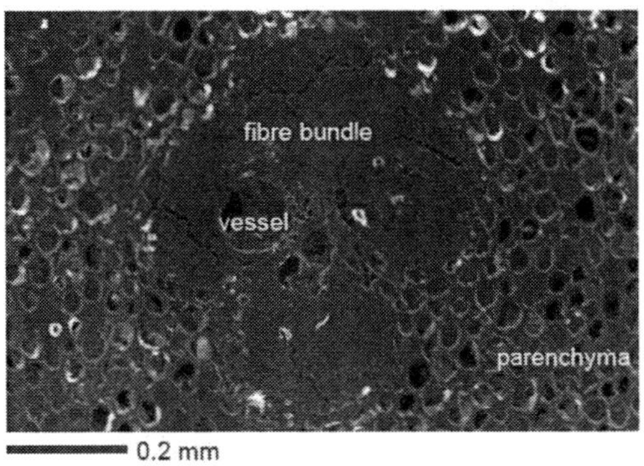

FIGURE 4 SEM IMAGE OF *P.EDULIS* VASCULAR BUNDLE

QUANTIFICATION OF MATERIAL VARIABILITY

As described above, the variation of bamboo through-culm wall properties is of interest. Such data will be used to quantify uncertainty in finite element modelling full-culm bamboo. The variation of constituent bamboo strips in glue-laminated bamboo application is also of interest to manufacturers of these products. Image analysis was used to quantify the distribution of fibre volume ratio, Vf, in strips of P. edulis bamboo used in commercially available glue-laminated bamboo structural elements.

High resolution images of cross sections of 60 glue-laminated bamboo beams were obtained. The images used in this study are those produced by Penellum et al. [17], an example is shown in Figure 5a. The commercially obtained beams – obtained from two different manufacturers (denoted M and P) – had been previously tested in flexure. The beams were fabricated from bamboo strips each of which was 19 mm wide and 6 mm thick in the direction through the culm wall. Over 3600 individual images of the 19 mm x 6 mm strips were extracted from the 60 beam section images available. Each image is approximately 900 x 300 pixels corresponding to a pixel resolution of approximately 2400 pixels/mm2, equivalent to 1200 dpi (this is available on a standard document scanner). Approximately 13% of the extracted images were excluded from further analysis due primarily to poor image quality. Additionally, approximately 3% of the strip sections were near the bamboo nodal region; the varying fibre orientation and bamboo morphology in this region is also unsuitable for the analysis conducted and so these strips were also excluded from further analysis.

Using a purpose-written MatLab script, each image was divided into ten equal sub images in the through-culm wall 6 mm direction (similar to the manner shown in 1). The fibre components were extracted from the images based on contrast and the fibre volume ratio determined. From this analysis the total fibre volume ratio of each strip, Vf, and the distribution as a function of location in the strip can be determined. Figures 5b and 5c shows examples of data obtained for M and P strips, respectively.

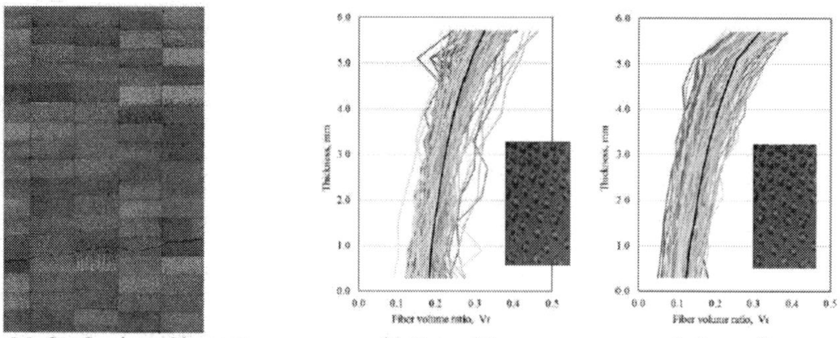

a) Group M glue-laminated beam cross section: each strip is 19 x 6 mm. b) Group M c) Group P

FIGURE 5 REPRESENTATIVE FIBER VOLUME DISTRIBUTIONS FROM 100 STRIPS.

Total fibre volume ratio, Vf obtained is given in Table 2. Distribution of fibre volume ratio through the culm wall thickness is not meaningful in so far as the actual location of the 6 mm sample within the culm wall is unknown. Nonetheless, the nature of the distribution and its variation is a measure of the uncertainty inherent in ascribing geometric or material properties to bamboo. Each acquired fibre distribution was fitted to both a linear (Vf = mx + b) and exponential relationship (Vf = bemx). The value of m, in each case, describes the variation of the fibre volume across the strip dimension x which ranges from 0 to 6 mm. The value b is not addressed as this is a function of the location of the 6 mm strip within the culm wall and is not uniquely defined in this study. The values of m determined are given in Table 2.

TABLE 2 SUMMARY OF FIBER VOLUME RATIO AND DISTRIBUTIONS DETERMINED FROM IMAGE ANALYSIS.

Group	n	fibre volume ratio		$V_f = mx + b$		$V_f = be^{mx}$	
		mean V_f	COV	mean m	COV	mean m	COV
M	2450	0.235	0.130	0.025	0.32	0.105	0.315
P	603	0.190	0.191	0.032	0.25	0.172	0.296
M+P	3053	0.226	0.160	0.026	0.31	0.118	0.387

Further work is necessary to automate the calculation of fiber volume ratio distribution in bamboo sections, particularly to conduct it directly on intact specimens. As of the writing of this article, another purpose-written MatLab script has been developed, but not yet verified, that can measure fiber volume ratio in each segment of an arbitrary "polar grid" that divides a full culm section circumferentially and through the thickness.

NUMERICAL MODELING

At the time of writing, limited progress had been made in modelling the bamboo culm wall. Linear finite element models (FEM) of the flat ring flexure specimen are developed using ABAQUS based on the results presented above. These have been validated using both flat ring flexure and circumferential compression tests [12] which use the same specimen geometry. Digital image correlation (DIC) is used in some tests (Figure 6a) to obtain full field strain data during testing. FEM complexity is also being considered with results for three modelling scenarios being considered.

Case 1: a model to which gross section material properties are assigned. While simple, such an approach does not capture the graded nature of the culm wall. An effect of this is seen in Figure 6c in which the strain gradient passes through zero at the section elastic neutral axis rather than being shifted toward the outer culm wall.

Case 2: divides the culm wall into ten concentric sections and assigns these material properties based on the approach described above in Figure 2 and Table 1.

Case 3: assigns material properties through the culm wall using similar distributions as Case 2 but using a continuum approach (graded elements) assigning properties to integration points through the culm wall thickness. This approach mitigates some of the error anticipated using discrete steps in Case 2 and will lead to smooth strain fields [19]. This approach is implemented using the ABAQUS USDFLD subroutine. Once validated, this routine will be made freely available to the bamboo research community.

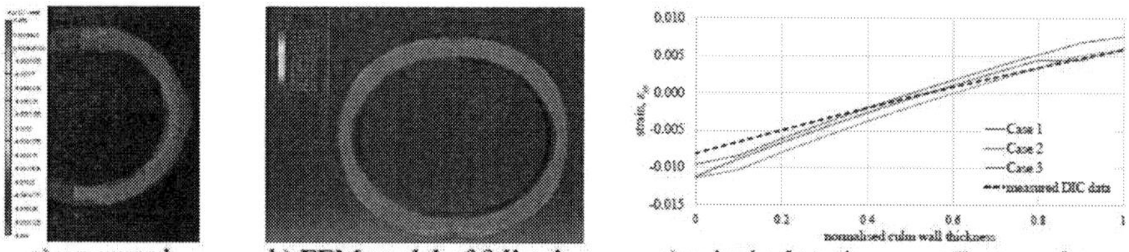

a) measured experimental DIC data b) FEM model of full culm section (Case 3 shown) c) principal strain, ε_{yy}, at East quadrant

FIGURE 6 NUMERIC MODELING OF BAMBOO CIRCUMFERENTIAL COMPRESSION TESTS SHOWING RESULTS FOR VERTICALLY ORIENTED STRAIN, ε_{yy}.

As the modelling tasks progress, the effects of uncertainty in bamboo microstructure will be integrated into the model. One objective of this activity is to begin to quantitatively understand the variation in tested bamboo materials properties – particularly why some test methods exhibit less variation than others [7, 9].

CONCLUSIONS

Performance of full-culm bamboo structures is dominated by longitudinal splitting behaviour which is a function of the transverse properties of this highly orthotropic material. Considerable study of the longitudinal properties of bamboo is available; a primary hypothesis of much of this work is that bamboo may be considered as a fibre-reinforced composite material and material properties may be assessed using rule-of-mixture methods. Nonetheless, few studies have addressed the transverse properties of the bamboo culm wall, despite these largely governing full-culm behaviour. This study investigates the transverse material property gradient through the culm wall and attempts to connect the mechanical behaviour with physical characteristics. Most importantly, the study demonstrates that the transverse behaviour is complex and that bamboo does not appear to behave as a classic fibre-reinforced composite material in the direction transverse to the fibres.

A primary objective of this project is to develop a framework and tools required to standardize the evaluation of the material and mechanical properties of full-culm bamboo. The variation inherent in bamboo from species to species and location to location is significant. The validated modelling tools developed will be shared and refined for use in a variety of international contexts so that they can be used to better interpret the considerable range of materials test data available. Coupled with a materials informatics approach [19], the opportunities for bamboo materials refinement and discovery are significant and – it is hoped – will help draw bamboo to the same degree of acceptance as comparable 'conventional' building materials.

ACKNOWLEDGEMENTS

This research was partially funded through National Science Foundation Awards 1634739 and 1634828 Collaborative Research: Full-culm Bamboo as a Full-fledged Engineering Material.

REFERENCES

[1] Liese, W. (1998) The Anatomy of Bamboo Culms, Brill. 204 pp.

[2] Amada, S., Munekata, T., Nagase, Y., Ichikawa, Y., Kirigai, A. and Zhifei, Y. (1996) The mechanical structures of bamboos in viewpoint of functionally gradient and composite materials, Journal of Composite Materials, 30, 800-819.

[3] Harries, K.A., Bumstead, J., Richard, M.J. and Trujillo, D. (2017) Geometric and Material Effects on Bamboo Buckling Behaviour, ICE Structures and Buildings Themed issue on bamboo in structures and buildings, 170(4), 236-249.

[4] Janssen, J.J.A., (2000) Designing and Building with Bamboo, INBAR Technical Report 20. International Network for Bamboo and Rattan, Beijing, China.

[5] Akinbade, Y., Harries, K.A., Flower, C., Nettleship, I., Papadopoulos, C., and Platt, S.P. (in review) Determining Through-Culm Wall Distribution of Mechanical Properties of Bamboo, Construction and Building Materials

[6] Amada, S., Ichikawa, T., Munekata, T., Nagase, Y., and Shimizu, H., (1997) Fiber texture and mechanical graded structure of bamboo, Composites: Part B, 28B, 13-20.

[7] Richard, M.J., Harries, K.A., (2015) On Inherent Bending in Tension Tests of Bamboo, Wood Science and Technology 49(1) 99-119.

[8] Janssen, J. (1981) Bamboo in Building Structures. Doctoral Dissertation, Eindhoven University of Technology, Netherlands.

[9] Richard, M., Gottron, J., Harries, K.A. and Ghavami. K. (2017) Experimental Evaluation of Longitudinal Splitting of Bamboo Flexural Components, ICE Structures and Buildings Themed issue on bamboo in structures and buildings, 170(4), 265-274

[10] Arce-Villalobos O.A. (1993) Fundamentals of the design of bamboo structures, Master's thesis, Eindhoven University of Technology, The Netherlands.

[11] Mitch, D., Harries, K.A., and Sharma, B. (2010) Characterization of Splitting Behavior of Bamboo Culms. ASCE Journal of Materials in Civil Engineering, 22, 1195-1199.

[12] Sharma, B., Harries, K.A. and Ghavami, K. (2012) Methods of Determining Transverse Mechanical Properties of Full-Culm Bamboo, Journal of Construction and Building Materials, 38, 627-637.

[13] Virgo, J., Moran, R., Harries, K.A., Garcia, J.J., and Platt, S. (2017) Flat Ring Flexure Test for Full-Culm Bamboo, Proceedings of 17th International Conference on Non-Conventional Materials and Technologies (17NOCMAT), Yucatán, México, November 2017.

[14] Halpin, J.C. and Kardos, J.L. (1976) The Halpin-Tsai Equations: A Review, Polymer Engineering and Science 16(5), 344-352.

[15] Yao, L.H., Wang, X.M. and Fei, B.H. (2011) Study on Permeability Coefficient of Different Bamboo/Fir Veneer Surface, Advanced Materials Research, 311-313, 1634-1637.

[16] Liese, W. and Weiner, W. (1996) Ageing of Bamboo Culms. A Review. Wood Science and Technology, 30, 77-89.

[17] Zheng, Q., Zhang, Q., Lu, Q., Zhou, Y., Chen, N., Rao, J. and Fan, M. (2019; in press) Wetting behavior and laminate bonding performance of profiled tangential surfaces of moso bamboo, Industrial Crops and Products

[18] Penellum A, Sharma B, Shah D.U, Foster R.M, Ramage M.H. (2018) Relationship of structure and stiffness in laminated bamboo composites. Construction and Building Materials 165, 241-246

[19] Silva, E,C,N, Walters, M.C. and Paulino, G.H. (2006) Modeling bamboo as a functionally graded material: lessons for the analysis of affordable materials, Journal of Materials Science, 41(21), 6991–7004.

[20] Zheng, C., Akinbade, Y., Nettleship, I. and Harries, K.A. (2019) The Potential Utility of Materials Informatics in Developing Non-Conventional Materials, 18th International Conference on Non-Conventional Materials and Technologies (18NOCMAT) Nairobi, Kenya, July 2019.

18th International Conference on
Non-Conventional Materials and Technologies
"Construction Materials & Technologies for Sustainability"
(18th NOCMAT 2019)
24th – 26th July 2019
Nairobi, Kenya

SCREW WITHDRAWAL CAPACITY IN FULL-CULM BAMBOO

Phoebe Morrill[1], Chase Rogers[1], Chelsea Flower[1], Yusuf Akinbade[1], Kent A. Harries[1]

[1] University of Pittsburgh, USA;

ABSTRACT

The use of screw connections in full-culm bamboo is often assumed to be limited, primarily due to the propensity for splitting of the culm. This study demonstrates that small diameter screws can be used effectively in full-culm bamboo. The study explores the withdrawal capacity of candidate screw types in order to identify those that may be used to achieve a high capacity while mitigating splitting failures. Twelve screw types of three standard sizes, ranging from hardwood screws, self-tapping wood screws and concrete anchors, are tested in conditions of both pre-drilled and self-tapping installation procedures. All tests are conducted on samples of P. edulis (Moso) having culm wall thickness on the order of 7 mm. The results of this study are intended to inform the applications for which screw connection to bamboo are viable.

Keywords: bamboo, bamboo connections, screws

INTRODUCTION

As interest in bamboo-based construction increases globally, significant efforts are underway to develop structural design codes and standards. Within this effort, developing efficient and economical connections for bamboo elements is of critical importance. While a large range of bamboo connections are possible [1], few use conventional dowel connectors (screws and nails) driven in to the culm wall. Indeed, in draft revisions to ISO 22156:2004 Bamboo Structural Design [2], the lead author of this paper proposed that screws and driven nails not be permitted in full-culm bamboo construction. This position, based on results presented by Trujillo and Malkowska [3] and those presented in this paper, has been reconsidered to permit screws but not driven nails. The study reported in this paper addresses the provision of screws for connecting bamboo in terms of the withdrawal capacity of screws embedded in the bamboo culm wall.

In full culm bamboo construction, screws may be used to secure gusset-plate [4] or pipe-sleeve [5, 6] connections. Using screws in place of through-bolts in some connections may make these connections easier to fabricate. As a secondary component of a bamboo connection, screws embedded in the culm wall may be used to secure lashing or other external confining or to secure a component of a connection. Helically-arranged screwed connections have also been proposed to transfer tension forces to a culm infill [1, 6].

Bamboo is also used in the form of strips split from a full culm requiring connections in which the use of screws would be beneficial. Connections in bahareque construction [7] are one example of this. Screwed connections to the exterior of glue-laminated bamboo also require quantification of screw withdrawal properties from the bamboo culm wall.

SCREW WITHDRAWAL

Screw withdrawal capacity derives from friction and/or mechanical interfaces for the transfer of axial (withdrawal) loads [7]. Screw (or nail or staple) withdrawal capacity from wood or wood-based engineered or composite materials can be determined based on ASTM D1761 [9] or EN 1382 [10] which are for most intents the same. In this paper, the screw withdraw capacity (in N) is denoted Fax and the 'withdrawal parameter' normalizes Fax by both screw diameter (d in mm) and depth of engaged threads (t in mm):

$$f_{ax} = F_{ax}/dt \quad (1)$$

Although many experimental programs are conducted otherwise, the thickness of the test specimen prescribed by ASTM D1761 is required to be greater than the depth of engaged threads t; that is, the screw does not protrude out the far side of the test specimen. There is limited data comparing the cases in which the screw protrudes through a thinner plywood or OSB specimen. Where such data exists [11] the withdrawal capacity for protruding cases is marginally greater than when the screw does not protrude. Such an effect is sufficiently small (and may not be statistically significant) that it should not be (and is not) considered in design.

Characteristic screw withdrawal capacity from wood is prescribed in EN 1995-1-1 [12] and, in North American practice, allowable capacity is prescribed by AWS [8]. Values of screw withdrawal are prescribed in terms of screw diameter, d, depth of embedment, t, and density of the material penetrated, ρ.

Screw Withdrawal from Bamboo

There are multiple studies which report the screw withdrawal capacity of various engineered bamboo materials [13-15] and hybrid wood/bamboo materials [14, 16]. All report a modest improvement in screw withdrawal capacity in bamboo materials as compared to comparable wood and engineered wood products.

There is only a single known study reporting screw withdrawal capacity from full-culm bamboo [3]. This study reports 240 screw withdrawal tests [10] from the walls of Guadua angustifolia Kunth culms having a reported density ρ = 755 kg/m3 (COV = 0.11) at MC = 8.6% and a wall thickness, t = 10.5 mm (COV = 0.24). Five screw types were included in the study; a summary of screw parameters and test results are given in Table 1.

TABLE 1 SCREW PARAMETERS AND SCREW WITHDRAWAL RESULTS [3] (COV IN BRACKETS IN ALL CASES)

test ID	screw type	d	n	F_{ax}	$f_{ax} = F_{ax}/dt$
		mm		N	N/mm^2
3.5-b1	self-drilling wood screw	3.5	60	1264 (0.28)	34.43 (0.17)
4.0-b1		4.0	60	1342 (0.29)	31.81 (0.11)
5.0-b1		5.0	30	1505 (0.29)	28.65 (0.12)
4.0-b2	serrated tip self-drilling wood screw	4.0	30	1296 (0.27)	30.87 (0.15)
4.8-b3	self-drilling roofing screw	4.8	60	1284 (0.24)	25.97 (0.17)

OBJECTIVE OF PRESENT STUDY

Among craftspeople working with bamboo, screws are used. These may be whatever screw type/size is available or the craftsperson may have a screw type/size which they (anecdotally) believe out performs others. The objective of this study is to survey screw withdrawal capacity in full culm bamboo based on screw type and size. The study considers twelve screw type-size combinations and considers whether these are predrilled or the screws are inserted without predrilling, resulting in 20 screw type-size-predrill combinations. The range of parameters selected recognizes that a) bamboo construction is often informal and therefore may utilize a range of screw types; and b) although predrilling may be good practice, it may not be followed if screws can be inserted without predrilling.

EXPERIMENTAL PROGRAM

A total of 216 screw withdrawal tests were conducted according to the method specified in ASTM D1761 [9] modified to account for the round geometry of the bamboo culm as shown in Figure 1. A round steel pipe section was used to support the bamboo culm wall section. The screw was centred in a 45 mm opening in the side of the pipe (Figure 1). The pipe was attached to the test machine in a manner that permitted the system to be self-centered upon application of load. For consistency, tests on plywood used a rectangular steel tube in the same test configuration as used for bamboo (Figure 1). All tests were conducted in a 4500 N capacity universal test machine equipped with a load cell providing precision of 0.1%. Tests were conducted in displacement control at a rate of 1.27 mm/min resulting in failures typically occurring in 2-3 minutes. Applied load and machine cross-head displacement was recorded.

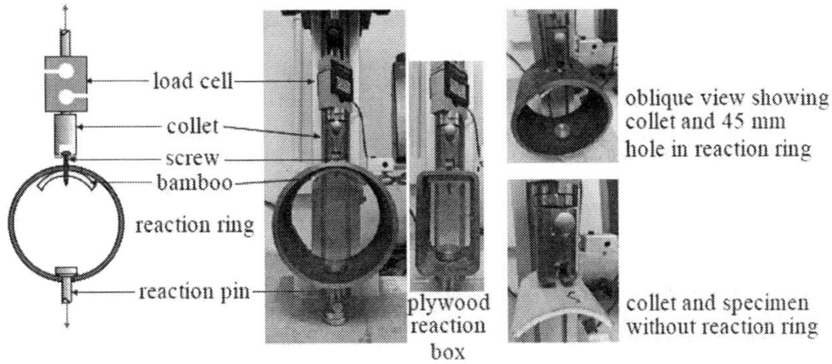

FIGURE 1 SCREW WITHDRAWAL TEST SET-UP.

All screws were installed from the outside of the culm wall, having to initially penetrate the tough bamboo epidermal layer. In a significant variation from ASTM D1761 practice, all tests were conducted with the screw completely protruding through the culm wall and engaged only over the region of the threaded full diameter of the screw. This condition would be typical of installations into bamboo culm walls.

An initial screening consisting of three tests of all 20 screw types (screw type and predrilling condition) was conducted to assess the performance of each screw type. Based on this screening, six types were selected for further testing of twenty samples each. Nineteen of the screw types were also tested using 3-ply 3/8 inch plywood to provide a comparison with typical results for wood.

Bamboo

All tests in this study were made with Phyllostachys edulis (Moso) bamboo. The culms were obtained from a commercial importer and were water treated and kiln dried. Samples ranged in diameter from approximately 65 mm to 100 mm. The culm wall thickness measured with a digital calliper, t, ranged from 4.8 mm to 10.4 mm having an average value of 7.0 mm (COV = 0.18). Moisture content determined by oven drying (ISO 22157-2019) at the time of screw insertion and testing was 7% with very little, if any, measureable variation. The bamboo was characterised as having a density, ρ = 730 kg/m3; compression strength [17], fc,0 = 48.1 MPa (COV = 0.20); a longitudinal shear strength [17], fv = 15.1 MPa (COV = 0.11) and transverse modulus of rupture [18], fr = 17.3 MPa (COV = 0.18).

Plywood

In order to provide a basis of comparison, companion tests using three-ply, 8.5 mm thick (supplied as 3/8 inch plywood) were also carried out. This material was selected as it has a similar thickness to the bamboo tested. The properties of this commercially obtained material were not experimentally determined; the allowable material properties [8] are given as: compression strength, Fc = 4.4 MPa; and 0o modulus of rupture, Fb,0 = 6.1 MPa.

Screws

Twelve commercially available steel screw types were used; these are identified as A through M (excluding I) as indicated in Table 2. Screws were selected to capture a range of thread geometry while being easily obtained in most areas of the world. Schematic drawings of each screw type are shown in Figure 2. Screws A, H and K are self-drilling versions of conventional screws B, J and L, respectively; the former were not tested in the predrilled condition. Screws G and M are concrete anchor screws; these were selected for investigation due to their deep and widely spaced threads. Screw type C has a smaller pitch than conventional wood screws (D) and has a very sharp auger tip for installation into hardwood without predrilling. Screw type C in this study is similar to the 4.0-b1 specimens reported by [3]. Screw Type F is intended for use in plastics and has a dual height thread pattern; this screw required predrilling.

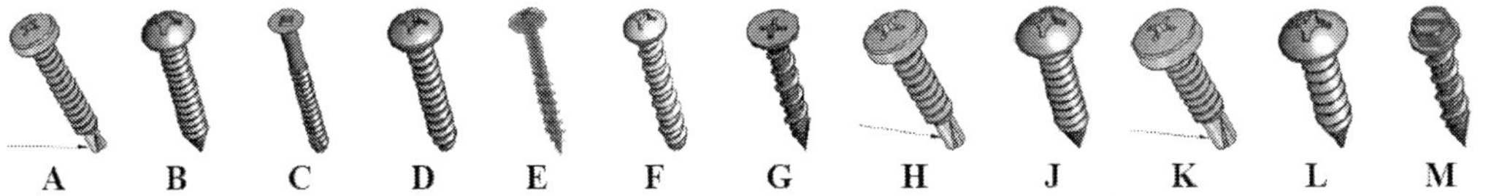

FIGURE 2 SCREW TYPES USED IN THIS STUDY (NOT TO SCALE).

Screening for Culm Splitting

Screws were inserted using a handheld rechargeable drill/screwdriver. Predrill diameters are given in Table 2. Initially, all screw types were inserted 25 mm from the cut end of a culm. There was no node between the screw and the end of the culm to arrest any splitting caused by screw insertion; inclusion of an end node is recommended in practice. The purpose of this was to screen the screw types for their likelihood of causing splitting of the culm upon insertion. None of the screw types caused splitting when they were inserted into predrilled holes. As expected, the larger diameter screws (J and L) resulted in bamboo splitting when inserted without predrilling. Interestingly, large diameter self-drilling screws (H and K) did not result in splitting.

SCREW WITHDRAWAL RESULTS

All data reported is in terms of the withdrawal parameter fax (Eq. 1). Figure 3 and Table 2 provide a summary of all tests conducted in this study. Figure 4 shows the initial three tests of each screw type (predrilled and non-predrilled); curves are offset 4 mm horizontally for clarity. The curves in Figure 4 may be considered surrogate for, and interpreted in a manner similar to, load-extraction displacement curves. As seen in Figure 4, the withdrawal behaviour is quite consistent for all screw types. The concrete anchor screws (G and M) both display more variable performance likely attributed to damage caused during insertion. The largest self-drilling screw (K) exhibits the least stiff response also suggesting damage during insertion. The withdrawal parameters for bamboo exceeded those determined for 3/8 inch plywood by between 130 and 270%, depending on screw type.

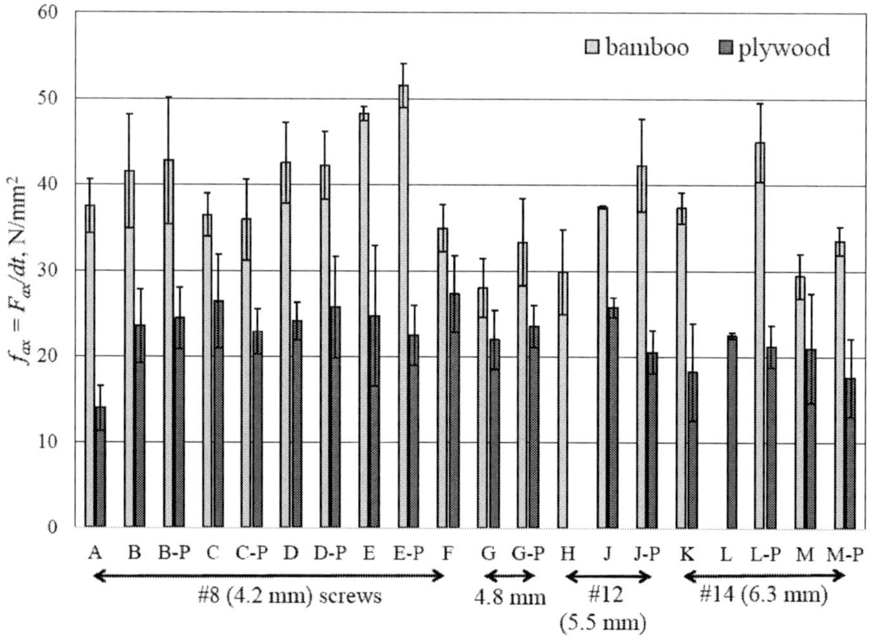

FIGURE 3 SUMMARY OF SCREW WITHDRAWAL TEST RESULTS.

TABLE 2 WITHDRAWAL TEST RESULTS.

ID	screw description	predrill dia.	d	pitch	*P. edulis*			3/8" plywood		
		mm	mm	mm	n	f_{ax} N/mm²	COV	n	f_{ax} N/mm²	COV
A	#8-18 – 1" self-drilling	none	4.166	1.41	3	37.5	0.08	3	13.9	0.19
B	#8-18 – 1" sheet metal	none	4.166	1.41	20	41.5	0.16	3	23.5	0.18
B-P		1.98			20	42.8	0.17	3	24.4	0.15
C	#8 – 1 5/8" auger point	none	4.166	1.69	3	36.5	0.07	3	26.4	0.21
C-P		1.59			3	35.9	0.13	3	22.9	0.11
D	#8 – 1" plywood/OSB	none	4.166	2.54	20	42.5	0.11	3	24.1	0.09
D-P		1.98			20	42.2	0.09	3	25.7	0.23
E	#8 – 1" sheetrock	none	4.166	2.54	4	48.2	0.02	3	24.7	0.33
E-P		1.98			3	51.5	0.05	3	22.5	0.15
F	#8 – 1" thread forming	2.38	4.166	1.41	3	34.9	0.08	3	27.3	0.16
G	3/16" – 1 ¼" concrete	none	4.775	2.54	3	28.0	0.12	3	22.0	0.16
G-P		2.38			3	33.3	0.15	3	23.5	0.10
H	#12 – 1" self-drilling	none	5.455	1.81	3	29.8	0.16	0	[2]	[2]
J	#12 – 1" sheet metal	none	5.455	1.81	3	37.4	0.00	3	25.7	0.05
J-P		3.18			20	42.3	0.13	3	20.5	0.12
K	¼" – 1 1/2" self-drilling	none	6.317	1.81	3	37.3	0.05	3	18.2	0.31
L	#14 – 1" sheet metal	none	6.317	1.81	3	[1]	[1]	3	22.4	0.02
L-P		3.18			20	45.0	0.10	3	21.2	0.12
M	¼" – 1 ¼" concrete	none	6.350	2.82	3	29.4	0.09	3	21.0	0.30
M-P		3.97			3	33.5	0.05	3	17.6	0.26

[1] screw insertion splits bamboo; no withdrawal test possible
[2] screw threaded length was inadequate to fully embed threads

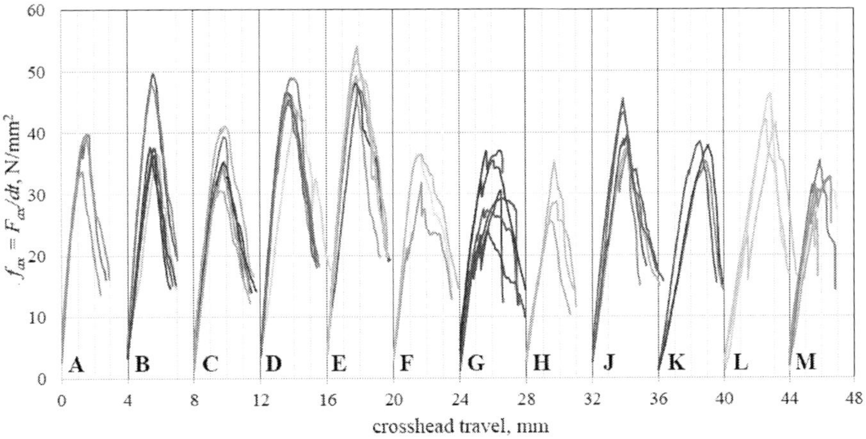

FIGURE 4 WITHDRAWAL PARAMETER VERSUS CROSSHEAD TRAVEL FOR THREE SCREENING WITHDRAWAL TESTS FOR ALL SPECIMENS. [HORIZONTAL AXES OFFSET 4 MM FOR EACH SERIES FOR CLARITY]

Characterization of Failures

Observed failures were dominated by a pull-out mode of failure (Figure 5). Due to the anisotropic nature of bamboo, this was typically followed by splitting following the peak capacity being achieved. In Figure 4, pull-out failures are characterized by a small drop in apparent stiffness prior to the peak capacity being reached and a somewhat ductile post-peak ascending curve. Splitting failures are brittle and characterized by little (if any) drop in stiffness prior to the peak load being achieved and an abrupt loss of load carrying capacity. Few specimens exhibited splitting dominated failures.

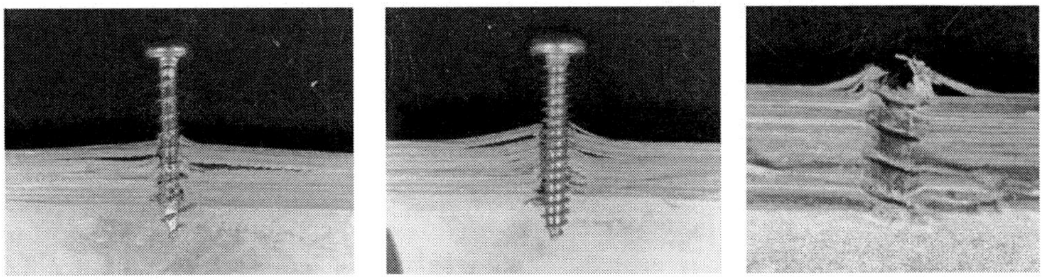

a) screw type D (t = 6.9 mm)	b) screw type B-P (t = 7.9 mm)	c) screw type D following removal

FIGURE 5 TYPICAL SCREW WITHDRAWAL TESTS (SAMPLES SPLIT FOLLOWING TEST TO SHOW EMBEDMENT).

The nature of the embedment and withdrawal behaviour is evident in Figure 5. Withdrawal capacity is primarily affected by mechanical interaction between the screw and the bamboo. The anisotropic nature of the bamboo results in interlaminar splitting, presumably initiated at the threads. Depending on thread engagement, the portions of the culm wall continue to delaminate (Figure 5a) or the outer laminates bend (Figure 5b) as the withdrawal progresses. Care should be taken examining Figure 5; these images were taken following continued post-peak displacement (Figure 4). The relatively 'soft' nature of the bamboo in terms of engaging threads is evident in Figure 5c. Here the threads are clearly seen as having engaged and cut across both the bamboo fibres and parenchymal matrix.

Discussion of Results

ANOVA evaluation of all test results was conducted. This clearly indicated that withdrawal parameter is affected by screw type. Two results in particular are informative: Firstly, there is no statistical difference between non predrilled and predrilled cases. This finding is useful in that not having to predrill screw connections into bamboo would be a preferable condition in field applications. Secondly, among the best performing screw types – those subsequently subject to 20 tests (B, B-P, D, D-P, J-P and L-P) – there is no significant difference in performance except when comparing L-P to D-P. In general, screw type L-P exhibits a greater characteristic withdrawal capacity than the other conditions tested.

Contrary to the trend reported for wood construction [19], self-drilling screws (types A, H and K) exhibited lower capacities than comparable regular-point screws (B, J and L) whether the latter was predrilled or not. This observation suggests that the drilling tip on self-drilling metal screws is inappropriate and may cause damage to the culm wall. The auger tip on screw type C, on the other hand appeared to perform well, easily penetrating the bamboo epidermis. The performance of screw type C, however, may have been limited by it relatively smaller thread depth.

Screw types D and E have identical geometry. Screw type D is zinc-plated steel to resist corrosion in wet environments while type E is black oxide steel suitable only for dry environments. The difference in performance between D and E is statistically significant ($p < 0.03$ for all variations) suggesting an effect of screw surface treatment on withdrawal capacity. This parameter requires further study.

When interpreting the results presented it is important to recall that these are presented for a single bamboo species (P. edulis) from a single batch having both uniform density (ρ = 730 kg/m3) and moisture content (MC = 7%). Greater withdrawal capacities are reported in the present study than were reported previously [3] for Guadua bamboo. Comparing essentially identical screw type C from this study and 4.0-b1 from [3] (Table 1), the former is observed to have a withdrawal parameter 115% the latter (t-test p-value = 0.03). The difference is hypothesized to be attributable to species morphology.

CONCLUSION

A series of screw withdrawal tests from P. edulis bamboo has demonstrated capacities and behaviour suitable for structural applications. For the better-performing screws, average withdrawal parameters exceeding fax = 40 N/mm2 were achieved; twice that determined for 3/8 inch, 3 ply plywood. Additionally, less variation was observed in bamboo withdrawal tests than in plywood tests. In order mitigate splitting upon screw insertion in to bamboo, screw diameters generally less than 6 mm were required unless the screws are inserted into predrilled holes. There was no advantage observed to using self-drilling screws; indeed, the self-drilling screw tip may cause more damage to the bamboo than a simple sharp screw of the same size. The lubricating and/or friction effect of the screw coating is a factor not considered widely in this study although results suggest that this is a factor that should be included in further study.

ACKNOWLEDGEMENTS

This research was partially funded through National Science Foundation Award 1634739 Collaborative Research: Full-culm Bamboo as a Full-fledged Engineering Material.

REFERENCES

[1] Widyowijatnoko, A. and Harries, K.A. (2019, in press) Chapter 20: Joints in Bamboo Construction in Nonconventional and Vernacular Construction Materials: Characterisation, Properties and Applications, 2nd edition. K.A. Harries and B. Sharma, editors. Woodhead (Elsevier) Publishing Series in Civil and Structural Engineering

[2] ISO 22156:2004 Bamboo Structural Design, International Organisation for Standardisation (ISO) [presently undergoing extensive revision chaired by Harries]

[3] Trujillo. D.J.A. and Malkowska, D. (2018) Empirically derived connection design properties for Guadua bamboo. Construction and Building Materials 163 (2018) 9–20.

[4] Morisco and Mardjono, F. (1996) Strength of Filled Bamboo Joint, 5th International Bamboo Workshop, Ubud, Bali (P. Ganapathy, J. Janssen and C. Sastry, editors) pp. 113-120.

[5] Forero, E. H. (2003) Uniones a Tension en Guadua con Mortero y Varilla [Mortar and Rod Bamboo Joints in Tension], Undergraduate Dissertation, Departemento de Ingeniera Civil y Agricola, Universidad Nacional de Colombia.

[6] Clavijo, O., and Trujillo, S. (2000) Evaluacion de Uniones a Traccion en Guadua [Evaluation of Bamboo Joints in Tension]. Santaf e de Bogota: Facultad de Ingenieria, Universidad Nacional de Colombia.

[7] Gonzalez G, Gutierrez J (2005) Structural performance of bamboo 'bahareque' walls under cyclic load. Journal of Bamboo and Rattan 4(4), 353–368

[8] American Wood Council (AWC) (2015) Allowable Stress Design (ASD)/LRFD Manual for Engineered Wood Construction, American Wood Council

[9] ASTM D1761-12 Standard Test Methods for Mechanical Fasteners in Wood. ASTM International.

[10] EN 1382:2016 Timber Structures –Test methods – Withdrawal capacity of timber fasteners, CEN (European Committee for Standardization).

[11] Erdil, Y.Z., Zhang, J. and Eckelman, C.A. (2002) Holding Strength of Screws in Plywood and Oriented Strand Board, Forest Products Journal, 52(6), 55-62.

[12] EuroCode 5: EN 1995-1-1 (2008) (English): Eurocode 5: Design of timber structures - Part 1-1: General - Common rules and rules for buildings

[13] Chen, Y., Zhu, S., Guo, Y., Liu, S., Tu, D. and Fan, H. (2016) Investigation on Withdrawal Resistance of Screws in Reconstituted Bamboo Lumber, Wood Research 61(5), 799-810.

[14] de Melo, R.R., Strangerlin, D.M., Santana, R.R.C and Pedrosa, T.D. (2014), Physical and Mechanical Properties of Particleboard Manufactured from Wood, Bamboo and Rice Husk, Materials Research 17(3), 682-686

[15] Guo, Y., Zhu, S., Chen, Y. and Qiqiqin, S. (2018) Contrastive Analysis of Screw Withdrawal Resistance Between Bamboo Oriented Strandboard and Conventional Particleboard. Wood Research, 63(6), 1071-1080

[16] Nurhazwami, O., Jawaid, M., Tahir Paridah, M. Juliana, A.H. and Hamid, S.A. (2016) Hybrid Particleboard Made from Bamboo (D. asper) Veneer Waste and Rubberwood (H. brasilienses), BioResources, 11(1), 306-323.

[17] ISO 22157:2019 Bamboo structures — Determination of physical and mechanical properties of bamboo culms — Test methods, International Organisation for Standardisation (ISO)

[18] Virgo, J., Moran, R., Harries, K.A., Garcia, J.J., and Platt, S. (2018) Flat Ring Flexure Test for Full-Culm Bamboo, Materials Research Proceedings, 7, 349-358.

[19] Soltis, L.A. (1999) Chapter 7: Fastenings in FPL-GTR-113 Wood Handbook, United Stated Department of Agriculture Forest Service, Forest Products Laboratory.

NON-CONVENTIONAL MATERIALS / TECHNOLOGIES & RESILIENCE

18th International Conference on
Non-Conventional Materials and Technologies
"Construction Materials & Technologies for Sustainability"
(18th NOCMAT 2019)
24th – 26th July 2019
Nairobi, Kenya

RWANDA INSTITUTE FOR CONSERVATION AGRICULTURE: LESSONS LEARNT DESIGNING EARTH WALLS TO CODE IN A MODERATE SEISMIC ZONE

Rosie Goldrick[1], James Kitchin[2], Aimable Mukire[3], Will Arnold[4], Jorge Lopez[5], Tim White[6]

[1] Engineering Director, MASS Design Group, Rwanda;

[2] Associate Engineer, MASS Design Group, Rwanda;

[3] Graduate Engineer, MASS Design Group, Rwanda;

[4] Senior Engineer, Arup, UK, seconded to MASS Design Group, Rwanda;

[5] Senior Engineer, Arup International Development, UK;

[6] Senior Consultant, Arup International Development, UK;

ABSTRACT

Due to open in September 2019, the Rwanda Institute for Conservation Agriculture will be a new 20,000m^2 institution in the Eastern province of Rwanda that will prepare future entrepreneurs with the technical and practical skills to create solutions for more sustainable food production in Rwanda. With 80% of the country's population relying on agriculture for their livelihood, yet 20% of the population being food insecure, there is a pressing need to promote more sustainable agricultural techniques. The campus design brief therefore demanded a high-quality institute that would push the boundaries of sustainable design. Consisting of buildings up to two storeys and having determined the site soil to be suitable for construction, the walls are built almost entirely from site produced reinforced earth blocks and reinforced rammed earth. Funded by the Howard G. Buffett Foundation and designed in collaboration by MASS Design Group and Arup, the team made reference to both Eurocodes and New Zealand Standards to ensure that the structures would resist the moderate seismicity present in Rwanda, whilst also aiming to minimise embodied carbon from the use of cement and steel reinforcement. The paper describes the project design process, highlighting the limitations of the codes and the challenges faced by engineers wishing to design sustainable, affordable, and buildable earth structures in moderate seismic zones. The paper concludes by suggesting areas for further work.

Keywords: Earth, CSEB, Rammed Earth, Seismic, Code

ACKNOWLEDGEMENTS

The authors would like to acknowledge the contribution of members of the design and site teams to this paper, including Nelson Habintwari, Harriet Kirk, Obed Sekamana, Shakira Nyiratuza, Sherryen Mutoka and Chris Hardy from MASS Design Group and Katherine Coates and Caroline Ray of Arup. This work has built on previous partnerships between MASS and Arup that have investigated earth block and rammed earth construction.

INTRODUCTION

Is it possible to design earth buildings to be resilient to moderate seismicity, whilst maintaining the earth's qualities as a sustainable, affordable, and buildable material? This was the question posed to the design team at the start of the project. Due to open in September 2019, the Rwanda Institute for Conservation Agriculture (RICA) will train the next generation of agricultural leaders to adopt and promote conservation agriculture to support sustainable food production. The campus construction and first five years of the institute's operations are being made possible by a $125 million commitment from the Howard G. Buffett Foundation as well as the donation of a 1500 hectare site by the Government of Rwanda. This paper describes the project design process, the challenges faced by the structural engineers aiming to meet this brief and the lessons learnt. It concludes by suggesting areas for further work.

MASS Design Group is a non-profit design collective that began in 2008 with a mission to research, build, and advocate for a built environment that promotes justice and human dignity. Arup International Development is a specialist team within Arup working exclusively with humanitarian and development organisations. MASS has recruited its 24 engineers through a combination of partnerships with Arup, Engineers Without Borders and local and regional hires. It currently boasts geotechnical, structural, civil and MEP expertise. The institute's structural design was a collaboration between MASS and Arup; Arup led the structural engineering during schematic design, and then provided a secondee structural leader and remote technical support to MASS during detailed design and construction. In addition to structural engineering, MASS also carried out architectural design, building services, and oversaw construction of the campus.

Land pressure, deforestation, degraded soils and climate change are creating a food shortage in Rwanda, Africa's most densely populated country. With 80% of the country's population relying on agriculture for their livelihood, there is a pressing need to promote more sustainable agricultural techniques. Through a curriculum based on Conservation Agriculture and One Health principles, the institute intends to teach students to be socially responsible, innovative and entrepreneurial, and provide them with the practical skills required to transform local and regional farming practices.

The design brief for the institute demanded high quality, highly sustainable buildings that would stand the test of time. It was decided to explore extensive use of natural materials, including locally and sustainably sourced timber for roofing, natural stone for the foundations and rammed earth (RE) and compressed stabilised earth blocks (CSEB) for the walls. Testing of the soil on site confirmed its suitability for CSEB and RE, thereby eliminating the cost of importing soil, reducing the carbon footprint and providing the local community the opportunity to benefit from training in earth construction. Whilst 90% of Rwandans live in earthen homes (EICV5 [1]) the use of earth was until recently restricted by the building code and may have contributed to it being under utilised in the formal construction sector. The current demand for housing has changed government perceptions on earth construction and it is now accepted that a range of material options will be required. Whilst earthen construction is affordable and sustainable it is also low strength, brittle and heavy, and performs notoriously poorly in earthquakes. With Rwanda exposed to moderate seismicity associated with the East African Rift System there is an immediate need for low cost, sustainable solutions that can be built locally and are designed to resist moderate seismicity.

DESIGN APPROACH

Codes

Aiming to build a world class facility, the design team needed to ensure that codified performance criteria such as a 50 year design life and seismic resistance were met. As a first step the team conducted a review of available codes, both locally and internationally. With the site in a zone of moderate seismicity it was necessary to use codes that would cover seismic design of earth construction as well as more commonly used structural materials such as timber and reinforced concrete.

The Rwanda Building Code (RBC) [2] is a performance-based design code, which defines the minimum requirements but allows alternative solutions to be developed by the designer. There is no Rwandan standard for seismic analysis and design, instead the RBC references the generic international standard ISO 3010 [3] for the seismic basis of design. ISO 3010, however, does not give specific values for certain factors to determine the seismic loading to be used in analysis and design. Therefore, it is not possible to design a structure to ISO 3010 alone.

The RBC references the Rwanda Standard RS 112 (Basis of structural design) [4] for load combinations and partial safety factors. RS 112 is largely based on the Eurocodes. While a number of other standards (for example British, American and South African) are accepted in Rwanda, it was decided to use Eurocodes as they provide a full suite of coordinated structural design standards, including seismic design. It was decided to use EN 1996 [5] for the design of CSEB walls, however earth construction is not explicitly covered, raising questions about its applicability. Although CSEB are covered in the NZS codes, provisions were found to be slightly less stringent and less detailed compared to EN 1998-1 [6] (Table 1). Given the intention to follow Eurocodes where ever possible it was decided to refer to EN 1998 1 for design of CSEB. For the design of RE walls, alternative design codes were sought as Eurocodes do not cover such construction methods. The New Zealand standards NZS 4297 [7], NZS 4298 [8] and NZS 4299 [9] were used for the design, testing and construction of RE walls, taking due consideration of the required consistency with the other Eurocodes used, in particular EN 1998-1.

TABLE 1 - MINIMUM CODE REQUIREMENTS COMPARISON

Minimum Code Requirement	EN 1998-1 - Reinforced masonry	NZS 4297 - Cinva Bricks (CSEBs)
Minimum thickness	240mm	130mm
Maximum slenderness ratio	15	16
Minimum compressive strength normal to face	$5N/mm^2$	$3.6N/mm^2$
Minimum horizontal bar diameter and spacing	B4-600	B6-600
Minimum horizontal reinforcement	0.05% of the gross wall section	N/A
Minimum vertical reinforcement	0.08% of gross plan area	N/A

Design hazard

The RBC defines a maximum credible peak ground acceleration (PGA) in the country equal to 0.163g for a 475 year return period whilst a recent seismic hazard assessment performed for East Africa by Arup (Lubkowski et al. [10]) identified that the PGA at the site is 0.15g. 0.163g was used in the design calculations. Ground investigation data identified the seismic site classification as Ground Type B. The buildings were classified as per their 'seismic' importance (Class I - III). Type 2 response spectrum was adopted at conceptual design stage given the moderate seismicity context.

Whilst the RBC PGA of 0.163g is just above the minimum for which EN 1998-1 Cl. 9.3(3) allows the use of unreinforced masonry, a PGA of 0.15g is just within. Recognising the project's aim to develop low cost, low carbon, locally buildable solutions it was decided that lower risk structures on site, such as single storey buildings with closely spaced walls, would be unreinforced. Any building of two storeys, with taller or longer walls would be reinforced with internal steel rebar as per the provisions of EN1998-1.

Behaviour (ductility) factors

At this stage it was also decided to assume a behaviour factor of 1, implying fully elastic design, for all earth construction across the project. NZS codes recommend 1 and up to 2 for unreinforced and reinforced Rammed Earth respectively. EN 1998-1 allows 1.5 and 2.5 for unreinforced and reinforced masonry respectively. The decision to use 1 was made for a number of reasons, namely uncertainty over the providence of the NZ factors, the fact that CSEB are not explicitly covered by Eurocodes and uncertainty over eventual material and workmanship quality. It also serves to provide additional capacity to unreinforced earth walls.

Strength and slenderness

The CSEB and RE walls were designed to have the same strength. Masonry panels are required by EN 1998-1 to have a minimum normalised compressive strength of 5N/mm2, with the same requirement placed on the RE. Both the CSEBs and RE were stabilised with Pozzolana Portland Cement in order to improve their strength and durability. Slenderness limits recommended in EN 1998-1 Table 9.2 and NZS 4297 Cl. 4.6.2 were used for CSEB and RE respectively. The Auram press used on site produces blocks 240mm wide, equal to the minimum allowed by EN 1998-1, and so most walls were a single block. For taller walls a second layer of CSEB was required (490mm thick in total) in order to satisfy the H/15 slenderness limit. According to NZS 4297 the site PGA approximately equated to an Earthquake Zone Factor of >0.6 (albeit the lower end), requiring use of the more onerous slenderness limits for RE according to Cl. 4.6.2. The minimum width of the RE walls was based on a practical minimum width of 500mm to allow comfortable forming and ramming. Unreinforced RE walls have a maximum slenderness ratio of H/6, which for a 500mm thick wall limits the height to 3m. Reinforced RE have a maximum slenderness ratio of H/10 and therefore a maximum height of 5m.

A cradle to site embodied carbon comparison of different 3.3m unreinforced wall types was performed and is presented in Figure 1. Fired brick and hollow concrete blocks are included in Figure 1 to represent typical walling materials. The embodied carbon of CSEB is less than half of fired brick or hollow concrete blocks, whereas RE is about 20% lower which is predominantly due to the thickness.

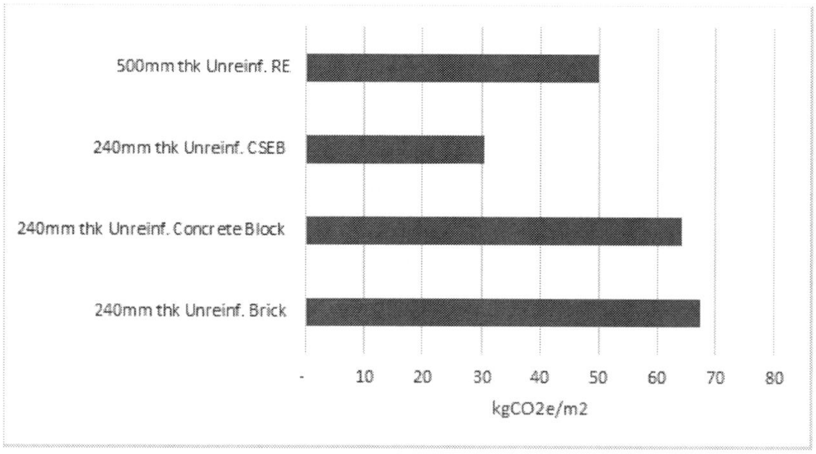

FIGURE 1 - COMPARISON OF CRADLE TO SITE EQUIVALENT EMBODIED CARBON FOR DIFFERENT WALL TYPES

Regular buildings and analysis

With detailed design to be carried out in MASS's Kigali office for 50+ buildings to a rapid design schedule it was important to develop a family of structural systems that could be replicated across the site. This robust design approach also recognises the lack of available material design data. The structural design team worked closely with the architects to 'design out' as much of the seismic risk as possible. Seismic joints were utilised to ensure regularity on plan and elevation. Approximate symmetry in both directions was determined by limiting the difference in stiffnesses between the two orthogonal directions to 30%. By adhering to the Criteria for Structural Regularity set out in EN 1998-1 Cl 4.2.3 the structures could be analysed using equivalent lateral force method, enabling checks to be understood by trainee engineers and performed quickly. This approach recognises that complex and lengthy analysis is not appropriate for cost effective and replicable buildings. The period of the buildings were approximated using the method in EN 1998-1 which led to the plateau of the elastic response spectrum being used for all building analysis (Figure 2).

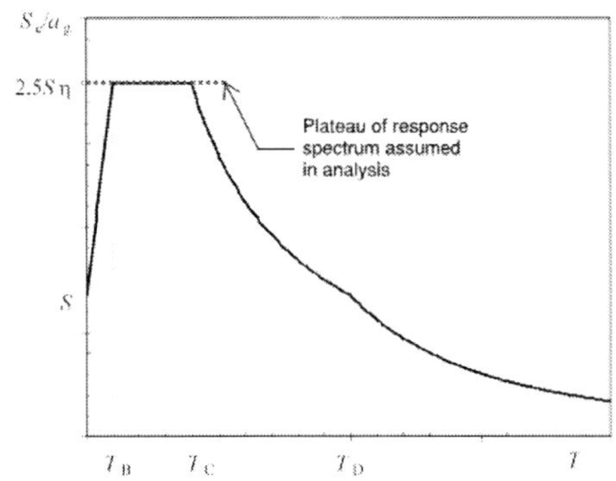

FIGURE 2 - ANNOTATED RESPONSE SPECTRUM FROM EN 1998-1

Design overview

With the design brief looking to minimise embodied carbon, local sustainably sourced timber was to be used for the roof system. Strip footings under walls were to be made from mortared stone masonry, also locally sourced and commonly available in Rwanda. All buildings were to have a reinforced concrete (RC) grade beam to raise the base of the earth walls clear of the ground for durability purposes. A roof overhang of H/3 was provided on all sides to protect the walls from driven rain. Where this could not be achieved or walls were exposed to other sources of water they are rendered or clad with a ventilated cavity between the cladding and the earth walls. RC was to be used sparingly, leading the team to test the limits of timber in its place, for example as a ring-beam at the tops of walls. Structures on the campus fell into two broad categories with structural stability and resistance of vertical and lateral loading provided by CSEB and Rammed Earth walls for both.

The first consisted of walls at regular spacings, typically less than 4m apart, single storey and around 3m tall - known as 'Residential Buildings'. Residential buildings were made from un-reinforced earth walls with timber ring beams provided at roof level. The roof diaphragm was provided by timber planks, laid diagonally and nailed into timber roof structure for improved stiffness using code AWC SDPWS [11].

The second category known as 'Enterprise Buildings' consisted of much larger spaces, with greater spans and taller walls, often two storey, containing a variety of agricultural and educational functions. The lower wall density and taller walls of the Enterprises necessitated that the walls be reinforced with steel bars tying into RC ring-beams provided at the base, roof and floor level (where present). Walls were designed to span vertically between ring-beams, allowing full height openings and greater freedom in where they are located. At the base the RC ring-beam serves to transmit shear into the base and tie the base of the walls together. At floor level RC ring-beams and RC floors provide a diaphragm. Roof level diaphragms are provided by steel bracing in each bay.

DETAILED DESIGN APPROACH

During detailed design, and following on from Criteria for Structural Regularity set out in EN 1998-1 Cl 4.2.3, the team conducted a number of more detailed checks. Shear was found to govern for RE and CSEB for all of the buildings. In plane checks relating to global stability are discussed, in particular with regard to overturning of shorter walls and the implications for un-reinforced walls. Walls were also checked locally for out of plane effects. Timber ring-beams at tops of walls are also discussed here in relation to the challenges they posed during detailing.

Shear analysis

Generally both the CSEB and RE walls were governed by shear. The characteristic shear strength of unreinforced masonry is assumed to be 0.2N/mm2 based on values provided in EN 1996-1 where the mortar strength governs. A small benefit in shear resistance is realised by accounting for the compressive stress perpendicular to the shear stress.

The unreinforced CSEB requires 750mm tall B12 dowels to anchor the wall into the grade beam, as the damp proof course at the bottom of the wall is assumed to have no frictional capacity. The shear in the unreinforced masonry wall was checked both at the base (resisted by the dowels) and in the wall just above the dowels.

For reinforced CSEB 2no. B8 horizontal reinforcement bars at 500mm vertical centres provide significant additional shear capacity, typically accounting for 72% of the total shear resistance. In no case did the shear resistance exceed the 2N/mm2 limit set in EN 1996-1.

A minimum of 5no. RE samples per mix, per day were tested for unconfined compressive strength in accordance with NZS 4298, enabling classification as special grade earth wall. According to NZS 4297 this enables shear strength of the RE to be taken as 7% of the compressive strength, 0.35N/mm2, compared to 0.08N/mm2 for a standard grade wall, that has undergone less rigorous testing.

In-plane bending analysis

For global stability analysis the forces at each level were distributed assuming a rigid diaphragm. An accidental torsion factor was applied to the seismic force on each wall based on the relationship between the centre of gravity and distance to bracing element as specified in EN 1998-1 Cl. 4.3.3.2.4. It was determined that nominally unreinforced RE and CSEB walls would still require reinforcement in order to prevent overturning and remain stable in a seismic event. To prevent this B16 'tie-downs' were used at either end of walls to act in tension and prevent overturning. Overall the nominally unreinforced walls contained significantly less reinforcement than the reinforced walls. The 'tie-downs' could have been reduced if wall lengths increased but this was limited by architectural requirements and movement joints.

FIGURE 3 - UNREINFORCED CSEB 'TIE-DOWNS'

For reinforced CSEB walls axial and major axis moment interaction curves were calculated for all wall types to allow the required reinforcement to be quickly determined. The curves for a 240mm thick CSEB wall are presented in Figure 4. Where possible design decisions were taken to limit the vertical reinforcement to 12mm diameter bars at 500mm centres, in line with EN 1998-1 minimum rebar, by lengthening walls or classifying walls as secondary seismic elements. Typically axial loads were less than 50kN/m, and moments less than 200kNm/m in the two storey buildings, showing this is not often the critical check for walls greater than 4m.

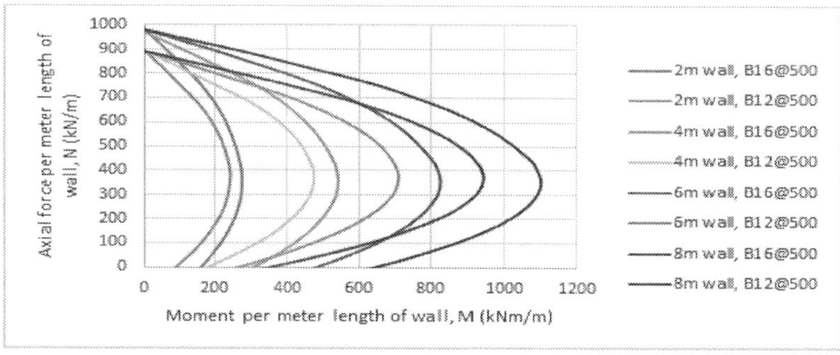

FIGURE 4 - N-M INTERACTION CURVES FOR 240MM THICK CSEB WALLS OF VARYING LENGTH AND VERTICAL REINFORCEMENT

Out of plane wall analysis

Walls were also checked to ensure they were stable out of plane under seismic and wind loading. The reinforced CSEB walls are designed to span vertically out of plane, whereas the unreinforced CSEB walls span in both directions and typically rely on return walls or grouted piers. Reinforced and unreinforced RE walls are both designed to span top to bottom, with seismic force generated from their own mass governing design.

Reinforced CSEB walls were designed using the approach in EN 1996 and ignores the contribution from masonry flexural strength or arching. The seismic force equates to 2.65kN/m2 for a 240mm thick CSEB wall. With minimum reinforcement the typical utilisation of CSEB out of plane failure was 35%.

Reinforced RE walls were designed to have B16 full height vertical reinforcement at 500mm centres and B12 horizontal reinforcement at 500mm centres. As no ductility was taken in the design of the walls, no minimum reinforcement was required by code, even for crack control, and instead the walls were designed from first principles with size and spacing based on out of plane bending calculations. There is limited history of engineered reinforced rammed earth, and so there is varied guidance on detailing requirements. NZS 4299 Cl. 5.9 recommends the reinforcement sits in a grouted tube. The intention of this was unclear, and there was concern that this could result in shrinkage of walls leaving a gap between the wall and the ring beam, and so bars were to be in direct contact with the earth instead. Whilst NZS 4297 Table 4.1 provides for bond between steel rebar and RE the relationship between bond strength and other factors such as soil mix, stabilisation content etc was unclear to the team. It was deemed prudent to assume no bond and design the wall as a tied arch instead. When coming to detail the reinforcement the absence of earth to rebar bond ruled out any laps, with mechanical couplers unavailable, 5m high bars were required. This approach was backed up somewhat by NZS 4297 Cl 8.3.4.1 which states that splices shall be welded or mechanical laps only, unless the reinforcement is in a grouted tube.

FIGURE 5 - FULL HEIGHT RAMMED EARTH WALL REINFORCEMENT

Unreinforced RE was checked against a method set out in NZS 4297. It involves plotting the response acceleration against wall displacement and using the equal energy principle to determine the equivalent elastic response acceleration at which failure is considered to occur. Practically this means calculating the horizontal wall deflection and associated acceleration in 4 situations: initial cracking when there is zero compression at one face, when wall is half cracked, when wall is three quarters cracked and when the wall is at failure. The wall displacement is calculated assuming a modulus of elasticity 300 times the compressive strength as stated in NZS 4297 which also suggests this will be a low average estimate. The wall is considered to have failed when the stress exceeds 85% of the compressive strength. The response acceleration calculated to induce failure is 1.74g compared to a design acceleration of 0.55g, giving a significant safety margin.

Timber ring beam

In order to simplify construction of residential buildings the option to omit an engineered diaphragm was explored, with NZS 4299 stating that this would necessitate an RC ring-beam, a provision verified somewhat by calculations determining that a timber ring beam would need to be 300mm x 300mm, an impractically large section. Given the cost and embodied carbon of RC it was ultimately decided to utilise diagonally placed timber ceiling boards as a diaphragm, enabling a reasonably sized timber ring beam.

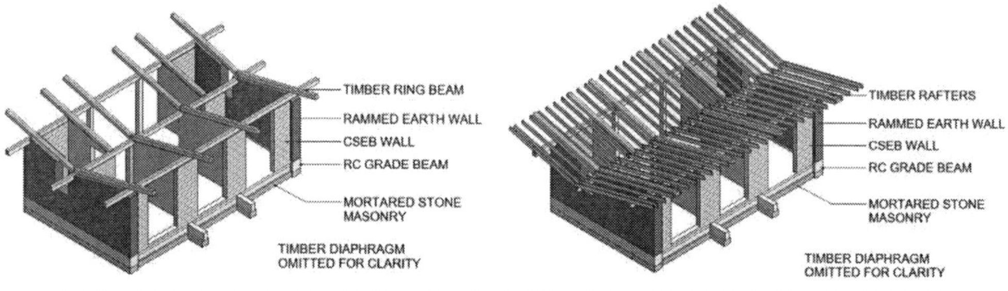

FIGURE 6 - ANNOTATED 3D IMAGE OF TYPICAL RESIDENTIAL BUILDING

CHALLENGES

Throughout the design and construction process the team faced a number of challenges which are detailed below. These include the inherent variability of earth as a construction material, the lack of available design data, uncertainty over what data is available and the infrequent use of the material in the formal construction sector.

Material properties

The availability of material data which engineers rely upon during design is taken for granted when designing in concrete, steel and timber. This is not the case for CSEBs and RE where a lack of data was a key source of uncertainty in the design. Whilst there is a significant body of academic research into earth construction, relatively limited codified guidance is available and where data does exist there was often uncertainty over its veracity/applicability to the project. Whilst testing laboratories do exist in Rwanda they are better equipped for testing concrete and road construction. In particular the design was affected by uncertainty over appropriate bond strength between rebar and earth, modulus of elasticity, movement joints in RE panels, and ductility behaviour factors.

Although the addition of cement to CSEBs and RE can significantly impact material properties the team found that available guidance would often only give a single value for a material property, and that it would not always be clear as to whether this applied to stabilised or unstabilised earth. This stands in contrast to materials such as concrete and timber, whose variability are recognised through different strength/stiffness grades.

Eurocode applicability for CSEB

The applicability of EN 1996 and EN 1998-1 to CSEB walls is not clear with the blocks not falling directly into any of the existing allowed units allowed by EN 1996-1. Whilst properties such as compressive strength and water absorption of CSEBs can be demonstrated to fall within EN 1996-1, literature suggests that other properties are less comparable, with earth construction having a significantly lower modulus of elasticity for the same compressive strength. EN 1996 recommends the modulus of elasticity to be 1000 times the characteristic compressive strength, equating to 5000N/mm2 for 5N/mm2 CSEBs. By comparison Auroville [13] and NZS give the modulus of elasticity of CSEBs as between 700N/mm2 and 1000N/mm2 for a block of similar strength. Uncertainty over the applicability of ductility behaviour factors, and decision to design fully elastically was to have limited impact on the design. For example, a ductility factor of 2.5 for reinforced CSEBs may have allowed (taller) walls that are two blocks wide to be reduced to a single block, but would then have exceeded slenderness limits.

Rebar Lapping in CSEB

Whilst masonry is common in Rwanda, the use of vertical reinforcement is not, requiring additional effort to co-ordinate its location when casting the RC ring-beams which sit at the base and at the top of the wall. CSEBs were formed in a way that avoided the need to thread each block down from the top of a vertical bar, improving buildability. In terms of continuity of vertical bars, the contractor's preference was to avoid a single full height bar, which is difficult to hold in place whilst casting the foundation and harder to build a wall around. With a mortar strength of 5N/mm2 this meant a lap length of 1.3m and so as one lap finished another would start almost immediately.

Timber Diaphragm Connections

Whereas the reinforcement in an RC ring-beam provides inherent continuity and is relatively straightforward to detail and construct, a timber ring-beam is made up of a number of discreet elements which must be connected together through bolts, screws or nails. In order to transfer seismic loads out of the unreinforced walls and into the timber ring-beam a series of 500mm B12 steel reinforcement anchors were required. They were installed with threaded ends so that they passed through holes drilled in the timber ring-beam, allowing it to be secured in place. Locating the anchor bars and the holes in the timber ring-beam so that they aligned required close attention on site. The timber ring-beam was then connected to the timber diaphragm through screws into timber blocking. This was another detail that proved tricky to detail and construct, in particular ensuring sufficient screws could be accommodated.

CSEB and RE Production

The scale of the project generated a number of challenges for the onsite manufacture and quality control of CSEBs and RE. With an estimated 700,000 CSEBs required to complete the project, it was determined that 6 block presses and a crew of 60 people were required to produce 4800 blocks per day. Blocks were then protected from rain and sun whilst they cure, requiring a significant amount of dry space. After curing they were transported to the building site for which they would be used, with a number being damaged in the process. Once CSEBs were brought to site the construction process proceeded in much the same way as for traditional masonry, a technique which is well established in Rwanda.

In contrast Rammed Earth is a new construction technique, so training and additional supervision were required in ramming as well as preparation of the soil. To attain the required surface finish, it was decided to import re-usable formwork, an additional expense and embodied carbon in comparison to CSEBs.

LESSONS LEARNT/ SUGGESTIONS

Codes and material data

Eurocodes are being adopted in some form or another by an increasing number of countries beyond Europe. Expanding this suite to cover earth construction may be fanciful but would serve to address many of the challenges faced by the team. This could increase the accessibility of earth construction and provide greater certainty for designers, especially if it were able to draw upon the significant body of academic research that is undertaken in this field.

Establishment of verified material design data for earth construction would ensure that designers are achieving a balance between affordability, sustainability and seismic resilience. This includes bond strength between rebar and earth, modulus of elasticity, movement joints in RE panels, and ductility behaviour factors. Where material data is given it would be useful to develop this to reflect the 'grade' of material being used, in particular to reflect whether it is stabilised or not.

Non-engineered building guidance

The aspiration for the residential buildings to be as locally buildable, sustainable and affordable as possible, as well as ensuring seismic resilience, was a mixed success, illustrating that trade-offs are required. The use of timber ring-beams did control cost and embodied carbon compared to RC, but the screw layouts and anchor rebar required impacted on their buildability. By contrast the construction of a concrete ring beam over the reinforced walls was much simpler to detail and construct.

Whilst the nominally unreinforced residential walls did contain significantly less steel reinforcement (and therefore reduced cost and carbon) compared to the enterprise buildings, full height steel rebar tie downs were required at the end of each panel to combat overturning, and steel rebar dowels were required at the base and at roof level, in order to tie the walls to the foundation and roof respectively. On average the 'unreinforced walls' contained approximately one quarter of the steel reinforcement provided in the reinforced walls.

Overall the detailing and construction of the residential buildings were not as simple as hoped. Whilst timber ring beams are often recommended in "non-engineered" seismic earth construction guidance, it was difficult to resolve by calculation, even for moderate seismicity. It may have been possible to reduce or eliminate the need for tie-downs at end of wall panels through adjustments to the plan layouts, but was deemed too restrictive for the architecture in this case.

Preparation

Engineered earth construction is unusual in Rwanda and training was essential in ensuring that processes for the mix, manufacture, curing and construction were established and followed. The scale of the project posed a challenge in adhering to quality standards but also an opportunity in that economy of scale enabled the contractor to procure specialist equipment and hire earth construction professionals to train staff and set up processes. Extensive testing of the site soil and various mixes were trialled for both CSEBs and RE giving the design team and contractor a good understanding of the material.

REFERENCES

1. NISR (2018). Integrated Household Living Conditions Survey 5 (EICV 5). [Online] Available at: http://www.statistics.gov.rw/datasource/integrated-household-living-conditions-survey-5-eicv-5?page=1

2. MININFRA, RHA, (2015). Rwanda Building Code (2015).

3. International Standard (2001) ISO 3010 2001 International Standard Based for Design of Structures - Seismic Actions on Structures.

4. RBS, RS 112: 2011 Basis of structural design, 2011

5. CEN, EN 1996-1-1:2005+A1:2012 Eurocode 6 - Design of masonry structures - Part 1-1: General rules for reinforced and unreinforced masonry structures, 2012

6. CEN, EN 1998-1:2004+A1:2013 Eurocode 8: Design of structures for earthquake resistance - Part 1: General rules, seismic actions and rules for buildings, 2013

7. SNZ, NZS 4297:1998 Engineering design of earth buildings, 1998

8. SNZ, NZS 4298:1998 Materials and workmanship for earth buildings incorporating amendment no. 1, 2000

9. SNZ, NZS 4299:1998 Earth buildings not requiring specific design incorporating amendment no. 1, 19995

10. Lubkowski Z, Villani M, Coates K, Jirouskova N, and Willis M. Seismic Design Considerations for East Africa. Second European Conference on Earthquake Engineering and Seismology, Istanbul, 2014.

11. AWC, ANSI/AWC SDPWS-2015 Special Design Provisions for Wind & Seismic, 2015

12. Walker, P., & Standards Australia. The Australian Earth Building Handbook, Sydney: Standards Australia International Ltd, 2002

13. Earth-auroville.com. Compressed stabilised earth block. [Online] available at: http://www.earth-auroville.com/compressed_stabilised_earth_block_en.php

18th International Conference on
Non-Conventional Materials and Technologies
"Construction Materials & Technologies for Sustainability"
(18th NOCMAT 2019)
24th – 26th July 2019
Nairobi, Kenya

ANALYSIS OF EARTHQUAKE TESTING TRENDS OF ALTERNATIVE BUILING MATERIALS

Elisabeth Vitswamba[1]

[1]Engineering for Change;

ABSTRACT

In a world that is experiencing rapid population growth, an increase in precarious building, and increase in the effects of climate change, promoting the use of alternative, sustainable building materials is a way to positively impact communities by providing affordable housing with materials that have a smaller carbon footprint. One challenge in the effort to make alternative building materials accepted and adopted by investors, designers, builders, and households, is the education needed to build well with the materials, and the need to have these methods accepted by governments and included in local building codes. An extra challenge exists in seismic regions where buildings must be designed to sustain additional forces. Several countries have seen extreme damage and loss of life due to earthquakes including Pakistan, Haiti, Nepal, and Chile. Designers and implementers of alternative building materials often do not have expertise in earthquake design, a complex area of study. However, even a basic knowledge of what is being done in earthquake design and testing, and where the bar has been set by way of testing of alternative building assemblies, can provide clearer direction to next steps in developing appropriate details using these materials, and provide the proof that these materials can be resilient to seismic forces to push for inclusion in local standards. An example of one such development is the work of Good Earth Global that have worked closely with the government of Nepal and have advocated for earthbag designs to be included in he local building code. The paper reviews existing literature to summarize the many trends in earthquake testing, determine the status of testing of specific alternative building materials, and based on the findings highlight where gaps may exist and what next steps may be taken in an effort to give additional credibility in the seismic resistance of alternative building assemblies.

Keywords: seismic resistance, earthquake simulation, earthquake testing, alternative building materials

INTRODUCTION

Earthquake engineering is the study of how a structure performs when subjected to the forces of an earthquake. Because earthquake forces are more complex than other structural forces, physical and digital simulation methods are used to validate the seismic performance of a structure. The physical method of testing is done with the use of a shake table or a set-up of horizontal or virtual actuators, and the numerical methods use computer model simulations using either static or dynamic methods of mathematical assessment (pseudo-static and pseudodynamic). These tests are performed on new materials, partial configurations, or full designs. Building codes often use the results of these tests when establishing design requirements for specific constructions. History has shown that buildings designed using earthquake engineering can save lives in seismically active zones. Earthquake simulations play an important role to protect the safety of the people that use or are in proximity of a structure. The following paragraphs provide an analysis of literature about shake table and pseudo-static/dynamic methods to assess the trends of how, when, and why specific methods are being used, with a focus on tests of alternative building materials including adobe, compressed earth blocks, cob, rammed earth, bamboo/wattle and daub, earth bags, and straw bale.

PHYSICAL TESTING

Quasi-static Testing

Quasi-static testing is a physical test that involves holding a building assembly in a static position while putting loads on particular points and analyzing the resulting behavior. It is used to analyze one or two particular characteristics of a material or assembly such as compression, tension, shear strength in or out of plane, or flexural strength. The loads can be monotonic, meaning the force is applied continuously on the assembly, or cyclic, meaning an oscillating load that is similar to what an assembly would experience in an earthquake. The physical simulation has hydraulic actuators that can push or pull at the structure. Compression and tension forces can be implemented simultaneously such as the pushing and pulling of a hydraulic actuator at the top of a structure while a continuous vertical deadload is simulated coming down on the structure [1]. Horizontal cyclic loading is applied and then is increased step by step until the specimen cracks, losing its resistance to horizontal load [2]. Sensors measure the structural performance including rate of propagation of cracks, hierarchy of collapse, and levels of damage. The advantage of quasi-static testing is that it is a more affordable option than dynamic shake table testing and can obtain focused results that are used as values and coefficients in numerical simulations. It can also measure stress-deformation conditions near to failure which is not always possible with shake-table testing [3]. The disadvantage of quasi-static testing is that it does not take into consideration the effects of inertia. Although it can be used as an alternative to shake table testing if the resources are unavailable, working under such controlled conditions is limited since earthquake motions are not static. Cyclic dynamic tests show strengths that can be double that of static tests because the specimens are able to function as full assemblies [4].

Quasi-static tests have been performed on adobe, compressed earth blocks, rammed earth, bamboo/wattle and daub, and straw bale assembles. No tests were found for cob construction. Adobe has been tested for compression, shear, bond, and diagonal shear of different sized assemblies ("wallets") of brick [5]. Static tilt tests of a larger assembly were performed to assess displacement with a known acceleration [6]. Researchers are exploring how adobe can be used as infill wall sections with wood boards on either side that can rock with in-plane motions and slide against one another to reduce the regular cracking and failure that commonly occurs. The infill walls performed satisfactorily to horizontal in-plane and out-of-plane loads, and more testing is needed [7]. Another alternative masonry is compressed earth blocks (CEBs). Normal CEBs would have a similar behaviour to adobe, but CEIBs (Compressed earth interlocking blocks) or ISSBs (Interlocking stabilized soil blocks) provide an opportunity to use less mortar in construction by dry stacking and allowing energy to be dissipated through the controlled movement of the blocks. Several quasi-static tests have been conducted on individual blocks, wallets, and wall panels seeking to understand the behaviour of the assemblies in compression and shear [8]. Some researchers are focusing on the shape of the individual modules to determine what allows for the most controlled movement with minimal shifting and damage to the bricks [8] [9].

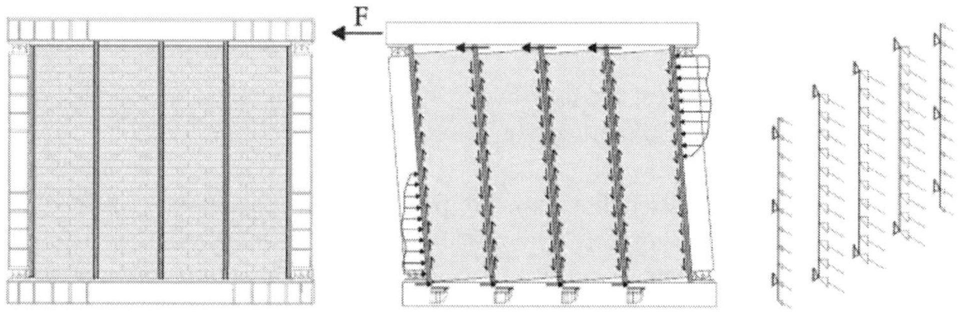

FIGURE 1 – DIAGRAMS SHOWING FORCES AND MOVEMENT OF ADOBE WALL PANELS USING WOODEN BOARDS AS SLIDING JOINTS [7].

Earthbags are also a type of modular construction but behave differently because of barbed wire that acts in tension between the layers as well as the deformation characteristics of the tamped earth. Earthbags have undergone study and testing for compressive strength, deformation characteristics, and general structural behavior [9] [10] [11] [12] [13] [14] [15] [16] [17]. Earthbag walls with returns were tested with quasi-static out-of-plane loading to analyze various reinforcing strategies at the corners. Several specimens were tested including one that followed standard procedure with the barbed wire between the withes bending and continuing at the corner, another that used ties binding courses together, another where the barbed wire was tied into ladder reinforcement at the corner, and a last one that had rebar on the outside anchored in a concrete footing. The walls tended to deform more than to detach, although some gapping occurred. Deformation was the least for the anchored rebar specimen. It was also found that tubing had less deformation than modular bags that tended to rotate and create gaps. The results of the test were tentative, and more testing is needed to more fully understand the behavior [4].

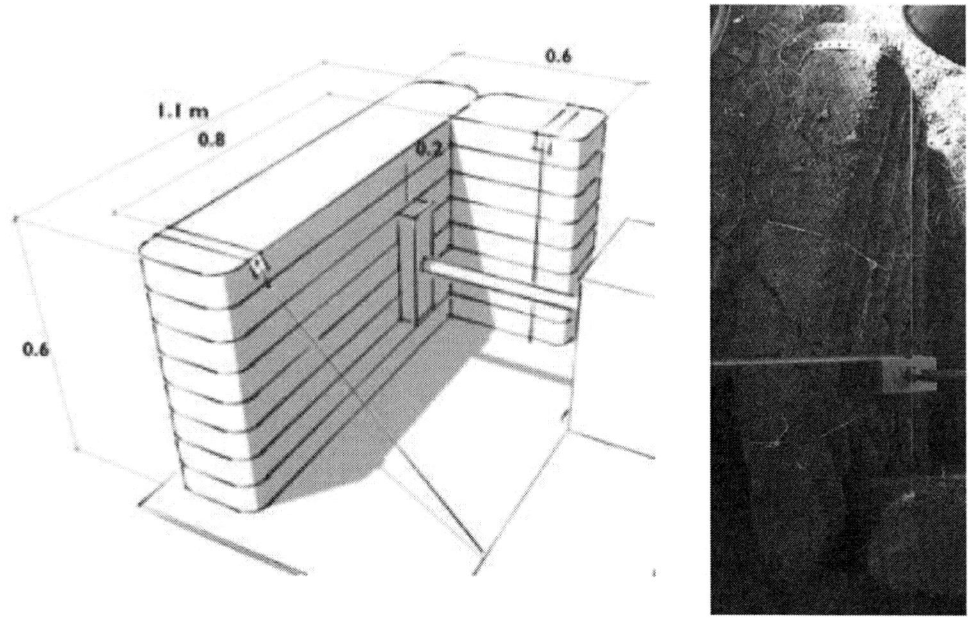

FIGURE 2 - QUASI-STATIC TESTING OF EARTHBAG CORNER [4].

Similar to that of adobe, rammed earth has undergone compression and shear testing. Sections of rammed earth were cut from a fully constructed wall because it was expected that results would differ from the middle to the top of the wall. Standard (10x10x3.5cm) and full size (50x50x45cm) blocks were tested, the results of the friction angles and cohesion to be used in later numerical tests. Next, a larger section of wall was constructed on a fixed concrete beam and subjected to horizontal loading to study the displacement and crack propagation, also to be compared with a numerical simulation [18].

Bamboo is used diversely in construction around the world, but it is primarily the wattle and daub type of construction that has undergone seismic testing because it is an affordable and simple way to construct a shear wall. In Costa Rica in-plane tests were performed on wattle and daub walls in the traditional bahareque style [19]. Researchers in El Salvador recently took it a step further by building walls in an engineered version of the bahareque that had 1m return walls with vertical ties to see how the assembly would fail in shear. They were tested in and out of plane and achieved the attended design load. The specimen that had mesh on both sides performed better, particularly for the out-of-plane loading [20].

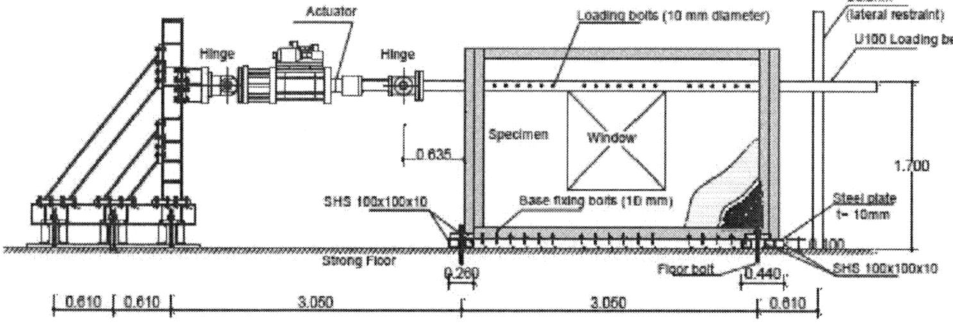

FIGURE 3 – DIAGRAM OF QUASI-STATIC TESTING OF BAMBOO WALL PANEL [20].

Lastly, the Ecological Building Network initiated a straw bale test program that performed several quasi-static tests on straw bale assemblies including compression, out-of-plan and inplane loading. Multiple constructs were built at full scale testing earth skins and cement stucco skins with light, medium, and heavy detailing. Walls with mesh (medium detailing) performed better than none. More tests need to be done to understand the behavior of assemblies with earth versus cement stucco skins [21].

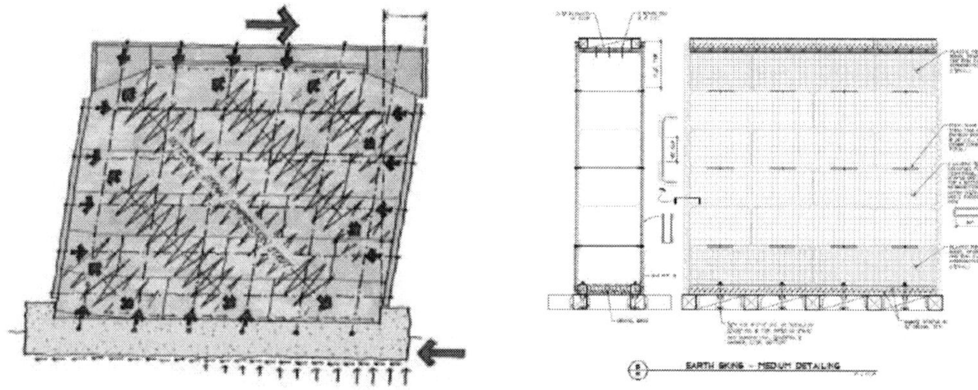

FIGURE 4 – DIAGRAM OF STRAW BALE FORCES AND SPECIMENT DETAILS [21].

Shake Table Testing

Shake table testing is a physical dynamic test that involves fixing a building assembly on a platform that moves on one or more axis and measuring and analyzing the resulting behavior. The shaking simulates an earthquake's movement and is able to accurately represent the forces involved. This is the most widely recognized test and is often used to validate other numerical analysis and confirm theoretical and partial-experimental behavior [22]. Since it is expensive, the computer models are generated and tested once to find the accuracy of the model. If it accurately predicts the result of the shake table test, it is used for future calculations instead of repeating the shake table test. The shake table test is the only way to study the performance of structural systems that are influenced by the dynamic and spatial characteristics of earthquake action, such as masonry.

Shake tables consist of a platform, a guiding system and actuators and can vary considerably in size. Some are relatively small – 3m by 3m [23] or 5.8m by 4.4m [24]. The biggest in the world, E-Defense, measures 20m by 15m. Others fall in between, such as NIED's Tsukuba Shaking Table, which is 14.5m by 15m. An extension frame can be added to make the platform larger [25]. A shake table with three axes (6 Degree-of-Freedom) is the most accurate [26] [22], but some tests require less, and other studies do not have the possibility of a shake table with three dimensions of movement and so researchers work with what they have. Some tests are performance only with horizontal motion, either in one direction or in the two orthogonal directions even when the shake table used is capable of moving vertically [27] [28]. Some shake tables are designed to operate in tandem as one larger shake table. The input motion can vary depending on what the test is meant to evaluate. The movements can be sinusoidal wave, real input ground acceleration, or ambient and forced vibration [3]. "Real input ground acceleration" means that in an earthquake event, the movement was measured, and it is reproduced in the simulation [26] [23] [29] [30] [24]. In some cases, loading history is used to expose the structure to a wide array of frequencies representing damage over time [25].

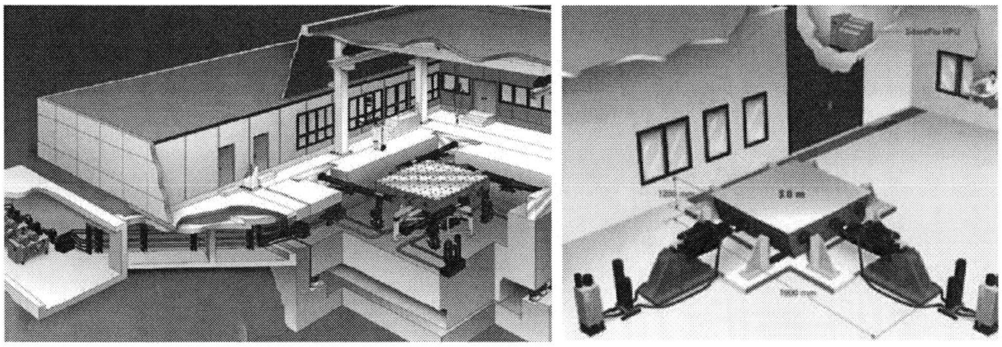

FIGURE 5 – CONCEPTUAL DRAWING OF 6 DEGREE OF FREEDOM SEISMIC SIMULATOR FACILITY AND A BI-AXIAL SIMULATOR [31]

The tested structure can be a full-scale model or sized down. The ratio of the model to prototype is generally determined according to the load capacity and dimension of the available shake table. [32] [33]. There are tests with 1:3 scale 4 story buildings [29], structures that are scaled down to 1:6 [33], and in the case of a massive structure such as a cooling tower, the model is scaled 1:30 [32]. If possible, full scale structures are tested because they are more accurate as real constructions rather than representations [27] [22] [34] [3] [35] [25] [23]. Tests have also been done where modular brick construction is scaled by one half [36]. It is important to note scaling removes mass that contributes to inertial forces. This loss must be made up with artificial mass and must be positioned carefully to make sure it does not affect other parameters of the structure [32]. The shake table method permits the testing of irregular shaped buildings that may be more difficult to model digitally [34]. Shake table tests are performed on almost every type of building assembly. Most tests do not include foundations and fix the superstructure directly to the shake table with bolts [34] [25]. Some exceptions to this general trend are shake table tests that have been performed on shallow foundations or retaining walls. In both cases the assemblies are placed in boxes of earth that has the desired soil mixture. The box is made of a steel frame with steel plates and sometimes glass panes are used on the sides to see the soil movement [37] [38].

Shake table tests have been performed on all of the alternative building materials researched, though some were only partial assemblies. Several shake table tests have been performed on adobe structures [39]. Most of experiments are performed to test the effectiveness of reinforcement strategies that include internal wood or steel reinforcement, interior steel mesh [40], PP-band (polypropylene band) meshes, and geogrid reinforcement [6]. In most cases the reinforcement strategies were satisfactory in preventing spalling and collapse. One exception was a test of internal steel reinforcement that gave unpredicted results [40]. Testing has also been done on roof connections to see if a diaphragm roof design can have a significant impact on the seismic response. This proved to be an accurate hypothesis [5]. The testing specimens focusing on reinforcement were scaled at 1:2 and 1:3. The test for roof designs was scaled at 1:4 [5]. Several groups are exploring dry stacked CEIBs and have conducted shake table tests. One was conducted in South Africa using blocks developed by "Hydraform" [41] and a second called "HiLoTec" for construction in Malawi [42]. More research needs to go to optimizing the shape and strength of the bricks to reduce cracking and gapping and testing the assemblies out of plane [43].

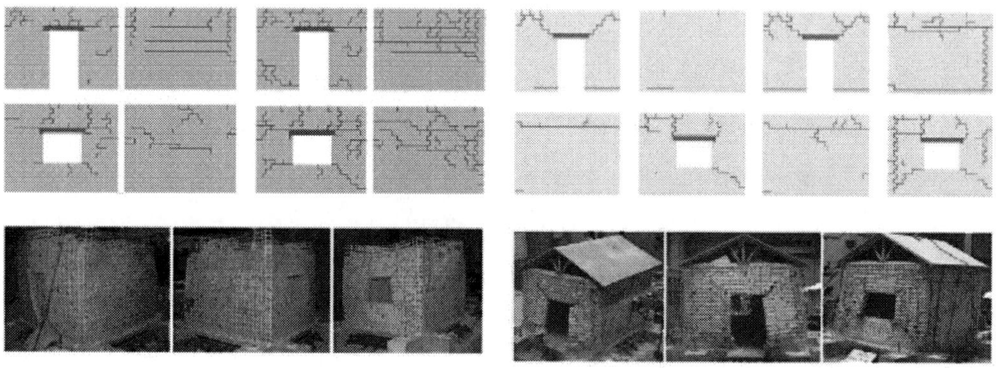

FIGURE 6 – SHAKE TABLE TESTING OF ADOBE HOUSE WITH AND WITHOUT DIAGRAIGM ROOF [5].

One shake table test has been performed on a cob house, but no academic reports were published about the findings. A video describing the project shows the building withstood a simulation of a 7.2 level earthquake on the Richter scale and then collapsed for the next one that was 9.0; however, the assembly contained no tensional reinforcing members beyond the hay strands within the clay-sand mixture [42]. There is more room for experimentation in the hybridization and detailing of cob similar to other earthen materials. A few u-shaped rammed earth wall assemblies have been tested on a shake table to determine assess a strategy of using wire mesh with a cement mortar finish applied to the walls similar to solutions that have been deemed effective for adobe [43].

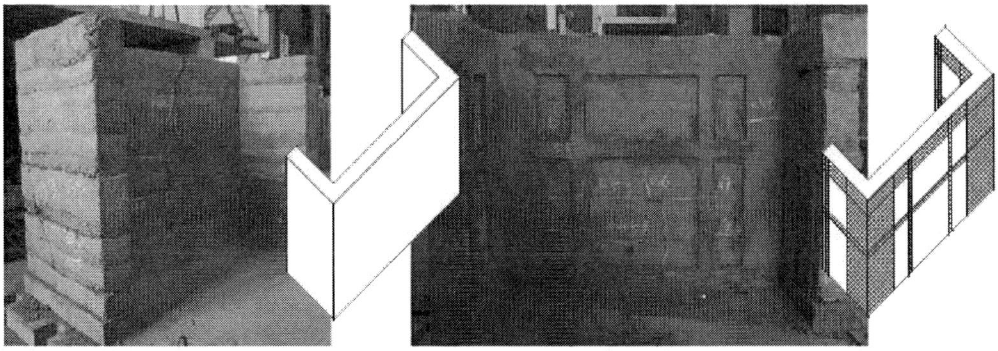

FIGURE 7 – RAMMED EARTH SPECIMENTS AFTER QUASI-STATIC TESTING [43].

There is one documented test of a bamboo house, also in the bahareque style that was tested in Guatemala. It had chicken wire placed on the interior and exterior of the walls. The house performed well and although it experienced cracking it met life safety requirements [20]. In Columbia there was a shake table test done of a two storey house that followed the Colombian seismic design code [44], but no report could be found with results from the test which was a collaboration between Arup and Imperial College. One shake table test has been performed on an earthbag assembly. The assembly was only 24in by 24in at 1:6 scale. Three types of corner configurations were tested [45]. Pakistan Straw Bale and Appropriate Building (PAKSBAB) along with the University of Navada tested a full-scale clay-plastered, thin, load-bearing straw bale house. Fishing net was used for reinforcement and the house was built on a gravel bag foundation that was encased in soil and cement and held in place on the shake table by formwork meant to simulate ground confinement. The house performed well, and damages were considered repairable. The gravel bags and straw bale were shown to dampen the movement [46].

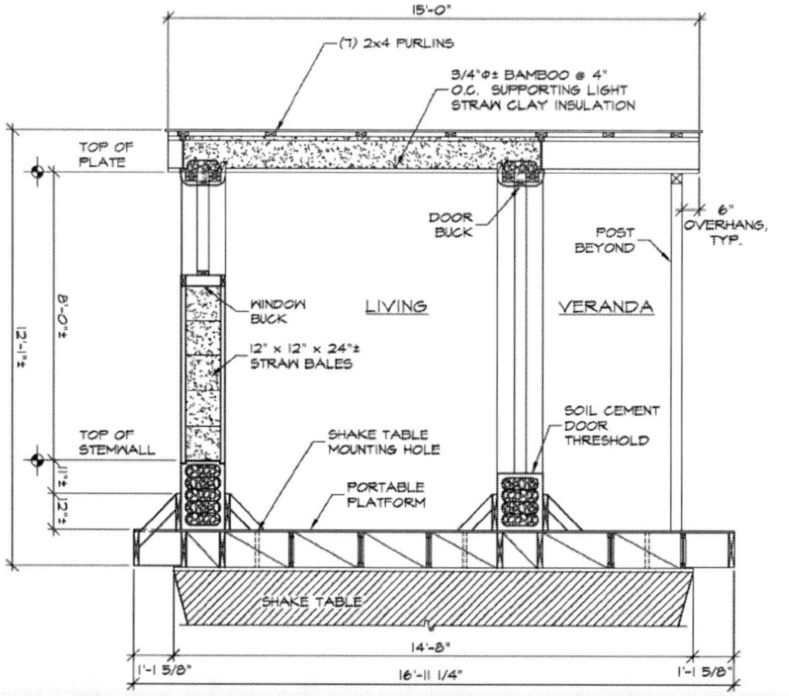

FIGURE 8 – SECTION OF PAKSBAB STRAW BALE HOUSE TESTED AT THE UNIVERSITY OF NEVADA [46]

Measurement and Analysis Tools

Various types of sensors and hardware are used to measure the movement of the structure during a quasi-static or shake table test. Acceleration and displacement are often measured with strain-gage acceleration sensors. Triaxial geophones or voltage displacement transducers (LVDT's) can be placed on the structure that convert movement into voltage that can be recorded. Video cameras are used for image processing for response displacements and retroreflective targets can be placed to be picked up by the cameras for higher accuracy [30] [25]. Optical laser sensors can also be used [22]. Sound and vibration can be measured with hardware such as the BBM data acquisition system [38]. The data collected is used for qualitative and quantitative analysis. The data is often used as inputs for pseudo-static or pseudo-dynamic simulations (outlined in the next section) and are often used to confirm results.

NUMERICAL TESTING

Pseudo-static testing

Pseudo-static testing is the representation (calculated by hand or in a computer program) of earthquake loads with "single constant unidirectional pseudo-static acceleration". Pseudo-static testing uses the "statics" mathematical method that assumes the structure is in a state of static equilibrium. A pseudo-static coefficient must be selected and is the most important aspect of the calculations because it is a measure of the amplitude of inertial force [47]. This method applies the force of a magnitude equal to the multiplication of the weight of the facility and the seismic coefficient. Earthquake loads are represented as a series of horizontal eigenvectors that are accumulated across the storeys of a building and must be balanced using equivalent storey stiffness and damping. Pseudo-static tends to agree with results evaluated by time history analysis [48].

Pseudo-static tests have been used in the calculation of earth pressures behind retaining walls [49], slopes [50] [51], earth dams [47], bricks [1] [52], concrete [53], pile foundations [54], and concrete blocks [55]. Pseudo-static is selected as testing method for large constructions where using a shake table is not an option such as earth dams or slopes. In the case of (Li et al., 2004), the research team had access to shake table, but it was too small for the structure they wanted to test [1]. An advantage of pseudo-static is that it is a simple and straightforward analysis, but the disadvantage is that it is crude and doesn't always accurately represent real affects [47]. No research involving pseudo-static testing was found for alternative building materials.

Pseudo-dynamic Testing

Pseudo-dynamic (PsD) testing is a more recently developed method that includes time in the calculation of applied earthquake loads and the effect of shear and primary waves. It incorporates computational modeling with the experimental testing that follows a feedback loop of the computer analyzing the resulting displacement and then using it to determine the response in the next time step. It is an experimental computer simulation that is creating new results and analyzing them as the program runs [56].

The pseudo dynamic is reliable, informative and economical technique for small scale specimens [33]. It can be used to analyze structures that are already in place and cannot be physically tested, such as internal steel frames [57]. This also includes structures like retaining walls that can't be analyzed any other way [58]. It can be used to study full size structures as well, without the need of constructing the whole thing in a test rig [59]. The pseudo-dynamic method has been found to be accurate when compared to codes and standards such as FEMA 273 [60] and GB50003-2001 [1].

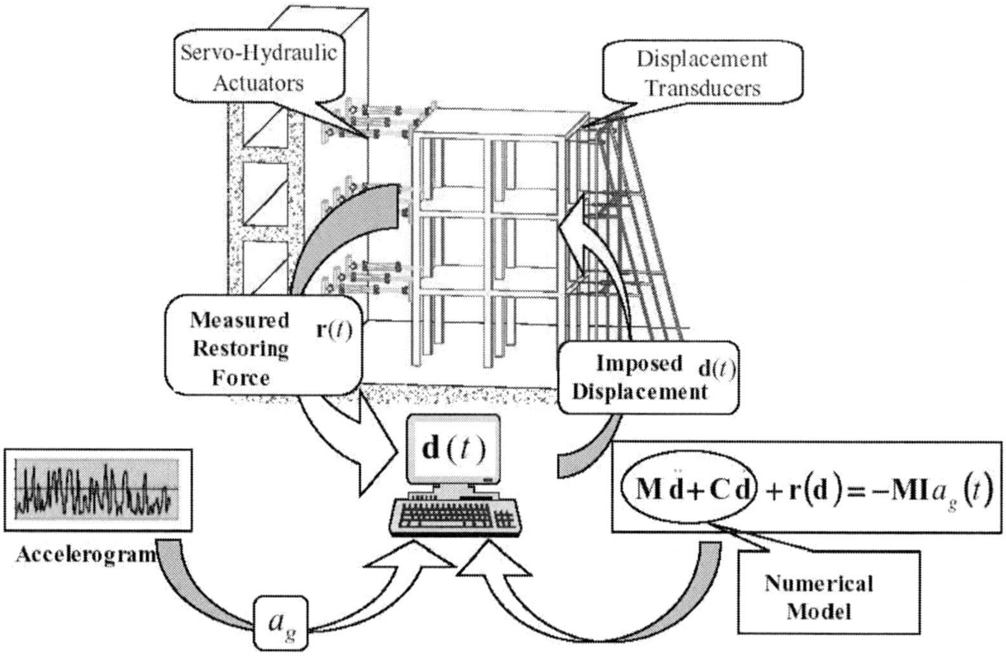

FIGURE 9 – DIAGRAM SHOWING FEEDBACK LOOP OF PSEUDO-DYNAMIC SIMULATION [56]

Pseudo-static and pseudo-dynamic tests can be done by hand for simple structures such as a single retaining wall [58] but can quickly become arduous for larger, complex structures necessitating the use of computers. Software used for pseudo-static and pseudo-dynamic testing include open-source, proprietary, and custom developed tools. Proprietary examples include ANSYS [32], Diana [18], Solidworks [45], and LS-DYNA [32]. ANSYS and Solidworks are engineering simulation software whereas LS-DYNA is a multi-physics simulation software. Tensor Flow (TF) Multi [46] and Code Aster [18] are multi-physics simulation software that are open source. An example of a custom software is FELISA//3M that was developed by the Institute of Earthquake Engineering and Engineering Seismology (IZIIS), Skopje [3].

The programs modeling dynamic behavior use a method of analysis called Finite Element Method (FEM) that breaks down forces into a field of nodes that relate to one another (Sattar, 2014). Wikipedia has a list of software packages that implement the finite element method for use by various applications. The calculations made from these programs often use inputs taken from shake table or quasi-static tests that represent the inertial forces, and the results are analyzed and compared [22]. For granular or discontinuous structures or particles, a numerical simulating method called DEM or Extended Distinct Element Method (XDEM) are used. Researchers (Hanazato et al.) outline the use of a new simulating method based on XDEM applied to the simulation of earthquake forces on a brick house. In the computer simulation 2600 brick elements are calculated in relation to one another [30].

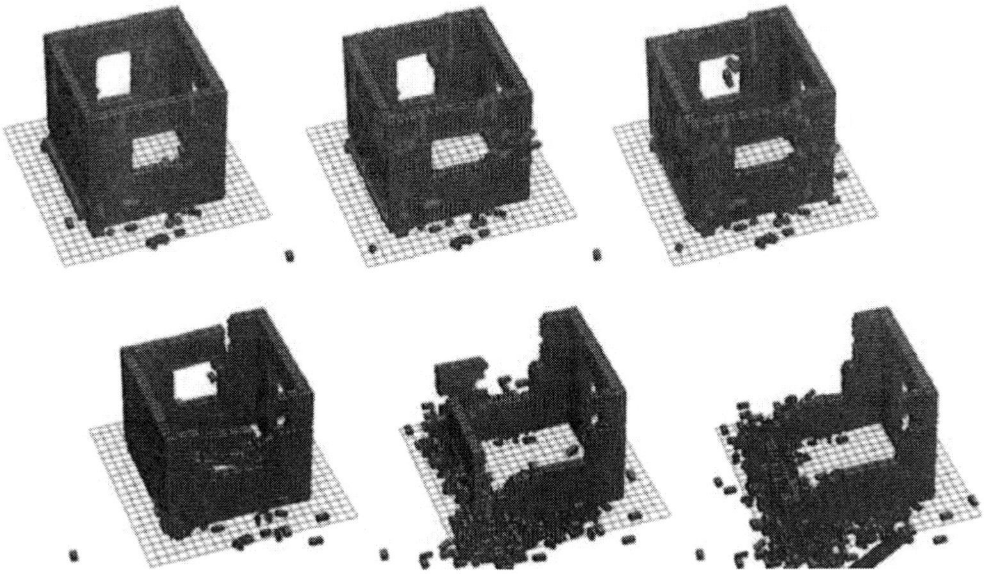

FIGURE 10 – DIAGRAM SHOWING FEEDBACK LOOP OF PSEUDO-DYNAMIC SIMULATION [56]

Research for alternative building materials using Pseudo-dynamic testing was found for dry stack masonry, rammed earth and straw bale brick infill panels. The rammed earth numerical model had intralayers and interlayers that represent the layers of the wall that have differing properties. The cohesion and friction angles were taken from the first shear tests and used as values in the numerical model. The model was then compared to the quasi-static wall test and it was found to be satisfactorily representative of what took place in the physical pushover test [18].

DISCUSSION

The shake table test and pseudo-static test give very close results [61] [38]. When compared with pseudo-dynamic, both gave similar results on earth pressure coefficients and expected start of failure [49]. The main difference found between testing with a shake table and quasistatic is the stiffness. Pseudo-static is more conservative- it was also found that the value calculated with this method for collapse, when validated with a shake table test, was much lower than the actual collapse [22]. Quasi-static provides information near to failure that is not possible with shake tests. Both are necessary to describe the behavior [3]. In general, it is important to note that no matter what test is used, it might prove to be different to actual collapse modes [30]. Because of differing focus of each type of test, they work effectively in conjunction with one another.

There has been promising headway in the development of earthquake testing research for alternative building materials although gaps and opportunities exist as well. This paper does not cover all ground but provides a comprehensive picture of the types of tests that are performed and what is being explored. Materials like adobe, CIEBs, and rammed earth have had more interest from researchers, whereas cob, and earthbags have gaps that could be further explored. Materials like cob and earth bags also face a challenge in numerical modelling as the deformation characteristics and tensile fibers would be difficult to represent numerically. They would also be more challenging to scale down. For this reason, full scale shake table testing offers the most insight into the design and detailing of seismic resistant assemblies using these materials.

APPENDIX

TABLE 1 – EARTHQUAKE TESTING TRENDS OF MODULAR EARTH ASSEMBLIES

	Adobe	Compressed Interlocking Earth Blocks (CIEBs)	Earth Bags
Quasi-static	Year: 2010 Group: University of Auckland [6] Construct: U-shaped assembly Load: Static tilt test Year: 2018 Group: University of Jaffna [5] Construct: Adobe blocks and wallets Load: Shear and compression Year: 2018 Group: Università degli studi de Brescia [7] Construct: Infill wall with vertical sliding joints Load: In-plane and out of plane cyclic loading	Year: 2014 Group: University of Minho [62] Construct: Masonry prisms and masonry wallets Load: Compression and horizontal loading Year: 2016 Group: Shenzhen Engineering Lab [8] Construct: Multiple sets of Light Aggregate Concrete (LAC) bricks with variously shaped interlocking joints Load: Compression and horizontal cyclic loading	Year: 2008 Group: Bryce Daigle – Queen's University [10] Construct: Unplastered earthbags with diferente types of fill Load: Compression Year: 2011 Group: Chris Croft – The University of Bath [9] Construct: Wall assemblies Load: horizontal flexural and shear tests Year: 2016 Group: Build Simple [4] Construct: Corner walls Scale: 60% Additions: Barbed wire, ladder reinforcement, vertical reinforcement Load: Horizontal out of plane

Shake Table	Year: 2005 Group: University of Technology Sydney [40] Construct: U-shaped assembles and one room house Size table: 3x3m Scale: 1:2 Additions: internal horizontal reinforcement (wire mesh), external vertical or horizontal reinforcement (bamboo), and timber ring beam Year: 2010 Group: University of Auckland [6] Construct: U-shaped walls Table size: 1 x 1.5m Scale 1:3 Additions: vertical steel bars at corners, structural bondbeam, and polysynthetic geogrid placed horizontally every third course, tied to vertical members Year: 2018 Group: University of Jaffna [5] Construct: One room house with and without roof Scale: 1:4 Additions: PP-band retrofitting	Year: 2011 Group: University of Witwatersrand [41] Construct: Full-scale drystack masonry house using Hydraform bricks Size table: 4x4m Scale: 1:1 Additions: Minimal rebar in mortar Year: 2013 Group: University of Minho [63] Construct: Full-scale drystack masonry house using HiLoTec system Size table: Unknown Scale: Unscaled Additions: Steel rebar	Year: 2013 Group: Awais Malik, University of Dartmouth [45] Construct: Partial wall construction Size table: 24x24" Scale: 1:6 Additions: Barbed wire and rebar
Pseudo-static	None found	None found	None found
Pseudodynamic	None found	None found	None found

TABLE 2 – EARTHQUAKE TESTING TRENDS OF MONOLITHIC EARTH ASSEMBLIES

Test types	Cob	Rammed Earth
Quasi-static	None found	Year: 2018 Group: University of Grenoble [18] Construct: Block samples and wall section Load: Shear and horizontal pushover
Shake Table	Year: Unknown Group: UBC Civil Engineering [42] Construct: One room house Size table: 3x3m Scale: 1:2 Additions: None	Year: 2010 Group: Xian University [43] Construct: Rammed Earth wall panels Scale: 1:3 Additions: Mesh reinforcement
Pseudo-static	None found	None found
Pseudodynamic	None found	Year: 2018 Group: University of Grenoble [18] Construct: Wall section Load: Horizontal pushover

TABLE 3 – EARTHQUAKE TESTING TRENDS OF PLANT-BASED MATERIAL ASSEMBLIES

Test types	Bamboo/wattle and daub	Straw bale
Quasi-static	Year: 2015 Group: Imperial College London, REDES and Arup [20] Construct: Engineered bahareque timber, cane and mortar wall with 1m return Scale: 1:1 Additions: Wire mesh	Year: 2003 Group: University of Illinois and Ecological Building Network Construct: Straw bale wall assemblies with light,

	Load: Horizontal in and out of plane	medium and heavy detailing Scale: 1:1 Additions: Wire mesh, staples, steel dowels
Shake Table	Year: 2013 Group: PRIRTI-TRADA housing system at Central Power Research Institute (CPRI), Bangalore [44] Construct: 2.7 x 2.7m house bamboo skeletal frame and infill panels Scale: 1:1 Year: 2015 Group: Imperial College London, REDES and Arup [20] Construct: Engineered bahareque timber, cane and mortar house Size table: 3x3m Scale: 1:1 Additions: Wire mesh	Year: 2010 Group: George Institute of Technology, University of Navada, and Pakistan Straw Bale and Appropriate Building (PAKSBAB) [46] Construct: Thin straw bale house with clay plaster and lime wash finish on gravel bag foundation Scale: 1:1
Pseudo-static	None found	None found
Pseudodynamic	None found	Year: 2018 Group: University of Ilorin [64] Construct: Straw bale infill panels Load: Accelerated history of uniaxial shake table tests

REFERENCES

[1] X. Li, S. X. Gong and X. L. Gu, "Interaction and seismic capacity of brick wall and supporting RC frame," in 13th International Brick and Block Masonry Conference, Amsterdam, 2004.

[2] A. P. Russell and J. M. Ingham, "Flange effects of an unreinforced masonry wall subjected to pseudo-static in-plane seismic forces," in The 14th World Conference on Earthquake Engineering, Beijing, 2008.

[3] V. Hristovski, M. Stojmanovska and B. Dujic, "Full-Scale Shaking Table Tests of XLam Panel Systems - Correlation With Cyclic Quasi-Static Tests," in 15 WCEE, Lisboa, 2012.

[4] P. Stouter, "Stronger earthbag corners: Test results," Build Simple Inc., 2016.

[5] S. Navaratnarajah, "Effect of Roof and Diaphragm Connectivity on Dynamic Behavior of the PPband Retrofitted Adobe Masonry Structures," Periodica Polytechnica Civil Engineering, 2018.

[6] J. F. Tipler, M. L. Worth, H. W. Morris and Q. T. Ma, "Shake table testing of scaled geogrid reinforced adobe wall models," in NZSEE Conference, 2010.

[7] M. Preti, M. Neffati and V. Bolis, "Earthen masonry infill walls: Use of wooden boards as sliding joints for seismic resistance," Construction and Building Materials, vol. 184, no. 2018, pp. 100- 110, 2018.

[8] H. Liu, P. Liu, K. Lin and S. Zhao, "Cyclic Behavior of Mortarless Brick Joints with Different Interlocking Shapes," Materials, no. 166, 2016.

[9] C. Croft, "Structural Resistance of Earthbag Housing Subject to Horizontal Loading," The University of Bath, 2011.

[10] B. C. Daigle, "Earthbag housing: Structural behaviour and applicability in developing countries," Queen's University, Kingston, Ontario, Canada, 2008.

[11] R. Dunbar, "Prism Testing of Polypropylene Earthbags," May 2006. [Online]. Available: www.earthbagbuilding.com/Testing/prismtest.htm. [Accessed 1 March 2019].

[12] T. N. Lohani, K. Matsushima, U. Aqil, Y. Mohri and F. Tatsuoka, "Evaluating the strength and deformation characteristics of a soil bag pile from full-scale laboratory tests," Geosynthetics International, vol. 13, no. 6, pp. 246-264, 2006.

[13] P. Stouter, "Friction and tensile strength of earthbag components," Wycliffe Associates and EarthBagStructures, 2010.

[14] R. Pelly, "Plastic Limit Analysis of Earthbag Structures," The University of Bath, 2009.

[15] S. F. Tantono, "The Mechanical Behavior of a Soilbag under Vertical Compression," TU Graz / Büroservice, Graz, 2008.

[16] N. Vadgama, "A material and structural analysis of earthbag housing".

[17] B. E. Ross, M. Willis, P. Datin and R. Scott, "Wind Load Test of Earthbag Wall," Buildings, vol. 3, no. 3, pp. 532-544, 2013.

[18] R. El-Nabouch, Q. B. Bui, P. Perrotin and O. Plé, "Shear Parameters of Rammed Earth Material: Results from Different Approaches," Hindawi: Advances in Materials Science and Engineering, 2018.

[19] G. Gonzàlez and J. Gutiérrez, "Structural performance of bamboo 'bahareque' walls under cyclic load," Bamboo and Ratan, vol. 4, no. 4, pp. 353-368, 2005.

[20] S. Kaminski, K. Coates, A. Lawrence and J. Aleman, "Seismic design of low-cost and sustainable cane, timber and mortar housing for El Salvador," in SECED 2015 Conference, Cambridge, 2015.

[21] C. Ash, M. Aschheim and D. Mar, "In-plane cyclic tests of plastered straw bale wall assemblies," Ecological Building Network.

[22] I. Ricci, G. Gasparini, S. Silvestri, T. Trombetti, D. Foti and S. Ivorra-Chorro, "Design of a shaking table test on a 3-story building composed of cast-in-situ concrete walls," in 15 WCEE, Lisboa, 2012.

[23] C. Petrone, G. Magliulo and G. Manfredi, "Shake table tests for the seismic assessment of hollow brick internal partitions," Engineering Structures, vol. 72, pp. 203-214, 2014.

[24] M. Urban and L. Stempniewski, "Comparison of different earthquake strengthening methods for masonry buildings," in 15 WCEE, Lisboa, 2012.

[25] D. J. Heath, E. F. Gad and J. L. Wilson, "Shaking table test of full-scale brick veneer house," Uminho Research Group.

[26] T. Nagao, H. Kagano and K. Hamaguchi, "Full-Scale Shaking Test on Furnitures subjected to Long-Period Earthquake Motions," in 15 WCEE, Lisboa, 2012.

[27] I. Nagashima, H. Hibino, R. Maseki, H. Narihara, E. Sato, M. Iiba, H. Kitamura, Y. Yamano and N. Kashimoto, "Multi-cyclic Dynamic Loading Experiment of Full-scale Devices for Seismic Isolation against Long-Period Earthquake Motions -Preliminary study -," in 15 WCEE, Lisboa, 2012.

[28] M. Branco, L. Guerreiro, A. Campos Costa and P. Candeias, "Shake Table Tset of Structure Reinforced with Superelastic Dampers," in 15 WCEE, Lisboa, 2012.

[29] H. Su, J. Wang, X. Jiang and Y. Tan, "Shaking table test on building in masonry structure with construction waste recycled brick," World Journal of Engineering, vol. 11, no. 4, pp. 357-364, 2014.

[30] T. Hanazato, C. Minowa, T. Narafu, H. Imai, Q. Ali, K. Kobayashi, Y. Ishiyama and T. Nakagawa, "Shaking table test of model house of brick masonry for seismic construction," in 14 WCEE, Beijing, 2008.

[31] MTS , "MTS Seismic Testing Solutions," [Online]. Available: https://www.mts.com/cs/groups/public/documents/library/dev_002361.pdf.

[32] J. Dai, Y. Yang and X. Weng, "Shake Table Test for Large Indirect-Air-Cooling Tower Structure of Fire Power Plant - Part 1," in 15 WCEE, Lisboa, 2012.

[33] R. Sharma, A. Sachan and D. C. Rai, "Correlation between Pseudo-Dynamic and Shake Table Test on Steel Truss Moment Frames with Hysteretic Dampers," in 15 WCEE, Lisboa, 2012.

[34] K. Fujii and T. Ikeda, "Shaking Table Test of Irregular Buildings under Horizontal Excitation Acting in an Arbitrary Direction," in 15 WCEE, Lisboa, 2012.

[35] H. Imai, H. Inoue, C. Minowa, A. G. Lanuza, H. C. Penarubia, I. C. Narag, R. U. Solidum Jr., K. Okazaki, T. Narafu, J. Sakuma, K. Kusunoki and T. Hanazato, "Shaking

Table Experiment for Philippine Full-Scale Concrete Hollow Blocks (CHB) Masonry Houses," in 15 WCEE, Lisboa, 2012.

[36] V. Singhal and D. C. Rai, "Suitability of Half-Scale Burnt Clay Bricks for Shake Table Tests on Masonry Walls," Journal of Materials in Civil Engineering, vol. 26, no. 4, 2014.

[37] R. E. S. Moss, S. Kuo and V. Crosariol, "Shake Table Testing of Seismic Soil-Foundation-Structure-Interaction," in Geo-Frontiers 2011 Conference Proceedings, Dallas, 2011.

[38] C. Yang, J. J. Zhang, Q. Honglue, B. Junwei and L. Feicheng, "Seismic Earth Pressures of Retaining Wall from Large Shaking Table Tests," Advances in Materials Science and Engineering, 2015.

[39] C. F. Oliveira, H. Varum and J. Vargas, "Earthen Construction: Structural Vulnerabilities and Retrofit Solutions for Seismic Actions," in 15 WCEE, Lisboa, 2012.

[40] D. Dowling, B. Samali and J. Li, "An improved means of reinforcing adobe walls – External vertical reinforcement," in SismoAdobe , Lima, 2005.

[41] A. Elvin and H. C. Uzoegbo, "Response of a full-scale dry-stack masonry structure subject to exprimentally applied earthquake loading," Journal of the South African Institution of Civil Engineering, vol. 53, no. 1, 2011.

[42] M. Sider, "A Creative Impact: The Stanley Park Earthen Architecture Project," 29 July 2011. [Online]. Available: www.youtube.com/watch?feature=player_embedded &v=ChbccUQhpJc#!. [Accessed 1 March 2019].

[43] R. Hu and Y. Dong, "Shake Table Test on Rammed Earth Wall Panels," Advanced Materials Research, Vols. 133-134, pp. 795-799, 2010.

[44] J. Vengala, B. N. Mohanthy and S. Raghunath, "Seismic performance of Bamboo housing – an overview," Indian Plywood Industries Research and Training Institute (IPIRTI), Bangalore.

[45] A. Malik, "Structural analysis of earth-bag systems," 2013.

[46] C. Champion, D. Donovan and S. Elfass, "Seismic Response of Innovative Straw Bale Wall Systems and System Identification," 2010.

[47] T. Akhlaghi and A. Nikkar, "Evaluation of the Pseudostatic Analyses of Earth Dams Using FE Simulation and Observed Earthquake-Induced Deformations: Case Studies of Upper San Fernando and Kitayama Dams," The Scientific World Journal, 2013.

[48] M. Saitoh, M. Ikegame and S. Tanamura, "A quasistatic analysis method to improve collapse mechanism analyses of multistory buildings," in 13 WCEE, Vancouver, 2004.

[49] P. K. Woodward and D. V. Griffiths, "Comparison of the pseudo-static and dynamic behavior of gravity retaining walls," Geotechnical and Geological Engineering, vol. 14, pp. 269-290, 1996.

[50] H. Elahi, H. G. Poulos, H. Hajimollaali and A. Elahi, "Pseudostatic Seismic Response Analysis of a Pile Group in a Soil Slope," Geotechnical and Geological Engineering, vol. 36, no. 2, pp. 855- 874, 2018.

[51] D. Lo Presti, D. Marchetti and T. Fontana, "Pseudo-static vs. pseudo-dynamic slope stability analysis in seismic areas of the northern Apnnines (Italy)," Rivista Italiana di Geotecnica, 2010.

[52] T. Wang, M.-r. Zhou and Y.-m. Ding, "A quasi-static test on a new type of masonry built of "I"- shaped closed-top hollow bricks," Australian Journal of Structural Engineering, vol. 18, no. 2, pp. 103-112, 2017.

[53] K. Twigden, R. S. Henry and Q. Ma, "Pseudo-static cyclic, snap back and shake table testing of PreWEC self-centering wall systems," in Tenth US National Conference on Earthquake Engineering, 2014.

[54] A. Kumar, D. Choudhury and R. Katzenbach, "Effect of Earthquake on Combined Pile-Raft Foundation," International Journal of Geomechanics, vol. 16, no. 5, 2016.

[55] R. J. Bathurst and Z. Cai, "Pseudo-Static Seismic Analysis of Geosynthetic-Reinforced Segmental Retaining Walls," Geosythetics International, vol. 2, no. 5, pp. 787-830, 1995.

[56] H. Jamal, "Pseudo Dynamic (PsD) Tests - Non Linear Structural Dynamics Techniques," 7 July 2017. [Online]. Available: https://www.aboutcivil.org/pseudo-dynamic-tests. [Accessed 1 March 2019].

[57] R. Ozcelik, B. Binici and O. Kurc, "Seismic Retrofitting of Deficient RC Structures with Internal Steel Frames by Using Pseudo Dynamic Testing Procedure," in 15 WCEE, Lisboa, 2012.

[58] M. S. Tafreshi and M. Rahimi, "A Simplified Pseudo-Dynamic Method of Reinforced Retaining-Wall Subjected to Seismic Loads," in 15 WCEE, Lisboa, 2012.

[59] P. Negro, D. A. Bournas and F. J. Molina, "Seismic Testing of the SAFECAST Three-Storey Precast Building," in 15 WCEE, Lisboa, 2012.

[60] J. Paquette and M. Bruneau, "Pseudo-dynamic testing of unreinforced masonry building with flexible diaphragm," in 13 WCEE, Vancouver, 2004.

[61] M. Oliva, P. Gavrilovic and R. W. Clough, "Seismic testing of large panel precast walls: Comparison of pseudostatic and shaking table tests," Earthquake Engineering & Structural Dynamics, 1990.

[62] T. Sturm, L. F. Ramos and P. B. Lourenço, "Characterization of dry-stack interlocking compressed earth blocks," Materials and Structures, 2014.

[63] HiLoTec, "HiLoTec: Development of a Sustainable Self-Construction System for Developing Countires," [Online]. Available: www.civil.uminho.pt/hilotec /deliverables.html.

[64] A. A. Adedeji, A. Adedeji and A. Ibiyeye, "Hysteresis Analysis of Prestressed Brick Frame with Strawbale Masonry Infill Subjected to Seismic Loads," in MATEC Web of Conferences, 2018.

18th International Conference on
Non-Conventional Materials and Technologies
"Construction Materials & Technologies for Sustainability"
(18th NOCMAT 2019)
24th – 26th July 2019
Nairobi, Kenya

SOCIAL HOUSING ON EARTH AND BIOENVIRONMENTAL RESILIENCE: TWO CASES IN TUCUMÁN, ARGENTINA

Matías Ortega[1], Beatriz Garzón[1]

[1] San Miguel de Tucumán, Tucumán, Argentine;

ABSTRACT

Resilience as a concept in architecture is relatively new, and it responds to the dizzying era of changes that characterizes the contemporary world. The objective of any sustainable development agenda seeks to minimize the anthropogenic impact of the prevailing development model, responding to a cultural transition and paradigms. In this context, it is seen the flourishing of sustainable architecture, comprehensive search of architectural objects with multidimensional references to environmental issues: materials of construction, bioclimatic design, collaborative ways of execution, included technologies, maintenance, etc. This work aims to get to know the experience of two houses generated in the implementation of a program of social housing that, in coordination with a local community, resorted to adobe for the materialization of the walls, serving to a technological syncretism. It is concluded that these contribute to the resilience in the province of Tucumán and the paradigm of sustainable development.

Keywords: resilience, sustainable development, technology, building materials, adobe

INTRODUCTION

"Energivorous" is a neologism which attempts to describe the pattern of dominant development in the world, subject to the paradigm of sustained growth by the availability of abundant natural resources; the enormous energy contained in the oil, coal and gas enables the frenetic pace of global civilization. There are several anthropogenic effects on the biosphere, standing out as the most dramatic climate change, both by its effects at a global level and the risk that implies for the reproduction and survival of different ecosystems, as we know them today. The prevailing state of things is a reflection of the hegemonic paradigm. Below it, or in parallel, live other paradigms, new or ancient, which collaborate in the configuration of alternatives that are opening more visible paths and that give an account of resilient ability of societies, then, if dimension is taken from the boundaries of the land to certain modes of humans being in it, new theoretical and methodological frameworks can then be set. It is understood that the concept of sustainable development is part of this new paradigm, where the resiliency approach provides key elements. In terms of Calvente, the concept of resilience "impacts in the way in which societies behave to carry out their activities" and "sustainability, therefore, is the ability of a complex system to keep in time despite the environmental volatility fostered by learning, transformation, renewal and continuous evolution" (Calvente, 2007). The cyclical nature of complex adaptive systems, such as societies, is one of the characteristics of resilience. So this term is symptomatic of the time of changes (or change of time), in which we live.

Here, we are interested in the search to build responses and resilient environments for sustainability from the disciplines of design and planning of habitat and architecture. Following Cortés Cely (2015), in order for resilience to be manifested at larger scales, such as the territory, it is necessary to do so previously on smaller scales, such as the architecture that configures it, and even before, in the constructive elements and materials that constitute this architecture, that is to say, the incorporated technology for both its production and its operation. This is of enormous relevance, since the construction industry has a great responsibility in the existing environmental imbalances, since it "directly impacts 35% of the demand for primary energy in the country", and if we consider "the urban sectors and the companies that produce construction materials, this value exceeds 50% of the total "(Czajkowski, 2016); It is then defined as one of the industries that generates a higher impact in the territories by the quantity of matter and energy that it consumes in the production of the built habitat, besides the persistent energy consumption generated during the useful life of buildings and the Impact at the end of the cycle by the nature of the materials used in their composition. At this point you can reflect on the culture of reuse of materials on the one hand, and on the other in the problem of high energy consumption due to the nature of the energy matrix at global level, whose composition ends up being highly contaminant. This is what led, since the early 60`s, in the search for the rational use of energy, and currently in energy efficiency, through the bioclimatic and sustainable design of buildings.

Thus, this is how contemporaneousness moves towards solutions of low impact, of efficient use of the consumed energy and of diversification of the energy production through the technologies ofuse of renewable energies; we will call this type of resilience as "technological", the one that seeks to make use of a responsible and sustainable consumption of technology, in order to achieve a sustainable development in its three dimensions: economic, social and environmental (García Morales y Gutierrez Colino, 2014). Although there are projects in Argentina to establish normative and technological guidelines for the construction with energetic efficiency and renewable energy throughout the country, aimed at reducing the demand of energy by consumers and the consequent reduction of emissions of greenhouse gases (GHG), improving even the levels of hygrothermal comfort, they are not implemented in a systematic way.

In spite of this, the resilient architecture it is no more than was done historically by the accumulated environmental knowledges from generation to generation, that we know as vernacular architecture, and that for Gonzalo (2003), it is generated in the conjunction of factors closely related, grouped in the sociocultural dimension, which include the social organization of the group, the family structure, religion, worldview, economic organization, productive activities, relationship with other groups with Identical or different conformation, cultural level of the people, technologies managed by the group, etc.; and in those of the physical dimension, which include all those factors given by the natural characteristics of the place, such as climate, geomorphological characteristics of the site, natural landscape, hydrography, fauna and vegetation, the existing materials, Etc.

In the northwestern region of Argentina (NOA), most commonly in rural areas, earth has been one of the main materials used to obey to the construction of vernacular housing. It is considered here that it is necessary to regain and visualize these experiences in order to reduce the environmental impacts that the prevailing architecture implies.

Figure 1. Adobe Production Center in the way to Las Carreras, Valle de Tafí

This work will address the resilience linked to the ways of producing the Rural and Aboriginal social housing program, which, in the articulation with a local community of indigenous identity, used the earth material for the construction of masonry in adobe, being a constructive technology resilient, both for its intrinsic characteristics and its suitability for the environment, which gives account of a "technological syncretism" (Berger de White and Azurdia Bravo, 2008). At the same time, the experience relieved through the case study accounted for the opportunity generated by the social housing program to create work and to boost local markets, which allowed the family to adapt and overcome the conditions of previous vulnerability. Being the problematic of the housing deficit a fact that characterizes the global South, these experiences are interesting by the articulation of the state with vulnerable social groups for the joint construction of their habitat.

METHODS

For this work the exploratory method was applied, since until here the factor "resilience" was not considered in the preceding approaches, which resulted in an enormous contribution to the field of study. The hypothesis raised is that the experiences of social housing executed on land from the Rural and Indigenous program, constitute a valuable contribution in the configuration of more resilient environments, both for the materiality of the houses and for the technological-social modes or processes implemented to build them.

A descriptive method was also implemented, since it is sought to specify the properties and characteristics of the phenomenon submitted to analysis: on the one hand, it will refer here to the intrinsic properties of the earth material, considering with emphasis the bioenvironmental questions, to define it as resilient; on the other hand, the most relevant part of the housing program will be considered, which allowed the beneficiaries to build their homes using adobe, a traditional material in the area. The analytical method followed consisted in observing and examining the concrete experience, manifested in two houses, as a particular fact. Finally, using the correlation method, the relationships between the variables considered are evaluated, and the conclusions are raised.

RESULTS

Earth as a building material for a resilient architecture.

The rediscovery of the earth material applied to construction poses a profound change in the classic paradigm of western development, since it hierarchies a constructive material of the traditional techno-science practically forgotten by the innovations of the 19th and 20th century technology. From the decade of the years 1960, the value of this material for constructions has been recovering, since it was a context in which the critique of modernity was generalized and the vernacular values and the local identities were rescued. We will evaluate here the intrinsic characteristics to the earth material to define it as resilient, understanding that the construction with earth and the sustainability are closely linked, due to the low environmental impact that this technology has with respect to other conventionally used and dominant materials in the market, and to its constitution as social technology.

For Cortés Cely (2015), the resilience in architecture can be expressed around guidelines determined by four criteria of edification and design, which provide security and adaptation to the transformation of the environment. These criteria are: 1) the safety provided to the occupants, in terms of resistance, stability and sanitation; 2) Easy reconstruction of structures after a catastrophic event; 3) low environmental impact caused by the generation of building materials; and 4) efficiency in terms of saving resources and energy. In turn, he notes that the variables contained in the properties of the materials are those that determine their resilience, and that this approach opens a new path for the architecture of the 21st century. For this work we will consider some of those variables that make the thermal and environmental properties of the earth.

The thermal properties of the materials are the answers that these give when being heated. Both the temperature and the dimensions of a solid increase as it absorbs energy in the form of heat. The variables that characterize these properties are the calorific capacity and the specific heat, the dilation or thermal expansion, the thermal conductivity and the refractoriness or resistance to fire.

We are interested here to specifically address the variable conductivity, as it is the most decisive when analyzing the behavior of heat transfer. For the case of adobe, its thermal conductivity is 0.46 W/m°C (Moscoso Cordero, 2015), a low value compared to other masonry (being the brick conductivity of 0.85 W/m°C and the concrete of 1.50 W/m°C); as a result the heat resistance, that is, the ability of this material to oppose the heat flow, is high, which explains its good "insulating" behavior.

As this is a material with high density values (physical property, whose values range from 1500 to 1700 kg/m3), specific heat of 840 J/Kg°K and calorific capacity (between 500 and 1000 Kcal/m³°C), has a high thermal inertia; this determines that it is very appropriate in climates with important daily thermal amplitudes (where the temperature difference between the day and the night is big, as is the case of the valleys in Tucumán), because it possesses the capacity to accumulate the energy received during the day and release it during the night. It is thus that these properties are constituted in resources to achieve the hygrothermal comfort inside passively, maintaining stable conditions throughout the day and reducing in this way the need to consume additional energy by mechanical means, resulting in energy and economic saving.

The environmental properties of a material are defined by the variables that make the toxicity caused and, evaluating the life cycle of materials, the capacities contained in recycling, reuse and biodegradation (Cortés Cely, 2015). Unlike the "sick building", those who undermine our health, a resilient building must not "cause internal, chemical or electromagnetic pollution and should minimize external impacts such as noise, radon gas, quality of air, or non-ionizing radiations (Alonso, 2017). The earth material, because it is of natural origin, it does not contain risks for the occupants of a house that was built with it. At the same time, it is perfectly reusable and does not require biodegradation, being just a natural compound.

Another variable considered here in relation to the environmental properties of the materials or constructive elements is determined by the quantity of its greenhouse gas emissions (GHG), for example, carbon dioxide (CO2), which are released in the production process. The emission levels of this gas with respect to different constructive materials are observed.

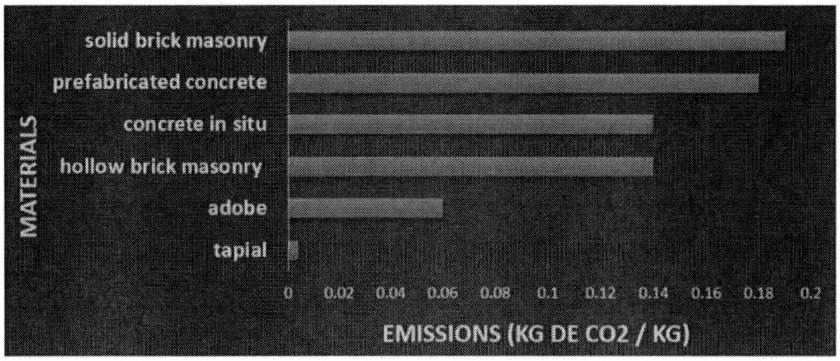

Figure 2. CO2 emissions of different materials (Bestraten; Hormías; Altemir, 2011, p. 18)

Working with techniques that incorporate earth is more energy efficient since the production of the constituent elements consumes less energy. That energy necessary to produce a material or a constructive element is considered as the primary energy content (CPE), and earth and other natural products have a low CPE (Minke, 2013). According to MacKillop (quoted by Garzón, 2005, p. 117), the energy required to manufacture adobe is 13 kcal/unit, a compressed earth block (BTC), stabilized with a 10 percent of cement requires 94 kcal/unit; while the brick will need 379 kcal/unit and the concrete block of 3830 kcal/unit.

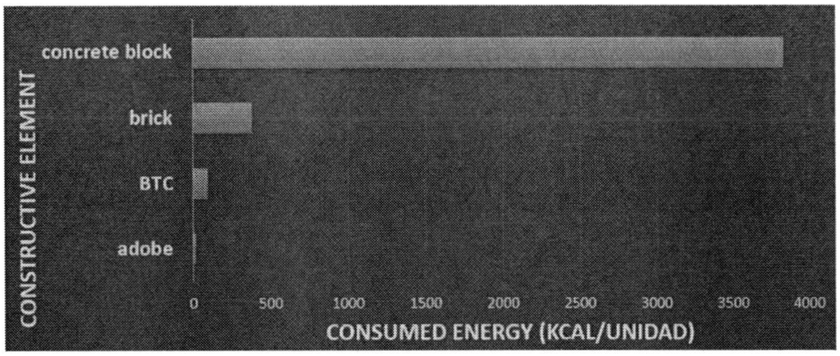

Figure 3. Necessary energy in production (MacKillop in Garzón, 2005, p. 117)

Finally, earth is considered as a resilient material because it is part of the constructive tradition of many peoples around the world, which defines it as a "social technology", as it constitutes the crystallization of local knowledge for solving shelter needs. Its normative regulation would enhance its development and promotion, collaborating in the resolution of the housing problem in an appropriate way. Although the country does not yet have specific norms that determine the art of good construction with techniques on earth, in recent years different territorial units have approved building ordinances with raw earth, as did the municipality of Santa Rosa, capital of the province of La Pampa, and the General Pueyrredón party in the province of Buenos Aires.

Synthetically, the characteristics that the earth material has, to define it as resilient are:

TABLE 1. EARTH MATERIAL CHARACTERISTICS TO CONSIDER IT AS RESILIENT

PROPERTIES	FEATURES
Thermoenergetic properties	It works as an insulator because of its resistance to the heat flow, collaborating in situations of interior comfort.
Environmental properties	It is not harmful to health. It is completely reusable. Its production releases low levels of greenhouse gases.
Technologic properties	It constitutes a social technology: community production, self-construction, environmental suitability.

Public policies as supports for the resilience of the built habitat.

If sustainability implies a social dimension, the design of public housing policies results in a thermometer to estimate the contained levels of resilience, since they reduce the housing deficit, so sensitive in countries of the global South, while they can promote the development of social structures or near communities, if they have the vocation to do so. Here we approach the experience of one of these programs implemented in the province that are particularly interested in how they addressed the problem in housing generation.

The year 2003 was of great relevance for the housing policy of Argentina, since many programs were activated since the Undersecretary of Urban Development and Housing, dependent of the Ministry of Federal Planning, Public Investment and Services, from the National Housing Fund (FONAVI), and from provincial channels, "tend to develop and organize a new programming of works according to the urgent needs" (Conavi, 2004). Product of the alarming social indices that were observed after the crisis of 2001, the Federal Housing Emergency Program (PFEH) was implemented, a framework agreement between the Ministry of Federal Planning, Public Investmentand Services, the Ministry of Social Development, the Ministry of Labour, Employment and Social Security and the provincial governments, with the objectives of advancing the housing emergency through the implementation of housing plans, and generating employment, promoting the formation of work cooperatives for construction.

In this context, the country's housing policy agenda became more complex by introducing progressive attention to relegated issues, such as articulation with local urban policies and other social policies, the participation of local and intermediates organizations, the development of the technical and human resources of local governments and the generation of program formats adaptable to local contexts, among others (Motta, 2016:145)

Thus, more integrative models were used for the construction of the habitat, but not without difficulties: in the articulation between the communities and the social housing programs, the constructive resolution expressed the tension between the local knowledges and practices and the regulations and recommendations of the control bodies. This reminds of Thomas, when he argues that "the resolution of the problems of poverty, exclusion and underdevelopment cannot be analyzed without taking into account the technological dimension" (Thomas, 2009:1).

Among the socio-economic organizations of the cooperative type with which articulated are co-operatives, present in the forms that the construction industry takes in the province of Tucumán; these have been the main piece in the whole process, and many of them were created to be able to give material response to the implementation of the programs. Here, the Provincial Institute for Cooperative and Mutual Action (IPACyM) had an important role, providing assistance and technical advice to co-operatives.

The experience of a social organization of the indigenous type that articulated with the Rural and Indigenous housing program in Valle de Tafí is developed here. In turn, the influences that they exerted to affirm their memories and identities and to preserve the constructive culture of the valley were examined.

The Monasterio houses as demonstrative models.

The department of Tafí del Valle is located in the west of the province of Tucumán, where its valleys are located, to an average of 2,100 M.A.S.L. It is an outstanding tourist destination because of the scenic beauty that characterizes it, Valle de Tafí is to the south of the mentioned department, in the point of contact g of two mountain ranges: Cumbres Calchaquíes and the chain of the Aconquija. The valley contains two important cities, Tafí del Valle and El Mollar, but in its territory different localities are distributed; some of them make up indigenous communities.

The Rural and Indigenous social housing program articulated with the indigenous community of El Mollar Pueblo Los Tafíes to collaborate in the situation of housing vulnerability of different families. This section explains the experience that allowed using adobe to build the houses of the Monasterio family, called "The twins", being a testimony that is sought to appreciate.

The houses are located in Ojo de Agua, very close to the cemetery of the valley, by the access from provincial Route N ° 307 (RP307), at KM 58. They constitute a group of two dwellings belonging each one to a member of the Monasterio family and his/her household. Both are differentiated in that they are mirrored with respect to an axis, being this the dividing line between the lots that contains each one of them.

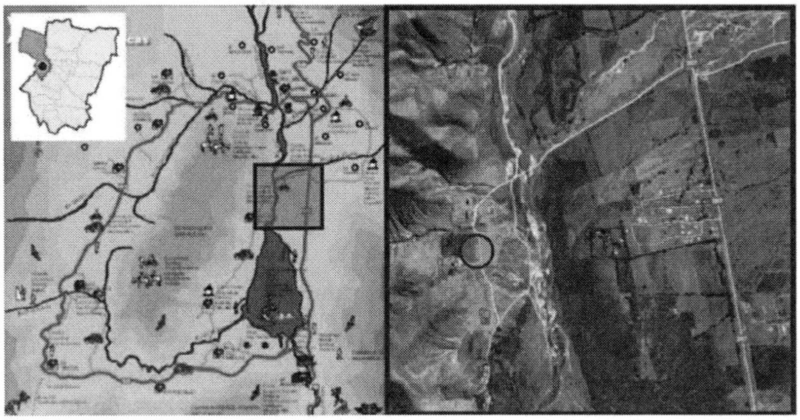

Figure 4. Location of the houses in the valley of Tafi

Figure 5. View of the Houses

The houses were executed thanks to the Federal Program of Housing and Habitat Improvement of Native and Rural Peoples (called synthetically "rural and indigenous"), between the years 2013 and 2015. This program provides materials and promotes self-construction, seeking to contain the most vulnerable population by improving their living conditions, as well as avoiding migration to urban centres (Ortega and Garzón, 2017). Among the objectives of the program are to incorporate the workforce of the beneficiaries who are familiar with the local uses and customs with the aim of respecting their cultural patterns, strengthening their family patrimony, enhancing the capacities of organization and self-management, taking advantage of existing skills to generate quality of life and root conditions, so that communities become agents of change for their own sustainable development.

The houses were built according to design guidelines that sought to provide solutions to specific needs, such as to contain an area for the productive function, while the bathroom has double access, both from the inside and from the outside; so the final design was agreed between the technicians and the beneficiaries. According to the singularity observed in these programs, the houses present elements of the local constructive tradition (vernacular architecture), while also observing the influence of the industry and the technical-constructive resolutions of the city. Although this is part of the transformation of applied technology and landscape that occurs in the valley (partly due to the great influence of tourism and the values they carry), it is a result of the articulation between the federal program and the target community. It was resolved through the cooperative construction, being built both simultaneously, and interspersing stages.

They are built with local resources and labor: foundations of stone and adobe walls. The roof is built with metal structure (C profile), and galvanized sheet cover. Inside it has an air chamber, an insulating membrane and a groove wood ceiling. As for the terminations, the houses have cement plaster, ceramic tiles inside, and the carpentry are metallic to the outside and wooden plate the interior doors.

Figure 6. View of one of the houses

According to estimates made with the Program for Calculating the Thermal Transmittance "K" (Garzón and Mendonça, 2013), this vertical enclosure in adobe has a value of 0.99 W/m2°C, compared to 1.76 W/m2°C of a hollow brick wall 18 cm wide with both sides revoked (resolution that predominates in the housing programs), which gives an account of the energy efficiency that implies the incorporation of the earth material in the construction of the walls.

Currently, these houses are being monitored with instrumental and simulated by means of software for a better analysis and understanding of their thermal-energetic behavior, whose results will be available at a later stage.

CONCLUSIONS

The constructive identity in these valleys is closely linked to adobe, a material revalued today from an approach of sustainability and design and production of environmentally conscious habitat. The characteristics of the Program for the Improvement of Rural and Aboriginal Habitat (Rural and Indigenous), constituted a significant contribution to diversity in the design of public social housing policies. They allowed to generate genuine and local labor by means of the cooperative construction, while contributing to valuing the constructive identities of the valleys of Tucumán, becoming promoters of a technological syncretism between the urban ways of building and vernacular knowledge, manifested in the close relationship between technology, community and environment. The referred cases become, for all this, in ecological and energetically efficient constructions, once they serve as prototypes, both for their conception and for the materials and constructive technologies incorporated, for the generation of affordable housing at low cost in the valleys of height of the province of Tucumán.

Although this program is not currently in force, the need to move forward to retake these initiatives with greater impetus is considered here, generating the necessary tools and relevant legal and regulatory frameworks. Understanding that the dwellings constitute the primary cells of the habitat in their different dimensions, land constructions promoted on a larger scale by social housing programs, with modalities such as those described, in which they articulated with communities would constitute very different human settlements to those currently promoted by housing policies in general.

It is thus concluded that the referred dwellings serve as testimony to a set of experiences that contribute to configure more resilient environments, in harmony with the paradigm of sustainability, serving to the recognition of the constructive diversity, collaborating in the reduction of emissions, and promoting local self-sufficiency and integration with the environment.

REFERENCES

1. Berger de White, M. y Azurdia Bravo, I. Resiliencia adaptativa y sincretismo tecnológico. Adaptación al cambio climático. Observatorio de la Sostenibilidad. Red Latinoamérica. [Consultado 3 agosto 2018]. In: http://www.acicafoc.org/wp-content/uploads/2017/08/CambioClimaticopopular.pdf

2. Bestraten, S., Hormías, E. y Altemir, A. (2010). Construcción con tierra en el siglo XXI. In: Informes de la Construcción. Barcelona, p. 18.

3. Calvente, A. Resiliencia: un concepto clave para la sustentabilidad. UAIS. 2007, junio, pp.1-4. In: http://capacitacionpedagogica.uai.edu.ar/pdf/cs/UAIS-CS-200-003%20-%20Resiliencia.pdf

4. CONAVI. Consejo Nacional de la Vivienda. Tucumán. Obras en la provincia: abril 2004. Buenos Aires: CNV, 2004.

5. Cortés Cely, O. Propiedades que definen los materiales resilientes en arquitectura. Revista de Tecnología I Journal of Technology. 2015, Volumen 14, Número 1, pp.117-126.

6. Czajkowski, J. La construcción debe pasar a un modelo de eficiencia energética y sustentabilidad [en línea]. [Consultado 5 agosto 2018]. In: https://www.revistapetroquimica.com/la-construccion-debe-pasar-a-un-modelo-de-eficiencia-energetica-y-sustentabilidad/

7. García Morales, L. y Gutierrez Colino, V. Resiliencia tecnológica. Arte y políticas de identidad. 2014, Volúmen 10-11, pp. 135-154.

8. Garzón, B. Variables bioclimáticas y uso de la energía en viviendas espontáneas y oficiales de interés social: análisis y propuestas. In: Red iberoamericana para el uso de energías renovables y diseño bioclimático en viviendas y edificios de interés social (San Martín de los Andes, Neuquén, Argentina, 2005): Los edificios bioclimáticos en los países de Iberoamérica. San Martín de los Andes, 2005, pp. 113-130.

9. Gonzalo, G. E. Manual de Arquitectura Bioclimática. Tucumán: CEEMA, [1998] 2003.

10. Minke, G. Manual de construcción con tierra. 4a ed. Bariloche: BRC Ediciones, 2013.

11. Moscoso Cordero, M. S. El adobe, sus características y el confort térmico. In: I Congreso Internacional online de Filosofía de la sustentabilidad de Vivienda Tradicional "Transformando comunidades hacia el desarrollo local", 2015. Disponible en: https://docplayer.es/73648337-El-adobe-sus-caracteristicas-y-el-confort-termico.html

12. Motta, JM. Proceso, participación, vivienda. Procesos de participación de la población en situación de pobreza en el mejoramiento del hábitat y la vivienda. Evaluación del Programa Federal de Mejoramiento de Viviendas "Mejor Vivir" en Chaco y Tucumán (2003-2011). Tesis doctoral inédita, Universidad de Buenos Aires, 2016.

13. Ortega, M. y Garzón, B. Construcción con tierra y sustentabilidad en Tucumán, Argentina. Su análisis desde las políticas públicas de vivienda. In: 17° Seminario Iberoamericano de Arquitectura y Construcción con Tierra, SIACOT (La Paz, 9-12 de octubre de 2017): Tierra - Identidades. La Paz: FAADU-UMSA / PROTERRA, 2017, pp. 407-416.

14. Thomas, H. Tecnologías para la inclusión social y políticas públicas en América Latina. Grupos de Estudios Sociales de la Tecnología y la Innovación. 2009. In: http://www.redtisa.org/Hernan-Thomas-Tecnologias-para-la-inclusion-social-y-politicas-publicas-en-America-Latina.pdf

CONCRETE & CEMENTITIOUS MATERIALS

IC-NOCMAT 2019

18th International Conference on
Non-Conventional Materials and Technologies
"Construction Materials & Technologies for Sustainability"
(18th NOCMAT 2019)
24th – 26th July 2019
Nairobi, Kenya

EFFECTS OF SAWDUST ADDITION AND PARTIAL REPLACEMENT OF CEMENT WITH CARBIDE WASTE ON STRENGTH AND SORPTION PROPERTIES OF HOLLOW CONCRETE BLOCKS

Abel O. Olorunnisola[1], Oludele O. Oladejo[1]

[1] Department of Wood Products Engineering, University of Ibadan, Ibadan, Nigeria

ABSTRACT

Housing shortage in Nigeria due to rising price of construction materials makes it logical to consider alternative building materials. Concrete blocks are generally used for building contructions in West Africa. Paradoxically, many African counries are grappling with the challenges of waste management, when some of the so called waste materials could be incorporated into building materials. The aim of this study was to investigate the possibility of incorporating sawdust and welder's carbide waste in the mix for the production of hollow concrete blocks. Representative samples of 100 x 100 x 100 mm Sawdust-Reinforced Concrete Blocks (SRCB) were produced using different percentages of Teak (*Tectona grandis*) sawdust (20%, 25%, 30%) and carbide waste (30%, 25%, 20%) in the mix. These were cured for 28 days, after which compressive strength, 24- hour Water Absorption (WA) and Thickness Swelling (TS) were determined. The mean compressive strength values ranged from 0.32 to 0.75 N/mm^2. The maximum compressive strength which was obtained in the SRCB mix containing 20% sawdust and partial replacement of Portland cement by up to 30% of carbide waste fell within the measured strength values (0.5 - 1 N/mm^2) reported for commercially produced hollow concrete blocks in Nigeria. The WA (6.1 - 10.3%) and TS (1.2 - 1.9%) were quite acceptable. The preliminary cost analysis indicated the possibility of a 20% reduction in the production of the concrete blocks with the incorporation of sawdust and carbide waste in the production mix.

Keywords: Concrete blocks, Sawdust, Carbide waste, Building materials

INTRODUCTION

Effective sustainable buildings require careful attention to the full life cycle impact of resources. Building materials, one of the key issues of sustainable architecture, should be "green". Such materials should not constitute environmental and human health risks. Other criteria for selection include rational use natural resources; energy efficiency; elimination or reduction of generated waste; low toxicity; water conservation and affordability [1,2]. Accumulation of industrial wastes in developing countries over the years has resulted in environmental pollution. Recycling of such wastes as building materials appears to be viable solution not only to such pollution problem but also to the problem of economic design of buildings. These materials apart from being locally available have easy workability and speedy construction hence reducing costs [3].

One of the common sources of industrial wastes in Nigeria is calcium carbide which is widely used in oxy-acetylene welding as well as by local hunters and miners to fuel acetylene lamps. At present, the use of carbide lamps is highly restricted in some countries (Australia, USA, New Zealand, etc.). However, such lamps are still used in many other countries, and large carbide dumps are present in many caves [4]. Spent carbide, a hazardous waste, is still commonly found in many dump sites in Nigeria. However it has been reported that percentage purity of slaked lime recoverably from spent carbide could be up to 88% [5]. Also, there currently over 1000 sawmills, largely of small scale size in Nigeria. Not been vertically integrated, these are mostly concerned with their lumber products therefore proper utilization of sawdust is seldom given due attention. It is often treated as a waste rather than re-processing it into particleboard. The sawdust thereby constitutes an environmental nuisance as they form refuse heaps in the areas where they are disposed. These refuse heaps tend to contribute harmful leachates into local water systems, thereby creating an environmental hazard.

The commonest materials for wall construction in Nigeria are concrete blocks that are made of coarse natural sand or crushed rock dust mixed with cement in certain proportion and water; and moderately compacted into shapes and cured. Cement is the most costly material in the production of concrete blocks. It therefore follows that a reduction in the quantity of cement (without severely compromising the strength and sorption properties of the end-product) will lead to the corresponding decrease in the cost of the building. In this connection, spent carbide has been reported as a suitable material for partial replacement Ordinary Portland cement in selected concrete products [6]. Also, the use of sawdust as a partial replacement for sand in concrete blocks may be helpful in reducing the cost of production of hollow concrete blocks and ultimately reducing the cost of building construction. Ducman and Kopar [7], Turgut and Algin [8] as well as Chemani and Chemani [9] had earlier reported the use of sawdust in making porous clay bricks. Besides, Adebakin et al. [10] investigated the effect of partial replacement of sand with sawdust on the strength properties of hollow concrete blocks. Al-Khaja [4]) also reported the use of carbide lime waste as an alternative material to conventional hydrated lime in cement-lime mortars. However no study was found on the combined use of sawdust and calcium carbide utilisation for cement block production.

The aim of this study was to investigate the possibility of incorporating sawdust and calcium carbide waste in the mix for the production of hollow concrete blocks for use in the building construction industry in Nigeria.

MATERIALS AND METHODS

Material collection and preparation

Sawdust of Teak (Tectona grandis) timber converted into lumber was collected from a sawmill in Ibadan, Oyo State, Nigeria. The sawdust was soaked in water for 24 hours, drained of excess water and air-dried for three. The essence of the cold water treatment was to remove soluble sugars that inhibit the compatibility between sawdust and cement, for good setting time. Sharp sand was procured from a sand seller. By visual observation and by rule of thumb the size of the sand was of regular size. Carbide waste collected from a mechanical workshop was air-dried for 7 days, pulverized using hammer mill and screened with 75nm British Standard sieve. The screened particles were further air-dried for 21 days to an average moisture content of about 15% (dry basis). Ordinary Portland cement was procured from a cement depot. It was stored in an air tight container and was used up soon after delivery to prevent strength degradation. Potable water was obtained from the University of Ibadan water supply system.

Mixing of the ingredients and molding of blocks

A set of 100 x 100 x 100mm of sawdust-reinforced concrete cubes were produced in three replicates for each ratio which were 20% sawdust, 25% sawdust, 30% sawdust, respectively were mixed with cement, carbide (20%, 25%, 30%), sand, and water. Cube production involved mixing together measured weights of sand, sawdust, carbide, cement, and water and moulding the mixture. After de-moulding the cubes were cured in water for 28 days.

Property tests

The tests were conducted in a room at an ambient temperature of 24 ± 2 0C and relative humidity of 65 ± 5%. The densities of the test cubes were computed as a ratio of their measured masses and calculated volumes. The compressive strength test was carried out on a 300 kN-capacity compression machine (Figure 1) at a loading rate of 6mm/min. The maximum loads that the samples could carry until there was failure by crushing were recorded and the compressive strength was computed using equation 1:

$$F_c = P/A \quad (1)$$

Where:
Fc = Compressive strength (N/mm2)
P= Maximum applied load (N)
A = Cross-sectional area (mm2)

FIGURE 1: COMPRESSIVE STRENGTH TEST IN PROGRESS

The water absorption test involved immersing the samples in water at room temperature for 24 hours after taking the initial weight (Wi) of the samples. The samples were withdrawn from the water to allow for water to drain before taking the final weight (Wf) of the sample. The water absorption (WA) was calculated using the equation 2:

$$WA = \frac{W_f - W_i}{W_i} \times 100 \quad (2)$$

Thickness swelling is a function of how much water is absorbed by the composite which results in swelling. The thicknesses of the test samples were measure with a venier caliper before (Ti) and after (Tf) immersion in water. The thickness swelling (TS) was then calculated using equation 3:

$$TS = \frac{T_f - T_i}{T_i} \times 100 \quad (3)$$

Triplicate samples were tested. Data were analysed using descriptive statistics and analysis of variance.

Trial production of hollow concrete blocks

For cost estimation and demonstration purposes, samples of rectangular sawdust-reinforced hollow concrete blocks, 45 cm wide, 15 cm thick, and 30 cm high were produced with a cement: sawdust: carbide: sand: water mixing ratio of 0.5: 0.3: 0.2: 8. Hand mixing was employed and the materials were turned over a number of times until an even colour and consistency was attained. The mixtures were rammed into the metal moulds, compacted and smoothed off with a wooden pallet, de-moulded as shown in Figure 2 and air- dried for7 days. Blocks produced with the mixture that gave the best results were then used to construct a demonstration building (Figure 3).The cost of production of the blocks was computed and compared with that of a typical hollow concrete block.

FIGURE 2: THE HOLLOW SAWDUST-REINFORCED CONCRETE BLOCKS

FIGURE 3: THE HOLLOW SAWDUST-REINFORCED CONCRETE BLOCKS USED TO CONSTRUCT A DEMONSTRATION BUILDING

RESULTS AND DISCUSSION

Densities of the blocks

The densities of the blocks produced with different mixtures are shown in Table 1. The values ranged between 1210 and 1884 Kg/m3. The addition of carbide waste and sawdust reduced the density of the blocks significantly ($p < 0.05$) to the lightweight block category. As noted by Ogah and Igbe [11], the densities of light weight blocks range between 625 and 1,500 Kg/m3, while that of dense blocks may be as high as 2,080 Kg/m3. The reduction in density of the blocks by calcium carbide waste and sawdust was expectedly so because of their bulk densities of between 540 and 650 Kg/m3 compared to 1270 Kg/m3 reported for cement [12].

TABLE 1: DENSITIES OF THE CONCRETE CUBES

Concrete Cube Sample	Mean Density (Kg/m^3)
Control (cement : sand ratio 1:8)	1884 ± 97.2
30% sawdust, 20% carbide in the mix	1427 ± 7.0
25% sawdust, 25% carbide in the mix	1210 ± 11.4
20% sawdust, 30% carbide in the mix	1530 ±7.2

Compressive strength

The results of the compressive strength test are presented in Table 2. Expectedly, given their lower densities, the compressive strength values of all the samples containing calcium carbide sawdust were lower than that of the control. The positive correlation between density and compressive strength is shown in Figure 4. The mean compressive strength of only the mix containing 20% sawdust and 30% of carbide waste (0.75N/mm2), representing 75 % of the mean compressive strength of the control samples, fell within the measured strength values (0.5 - 1 N/mm2) of commercially available concrete hollow blocks in Nigeria, although The range of compressive strength values specified in the Nigeria Industrial Standard NIS 87:2000 for such blocks is between 2.5N/mm2 to 3.45N/mm2. What the result suggests is that the blocks are mainly suitable for use as partition walls in buildings. The test values, however, compare favourably with those reported by Oyetola and Abdullahi [13] for concrete blocks produced by partial replacement of cement with rice husk ash, i.e., 0.56 – 0.97 N/mm2; and Adebakin et al. [10] for hollow sawdust-reinforced concrete blocks (<3 N/mm2).

TABLE 2: COMPRESSIVE STRENGTH OF THE SAMPLES

Samples	Mean Load (KN)	Compressive Strength (Nmm^{-2})
Control (cement : sand, 1:8)	9.4	1.00
30% sawdust, 20% carbide	3.2	0.32
25% sawdust, 25% carbide	4.5	0.45
20% sawdust, 30% carbide	7.4	0.75

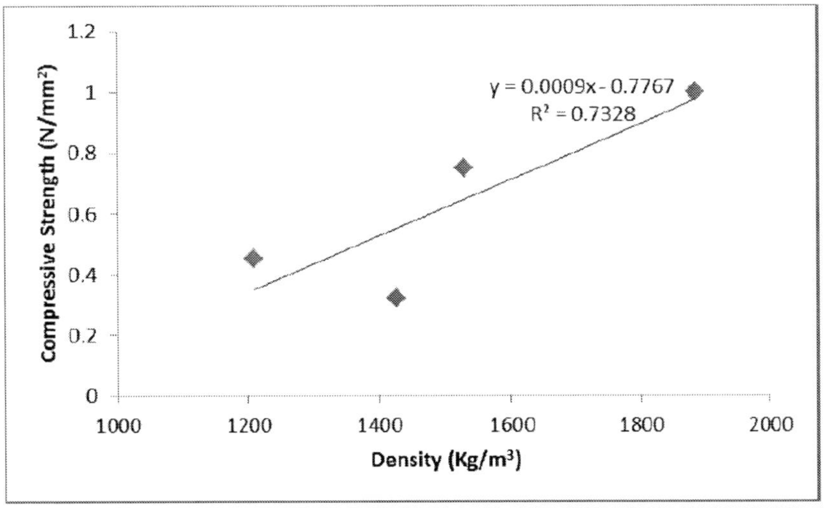

FIGURE 4: CORRELATION BETWEEN DENSITY AND COMPRESSIVE STRENGTH OF THE CONCRETE CUBES

Water absorption

Table 3 shows the mean values of water absorption (WA) which ranged from 6.1 to 10.3%, for the experimental samples, compared with 3.2 % for the control. The presence of sawdust and calcium carbide waste contributed to the WA. Samples containing higher percentages of sawdust had the highest WA. Incidentally, the samples containing 20% sawdust and 30% calcium carbide waste which exhibited the greatest compressive strength among the experimental samples also exhibited the lowest WA of 6.1%. As shown in Figure 5, there was a positive correlation between the density and the WA.

TABLE 3: WATER ABSORPTION TEST RESULTS

Sample	Mean WA (%)	Mean TS (%)
Control (cement : sand, 1:8)	3.2 ± 0.2	0.7 ± 0.2
30% sawdust, 20% carbide	10.3 ± 0.4	1.9 ± 0.6
25% sawdust, 25% carbide	9.2 ± 01.1	1.7 ± 0.5
20% sawdust, 30% carbide	6.1 ± 1.0	1.2 ± 0.9

Thickness swelling

The results of the thickness swelling test are presented in the Table 3. The values ranged between 07 % (control) and 1.9 %. There was an increase in TS with the addition of sawdust and calcium carbide. However, the sawdust appeared to largely responsible for the swelling because the thickness swelling values increased with percentage increase in sawdust content of the blocks. A positive linear correlation was also observed between the WA and TS of the samples as shown in Figure 6.

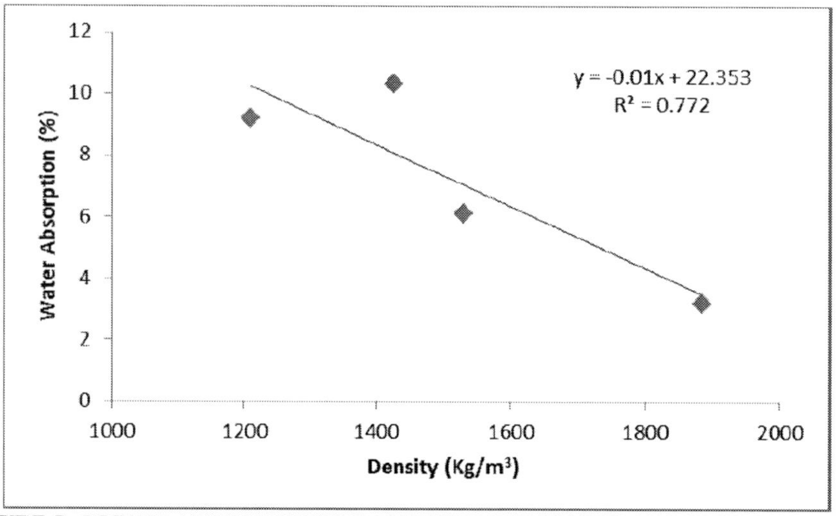

FIGURE 5: CORRELATION BETWEEN DENSITY AND WATER ABSORPTION BY THE CONCRETE CUBES

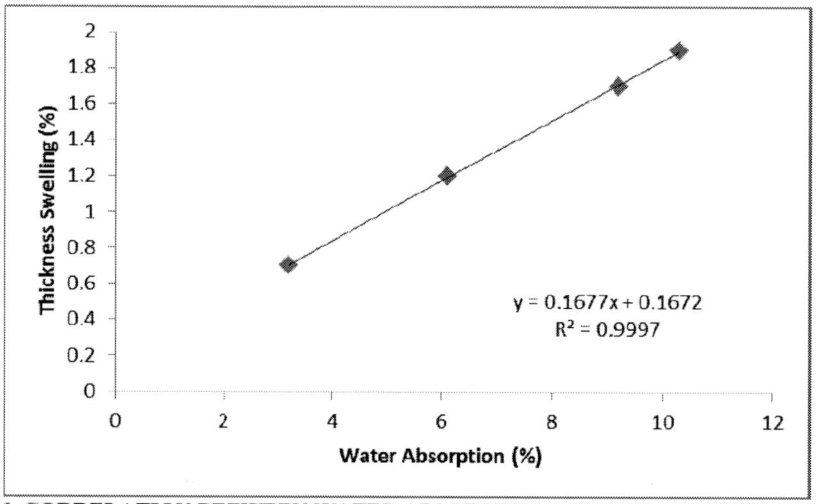

FIGURE 6: CORRELATION BETWEEN WATER ABSORPTION AND THICKNESS SWELLING OF THE CONCRETE CUBES

Estimated cost of production

The cost of production of a sawdust-reinforced block estimated at N115.40 was computed as shown in Table 4. This is about 23 % cheaper than a conventional hollow concrete block of the same size which was being sold at N150.00k per unit at time of the study.

TABLE 4: COST OF PRODUCTION OF A SAWDUST-REINFORCED BLOCK

S/No.	Cost Item	Quantity required per block	Cost Naira(₦)*
1	**Cement**	1.8 Kg	50.40
2	**Calcium Carbide waste**	0.54 Kg	5.40
3	Water	990 ml	0.80
4	Sand	¾ of Head pan	37.50
5	Labour		10.00
6	Miscellaneous (10%)	-	10.40
Total Cost per unit block			**115.4**

*1 Naira = 305 USD

CONCLUSIONS

The possibility of incorporating sawdust and calcium carbide waste in the mix for the production of hollow concrete blocks was investigated. It was concluded that blocks acceptable for low-stress applications could be produced based on partial replacement of Ordinary Portland cement with about 20 % of sawdust and 30% of calcium carbide waste

REFERENCES

1. Calkin, M. Materials for sustainable sites: a guide to the evaluation, selection and use of sustainable construction materials. Hoboken, NJ, USA: John Willey & Sons, 2009.

2. Haggag MA, SK Elmasry. Integrating passive cooling techniques for sustainable building performance in hot climate with reference to UAE. In C.A. Brebbia & Berlatos (ed.) Sustainable Development and Planning V WIT Press, 2011

3. Paki T, MA Halil. Limestone dust and wood sawdust as brick material. Department of Civil Engineering, Harran University, Osmanbey Campus, Sanliurfa, Turkey. 2006

4. Al-Khaja WA. Potential use of carbide lime waste as an alternative material to conventional hydraded lime of cement-lime mortars. Engineering Jornal of Qatar University, 1992, 5: 57-67.

5. Chukwudebelu JA, Igwe CC, Taiwo OE, OB Tojola.Recovery of pure slaked lime from carbide sludge: case study of Lagos state, Nigeria. African Journal of Environmental Science and Technology, 2013. 7(6):490 -495.

6. Olorunnisola AO, A Ogundipe. Impact and water resistance of rattan composites produced with rice husk ash and welder's carbide waste as partial replacement for Portland cement, in Fernando Caldeira (ed.), Towards Forest Products and Processes with

Lower Environmental Impact, Published by University of Fernando Pessoa, Porto, Portugal, p2014, p. 199-218.

7. Ducman V, T Kopar. Sawdust and paper-making sludge as pore forming agents for lightweight clay bricks. 2001, Ind. Ceram. 21(2):81-86.

8. Turgut P, HM Algin Limestone dust and wood sawdust as brick material. Build Environ. 2006, 42:3399-3403

9. Chemani H, B Chemani. Valorization of wood sawdust in making porous clay brick. Scientific Research and Essays 2013, 8(15):609 -614.

10. Adebakin IH, Adeyemi AA, Adu JT,.Ajayi FA, Lawal AA, OB Ogunrinola. Uses of sawdust as admixture in production of lowcost and light-weight hollow sandcrete blocks. American Journal of Scientific & Industrial Research, 2012 3(6):458-463.

11. Ogah O, EL Igbe 2016. Strength characteristics of hollow sandcrete blocks produced in Wadata-Makurdi, Nigeria. International Journal of Current Trends in Engineering & Research, 2(5): 181 – 187

12. Adefemi A, Muhammad U, Kebbi UM, Olugbenga S. Effect of admixture on fire resistance of ordinary Portland cements Concrete. 2013; Civil and Environmental Research, 3,(1) ISSN 2222-1719.

13. Oyetola EB, M Abdullahi. The use of rice husk ash in low-cost sandcrete block production. Leonardo Electronic Journal of Practices and Technologies 2006, 8:58-70

18th International Conference on
Non-Conventional Materials and Technologies
"Construction Materials & Technologies for Sustainability"
(18th NOCMAT 2019)
24th – 26th July 2019
Nairobi, Kenya

SELF-COMPACTING CONCRETES USING CALCIUM-RICH ASHES AS ALTERNATIVE FILLERS

Rodrigo Araújo Pereira Lima[1], Erika Pinto Marinho[1], Ana Cecilia Vieira da Nóbrega[1]

[1] Postgraduate Program in Civil and Environmental Engineering, Agreste Academic Center, Federal University of Pernambuco, PE, BR;

ABSTRACT

Algaroba wood is a usual energy source used in laundry-boilers in the Local Productive Arrangement (LPA) of Clothes, Pernambuco – Brazil. However, these algaroba wood ashes are constituted primarily by calcium carbonate (CaCO3), which can act as an alternative filler. The limestone fillers are widely used as fines in self-compacting concretes (SCC), being a natural resource that can be extinguished in the future. In order to verify the feasibility of the incorporation of the algaroba ashes in the SCC, self-compacting concretes (SCC) were evaluated with partial (0, 25, 50, 75%) and total replacement of the conventional limestone by the alternative filler. Cohesion, fluidity, and resistance to segregation tests were carried out according to the Brazilian Standards. Also, other fresh and hardened SCC properties were analyzed. The mixes were optimized using raw-materials compactness experimentally measured. With this research, it was possible to confirm that the algaroba wood ash is a viable material for incorporation in the SCC up to 50% (by cement weight), keeping the self-compactness ability, as well as providing compressive resistance gains in the early ages.

Keywords: algaroba wood ash, self-compacting concrete, filler.

INTRODUCTION

In the clothing local productive arrangement (LPA) in the state of Pernambuco, Brazil, one of the primary energy sources used in the water heating process for the dyeing of jeans is the algaroba wood (Prosopis julifora). This specimen is freely commercialized because it is not native to the Brazilian flora, as well as it has a high calorific performance [1], and low-cost compared to other fuel sources [2]. More than 200 industrial laundries registered in the region use it [3], generating around 1560 tons/month of ashes [2, 4]. Although some of these ashes are used informally to correct the soil acidity, most of this by-product is discarded without any criterion, being an environmental problem [1, 4].

On the other hand, these ashes can be considered a natural alternative filler, once they are rich in calcium carbonate (CaCO3) [5, 6]. Also, their grains have a spherical shape and irregular surface, with a low modulus of fineness, low magnesium oxide content, similar coloration of the Portland cement and no expansive oxides. The fact that explains their previous use as filler in cementitious materials [1, 2, 5-7].

Concerning self-compacting concretes, according to Silva and Brito [8], the presence of fillers improves the cementitious matrix by filling it. In the fresh state, the rheology suitable for molding is achieved due to its interaction with the superplasticizer [8]. The cohesion is increased, as well as the exudation and segregation of the mixture are reduced, allowing better flow conditions [9]. When the filler has a specific surface area smaller than that of the cement, there are still improvements in the passing ability verified by the L-box and Ring-J tests [10].

In the hardened state, the limestone filler in the SCC increases the density of the mixture due to the packing and the nucleation points, favoring resistance gains in the early and advanced ages, reducing porosity [9], making it difficult to penetrate aggressive agents and improving the transition zone [9, 11].

Thus, as the limestone filler is a natural material that may be extinguished in the future, it was sought to study the SCC using an alternative solution of partial and total replacement of the conventional limestone filler by algaroba wood ashes, observing the properties of the SCC.

The possibility of using these ashes in cement materials contributes to the sustainable development of the construction sector, in order to reduce environmental impacts produced by the inadequate disposal of this solid waste, and still generating employment and income opportunities, by adding commercial value to a residue and provide technological studies for the use of the material in construction.

EXPERIMENTAL PROGRAM

Materials

Table 1 presents the main physical characteristics and the chemical composition of the cement used, which complies with the Associação Brasileira de Normas Técnicas EN 197-1 [12].

TABLE 1 - CHARACTERISTICS AND COMPOSITION OF THE CEMENT USED.

Physical properties							
Designation	Type	Clinker (wt. %)	Density (g/cm^3)	Blaine/BET (cm^2/g)		d_{50} (μm)	Addition (wt. %)
C	CP V ARI	98.5	3.09	4692/20956		15.9	0-10 (Limestone)
Chemical composition							
Oxides (wt. %)	SiO_2	Al_2O_3	Fe_2O_3	CaO	SO_3	MgO	K_2O	LOI
	20.43	4.90	1.83	65.40	3.60	1.06	0.27	1.44

A calcite sedimentary limestone filler with a mesh size of 200 mesh (0.075 mm) was used. The algaroba wood ashes were collected in a laundry located in the city of Caruaru-PE, which uses algaroba firewood exclusively as the energy source. The material was ovendried at 105 °C, then sieved into 200 mesh (0.075 mm) mesh. The chemical analyze was performed using an X-ray fluorescence through the Rigaku spectrometer, model

Primini, with tube tension of 50W (Na to U) using pressed samples. The physicochemical properties of the limestone filler are shown in Table 2.

TABLE 2 - PHYSICAL-CHEMICAL CHARACTERIZATION OF THE FILLER AND ALGAROBA ASHES USED

Chemical composition								
Oxides (wt. %)	MgO	Al_2O_3	SiO_2	P_2O_5	Fe_2O_3	SO_3	K_2O	CaO
Calcareous filer	5.11	1.49	3.23	0.22	1.9	0.12	0.28	87.65
Calcareous ashes	2.44	0.37	0.83	1.99	0.35	1.33	12.30	77.52

Physical properties	Calcareous filler	Algaroba ashes
Blaine (cm^2/g)	5096	6857
BET (cm^2/g)	30045	47142
Density (g/cm^3)	2.71	2.76

The X-ray diffraction used a Bruker D8 Advanced Davinci diffractometer with CuKα radiation (40 kV/40 mA), and 0.02° step every 0.5 second in the interval of 5° <2θ <70 °. Figure 1 shows the X-ray diffractogram of the limestone filler and algaroba ashes. Analyzing the diffractograms (Figure 1), both the limestone filler and the algaroba ash present calcium carbonate as the peak of greater intensity near 30°.

The microscopy study of the calcareous ash was carried out on an SSX - 500 Shimadzu scanning electron microscope. The samples were fixed in a sample port with carbon tape after passing through a metallization process on Shimadzu's IC-50 gold metallizer for 5 minutes with a potential variation of 1.8 kV [5].

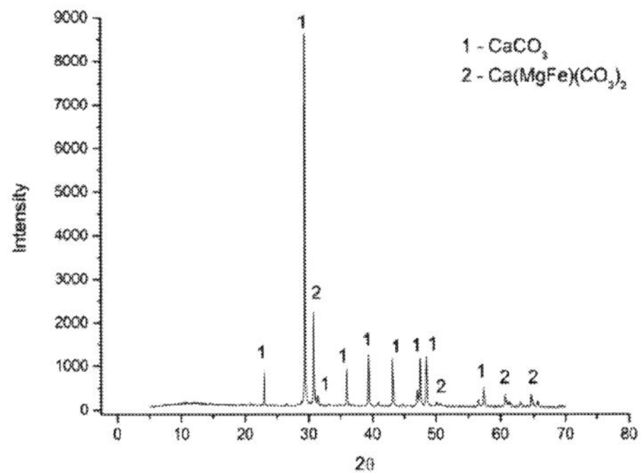

(a)

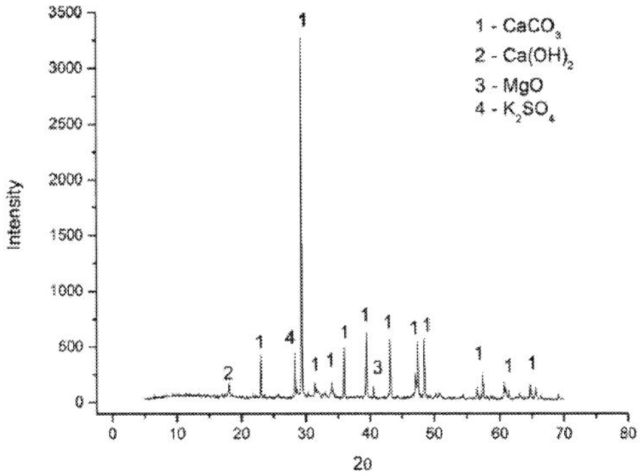

(b)

FIGURE 1 - X-RAY DIFFRACTOGRAM OF THE LIMESTONE FILLER (A) AND ALGAROBA ASHES (B) USED.

Granitic rounded coarse aggregate (designation: G) 0/9.5 mm (density = 2630 kg/m3; specific mass = 1,360 kg/m3) and silicous sand (designation: S) 0/4.75 mm (density = 2,580 kg/m3; specific mass = 1,470 kg/m3) were used. The granular skeleton was siliceous-granitic (designation: GE) 0/9.5 mm (density = 2,605 kg/m3; specific mass = 1,415 kg/m3). All aggregates were oven-dried at 105 °C. The granulometric distribution of aggregates (G, S) and the GE can be seen in the Figure 2.

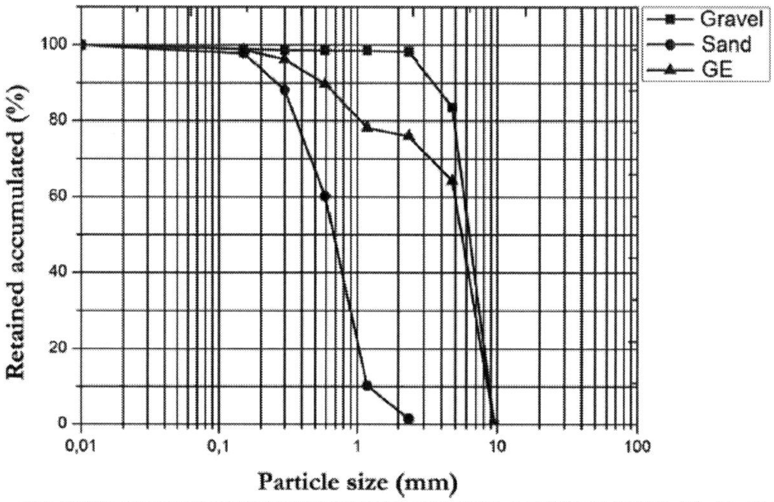

FIGURE 2 - GRANULOMETRIC DISTRIBUTION OF THE AGGREGATES (G, S) AND THE GE.

The superplasticizer employed was a modified eter-polycarboxylate type available as a commercial solution (specific gravity =1.067-1.107; active solid content by weight = 28.5-31.5%; viscosity<150; pH = 5-7). Distilled water (pH = 6.55) was used in all mixtures.

Methods

Composition, mixing sequence, and curing

The concrete compositions tested are presented in Table 2. A cement consumption of 360 kg/m³ with a content of 25% of limestone filler was added to determine the traces of the SCC. The water/fines ratio was fixed at 0.5. The superplasticizer content was achieved in each mixture to obtain an SF2 class SCC (spread from 660 to 750 mm) to observe the ashes influence on the compressive strength.

Table 2 presents the consumption of each mixture produced with the substitution of the limestone filler by the alternative filler (algaroba ashes) in the following contents: 0 (I), 25 (II), 50 (III), 75 (IV) and 100% (V). They were defined using a diphasic design approach combining a paste with the optimized flow and a wet aggregate skeleton [13, 14]. As can be seen in Table 3, the increase in the demand for superplasticizer occurred with the increase of the incorporation of the algaroba ashes (1.575 - I (0%), 2.5425 - II (25%), 3.375 – III (50%), 3.825 - IV (75%), and 4.05 - V (100%). This behavior is due to the progressive increase in the specific surface of the fines since the ashes have a BET specific area of 47,142 cm2/g against 30,045 cm2/g of the limestone filler.

The mixing sequence comprised of three stages [14]: first, the introduction of gravel, cement, limestone filler and sand followed by 60 s of mixing; second, the introduction of water and 1/3 of the mass of the superplasticizer followed by 90 s of mixing; third, the introduction of 2/3 of the superplasticizer followed by 90 s of mixing. The procedure of molding the specimens followed the parameters assigned by ABNT NBR 5738: 2016 [15], where cylindrical molds of 10 cm x 20 cm (diameter x height) were used. After 24h, all the samples were immersed in water at room temperature (20 °C ± 1 °C) until the tests were performed.

TABLE 3 - SELF-COMPACTING CONCRETE COMPOSITION TESTED.

Materials (kg/m³) / Nomenclature	I (0%)	II (25%)	III (50%)	IV (75%)	V (100%)
Cement	360	360	360	360	360
Calcareous filler	90	67.5	45	22.5	-
Algaroba ashes	-	22.5	45	67.5	90
Sand	784.85	785.01	785.16	785.32	785.47
Grave	784.85	785.01	785.16	785.32	785.47
Superplasticizer	1.575	2.5425	3.375	3.825	4.05
Water	225	225	225	225	225

Fresh state tests

The self-compacting ability of the concretes was assessed using various standardized methods and their criteria. Taking the time at the end of mixing as t0, the same test sequence was applied to all the concretes: spread (ABNT NBR 15823-2: 2017 [16]) and t500 (ABNT NBR 15823-2: 2017 [16]) at t0, L-box (3 rebars, ABNT NBR 15823-4: 2017 [17]) and Vfunnel (ABNT NBR 15823-5: 2017 [18]) at t0+5'. In parallel with these tests, the and visual stability index (IEV) (ABNT NBR 15823-2: 2017 [16]) was also measured.

Hardened state tests

Compressive strength values were measured at 1, 7 and 28 days of age according to Brazilian standard ABNT NBR 5739:2007 [19] using a Shimadzu hydraulic press with a capacity of 2000 kN. Total water absorption (ABNT NBR 9778:2009 [18]) and capillary water absorption (ABNT NBR 9779:2012 [20]) were carried out at 28 days. Each test result reported is the average of three values.

To determine the electrical resistivity was carried out the method adopted by Lavagna et al. [21]. A water-saturated sponge was used between the concrete sample and the copper plate to improve the electrical contact. The electrical resistance was measured by applying an alternating current of 300 mV at 200 kHz. In this test, oven-dried specimens (80 ºC, 24 hours) with dimensions of 10 cm x 10 cm (diameter x height) were adopted.

Also, the SCC resistance was verified to the attack of three different types of acids. Sulfuric acid, acetic and hydrochloric acid, with a molar concentration of 0.1 N each. The influence of the addition of algaroba ashes on the acid attack resistance was analyzed.

Sulfuric acid, acetic acid and hydrochloric acid, with a molar concentration of 0.1 N each were applied. For this test three cube samples, 2 cm in size, were extracted from the central part of the cylindrical specimens, which were oven-dried at 65 ± 5 °C for 24 hours to remove moisture.

After drying, the dry mass of the cubes was measured. Each of the three samples was placed in a jar containing 500 ml of the acid. Then the vials were subjected to a constant rotation for 24 hours on a rotary flask shaker. After the 24 hours rotation, the samples were taken out of the acid and washed with distilled water, then passed through the procedure drying at 65 ± 5 ° C for 24 hours. Finally, the dry samples were weighed, and the difference in mass was verified.

RESULTS ANS DISCUSSION

Fresh State Tests

Table 4 gathers together the results (average and experimental scatter on two batches) corresponding to the slump flow (spread and t500), the filling ratio (H2/H1) measured in the L-box test, J-ring, V-funnel, and IEV according to the ABNT NBR 15823-1:2017 [21].

TABLE 4 - SPREAD, T_{500}, H_2/H_1, J-RING, V-FUNNEL, IEV, AND RESPECTIVE CLASSIFICATIONS ACCORDING TO THE ABNT NBR 15823-1:2017 [21].

	Spread (mm)	t_{500} (s)	H_2/H_1	J-ring (mm)	V-funil (s)	IEV (visual)	Classification based on the average value (ABNT NBR 15823-1:2017 [21])
0AFM	700	1.6	0.92	15	4.5	IEV 1	SF2, VS1, PJ1, VF1, IEV 1
25AFM	705	1.1	0.90	10	3.9	IEV 0	SF2, VS1, PJ1, VF1, IEV 0
50AFM	665	1.3	0.81	22	5.2	IEV 0	SF2, VS1, PJ1, VF1, IEV 0
75AFM	680	2.9	0.75	23	7.1	IEV 0	SF2, VS2, PJ1, VF1, IEV 0
100AFM	695	2.3	0.76	15	6.8	IEV 1	SF2, VS2, PJ1, VF1, IEV 1

All the properties of fluidity, passing ability, and resistance to segregation were reached when the limestone filler was partially replaced by algaroba ashes in contents of up to 50%, obtaining self-compacting concretes with classes similar to the standard: SF2, VS1, PJ1, and VF1. Only the mixtures 0AFM (0%), 25AFM (25%) and 50AFM (50%) passed in the L-box test, mixtures 75AFM (75%) and 100AFM (100%) were below the allowable limit. Thus, substitution rates of the lime filler by the algaroba ash above 50% are not recommended. Also, the visual stability was improved when the limestone filler was replaced by 25% and 50% from IEV classification (stable) to IEV 0 (highly stable).

Loss of passing ability with an increasing amount of algaroba ashes incorporated is closely related to the increase in viscosity related to the larger surface area of the algaroba ash. There is thus a larger wetting area due to the use of a material with more roughed characteristics, which will rheologically block the mixture. Thus, although all the blends have the same class of spreading, they have different flows and different passing abilities due to the increase of the viscosity due to the increase of the surface area of the algarroba ashes.

Hardened State Tests

Compressive strength values measured at 1, 7 and 28 days of age and the respective standard deviation are exposed in Figure 3.

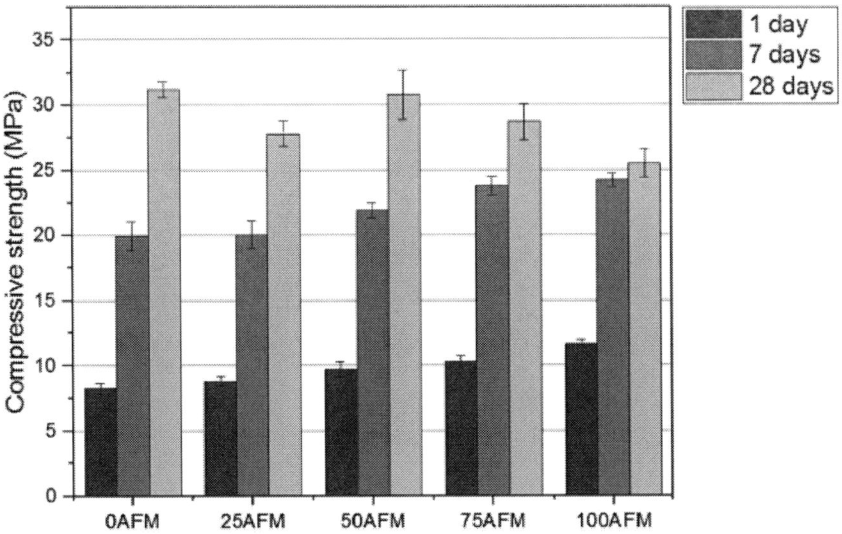

FIGURE 3 - COMPRESSIVE STRENGTH AT ONE DAY, SEVEN DAYS AND 28 DAYS.

In the hardened state, it was observed that the presence of algaroba ashes accelerated the development of the compressive resistance in the first ages (1 and seven days). That is an indication that the presence of algaroba ashes in the system can accelerate the initial hydration reactions of the cementitious compounds.

At the 28 days, the proportion that came closest to the standard was trace III (50% replacement), with average compressive strengths of 31.19 MPa to 30.77 MPa. It should be noted, however, that the average compressive strength decreases around 18% (30 MPa to 25 MPa) when replacing 100% by the alternative filer, while still maintaining an average value for concrete of 25 MPa.

According to Sua-Iam et al. [22], the use of limestone in the production of concrete improves the distribution of the particles of the granular skeleton of the mixture, guaranteeing a better arrangement of the system [10, 23, 24], as well as a reduction in the friction between the grains. In addition to the packing [25], the limestone filler also influences the cement hydration process through the interaction between tricalcium silicate (C3S) and calcium carbonate (CaCO3), acting as a nucleating agent [26]. Recent research indicates that the filler also has reactivity, with the formation of carboaluminates and monocarboaluminates, positively influencing the microstructures of cementitious materials [27, 28], favoring resistance gains in the early and advanced ages, reducing porosity [9], making it difficult the penetration of aggressive agents and improving the transition zone [9, 11].

Concerning the ashes, besides the increase in the nucleation points due to the increase of the specific surface area, the demand for more superplasticizer improves the dispersion in the cement, improving the resistance in the early ages. Also, the ashes seem to present a better nucleation behavior than the calcareous filer due to their morphology, as can be seen in the electron micrographs of the ashes below 75 μm, Figure 4. It is verified that the ash presents irregular and porous appearance, being formed by agglomerates with a predominantly rounded form resulting from the agglomeration of particles. That is, the ash does not appear as separate particles, but rather as a synthesized product and slightly porous in its structure.

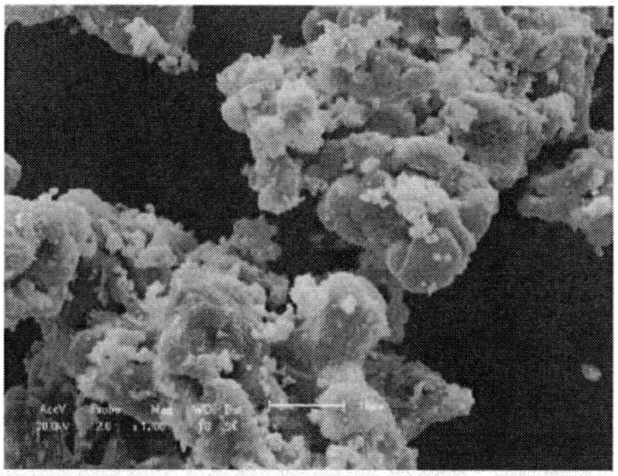

FIGURE 4 - SCANNING ELECTRON MICROGRAPH OF THE ALGAROBA ASHES USED.

Results about total water absorption indicated: 6.3% (0AFM, 0%), 7.5% (25AFM, 25%), 5.9% (50AFM, 50%), 7.6% (75AFM, 75%), and 8% (100AFM, 100%). The lowest value was to the mixture with 50% of algaroba ashes, ratifying the best compression strength results at 28 days to this mixture. Also, the capillary water absorption results corroborate with the total water absorption ones, as can be seen in Figure 5. It should be noted that many phenomena govern the responses of a capillary absorption test, from the granulometry of the fillers, capillary size, and diameter, tortuosity effects, molding failures, among others. Thus, it is difficult to recommend a single explanation for the observed behaviors.

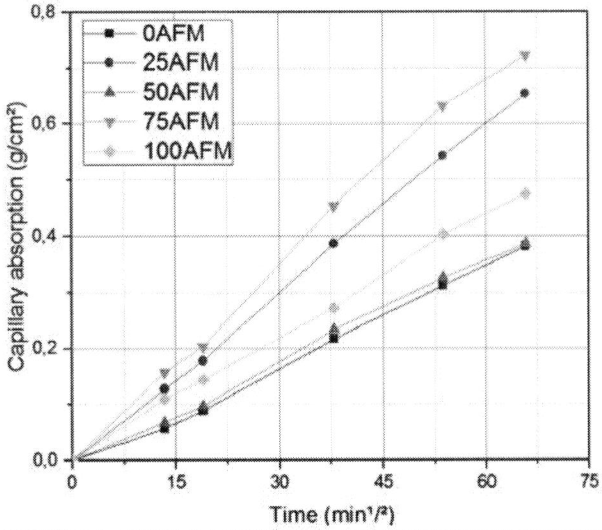

FIGURE 5 - CAPILLARY WATER ABSORPTION AT 28 DAYS.

Regarding the electrical resistivity, the mixture with 50% substitution of the limestone filler by the ashes (mix III, 50%) presented greater electrical resistivity to the other traces, 12.5% more than the mix 0AFM (0%) and 24.1% more than the mix 100AFM (100%). Providing, therefore, less ionic mobility; in this way, it can be more resistant to the process of reinforcement corrosion in the structures of reinforced concrete, benefiting the durability.

Concerning the acid attack behavior, Table 5, can be seen that the mixture 50AFM shows the lowest loss of weight when exposed to the sulfuric and chloridric acid when compared to the reference and the 100% algaroba ashes mixtures. When exposed to the acetic acid, the lowest loss of weight was achieved to the reference mix.

TABLE 5 - LOSS OF WEIGHT DUE TO THE ACID ATTACK.

Mixture	Acids / Loss of weight (%)		
	Acetic	Sulfuric	Chloridric
0AFM	9.97	6.13	28.76
50AFM	19.91	4.49	25.04
100AFM	21.64	6.04	29.82

CONCLUSIONS

The replacement of the calcareous filler by the algaroba ashes is feasible at contents up to 50% so that the SCC does not lose its self-compacting properties. Ash has proved to be an addition that can be successfully incorporated, providing compressive resistance gains in the early ages, as well as keeping it at 28 days and improving the water transport properties and some durability indicators.

REFERENCES

1. Mota, M.H.A., Concreto seco com incorporação de cinzas de madeira de algaroba (prosopis juliflora) moldado sob pressão, in Mestrado em Engenharia Civil e Ambiental. 2014, Universidade Federal de Pernambuco: Caruaru, PE, Brazil. p. 122.

2. Melo, M.C.S., Estudo de argamassas adicionadas de cinzas de algaroba geradas no arranjo produtivo local de confecções do agreste pernambucano, in Mestrado em Engenharia Civil e Ambiental. 2012, Universidade Federal de Pernambuco: Caruaru, PE, Brazil. p. 122.

3. Melo, M.C.S., et al., Cal produzida a partir de cinza de biomassa rica em cálcio. Cerâmica, 2018. 64: p. 318-324.

4. Moura, M.H.A., Incorporação de cinzas de algaroba geradas no APL de confecções do agreste pernambucano em concreto betuminoso usinado a quente – CBUQ, in Mestrado em Engenharia Civil e Ambiental. 2017, Universidade Federal de Pernambuco: Caruaru, PE, Brazil. p. 97.

5. Leloup, W.A., Efeito da adição de lodo têxtil e cinzas de lenha gerados no APL de confecções pernambucano em argamassas de cimento Portland, in Mestrado em Engenharia Civil e Ambiental. 2013, Universidade Federal de Pernambuco: Caruaru, PE, Brazil. p. 122.

6. Nascimento, J.E.M.F., Avaliação dos efeitos da substituição da cal hidratada por cinzas de algaroba em argamassas de revestimento, in Mestrado em Engenharia Civil e Ambiental. 2014, Universidade Federal de Pernambuco: Caruaru, PE, Brazil. p. 99.

7. Pires, D.R.P., Desenvolvimento de argamassas com substituição parcial do cimento Portland por cinzas de algaroba geradas do APL (Arranjo Produtivo Local) de confecções pernambucano), in Mestrado em Engenharia Civil e Ambiental. 2016, Universidade Federal de Pernambuco: Caruaru, PE, Brazil. p. 96.

8. da Silva, P.R. and J. de Brito, Experimental study of the porosity and microstructure of self-compacting concrete (SCC) with binary and ternary mixes of fly ash and limestone filler. Construction and Building Materials, 2015. 86: p. 101-112.

9. Tutikian, B.F. and D.C. Dal Molin, Concreto Auto-Adensável. 1 ed. 2008, São Paulo: PINI.

10. Uysal, M., K. Yilmaz, and M. Ipek, The effect of mineral admixtures on mechanical properties, chloride ion permeability and impermeability of self-compacting concrete. Construction and Building Materials, 2012. 27(1): p. 263-270.

11. Bosiljkov, V.B., SCC mixes with poorly graded aggregate and high volume of limestone filler. Cement and Concrete Research, 2003. 33(9): p. 1279-1286.

12. British Standard Institution. BS EN 197-1. Cement – Part 1: Composition, specifications and conformity criteria for common cements, . February 2001, London.

13. Diederich, P., M. Mouret, and F. Ponchon, Simple tools for achieving selfcompacting ability of concrete according to the nature of the limestone filler. Construction and Building Materials, 2013. 48: p. 840-852.

14. da Nóbrega, A.C.V., et al., Improved 28-day compressive strength of SCC mixed and cured in hot weather: Non-conventional behavior. Construction and Building Materials, 2018. 173: p. 650-661.

15. Breintebach, S.B., Desenvolvimento de argamassas para restauração utilizando resíduo do polimento de porcelanato, in Programa de pós-graduação em Ciência e Engenharia dos Materiais. 2013, Universidade Federal do Rio Grande do Norte: Natal.

16. Souza, P.A.B.F., Estudo do comportamento plástico, mecânico, microestrutural e térmico com concreto produzido com polimento de porcelanato, in Programa de pós-graduação em ciência e engenharia dos materiais. 2007, Universidade Federal do Rio Grande do Norte: Natal.

17. Lee, C.Y. and S.R. Wang, Application of four-electrode method to analysis resistance characteristics of conductive concrete. World Acad Sci Eng Technol, 2010. 4: p. 77-80.

18. Wen, S. and D.D.L. Chung, Effect of stress on the electric polarization in cement. Cement and Concrete Research, 2001. 31(2): p. 291-295.

19. Commitee, R.-R.C.P., Draft Recommendation: measurement of hardened concrete carbonation depth CPC-18, in Materials and Structures. 1984, RILEM – Rilem Concrete Permanent Commitee. p. 435-440.

20. Zhu, X., et al., Characterisation of pore structure development of alkaliactivated slag cement during early hydration using electrical responses. Cement and Concrete Composites, 2018. 89: p. 139-149.

21. Lavagna, L., et al., Cement-based composites containing functionalized carbon fibers. Cement and Concrete Composites, 2018. 88: p. 165-171.

22. Sua-iam, G., P. Sokrai, and N. Makul, Novel ternary blends of Type 1 Portland cement, residual rice husk ash, and limestone powder to improve the properties of selfcompacting concrete. Construction and Building Materials, 2016. 125: p. 1028-1034.

23. De Schutter, G., Effect of limestone filler as mineral addition in self-compacting concrete, in 36º Conference on Our World in Concrete & Structures. 2011: Singapore.

24. Gesoğlu, M., et al., Fresh and hardened characteristics of self compacting concretes made with combined use of marble powder, limestone filler, and fly ash. Construction and Building Materials, 2012. 37: p. 160-170.

25. Moraes, K.A.M., Otimização do uso de adições minerais para a produção de concreto auto-adensável, in Doutorado em Engenharia Civil. 2010, Universidade Federal de Pernambuco: Recife, PE, Brazil.

26. Péra, J., S. Husson, and B. Guilhot, Influence of finely ground limestone on cement hydration. Cement and Concrete Composites, 1999. 21(2): p. 99-105.

27. Schöler, A., et al., Early hydration of SCM-blended Portland cements: A pore solution and isothermal calorimetry study. Cement and Concrete Research, 2017. 93: p. 71-82.

28. Moon, G.D., et al., Effects of the fineness of limestone powder and cement on the hydration and strength development of PLC concrete. Construction and Building Materials, 2017. 135: p. 129-136.

18th International Conference on
Non-Conventional Materials and Technologies
"Construction Materials & Technologies for Sustainability"
(18th NOCMAT 2019)
24th – 26th July 2019
Nairobi, Kenya

EFFECTS OF TEMPERATURE IN SELF-COMPACTING CONCRETES USING CALCIUM-RICH ASHES AS ALTERNATIVE FILLERS

Jonatércio Florêncio de Vasconcelos Marinho[1], Erika Pinto Marinho[1], Ana Cecilia Vieira da Nóbrega[1]

[1] Postgraduate Program in Civil and Environmental Engineering, Agreste Academic Center, Federal University of Pernambuco, PE, BR;

ABSTRACT

The use of wood as an energy matrix still prevails in industrial scenarios, generating by-products. The ashes sometimes are discarded in nature without an adequate destination. Algaroba trees are in the northeast part of Brazil, India and other places. Algaroba tree ashes are rich in calcium carbonate ($CaCO_3$) and can be applied instead of limestone. Limestone fillers are commonly used as fines in cementitious materials. The use of fines is essential to ensure the maintenance of the fresh-state behavior in self-compacting concretes (SCC). Also, mixing and casting concrete in hot climates (for example, in tropical regions) can affect the fresh state conditions, due to the evaporation of the mixing water. Self-compacting concretes are even more susceptible due to their high-water demand. Thus, the present research aimed to analyze the properties in the fresh and hardened state of self-compacting concretes made incorporating the algaroba trees ashes produced in hot weather conditions. For this, an optimized compact granular skeleton was developed based on the constituent materials. The raw-materials were preheated simulating a hot climate region. The use of algaroba ashes improved the compressive strength and decreased the permeability of the SCC, especially in hot weather conditions. To sum up, this material can be an excellent alternative to replace the calcareous filler for SCC production.

Keywords: self-compacting concrete, algaroba trees ashes, temperature.

INTRODUCTION

In the northeastern region of Brazil, the predominance of tropical climate and the influence of the temperature derived from this climate are practically ignored in the scope of construction, especially about concreting in hot weather conditions. Concreting in hot climates may impair the properties of cementitious systems [1, 2]. Pineaud et al. [3] emphasize that the effects of temperature on SCC's have been little studied and have verified that this factor interferes in the mechanical properties of the SCC.

Studies show that the main consequences of the influence of hot weather on concrete are: high temperature during mixing and/or curing, observing a maximum value of 50 ºC; loss of workability and increased water consumption [4-8] extending to losses in compressive strength at ages of 28 days or more [6, 8-12], although compressive strength increases at early ages [6, 8, 9, 11, 12]. Also, according to Paul [4] and ACI 305.1: 2014 [13], for concrete produced with ordinary Portland cement, water consumption increases significantly when the initial temperature of the mixture is above 30 °C. Zhang et al. [14] observed that there is a decrease in compressive strength at the age of 28 days when the temperature of the concrete during the first 24 hours is 38 °C.

Nóbrega et al. [15] analyzed the influence of temperature on SCC's and found that concrete mixed at about 50 °C required correction of water to maintain the properties selfcompacting, and that this correction can influence negatively, having a decrease in compressive strength, durability, passing ability, fluidity, flow, and plastic viscosity. Although research has indicated a better mechanical behavior of self-compacting concretes compared to traditional vibrated concretes in terms of mechanical strength development [15], behavioral assessments in hot weather are essential.

On the other hand, it is known that the use of fines (fillers) combined with the use of superplasticizer [16-18] allows adequate rheological properties and non-segregation in self-compacting concretes. The filler is a finely ground mineral addition obtained from the comminution of limestone, which can act both with a physical aspect of grain size packing and also as a nucleating agent, accelerating the hydration of the grains of the cement [16, 17, 19-26]. Recent researches indicate that the fillers also have reactivity, with the formation of carboaluminates and monocarboaluminates, positively influencing the microstructures of the cementitious materials [24, 26].

As the limestone filler is a non-renewable source, it is evaluated the possibility to use the algaroba ashes, once they are primarily composed of calcium carbonate (CaCO3) [27, 28]. Also, due to the similarity of this ash with natural limestone filler in relation to its chemical composition and possible treatment in the physical aspect, and the potential of application in the civil construction industry, it is observed the possibility of the SCC in incorporating these ashes of calcium-rich wood as an alternative to guarantee fluidity in the SCC.

They have a representative production, once whose wood is the fuel used by about 240 laundries in Agreste Pernambucano [29]. For only a medium-sized laundry, the residue generated by the burning of algaroba wood is about 6,500 kg/month of ash [30]. Informally, these ashes are used for soil pH correction, but most of the waste is still disposed of without environmental control or criteria [31]. Therefore, based on the environmental conditions of the region, this research aims to evaluate the effects of temperature in SCC's produced with the incorporation of algaroba ash as alternative filler.

EXPERIMENTAL PROGRAM

Materials

Table 1 presents the main physical characteristics and the chemical composition of the cement used, which complies with the ABNT EN 197-1 [32].

TABLE 1 - CHARACTERISTICS AND COMPOSITION OF THE CEMENT USED.

Physical properties							
Designation	Type	Clinker (wt. %)	Density (g/cm^3)	Blaine/BET (cm^2/g)	d_{50} (μm)	Addition (wt. %)
C	CP V	98.5	3.09	4692/20956	15.9	0-10
Chemical composition						
Oxides (wt. %)	SiO_2	Al_2O_3	Fe_2O_3	SO_3	K_2O	LOI
	20.43	4.90	1.83	3.60	0.27	1.44

A calcite sedimentary limestone filler with a mesh size of 200 mesh (0.075 mm) was used. The algaroba wood ashes were collected in a laundry located in the city of Caruaru-PE, which uses algaroba firewood exclusively as the energy source. The material was ovendried at 105 °C, then sieved into 200 mesh (0.075 mm) mesh. For characterization, chemical analyses were performed as a) X-ray fluorescence through the Rigaku spectrometer, model Primini, with tube tension of 50W (Na to U) using pressed samples; b) X-ray diffraction using the Bruker D8 Advanced Davinci diffractometer, adopting the following input data: CuKα radiation (40 kV/40 mA), 0.02° step every 0.5 second, in the interval of 5° <2θ <70 °. The physicochemical properties of the limestone filler are shown in Table 2. Figure 1 shows the X-ray diffractogram of the limestone filler and algarroba ashes, both the limestone filler and the algaroba ash present calcium carbonate as the peak of greater intensity near 30°.

TABLE 2 - CHARACTERISTICS AND COMPOSITION OF THE FILLER AND ALGAROBA ASHES USED.

Chemical composition								
Oxide wt.%	MgO	Al_2O_3	SiO_2	P_2O_5	Fe_2O_3	SO_3	K_2O	CaO
LF	5.11	1.49	3.23	0.22	1.9	0.12	0.28	87.65
AF	2.44	0.37	0.83	1.99	0.35	1.33	12.30	77.52
Physical		LF		AF				
Blaine (cm^2/g)		5096		6857				
BET (cm^2/g)		30,045		47,142				
Density (g/cm^3)		2.71		2.76				

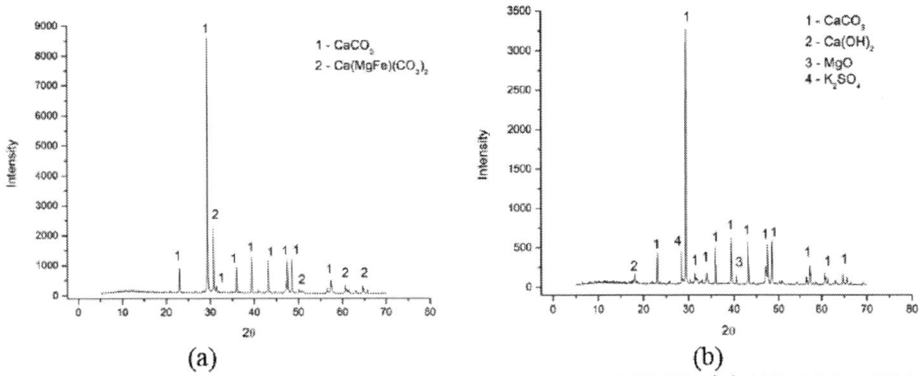

(a) (b)

FIGURE 1 - X-RAY DIFFRACTOGRAM OF THE LIMESTONE FILLER (A) AND ALGAROBA ASHES (B) USED

Granitic rounded coarse aggregate 0/9.5 mm (density = 2630 kg/m3; specific mass = 1360 kg/m3) and silicous sand 0/4.75 mm (density = 2580 kg/m3; specific mass = 1470 kg/m3) were used. The granular skeleton was siliceous-granitic (designation: GE) 0/9.5 mm (density = 2605 kg/m3; specific mass = 1415 kg/m3). All aggregates were oven-dried at 105 °C. The granulometric distribution can be seen in the Figure 2.

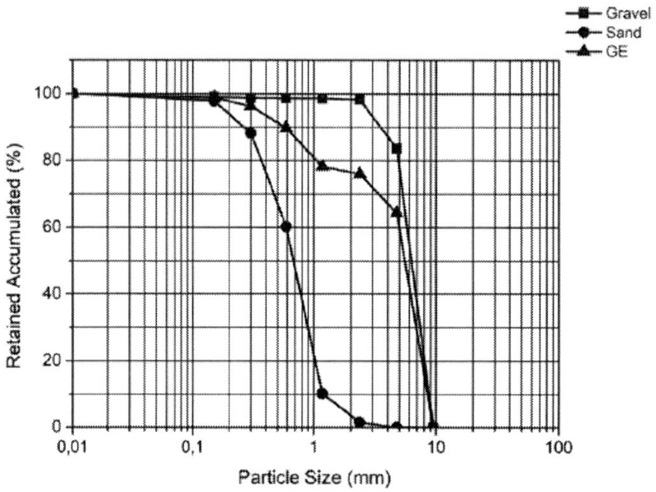

FIGURE 2 - GRANULOMETRIC DISTRIBUTION OF THE AGGREGATES (G, S) AND THE GE.

The superplasticizer (designation: SP) employed was a modified eter-polycarboxylate type available as a commercial solution (specific gravity =1.067-1.107; active solid content by weight = 28.5-31.5%; viscosity<150; pH = 5-7). Distilled water (designation: W, pH = 6.55) was used in all mixtures.

The microscopy study of the calcareous ash was carried out on an SSX - 500 Shimadzu scanning electron microscope. The samples were fixed with carbon tape after passing through a gold metallization for 5 minutes with a potential variation of 1.8 kV [27].

Methods

The concrete compositions tested are presented in Table 2. A cement consumption of 360 kg/m³ with a content of 25% of limestone filler was added to determine the traces of the SCC. The water/fines ratio was fixed at 0.5. The superplasticizer content was achieved in each mixture to obtain an SF2 class SCC (spread from 660 to 750 mm) to observe the ashes influence on the compressive strength.

Table 3 presents the consumption of each mixture produced with the substitution of the limestone filler by the alternative filler (algaroba ashes) in the following contents: 0 (LFM), and 50% (LFAFM). They were defined using a diphasic design approach combining a paste with the optimized flow and a wet aggregate skeleton [15, 33]. As can be seen in Table 2, the increase in the demand for superplasticizer occurred with the increase of the incorporation of the algaroba ashes (1.575 - I (0%), and 3.375 - III (50%)). This behavior is due to the progressive increase in the specific surface of the fines since the ashes have a BET specific area of 47,142 cm2/g against 30,045 cm2/g of the limestone filler.

TABLE 3 - SELF-COMPACTING CONCRETE COMPOSITION TESTED

Materials (kg/m³) / Nomenclature	LFM (0%)	LFAFM (50%)
Cement	360	360
Calcareous filler	90	45
Algaroba ashes	-	45
Sand	784.85	785.16
Gravel	784.85	785.16
Superplasticizer	1.575	3.375
Water	225	225

In order to verify the properties of the SCC in hot weather conditions, a temperature closed to the reality was studied, where they can quickly achieve more than 70 °C in the tropical climate region in the summer. Thus, the final temperature of concrete mixing can be calculated by the specific heat and mass of each material, obtaining the temperature of approximately 50 °C, independent of the heat generated by the chemical reactions generated by the cement in contact with water [15]. Finally, in the laboratory context, the temperature values of the cement, filler, and aggregates are raised to 80 °C to maintain the temperature at the end of the blend at 50 °C [15]. Water and superplasticizer at 20 °C, an acceptable value for the conditions of distribution and storage.

The study evaluated two different types of correction to maintain the self-compacting properties: by water increment and by superplasticizer increment, as in [15]. The addition of water or superplasticizer was done using the spread measurement by the slump-flow test (ABNT NBR 15823-2: 2017 [34]), maintaining it between 660 and 750 mm. LFM (mixture with calcareous filler – reference - molded in ordinary climate conditions); LFM-W (mixture with calcareous filler – reference - molded in hot climate conditions and watercorrected); LFM-S (mixture with calcareous filler – reference - molded in hot climate conditions and superplasticizer-corrected); LFAFM (50% calcareous filler and 50% algaroba ashes alternative filler molded in hot weather conditions); LFAFM-W (50% calcareous filler and 50% algaroba ashes alternative filler molded in hot weather conditions and water-corrected), LFAFM-S (50% calcareous filler and 50% algaroba ashes alternative filler molded in hot weather conditions and superplasticizer-corrected).

The mixture was carried out in an air-conditioned room with a temperature of 20 ± 2 °C, using a stationary mixer with a total volume of 120 liters [15]: first, the introduction of gravel, cement, limestone filler and sand followed by 60 s of mixing; second, the introduction of water and 1/3 of the mass of the superplasticizer followed by 90 s of mixing; third, the introduction of 2/3 of the superplasticizer followed by 90 s of mixing. For the temperature of 50 °C, 30 additional seconds step was added to increase of water, or superplasticizer in the concretes, not exceeding the time limit of 5 minutes of complete mixing, as in [15]. If this time was exceeded, the sample was discarded.

The procedure of molding the specimens followed the parameters assigned by ABNT NBR 5738: 2016 [35], where cylindrical molds of 10 cm x 20 cm (diameter x height) were used. After 24h, all the samples were immersed in water at room temperature (20 °C ± 1 °C) until the tests were performed.

Fresh state tests: the self-compacting ability of the concretes was assessed using various standardized methods and their criteria. Taking the time at the end of mixing as t0, the same test sequence was applied to all the concretes: spread (ABNT NBR 15823-2: 2017 [34]) and t500 (ABNT NBR 15823-2: 2017 [34]) at t0, L-box (3 rebars, ABNT NBR 15823-4: 2017 [36]) and V-funnel (ABNT NBR 15823-5: 2017 [37]) at t0+5'. In parallel with these tests, the and visual stability index (IEV) (ABNT NBR 15823-2: 2017 [34]) was also measured.

Hardened state tests: compressive strength values were measured at 1, 7 and 28 days of age according to ABNT NBR 5739:2007 [38] using a Shimadzu hydraulic press, 2000 kN. Total water absorption (ABNT NBR 9778:2009 [37]) and capillary water absorption (ABNT NBR 9779:2012 [39]) were carried out at 28 days. Each test result reported is the average of three values.

To determine the electrical resistivity was carried out the method adopted by Lavagna et al. [40]. A water-saturated sponge was used between the concrete sample and the copper plate to improve the electrical contact. The electrical resistance was measured by applying an alternating current of 300 mV at 200 kHz. In this test, oven-dried specimens (80 ºC, 24 hours) with dimensions of 10 cm x 10 cm (diameter x height) were adopted.

RESULTS ANS DISCUSSION

Regarding the evaluation of temperature during mixing of SCC, the average temperature behavior for the molded SCC's simulating hot weather conditions is shown in Figure 3. The temperature is around 50 °C at the end of the mix.

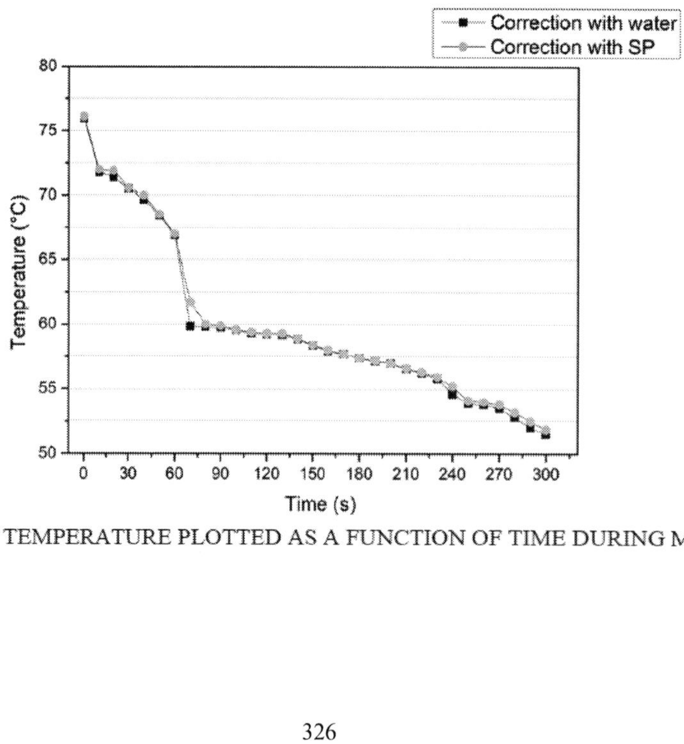

FIGURE 3 - TEMPERATURE PLOTTED AS A FUNCTION OF TIME DURING MIXING.

The volumes of water required to maintain the self-compacting properties in the concrete produced under hot weather conditions for the proposed features were 225 kg/m³ - LFM, 265 kg/m³ - LFM-W, 225 kg/m³ - LFAFM, and 238 kg/m³ - LFAFM-W. The volumes of superplasticizer required to maintain the self-compacting properties in the concrete produced under hot weather conditions for the proposed features are 1.575 kg/m³ - LFM, 2.750 kg/m³ - LFM-S, 3.375 kg/m³ - LFAFM, and 3.625 kg/m³ - LFAFM-S. All the SCC mixtures evaluated, after corrections, were classified in the same spreading class, SF2 (suitable for most current applications such as walls, beams, pillars, and others).

The lower water demand when replacing the limestone filler by algaroba ash is favorable to the microstructural development. It is due to the lower loss of water by evaporation, and it is closely related to the morphology of the ashes (the electron micrographs of the ashes below 75 μm, Figure 4). Although its greater specific surface area compared to the limestone filler, it has not demanded more water specifically due to their agglomeration process. Also, it is believed that its irregular and porous appearance and in agglomerates facilitated the imprisonment of the water, avoiding its evaporation in hot weather. Also, the accommodation between the materials and their physicochemical characteristics may facilitate or hinder the temperature dissipation, influencing the evaporation of water during the mixing process. Another factor to consider is the time between the removal of the heated materials from the greenhouse to the actual mixing process, as well as the distance between these two types of equipment.

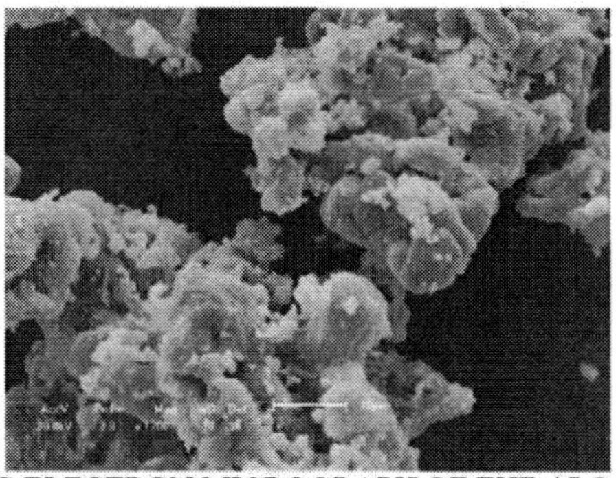

FIGURE 4 - SCANNING ELECTRON MICROGRAPH OF THE ALGAROBA ASHES USED.

When correcting with superplasticizer for mixing and molding in hot weather conditions (Figure 6b), it is verified that the partial replacement of the limestone by the algaroba ash, despite having demanded more superplasticizer under environmental conditions, demanded less superplasticizer in hot weather conditions. The increase in superplasticizer needed under environmental conditions is due to the increase of the specific surface area of the algaroba ash. However, the ash morphology resulted in the lower evaporation of the water during mixing. It is also worth noting that the dispersion of the cement particles increases with the amount of superplasticizer. So, the cement dispersion is improved in the mixtures with algaroba ashes.

Table 4 gathers together the results (average) corresponding to the slump flow (spread and t500), the filling ratio (H2/H1) measured in the L-box test, J-ring, V-funnel, and IEV according to the ABNT NBR 15823-1:2017 [40].

TABLE 4 - SPREAD, T_{500}, H_2/H_1, J-RING, V-FUNNEL, IEV, AND RESPECTIVE CLASSIFICATIONS.

	Spread (mm)	t_{500} (s)	H_2/H_1	J-ring (mm)	V-funil (s)	IEV (visual)	Classification based on the average value (ABNT NBR 15823-1:2017 [40])
LFM	700	1.6	0.92	15	4.5	IEV 1	SF2, VS1, PJ1, VF1, IEV 1
LFM-W	660	1.8	0.85	50	4.3	IEV 0	SF2, VS1, PJ2, VF1, IEV 0
LFM-S	700	2.0	0.71	20	5.1	IEV 1	SF2, VS1, PJ1, VF1, IEV 1
LFAFM	665	1.3	0.81	22	5.2	IEV 0	SF2, VS1, PJ1, VF1, IEV 0
LFAFM-W	685	1.5	0.83	30	3.0	IEV 1	SF2, VS1, PJ2, VF1, IEV 1
LFAFM-S	665	1.7	0.80	24	6.0	IEV 0	SF2, VS1, PJ1, VF1, IEV 0

All the properties of fluidity, passing ability, and resistance to segregation were reached when the limestone filler was partially replaced by algaroba ashes in contents of up to 50%, obtaining self-compacting concretes with classes similar to the standard: SF2, VS1, and VF1 (ABNT NBR 15832-1: 2017 [40]). After corrections when submitted to the hot weather conditions, positively, all the SCC mixtures evaluated were classified in the same spreading and passing ability, and resistance to segregation class. Regarding the passing ability skills, to add additional water to improve the J-Ring results and the LFM-S mixture did not pass on the L-Box test. The T500 and V-funnel results indicate that the corrections with water or superplasticizer increase the time of flow. Also, through this essay, LFAFM, LFAFM-W, and LFAFM-S are found to decrease the flowing time, indicating a positive effect of the alternative filler in the SCC, for both the reference mixture and the mixture under hot climate conditions.

Regarding to the hardened state, compressive strength values measured at 1, 7 and 28 days of age and the respective standard deviation are exposed in Figure 5.

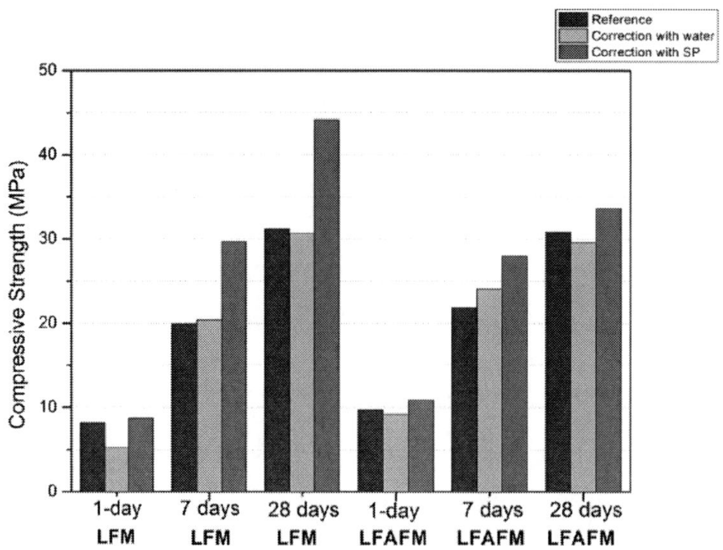

FIGURE 5 - COMPRESSIVE STRENGTH AT ONE DAY, SEVEN DAYS AND 28 DAYS.

The replacement of 50% of limestone filler by 50% of algaroba ash improve the compressive strength values at one day in all situations, to the reference, to the water corrected and the superplasticizer-corrected. It is believed that this behavior is due to the improvements on the superplasticizer values, better dispersing the cement particles, as well as increasing the nucleation sites due to the algaroba ashes morphology compared to the limestone filler (Figure 5). The early-age compressive strengths (1 and seven days) are more susceptible than the 28-days values in front of the proposed replacement. That is an indication that the presence of algaroba ashes in the system can accelerate the initial hydration reactions of the cementitious compounds.

At the literature about concretes in hot weather environments, it is related that the compressive strength increases at early ages to usual vibrated concretes [6, 8, 9, 11, 12]. It was observed in both water and superplasticizer correction to the reference, especially to the superplasticizer correction and at 7 days. At one day, the values were not prejudiced. When the ashes were incorporated into the system, similar behavior was observed.

Concerning the compressive strengths at 28-day or more, the literature referred losses when concreting in hot weather conditions to normal vibrated concrete [6, 8-12]. On the other hand, Nóbrega et al. [15] did not observe this behavior to SCCs. Equally, it was not observed losses in the compressive strengths to the studied SCCs when exposed to the hot weather mixing and molding conditions. Positively, the use of the algaroba ashes as alternative calcium carbonate filler keeps the compressive resistance values. When the mixture was superplasticizer-corrected, the responses were better.

Results about total water absorption indicated: 6.7% (LFM), 7.1% (LFM-W), 4.5% (LFMS), 5.9% (LFAFM), 6.3% (LFAFM-W), and 5.3% (LFAFM-S). Ratifying the results of compressive strength, the correction with water does not only recomposes the water evaporated by the temperature, but it negatively influences the compactness of the system. So, it increases the number of voids, leaving it less resistant and more permeable. Oppositely, it is the superplasticizer correction, which improves the arrangement of the granular arrangement and promotes lower permeability of the system, caused by the reduction of the water/fines ratio. The lowest value was to the mixture LFM-S, ratifying the best compression strength results at 28 days to this mixture.

In agreement with the compressive strength tests and total water absorption results, the capillary water results showed that the LFM-W (0.7 g/cm^3 at 65 min) and LFAFM-W (0.5 g/cm^3 at 65 min) are more permeable, and the LFM-S (0.29 g/cm^3 at 65 min) and LFAFM-S (0.31 g/cm^3 at 65 min) are the most compacts.

The electrical resistivity test can be evaluated in an inverse way to the total/capillary absorption, evaluating the permeability of the sample. In this case, the higher the electrical resistivity, the less permeable the sample. LFM (320 Ω.m), LFM-W (310 Ω.m), LFM-S (380 Ω.m), LFAFM (360 Ω.m), LFAFM-W (340 Ω.m), and LFAFM-S (370 Ω.m). Therefore, LFM-W and LFAFM-W are more permeable, whereas LFM-S and LFAFM-

S are those with a more compact granular arrangement, providing less ionic mobility; in this way, it can be more resistant to the reinforcement corrosion in the structures.

CONCLUSIONS

It was observed that algaroba ash, alternative filler, due to its specific surface area higher than to the limestone filler, demands more superplasticizer additive for the reference mixture. However, water correction or superplasticizer demands in hot weather conditions decreases. It was possible to obtain SCCs with the incorporation of algaroba ash with similar compacting results to SCCs produced only with limestone, maintaining the compressive strength and indicative of porosity-permeability. Thus, the potential reuse of this waste for use in SCCs in hot weather conditions is highlighted.

REFERENCES

1. Mouret, M., A. Bascoul, and G. Escadeillas, Strength impairment of concrete mixed in hot weather: analysis in relation to physical and chemical properties of hardened concrete. Magazine of Concrete Research, 2005. 57(5): p. 301-308.

2. M., M., Contribution à l'étude des phénomènes d'alteration des résistances des bétons confectionnés par temps chaud. 1997, Université Paul Sabatier.

3. Pineaud, A., et al., Mechanical properties of high performance self-compacting concretes at room and high temperature. Construction and Building Materials, 2016. 112: p. 747-755.

4. Paul, K., Effect of Mixing and Curing Temperature on Concrete Strenght. Journal Proccedings, 1958. 54(6): p. 1063-1081.

5. Hampton, J.S., Extended Workability of Concrete Containing High-Range Water-Reducing Admixtures in Hot Weather, in Developments in the Use of Superplasticizers V.M. Malhotra, Editor. 1981, American Concrete Institute: Detroit. p. 409-422.

6. Mouret, M., A. Bascoul, and G. Escadeillas, Strength impairment of concrete mixed in hot weather: relation to porosity of bulk fresh concrete paste and maturity. Magazine of Concrete Research, 2003. 55(3): p. 215-223.

7. Soroka, I., Concrete in Hot Environments. 2003: CRC Press. [S.l.].

8. Ortiz, J., et al., Influence of environmental temperatures on the concrete compressive strength: Simulation of hot and cold weather conditions. Cement and Concrete Research, 2005. 35(10): p. 1970-1979.

9. Burg, R.G., The Influence of Casting and Curing Temperature on the Properties of Fresh and Hardened Concrete, in Portland Cemente Association, R.D.B.R. 113, Editor. 1996.

10. Kim, J.-K., S.H. Han, and Y.C. Song, Effect of temperature and aging on the mechanical properties of concrete: Part I. Experimental results. Cement and Concrete Research, 2002. 32(7): p. 1087-1094.

11. Kanstad, T., et al., Mechanical properties of young concrete: Part I: Experimental results related to test methods and temperature effects. Materials and Structures, 2003. 36(4): p. 218-225.

12. Bascoul, A., M. Mouret, and G. Escadeillas, Strength impairment of concrete mixed in hot weather: Analysis in relation to physical and chemical properties of hardened concrete. Magazine of Concrete Research, 2005. 57: p. 301-308.

13. Cascudo, O. and H. Carasek, Ação da carbonatação no concreto, in Concreto: Ciência e Tecnologia, G.E. ISAIA, Editor. 2011, Instituto Brasileiro do Concreto: São Paulo. p. 1984.

14. Zhang, X., J.H. Li, and X.F. Li, Influence of slag powder on the water reducing effect of admixtures and its mechanism. Concrete, 1999. 6: p. 34-36.

15. da Nóbrega, A.C.V., et al., Improved 28-day compressive strength of SCC mixed and cured in hot weather: Non-conventional behavior. Construction and Building Materials, 2018. 173: p. 650- 661.

16. Tutikian, B.F. and D.C. Dal Molin, Concreto Auto-Adensável. 1 ed. 2008, São Paulo: PINI.

17. Tutikian, B.F., Proposição de um método de dosagem experimental para concretos autoadensáveis. 2007, Universidade Federal do Rio Grande do Sul.: Porto Alegre. p. 163.

18. Bradu, A. and N. Florea, Workability and compressive strength of self compacting concrete containing different levels of limestone powder, in Bulletin of the Transilvania University of Brasov, Special Issue, Nº 1, v. 8, n. 57. 2015: Brasov.

19. Dal Molin, D.C.C., Adições minerais para concreto estrutural, in Concreto: Ensino, Pesquisa e Realizações, G.C. Isaia, Editor. 2005, IBRACON: São Paulo. p. 345-379.

20. Silva, M.G., Cimentos portland com adições minerais, in Materiais de construção civil e princípios de ciências e engenharia de materiais, G.C. Isaia, Editor. 2007, IBRACON: São Paulo. p. 761-793.

21. C., D.M.D.C., Adições minerais, in Concreto: ciência e tecnologia, G.C. Isaia, Editor. 2011, IBRACON: São Paulo.

22. De Schutter, G., Effect of limestone filler as mineral addition in self-compacting concrete, in 36º Conference on Our World in Concrete & Structures. 2011: Singapore.

23. Ghomari, F., S.B. Hacene, and O. Taleb, Study of Limestone Addition on the Mechanical and Rheological Characteristics in the SCC. Jordan Journal of Civil Engineering, 2011. 5(3): p. 412-423.

24. Moon, G.D., et al., Effects of the fineness of limestone powder and cement on the hydration and strength development of PLC concrete. Construction and Building Materials, 2017. 135: p. 129-136.

25. Shadkam, H.R., et al., An investigation of the effects of limestone powder and Viscosity Modifying Agent in durability related parameters of self-consolidating concrete (SCC). Construction and Building Materials, 2017. 156: p. 152-160.

26. Schöler, A., et al., Early hydration of SCM-blended Portland cements: A pore solution and isothermal calorimetry study. Cement and Concrete Research, 2017. 93: p. 71-82.

27. Leloup, W.A., Efeito da adição de lodo têxtil e cinzas de lenha gerados no APL de confecções pernambucano em argamassas de cimento Portland, in Mestrado em Engenharia Civil e Ambiental. 2013, Universidade Federal de Pernambuco: Caruaru, PE, Brazil. p. 122.

28. Nascimento, J.E.M.F., Avaliação dos efeitos da substituição da cal hidratada por cinzas de algaroba em argamassas de revestimento, in Mestrado em Engenharia Civil e Ambiental. 2014, Universidade Federal de Pernambuco: Caruaru, PE, Brazil. p. 99.

29. Moura, M.H.A., Incorporação de cinzas de algaroba geradas no APL de confecções do agreste pernambucano em concreto betuminoso usinado a quente – CBUQ, in Mestrado em Engenharia Civil e Ambiental. 2017, Universidade Federal de Pernambuco: Caruaru, PE, Brazil. p. 97.

30. Melo, M.C.S., Estudo de argamassas adicionadas de cinzas de algaroba geradas no arranjo produtivo local de confecções do agreste pernambucano, in Mestrado em Engenharia Civil e Ambiental. 2012, Universidade Federal de Pernambuco: Caruaru, PE, Brazil. p. 122.

31. Mota, M.H.A., Concreto seco com incorporação de cinzas de madeira de algaroba (prosopis juliflora) moldado sob pressão, in Mestrado em Engenharia Civil e Ambiental. 2014, Universidade Federal de Pernambuco: Caruaru, PE, Brazil. p. 122.

32. British Standard Institution. BS EN 197-1. Cement – Part 1: Composition, specifications and conformity criteria for common cements, . February 2001, London.

33. Diederich, P., M. Mouret, and F. Ponchon, Simple tools for achieving self-compacting ability of concrete according to the nature of the limestone filler. Construction and Building Materials, 2013. 48: p. 840-852.

34. Souza, P.A.B.F., Estudo do comportamento plástico, mecânico, microestrutural e térmico com concreto produzido com polimento de porcelanato, in Programa de pós-graduação em ciência e engenharia dos materiais. 2007, Universidade Federal do Rio Grande do Norte: Natal.

35. Breintebach, S.B., Desenvolvimento de argamassas para restauração utilizando resíduo do polimento de porcelanato, in Programa de pós-graduação em Ciência e Engenharia dos Materiais. 2013, Universidade Federal do Rio Grande do Norte: Natal.

36. Lee, C.Y. and S.R. Wang, Application of four-electrode method to analysis resistance characteristics of conductive concrete. World Acad Sci Eng Technol, 2010. 4: p. 77-80.

37. Wen, S. and D.D.L. Chung, Effect of stress on the electric polarization in cement. Cement and Concrete Research, 2001. 31(2): p. 291-295.

38. Commitee, R.-R.C.P., Draft Recommendation: measurement of hardened concrete carbonation depth CPC-18, in Materials and Structures. 1984, RILEM - Rilem Concrete Permanent Commitee. p. 435-440.

39. Zhu, X., et al., Characterisation of pore structure development of alkali-activated slag cement during early hydration using electrical responses. Cement and Concrete Composites, 2018. 89: p. 139-149.

40. Lavagna, L., et al., Cement-based composites containing functionalized carbon fibers. Cement and Concrete Composites, 2018. 88: p. 165-171.

18th International Conference on
Non-Conventional Materials and Technologies
"Construction Materials & Technologies for Sustainability"
(18th NOCMAT 2019)
24th – 26th July 2019
Nairobi, Kenya

FROM PRESCRIPTIVE TO PERFORMANCE BASED DURABILITY DESIGN AND SPECIFICATION OF CONCRETE STRUCTURES

Mike Otieno[1]

[1] School of Civil and Environmental Engineering University of the Witwatersrand, Johannesburg

ABSTRACT

Africa is experiencing an increased growth in infrastructural development, with concrete being the most commonly used construction material. Concrete structures deteriorate with time at a rate dependent on both the quality of the concrete and the aggressivity of the exposure environment. A lot of evidence is available in the literature of premature deterioration (and durability failure) of concrete structures due to, for example, steel corrosion. It is therefore important that engineers are acutely aware of the need to incorporate durability design and specification in their day to day design process; focus should not only be placed on structural strength and stability. This paper presents fundamental concepts of durability design and specification of concrete. Two approaches prescriptive and performance-based are discussed. In general, prescriptive approaches present a number of deficiencies which can be overcome by using performance-based approaches. In Africa, South Africa has been on the forefront in developing and implementing performance-based durability design and specification using the durability index (DI) tests. The DI tests are simple and relatively inexpensive and have been briefly presented inn this paper with the intention of encouraging other African countries to adopt the tests for durability design and specification. This paper puts forward a case for the need to consistently incorporate durability in the design and construction stages in order to ensure that RC structures meet their design life. This is important for Africa if the infrastructural developments are to be worthwhile in the long-term. In order to implement concrete durability design and specification, civil engineering scholars in tertiary institutions and practicing engineers need adequate knowledge of concrete as a construction material.
Keywords: Concrete, Durability, Prescriptive, Performance-based, Service life prediction.

INTRODUCTION

Durability design and specification are now important components of design processes, construction, and maintenance and repair of concrete structures due to the increased premature deterioration of these structures. Engineers are now required by clients to provide high levels of confidence in the serviceability of concrete structures with almost maintenance-free service life. Furthermore, considering that (i) concrete is currently the most commonly used construction material not only in Kenya but all over the world [1], and that (ii) the use of concrete is associated with negative environmental impacts, engineers are constantly being challenged to incorporate sustainability demands in their day to day designs and specifications in order to reduce the negative environmental impact. Over the years, a lot of practical research developments have been made in the field of concrete durability innovative cements, concretes and design approaches have been developed and are being used. However, the suitability of these developments has not been actively explored and fully exploited in the majority of developing [African] countries. The new developments relate not only to the adoption of new cement and concrete standards but also to the drive towards the adoption of performance based specification approaches to optimize on strength, durability and sustainability requirements.

In the recent past, Africa has experienced increased reinforced concrete (RC) infrastructural investments such as roads, bridges, housing developments, etc, in a bid to unlock its economic growth. The increasing demand for cement in the continent is a clear indication of this [2-4]. In Kenya, typical examples include the recent and ongoing boom in the real estate sector (urbanization) and large infrastructural investments such as the LAPSSET (Lamu Port Southern Sudan Ethiopia Transport) project, and the Mombasa Nairobi Standard Gauge Railway project. It is without doubt that we will continue to experience this increase in infrastructural development in the coming years so long as there is a political goodwill, and the county and national governments are committed to investing in large infrastructural projects. However, for these RC infrastructural developments to be long-term investments as they are intended to be (or to maximize on the return on investment from both economic and functional viewpoints), it is important that civil engineers, especially those dealing directly with their design, specification and supervision during construction, ensure that they are not only strong enough and stable to carry the expected service loads but are also be durable enough to resist deterioration such as reinforcement corrosion in their service environments. This paper presents some of the fundamental principles of durability design and specification of concrete structures, which if applied, less capital investment will be spent in the near future in maintaining, repairing and retrofitting the structures being built now; this capital can better be spent on investing in new much needed infrastructure, and/or diverted to other well-deserving sectors of the economy such as education and health.

DURABILITY AND SERVICE LIFE PREDICTION OF CONCRETE STRUCTURES

Durability is fundamentally a function of the concrete quality (which is a function of intrinsic and extrinsic concrete properties) and the aggressivity of the exposure environment (which is a function of both physical and chemical deterioration mechanisms). It is defined as the ability of a structure or component to withstand its design environment over the design life, without undue loss of serviceability and need for major repair. It relates to the concept of material performance and cannot be regarded as an intrinsic material property. Durability is thus associated with the deterioration of the material over the intended service life of the structure in a given environment; a concrete that is durable in one environment may not be durable in another [5].

Durability is mainly related to the transport properties and chemical composition of concrete.Transport properties are a function of concrete penetrability. Penetrability refers to the ease with which an ion, molecule or fluid can move through concrete [6]. It is a function of interconnection between the pores, the pore size distribution and its tortuosity, and is a major factor responsible for the rate of deterioration of concrete structures. It is affected by several parameters such as the water/binder ratio and the degree of hydration of the cement paste.

Factors affecting durability of concrete

Figure 1 presents a summary of the two main factors that affect concrete durability viz (i) the aggressiveness of the exposure environment, and (ii) the quality (or penetrability) of the concrete. It is clear from this figure that concrete durability design, specification and construction will involve various players client, engineers, materials suppliers and contractors. For a given construction project, the responsibilities of these players have to be clearly defined. This is important not only to ensure each player clearly understands their role(s) but also for future arbitration purposes, if the need arises. These roles should be agreed upon and clearly defined in the contract tender documents. Further information on the roles of each of the players can be found in [1].

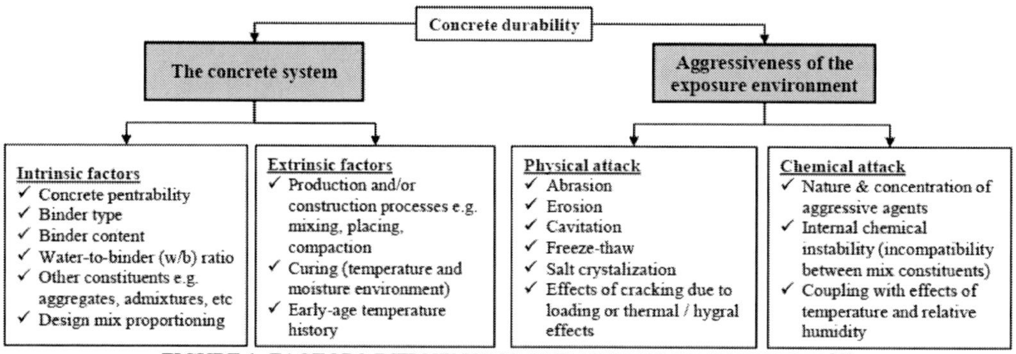

FIGURE 1: FACTORS INFLUENCING DURABILITY OF CONCRETE [5]

DESIGN AND SPECIFICATION OF DURABILITY

Service life

Service life refers to the assumed period for which a structure or part of it is to be used for its intended purpose, and in which [7]:

i. deterioration and material performance are quantified as far as possible;
ii. a suitable "risk" approach is adopted, usually based on probability which leads to measures of reliability;
iii. quantifications, costs, interventions (e.g. maintenance) and the like can be rationally considered.

Durability design

Durability design requires a proper understanding of deterioration mechanisms on the part of designers but also (i) practical Service Life Models that predict with sufficient accuracy the potential durability performance of the structure over its design life, (ii) proper selection of materials at the design stage with regard to foreseen exposure conditions to which the structure will be subjected during its service life, (iii) the means of characterising the potential durability of the concrete to be used in the structure, and (iv) the means of assessing the actual quality of the construction [5, 8, 9]. It therefore means that ensured durability may require higher initial costs.

Design for concrete durability involves service life modelling (SLM) and service life prediction (SLP), both of which are closely related:

i. SLM involves using tools to predict the time-based deterioration of concrete structures over their service life to quantify the design life of the structure. This is done for purposes of economic optimization, operational efficiency, and sustained structural and aesthetic performance [9].
ii. SLP requires development of deterioration models (e.g. South African chloride prediction model [10], ClinConc [11], Life-365 [12], DuraCrete [13-16], etc) that are sensitive to the varying exposure environments, construction materials, construction quality and definition of service life. The prediction models are related to specific deterioration mechanisms such as corrosion of steel, alkali-silica reaction, soft water attack, etc. These models need to be updated and/or calibrated from time to time to improve on their reliability in SLP.

When designing for concrete durability, it is important to remember that even though improved durability may be achieved with increase in concrete strength, this is not always the case unless the two parameters are carefully designed for. Furthermore, it is also important to acknowledge that increased cement contents do not necessarily lead to improved concrete durability. In fact, the contrary may be the case. For example, increased cement content leads to (i) high heat of hydration which can cause internal cracking of concrete, and (ii) increased past content which is highly penetrable [17].

Durability specification

Adequate durability design needs to be complimented by proper specification. Durability specification is carried out by the engineer based on the outcome(s) of the SLP in the design stage. It should focus on measurable and verifiable concrete durability parameters in the as-built structure. In the majority of cases, these parameters relate to the relevant transport properties such as permeation, diffusion and sorption which can be measured using [standard] tests; the transport properties are associated (directly or indirectly) with different deterioration mechanisms in the concrete. In South Africa, for example, a suite of three durability index tests oxygen permeability, water sorptivity and chloride conductivity are used [18-21].

Approaches for durability design and specification

Durability design and specification has evolved over the years. In this regard, two approaches have been used traditional prescriptive and modern performance-based approaches. These are briefly discussed in the following sections.

Prescriptive approach

Prescriptive durability design and specification approaches are of the recipe type, setting limits on water-to-binder ratio, cement type, compressive strength, etc for different exposure classes (such as those in EN 206 see Table 1[22]). They focus on the characteristics of raw materials, mix proportions, batching, mixing, and transport of fresh concrete, and a range of construction operations from placing to curing. Some of the limitations of prescriptive approaches include [5, 23]:

i. they fail to capture the material performance concept of a structure in a given environment;
ii. they do not take into account the effects of in-situ handling of concrete such as compaction and curing which impact on the durability of concrete;
iii. the limits (e.g. minimum cementitious contents, minimum compressive strength and maximum water/cement ratios) have tenuous or little correlation with properties that actually affect the durability of concrete;
iv. they stifle innovation in concrete mix design and construction;
v. they give no opportunity for effective quality control and conformity assessment;
vi. they may result in uneconomical and unsustainable concrete construction;
vii. the use of higher cement contents and/or lower water/cement ratios can result in increased shrinkage and increased risk of thermal cracking, which in turn negatively affects durability;
viii. they fail to account for the potentially positive effects of chemical admixtures, mineral additions and construction procedures on concrete durability.

It is therefore clear that prescriptive durability design and specification approaches are obsolete and do not resonate positively with the current needs and challenges of the concrete industry such as reduction of cement contents, and increase in the use of blended Portland cements [24].

TABLE 1: ENVIRONMENTAL EXPOSURE CLASSSES ACCORDING TO EN 206 [22]

Class designation	Description of the environment
X0	No risk of steel corrosion or attack
XC	Steel corrosion induced by carbonation
XD	Steel corrosion induced by chlorides other than from sea water
XS	Steel corrosion induced by chlorides from sea water
XF	Freeze/thaw attack with or without de-icing agents
XA	Chemical attack

Performance based approach

Performance-based approaches are based on quantitative prediction for durability (or service life) from exposure conditions and measured material parameters. They rely heavily on performance simulation test methods; a typical example is the suite of three simple and relatively inexpensive durability index tests that are used in South Africa [18, 25]. In applying performance-based durability design and specification, engineers need to appreciate that concrete durability is an interaction between the exposure environment and concrete system (quality, composition and condition).

It is clear that performance-based durability design and specification approach is superior to the prescriptive approaches, and should be adopted.

The following sections will briefly present the South African durability index tests [18-20, 26] that can easily be adopted by African countries for durability design and specification. The tests are carried out after 28 days of water-curing using 70 ± 2 mm diameter × 30 ± 2 mm thick discs cored and cut from companion 100 mm cube specimens. The discs are then oven-dried at 50 ± 2 C for a minimum of 7 days and a maximum of 8 days, after which they are air-cooled to a controlled room temperature (23 ± 2 C) before testing.

a) Oxygen permeability index test

The oxygen permeability index (OPI) is defined as Log k where k (m/s) is the coefficient of oxygen permeability determined from the rate of pressure decay through a sample when placed in a falling head permeameter, using the Darcy equation [20, 27, 28]. The coefficient of permeability k is defined as ( )×(ln(Po/P))×( ) 1 where is the molecular mass of the permeating gas (kg/mol), V is the volume of the pressure cylinder (m3), g is the acceleration due to gravity (m/s2), d is the sample thickness (m), R is the universal gas constant (Nm/Kmol), A is the cross sectional area of specimen (m2), = absolute temperature (K) and Po, P are the pressures at start of test and at time t (sec) respectively. The higher the index value, the less permeable is the concrete. Oxygen permeability indices range from 8 to 11, i.e. three orders of magnitude. The higher the index, the less permeable is the concrete. The test therefore gives an indication of the gas permeation properties of the concrete and a rapid estimation of the resistance of a particular concrete to the transport of gas, especially with respect to carbon dioxide diffusion into the concrete (Alexander et al., 1999).

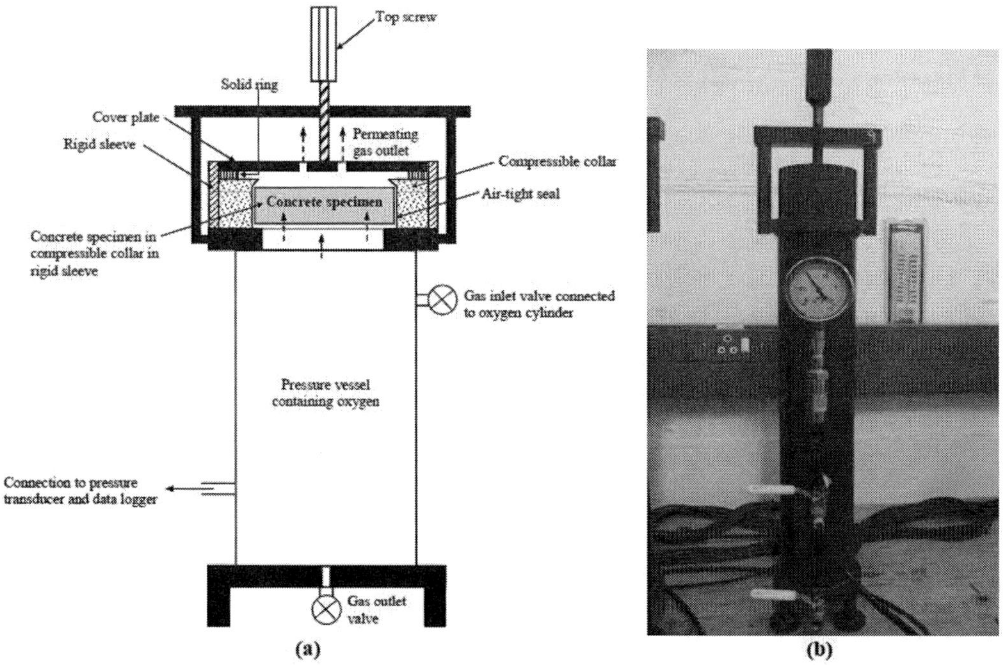

FIGURE 2: (A) SCHEMATIC AND (B) PHOTOGRAPH OF OXYGEN PERMEABILITY TEST SET-UP [20]

b) Chloride conductivity index test

The chloride conductivity is used as an index to characterize the intrinsic potential of a given concrete to resist chloride penetration [29-31]. To carry out the test, the oven-dried specimens are further preconditioned by vacuum saturation (between 75 and 80 kPa) with 5M NaCl solution for 18 hours before being placed in a test rig with a cell filled with the same salt solution on either side (see Figure 3). A 10 V potential difference is applied across the specimen and the current through the specimen measured, in a very short time (typically 10 seconds). The chloride conductivity index (CCI) is then calculated using the formula = (id)/(VA) [26]where dis the chloride conductivity index of the specimen (mS/cm), i is the electric current through the specimen (mA), V is the voltage difference across the specimen (V), d is the average specimen thickness (cm) and A is the cross-sectional area of the specimen (cm2). Chloride penetration resistance of a given concrete increases with decrease in and vice versa. Values of typically vary from < 0.5 mS/cm for dense chloride-resistant concretes to > 2.5 mS/cm for very penetrable concretes. CCI values depend both on the physical pore structure of the concrete and the pore water chemistry, such that the results are affected by the general microstructure and also the chemical composition of the binder type used. The chloride conductivity is fundamentally related to steady state diffusivity (Ds) by the Nernst Einstein equation which relates the conductivity of a bulk material to the diffusion coefficient [32]. Further information on the South African Durability Index tests can be found in the literature [30, 33, 34].

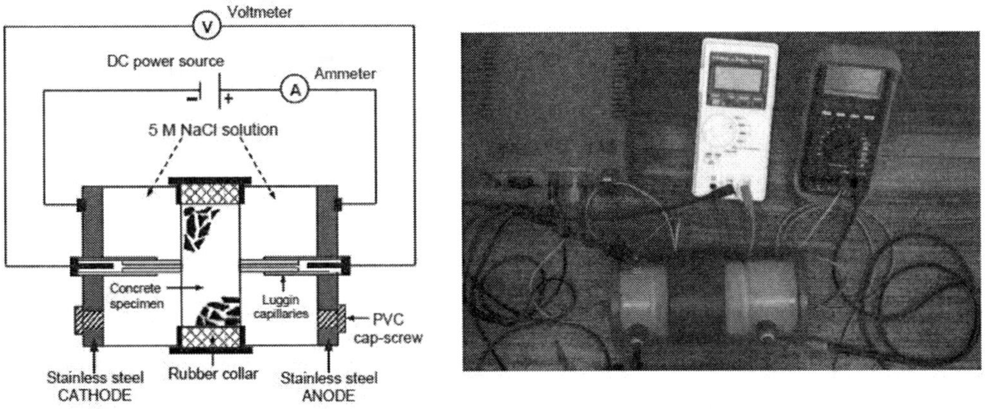

FIGURE 3: (A) SCHEMATIC AND (B) PHOTOGRAPH OF CHLORIDE CONDUCTIVITY TEST SET-UP [26]

c) Water sorptivity index test

Sorptivity is the rate of movement of a water front through a porous material under capillary action. This test is used to measure the rate of movement of a water front through a concrete by capillary suction. The test is a useful measure of the near surface characteristics of the material [10, 18]. The lower the water sorptivity index, the better is the potential durability of the concrete.

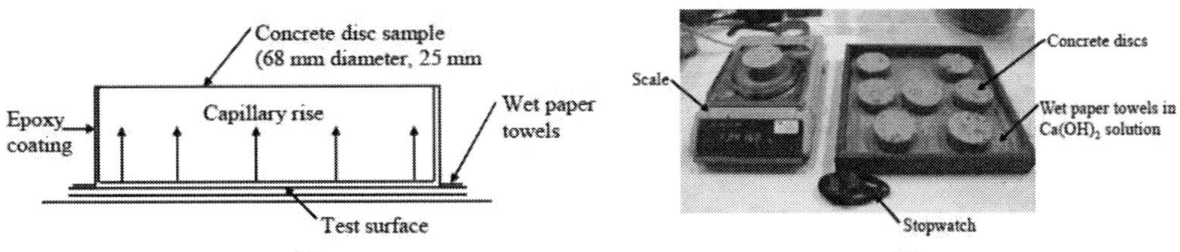

Typical ranges for durability classification of concretes for the three durability index (DI) tests, based on on-site and laboratory data, are shown in Table 2.

TABLE 2: TYPICAL RANGES FOR DURABILITY CLASSIFICATION USING DURABILITY INDEX VALUES [10]

Durability class*	Oxygen permeability index (log scale)	Water sorptivity index (mm/√hr)	Chloride conductivity index (mS/cm)
Excellent	>10	<6	<0.75
Good	9.5 – 10	6 – 10	0.75 – 1.50
Poor	9.0 – 9.5	10 – 15	1.50 – 2.50
Very poor	<9.0	>15	>2.50

* **Note:** This classification is relative to the aggressivity of the exposure environment and design life. For example, a CCI of 1.0 mS/cm may be classified as "*Good*" for a given exposure environment and design life, but may be "*Poor*" if the design life cannot be achieved in a more aggressive exposure environment.

Current hybrid approach to durability design and specification

Even though there is a universal consensus to move away from the traditional prescriptive durability design and specification approaches, there are still gaps in Service Life Modeling and Service Life Prediction that hinder this. These gaps relate to the understanding of the deterioration mechanisms, quantification of the aggressivity of the exposure environment, and reliability of the existing prediction models. Ongoing research efforts by various researchers around the world have made positive developments in all these areas but there is still much more work to be done. Consequently, current durability guidelines and standards are a hybrid of both prescriptive and performance-based approaches. Examples of hybrid durability design and specification standards/guidelines/codes include fib Model Code 2010 [35] and Canadian Standard A23.1-09 [36]. In South Africa, the durability design is still largely based on prescriptive requirements, but performance-based specifications and quality control for concrete durability have already been implemented on a large, industry-wide scale, about 10 years ago [23].

CLOSURE

Concrete was in the past perceived as a material whose main contribution to the structural system is compressive strength this is undoubtedly not the case now! The premature deterioration of concrete structures, together with the need to incorporate sustainability demands in concrete practice and the demand for maintenance-free structures by clients support the urgent need for concrete practitioners to incorporate durability design in the overall design process. Design for durability in concrete structures will lead to significant capital gains especially in Africa where limited capital is available for infrastructural development. A lot of strides have already been made towards durability design, and design approaches have been developed and adopted in a number of countries around the world. However, in Africa where a lot of infrastructural development is currently taking place, the uptake of these approaches has been relatively poor. Very few countries in Africa such as South Africa have actively embraced the concept of durability design and specification for concrete structures.

This paper has presented the fundamental concepts of concrete durability design and specification. The approaches for durability design and specification (i.e. prescriptive and performance-based) are also discussed. More importantly, some detail has been provided on the [simple and relatively inexpensive] South African durability design and specification approach in an attempt to encourage other African countries to adopt and implement it. In summary, African countries which are currently experiencing increased infrastructural developments need to quickly move away from prescriptive approaches and embrace performance-based durability design and specification of concrete structures in order to ensure that the investments are worthwhile in the long-term.

In order to implement concrete durability design and specification, both civil engineering scholars and practicing engineers need adequate knowledge of concrete as a construction material. More emphasis should be placed on the teaching of [modern] concrete technology in tertiary institutions. At the same time, concrete practitioners should constantly update their knowledge in this field; this is the intension of continuous professional education/development programmes. This requires the input of academics in our tertiary institutions to work closely with the engineering bodies to run conferences, workshops and seminars targeted towards updating practicing engineers with new and practical concrete practices.

ACKNOWLEDGEMENT

The authors acknowledge the financial support of The Concrete Institute (TCI), South Africa and the National Research Foundation (NRF). Gratitude is also extended to PPC Pty Ltd, AfriSam and Sika (SA) Pty Ltd. for their continued support.

REFERENCES

[1] Otieno, M. (2018) A case for the need to incorporate durability in the design and construction of RC structures. Kenya Engineer - Journal of the Institution of Engineers of Kenya, Vol. 39(1), pp. 32-38.

[2] KNBS (Kenya National Burueau of Statistics) Cement production and consumption. 1 pp (Accessed on 4th January 20118).

[3] Global-cement (2017) http://www.globalcement.com, 1 pp (Accessed on 4th January 20118).

[4] CW-Group, www.cwgrp.com.

[5] Ballim, Y., Alexander, M. & Beushausen, H. (2009) Chapter 9: Durability of Concrete - Fulton's Concrete Technology (9th Edition). Edited by Owens, G., Cement & Concrete Institute, Midrand, South Africa, pp. 155-188.

[6] Richardson, M. G. (2002) Fundamentals of durable concrete. Modern Concrete Technology, Spon Press, London.

[7] fib-Bulletin-No.-34 (2006) Model code for service life design. ISBN: 978-2-88394-074-1, 116 pp.

[8] Alexander, M. G. & Beushausen, H. (2007) Performance-based durability design and specification in South Africa. Proceedings of the International Concrete Conference and Exhibition (ICCX - Concrete Awareness), 14-16 February, 2007, Cape Town, South Africa.

[9] Alexander, M. G. (2018) Service life design and modelling of concrete structures background, developments, and implementation. Revista ALCONPAT International, Vol. 8(3), pp. 224-245.

[10] Alexander, M. G., Mackechnie, J. R. & Ballim, Y. (1999) Guide to the use of durability indexes for achieving durability in concrete structures. Universities of Cape Town and Witwatersrand, Research Monograph No.2.

[11] Tang, L. (2008) Engineering expression of the ClinConc model for prediction of free and total ingress in submerged marine concrete. Cement and Concrete Research, Vol. 38(8-9), pp. 1092- 1097.

[12] LIFE-365 (2005) ACI-Committee-365, Service life prediction model, Computer program for predicting the service life and life-cycle costs of reinforced concrete exposed to chlorides. American Concrete Institute.

[13] DuraCrete (1998) Probabilistic performance-based durability design: modelling of degradation. Document No. BE95-1347/R4-5, The Netherlands.

[14] DuraCrete (1998) Modelling of degradation. The European Union - Brite EuRam III Project BE95-1347, Document BE95-1347/R4-5.

[15] DuraCrete (1999) Statistical quantification of the variables in the limit state functions: summary.Brite/EuRam Project BE95-1347, Report No. BE95-1347-R9.

[16] DuraCrete (2000) Probabilistic performance-based durability design of concrete structy – Final Technical Report. Report No. BE95-1347/R17.

[17] Angelucci, M., Beushausen, H., Alexander, M. & Mackechnie, J. R. (2017) Specifying cement content for concrete durability: why less is more. Concrete Beton, No. 150, pp. 12-17.

[18] DI-Manual (2018) Durability index testing procedure manual, Version 4.5. www.theconcreteinstitute.org.za/durability, 43 pp.

[19] SANS-3001-CO3-1 (2015) Civil engineering test methods: Part CO3-1: Concrete durability index testing - Preparation of test specimens. South African Bureau of Standards – Standards Division, Pretoria, South Africa, ISBN 978-0-626-32799-6.

[20] SANS-3001-CO3-2 (2015) Civil engineering test methods: Part CO3-2: Concrete durability index testing - Oxygen permeability test. South African Bureau of Standards – Standards Division, Pretoria, South Africa, ISBN 978-0-626-32800-9.

[21] SANS-3001-CO3-3 (2015) Civil engineering test methods: Part CO3-3: Concrete durability index testing - chloride conductivity test. South African Bureau of Standards, ISBN 978-0-626- 32801-6.

[22] EN-206-1 (2000) Concrete - Part 1: Specification, performance, production and conformity, European Standard.

[23] Beushausen, H., Torrent, R. & Alexander, M. (2019) Performance-based approaches for concrete durability: State of the art and future research needs. Cement and Concrete Research, Vol. 119, pp. 11-20.

[24] Otieno, M. (2016) Mitigating against corrosion of steel in concrete a case for blended cement concretes. Proceedings of the 2nd Knowledge Exchange for Young Scientists (KEYS) Symposium, 6-10 June 2016, Accra, Ghana, ISBN 978-3-9817853-1-9.

[25] Ballim, Y. & Alexander, M. G. (2018) Guiding principles in developing the South African approach to durability testing of concrete. Proceedings of the 6th International Conference on Durability of Concrete Structures, 18-20 July, 2018, Leeds, West Yorkshire, United Kingdom, ISBN-13: 978-1849953948, pp. 36-45.

[26] SANS-3001-CO3-3 (2015) Civil engineering test methods: Part CO3-3: Concrete durability index testing - Chloride conductivity test. South African Bureau of Standards – Standards Division, Pretoria, South Africa, ISBN 978-0-626-32801-6.

[27] Ballim, Y. (1993) Towards an early age index for the durability of concrete. Proceedings of international conference, Concrete 2000 - Economic and durable construction through excellence. 7-9 September, 1993, Dundee, Scotland, Volume 2, E & FN Spon, London. pp. 1003-1012.

[28] Alexander, M. G., Ballim, Y. & Stanish, K. (2008) A framework for use of durability indexes in performance-based design and specifications for reinforced concrete structures. Materials and Structures, Vol. 41(5), pp. 921-936.

[29] Streicher, P. E. (1997) The development of a rapid chloride test for concrete and its use in engineering practice. PhD Thesis, Department of Civil Engineering, University of Cape Town.

[30] Streicher, P. E. & Alexander, M. G. (1995) A chloride conduction test for concrete. Cement and Concrete Research, Vol. 25(6), pp. 1284-1294.

[31] Alexander, M. G., Ballim, Y. & Mackechnie, J. R. (2009) Durability Index Testing Procedure Manual (Ver 1.0, 25 Feb 2009) - Revised Research Monograph No. 4 (1999), Department of civil engineering, University of Cape Town. Concrete Materials and Structural Integrity Research Unit - University of Cape Town, 30 pp.

[32] Garboczi, E. J. & Bentz, D. P. (1992) Computer simulation of the diffusivity of cement-based materials. Journal of Materials Science, Vol. 27(8), pp. 2083-2092.

[33] Beushausen, H., Alexander, M. G., Basheer, M., Baroghel- A., Gulikers, J., Jacobs, F., Khrapko, M., Monteiro, A. V., Nanukuttan, S. V., Otieno, M., Polder, R. & Torrent, R. (2016) Chapter 5: Principles of the performance-based approach for concrete durability. State-of-the-art Report: Performance-based specifications and control of concrete durability, RILEM TC 230-PSC (Editors: Beushausen, H. and Luco, L. F.), ISBN: 978-94-017-7308-9, Springer, pp. 107-132.

[34] Otieno, M. & Alexander, M. (2015) Chloride conductivity testing of concrete past and recent developments. Journal of the South African Institution of Civil Engineering (SAICE), Vol. 57(4), pp. 55-64.

[35] fib-Model-Code (2010) 3rd FIP/CEB Model Code for concrete structures. Comite Euro-International du Beton and Federation International de Precontrainte.

[36] CSA-A23.1-09 (2009) Concrete materials and methods of concrete construction, Canadian Standards Association, Mississauga, Ontario, Canada, L4W 5N6.

18th International Conference on
Non-Conventional Materials and Technologies
"Construction Materials & Technologies for Sustainability"
(18th NOCMAT 2019)
24th – 26th July 2019
Nairobi, Kenya

ASSESSMENT OF THE EVOLUTION OF THE COMPRESSIVE STRENGTH OF STRUCTURAL CONCRETE WITH RECYCLED COARSE AGGREGATES AND RICE HUSK ASH

Camila Lacerda Gomesa[1], Jerônimo Coura-Sobrinho[1], Conrado de Souza Rodrigues[1], Flávia Spitale Jacques Poggiali[1]

[1] Centro Federal de Educação Tecnológica de Minas Gerais (CEFET-MG), Brazil;

ABSTRACT

Concrete production is one of the activities that most emits pollutants in the atmosphere. The demolition of existing structures is also a major environmental problem, because it generates waste that needs appropriate disposal. Industrial and agricultural waste can be used in concrete production, reducing the use of cement and eliminating the need for disposal sites. From these considerations, it is possible to perceive the importance of studying the use of such materials in the production of concrete. The objective of this paper is to assess the evolution of compressive strength of concrete made from recycled aggregates (RA), produced from construction and demolition waste (CDW), and with rice husk ash (RHA) as a partial replacement for cement. For the accomplishment of the study, the characterization of the aggregates and the RHA is carried out in order to assess its properties, composition and pozzolanic activity. For this study, concrete mixtures were produced in two stages. In the first stage, concrete was produced with total substitution of the coarse aggregate by CDW and compressive strength tests were performed at 7 and 28 days. In the second stage, besides the substitution of natural aggregates (NA), the cement was replaced by RHA in 10% and 15% and the compression assays were performed at 7, 28 and 91 days, in order to assess the resistance gain over time. The mixture containing 15% of RHA presented the highest resistance gain over time, obtaining an increase of 34% from 7 to 91 days, while the mixture with 10% of RHA obtained an increase of 18%.

Keywords: Mineral admixture, Recycled aggregate, Recycled concrete, Construction and demolition waste.

INTRODUCTION

Aggregates resulted from construction and demolition waste (CDW) do not yet have standardization for use in concrete for structural purposes in Brazil. This is due to the inferior performance results when compared to concretes with natural aggregates (NA) and to the limited knowledge of their impact on long-term durability [1, 2]. Since the collection of waste is not standardized, its composition varies, producing aggregates with different properties among distinct batches, affecting the reliability of results and, consequently, their utilization.

To try to mitigate the negative effects, some industrial by-products can be used in concrete mixtures as mineral additions, improving their performance. This depends mainly on the physical and chemical characteristics presented by these by-products, and can also reduce the cost and negative environmental effects [3-5]. The combination of mineral additions with the use of CDW aggregates can generate concrete with acceptable properties, including for structural purposes [6].

The use of waste as a substitute for raw materials can minimize the emissions generated in the manufacture of new products and also reduce or even eliminate the need for landfill to dispose of these wastes [7].

The disposal of CDW is a major economic and environmental problem for the construction industry. Re-use of CDW can generate savings by reducing disposal costs and generating cheaper recycled materials in comparison with natural ones. The use of CDW as aggregates also reduces the extraction of raw materials, reducing the pollution generated by this activity [2, 8-10].

The agricultural industry produces a large quantity of by-products that can be used in civil construction, such as rice husk ash (RHA) that is already marketed for this purpose. Rice husks remaining from grain beneficiation are used as fuel for energy production in thermoelectric plants. The ashes generated in this process of controlled firing are usually rich in amorphous silica, being used as mineral addition to the production of concrete [11].

Based on the need to enable the use of CDW in concrete for structural purposes, obtaining a more sustainable concrete, this research assessed the performance of the axial compressive strength of concrete made of coarse aggregates of CDW in total substitution to natural coarse aggregate and RHA in partial substitution of Portland cement by 10% and 15%.

MATERIALS

The cement used to manufacture the concrete was the CPV-ARI of the company Lafarge/Holcim. The mineral addition used was the rice husk ash of Silcca Nobre, commercially produced by Pilecco Nobre. The physical and chemical properties of cement and RHA are presented in Tables 1 and 2. The pozolanicity index of the RHA, obtained by the Modified Chapelle´s method [12], was 456 mgCa(OH)2/gRHA.

TABLE 1 - PHYSICAL PROPERTIES OF PC AND RHA.

Physical properties	PC	RHA
Sample Average particle size D_{50} (μm)	17,28	22,46
Specific gravity	3,06	2,20
Surface area (BET)	1,37	12,31

TABLE 2 - CHEMICAL COMPOSITION OF RHA.

Chemical compositions	PC (%)	RHA (%)
SiO_2	18,85	93,58
Al_2O_3	4,67	1,87
Fe_2O_3	3,07	0,09
MnO	-	0,51
CaO	64,82	0,77
MgO	0,70	-
SO_3	2,90	-
K_2O	0,77	2,68
MnO	-	0,51
P_2O_5	-	0,47
CuO	-	0,01
Rb_2O	-	0,01
ZnO	-	0,01
Loss on ignition (LOI)	4,03	11,00

The fine natural aggregate used was mixed washed sand, with specific mass 2,57 g/cm³, unitary mass of 1,45 g/cm³, empty index of 43,5% and fineness module of 2,4.

The coarse natural aggregate was the gneiss crushed N° 1, classified according to NBR 7211 [13]. Its physical properties are shown in Table 3.

TABLE 3 - PHYSICAL PROPERTIES OF COARSE AGGREGATES

Physical properties	NA	RA
Specific gravity (g/cm³)	2,64	2,26
Absortion (%)	0,33	9,32
Los Angeles abrasion loss (%)	22,0	47,6

The recycled coarse aggregate of construction and demolition residue was obtained from the waste recycling station of Belo Horizonte (Minas Gerais, Brazil). Samples were obtained in September 2017. The RA has undergone a process of cleaning and separating contaminants before being used. The composition of the aggregate was 81,1% of natural rocks and cementitious materials, 16,1% of ceramic materials and 2,8% of contaminants. The physical properties are in Table 3.

The chemical additive used was the Superplasticizer (SP) based on polycarboxylates Fluxer RMX 7350.

The water used to manufacture the concrete was obtained from the local water supply system of the city of Belo Horizonte, under the responsibility of Copasa (Sanitation Company of Minas Gerais).

METHODOLOGY

The concrete was dosed following the methodology proposed by the American concrete Institute (ACI). The preparation was performed in a concrete mixer using the Double mixing method (or Two-Stage mixing). The specimens were molded with a diameter of 10 cm and height of 20 cm. After the demolding, a submerged cure was performed until the day before the assay.

The workability of all concrete mixtures was assessed by Slump test for determination of the consistency, specified by NBR NM 67 [14].

In the first stage of research, natural coarse aggregates were fully replaced by recycled aggregates. Compressive strength was determined at 7 and 28 days, following the recommendations of NBR 5739 [15].

In the second stage of the research, besides the total substitution of the coarse aggregates, the cement was partially replaced by the RHA in 10% and 15% by weight. The compressive strength of these mixtures was tested at 7, 28 and 91 days, following the recommendations of NBR 5739 [15].

The proportions of mixtures performed in the two research stages are presented in Table 4.

TABLE 4 - MIXTURE PROPORTION

Mix	Cement (kg/m³)	RHA (kg/m³)	Sand (kg/m³)	Coarse aggregates (kg/m³)		w/b	SP (%)
				Natural	Recycled		
CONTROL	364	0	639	1148	0	0,55	0,3
CDW100	364	0	639	0	983	0,55	1,1
CDW100RHA10	328	36	639	0	983	0,55	1,1
CDW100RHA15	309	55	639	0	983	0,55	1,1

RESULTS AND DISCUSSION

The results of the slump test [Figure 1] show that RA considerably influences the workability of the mixtures. Even the Control mixture having a lower percentage of SP it still has considerably higher abatement than the other mixtures. The highest water absorption capacity of the RA (9,32%) in relation to NA (0,33%) causes reduction in water free of mixtures, reducing the workability of the mixtures.

FIGURE 1 – SLUMP TEST

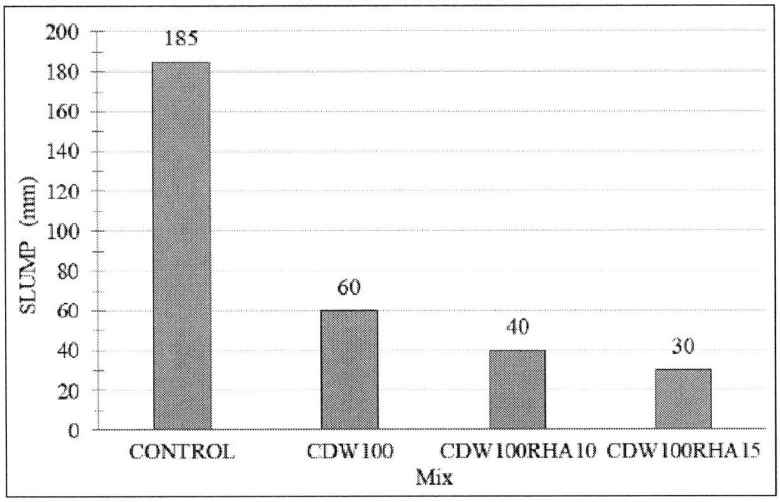

Previous studies also reported a reduction in the workability of concrete with the incorporation of CDW and/or RHA. Adnan et al. (2010) [16] reported 59,5% reduction in the mixture with 100% of CDW and 12% of RHA in relation to the mixture without substitutions. Padhi et al. (2018) [17] observed reductions of 16% and 30% in the workability of mixtures with 100% of CDW combined with 10% and 20% of RHA, respectively. The workability reductions are attributed to the greater water absorption capacity of CDW and RHA.

The results of axial compressive strength of the first stage of the study [Figure 2] show a reduction of 13.8% of the CDW100 mixture compared to the Control mixture at 28 days. The properties of RA and the creation of a new interfacial transition zone (ITZ) caused less resistance of the mixture produced. Poon et al (2004) [18] observed a 20% decrease in resistance at 28 days due to lower resistance to RA crushing in comparison to NA.

FIGURE 2 – COMPRESSIVE STRENGTH

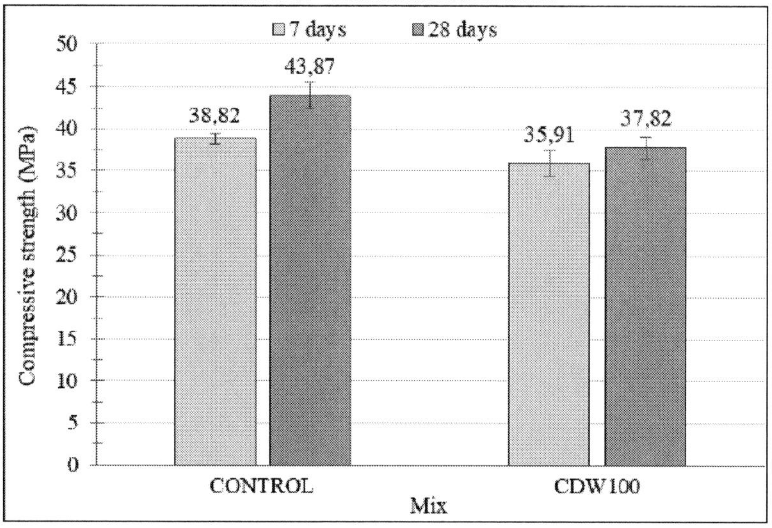

The results of compressive strength for the mixtures performed in the second stage of the study are presented in Figure 3. At 28 days it is possible observed a greater resistance of the CDW100RHA10 mixture compared to the CDW100 mixture. At 91 days, mixtures containing mineral addition have increased resistance due to the pozzolanic reactions of the RHA, which occur at ages above 28 days and improve the microstructure of the mixtures. The silica present in the RHA reacts with the calcium hydroxide formed in the hydration of the cement, forming an additional C-S-H gel that fills the pores existing in the recycled aggregate conferring greater resistance [19]. The surface area of the matrix is greater with the use of RHA, improving the bonding between aggregates and providing greater resistance to mixtures [20].

FIGURE 3 – COMPRESSIVE STRENGTH

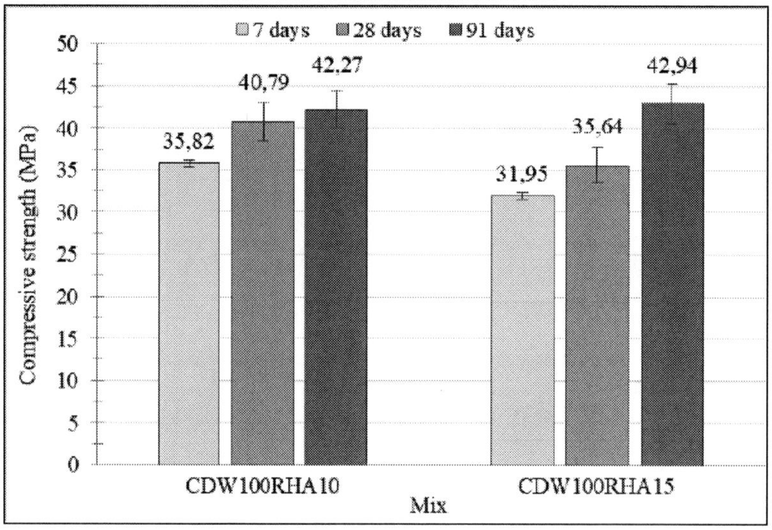

CONCLUSIONS

Recycled aggregates have lower specific mass, higher water absorption and greater loss of abrasion than natural aggregates. These differences in the properties of the aggregates influenced the workability of the concrete and its resistance at 28 days.

The incorporation of RHA in mixtures with CDW resulted in a lower workability of the concrete. The compressive strength showed a significant increase at 91 days, resulting from the improvement in the microstructure of the concrete provided by the RHA.

The combination of CDW and RHA, in 10% and 15%, proved to be efficient to reduce the losses in the axial compressive strength caused by the use of CDW in concrete mixtures. These results indicate that the incorporation of RHA may be a way to mitigate the damage produced by the use of high CDW content without segregation in concrete, and it is still necessary to assess other mechanical properties and durability for these mixtures.

REFERENCES

1. Marinkovic S, Radonjanin V, Malesev M and Ignjatovic I. Comparative environmental assessment of natural and recycled aggregate concrete. Waste Management 2010; 30(11):2255- 2264.

2. Lotfy A and Al-Fayez M. Performance evaluation of structural concrete using controlled quality coarse and fine recycled concrete aggregate. Cement and Concrete Composites 2015; 61:36-43.

3. Safiuddin M, West JS and Soudki KA. Hardened properties of self-consolidating high performance concrete including rice husk ash. Cement and Concrete Composites 2010; 32(9):708-717.

4. Madandoust R, Ranjbar MM, Moghadam HA and Mousavi SY. Mechanical properties and durability assessment of rice husk ash concrete. Biosystems Engineering 2011; 110(2):144-152.

5. Saravanan MM and Sivaraja M. Mechanical Behavior of Concrete Modified by Replacement of Cement by Rice Husk Ash. Brazilian Archives of Biology and Technology 2016; 59(spe2).

6. Radonjanin V, Malešev M, Marinković S and Al Malty AES. Green recycled aggregate concrete. Construction and Building Materials 2013; 47:1503-1511.

7. Ramasamy V. Compressive strength and durability properties of Rice Husk Ash concrete. KSCE Journal of Civil Engineering 2011; 16(1):93-102.

8. Nixon PJ. Recycled concrete as an aggregate for concrete - a review. Matériaux et Construction 1978; 11(65):371-378.

9. Oikonomou ND. Recycled concrete aggregates. Cement and Concrete Composites 2005; 27(2):315-318.

10. Malešev M, Radonjanin V and Marinković S. Recycled Concrete as Aggregate for Structural Concrete Production. Sustainability 2010; 2(5):1204-1225.

11. Fernandes IJ, Calheiro D, Kieling AG, Moraes CAM, Rocha TLAC, Brehm FA and Modolo RCE. Characterization of rice husk ash produced using different biomass combustion techniques for energy. Fuel 2016; 165:351-359.

12. ABNT. NBR 15895. Pozzolanic materials - Determination of calcium hydroxide fixed - Modified Chapelle´s method. Brazilian Association of Technical Standards 2010.

13. ABNT. NBR 7211. Aggregates for concrete - Specification. Brazilian Association of Technical Standards 2009.

14. ABNT. NBR NM 67. Concrete - Slump test for determination of the consistency. Brazilian Association of Technical Standards 1998.

15. ABNT. NBR 5739. Concrete - Compression test of cylindrical specimens. Brazilian Association of Technical Standards 2007.

16. Adnan SH, Rahman IA and Loon LY. Performance of Recycled Aggregate Concrete Containing Micronised Biomass Silica. International Journal of Sustainable Construction Engineering and Technology 2010; 1(2):1-10.

17. Padhi RS, Patra RK, Mukharjee BB and Dey T. Influence of incorporation of rice husk ash and coarse recycled concrete aggregates on properties of concrete. Construction and Building Materials 2018; 173:289-297.

18. Poon CS, Shui ZH, Lam L, Fok H and Kou SC. Influence of moisture states of natural and recycled aggregates on the slump and compressive strength of concrete. Cement and Concrete Research 2004; 34(1):31-36.

19. Prameetthaa J, Bharatkumar BH and Iyer NR. Investigation on micronized biomass silica as a sustainable material. Cement and Concrete Composites 2015; 60:25-33.

20. A. Jindal GD and Ransinchung RN. Behavioural study of pavement quality concrete containing construction, industrial and agricultural wastes. International Journal of Pavement Research and Technology 2018; 11(5):488-501.

18th International Conference on
Non-Conventional Materials and Technologies
"Construction Materials & Technologies for Sustainability"
(18th NOCMAT 2019)
24th – 26th July 2019
Nairobi, Kenya

COMPARATIVE STUDY BETWEEN PROPERTIES OF GEOPOLYMERIC CEMENTS PRODUCED WITH DIFFERENT TYPES OF METAKAOLIN

Paloma Santos Xavier de Alcantara[1], Normando Perazzo Barbosa[2], Primo Fernandes Filho[2], Letícia Cândido Mariz[3], Maria Laiz de Fátima Cabral Pontes[4]

[1] PhD Student, Federal University of Paraiba, João Pessoa, Brazil;

[2] Prof. Center of Technology of Federal University of Paraíba, João Pessoa, Brazil;

[3] Undergraduate Civil Engineering Student, Faculty of Maurício de Nassau, João Pessoa, Brazil;

[4] Master Degree Student, Program on Science and Materials Engineering, Federal University of Paraiba;

ABSTRACT

Geopolymers are produced by the alkaline activation of aluminosilicates. They have cementitious properties, excellent durability and dimensional stability. There are several applications on industry: fire-resistant composites, low-CO2 concrete, waste incorporation and metal encapsulation. One of the most used aluminosilicates (precursor material) for the production of geopolymers is metakaolin (MK). This research aims to compare properties of geopolymer cements produced with two different commercial MKs (metakaolin 1 – MK1 and metakaolin 2 – MK2). Alkaline activator solution was based on sodium hydroxide (NaOH) and sodium silicate (Na2OsiO3). Precursors were physically characterized by Blaine fineness, specific mass, as well as chemically and microstructurally through modified Chapelle, X-ray fluorescence (XRF) and X-ray diffraction (XRD). Property tested on pastes on fresh state was setting time through the Vicat apparatus. For the hardened state, compressive strength was tested. The analyzed metakaolins have different chemical reactivities. Compressive strength of the geopolymer cements produced with MK2 were 98.9% higher than those produced with MK1. Setting times presented big differences, of the order of hours. Thus, it can be stated that the properties of geopolymer cements are significantly altered when using different MKs.

Keywords: Geopolymeric cements. Metakaolin. Geopolymer.

INTRODUCTION

Geopolymers are produced by the alkaline activation of aluminosilicates. They have cementitious properties, excellent durability and stability. In addition, there are several applications on industry: from production of high-tech materials, fire-resistant composites, low-CO2 concrete, waste incorporation, to encapsulation of heavy metals. Geopolymers have been around for awhile. The main barriers to their widespread use are lack of standardization, high variability of raw materials - which causes a limitation of information exchange – and need to apply greater safety control in production of geopolymers, facilitating their use in precast concrete blocks. The geopolymerization reaction occurs by mixing a precursor (material rich in reactive silica and alumina) with an alkaline compound (called activator), resulting in a paste capable of agglomerating and hardening like Portland cement. Common examples of precursors are: construction residues, slag, fly ash, pozzolans and kaolin. A widely used precursor due to its disponibility and commercial consolidation is metakaolin (MK). Activating alkaline solutions are: sodium silicate (Na2OSiO3), potassium silicate (K2OSiO3), sodium hydroxide (NaOH) and potassium hydroxide (KOH) [1]. Granizo, Blanco-Varela and Martínez-Ramírez [2] defend the idea that the combined use of sodium silicate and sodium hydroxide results in geopolymers with better mechanical properties than those made only with sodium hydroxide.

The fundamental unit of the geopolymer is a tetrahedral complex of cations, such as Silicon (Si4+) and Aluminum (Al3+), in tetravalence, with four oxygens [3]. The atoms of Silicon alternate with those of Aluminum, sharing only one electron of the oxygen [3, 4], forming large inorganic molecular chains. According to the Si / Al molar ratio of the precursor oxides, the geopolymer arrangements are formed as can be seen in Figure 1.

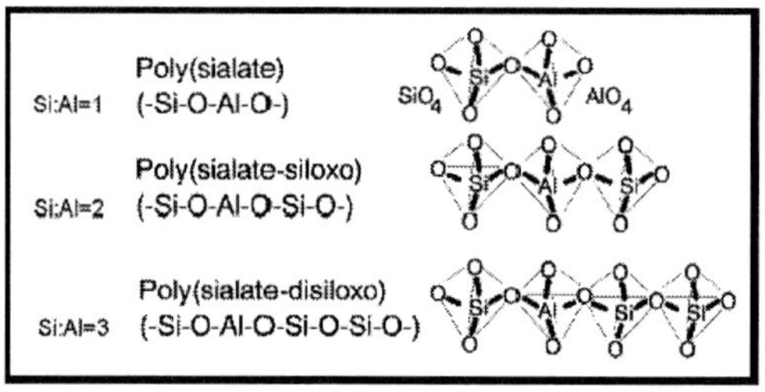

FIGURE 1. MOLECULAR ARRANGEMENTS OF GEOPOLYMERS IN FUNCTION OF SI / AL ratio [3].

It can be seen in Figure 1 that, for a Si / Al ratio equal to 1, the polysialate (-Si-O-Al-O-) is formed; for Si / For equal to 2, polysialato-siloxo (-Si-O-Al-O-Si-O-) is formed; for Si / Al equal to 3, the polysialato-disiloxo (-Si-O-Al-O-Si-O-Si-O-).

Papers have shown that the properties of the geopolymers are affected by changes in the amount of Si and Al available in the geopolymerization reaction [5] as well as by the molar ratios of the precursors and activators [6]. When using MK as a precursor material, it is worth pointing out that manufacturing conditions also directly influence dosage procedure and properties of geopolymers.

In the production of MK, after exploitation of quarries and extraction of kaolinite clays, processing and drying of the material, there is the calcination step at temperatures ranging from 600°C to 900°C. At this point the dehydroxylation of kaolinite (Al2O3.2SiO2.2H2O) - stable and less reactive compound - occurs, turning into metakaolin (Al2O3.2SiO2) - amorphous and more reactive compound [7]. Depending on duration and temperature firing, as well as the type of furnace [8], hydroxyl (functional group -OH) may not been totally eliminate, generating metakaolins with different reactivities with possible residual kaolinite in their composition. Ideal calcination temperature is between 700 and 800°C [8], however, it is known that due to lack of modernization and quality control of many factories, control of this parameter is not always respected. Thus this research aims to compare properties of geopolymer cements produced with two different commercial MKs.

Experimental Procedures

Two different commercial MKs (metakaolin 1 - MK1 and metakaolin 2 - MK2) were analyzed. MK1 is from northeast region, and MK2 is from southeast region of Brazil. Alkaline activator was based on sodium hydroxide solution (NaOH) and sodium silicate (Na2OSiO3). For preparation of the two geopolymers, quantitative material was fixed in the following mass ratio in relation to MK: 1: 0.992: 0.107: 0.25 (metakaolin: Na2OSiO3: NaOH: water). Paste 1 (P1) was the one prepared with the MK1, and paste 2 (P2), with the MK2. In order to comparison properties, all parameters were kept constant except MK type.

Precursors (MKs) were physically characterized by Blaine fineness, specific mass, as well as chemically and microstructurally through modified Chapelle, X-ray fluorescence (XRF) (Table 1) and X-ray diffraction (XRD) (Figure 2). Property tested on geopolymeric pastes on fresh state were set time through the Vicat apparatus according to the methodology adapted from Cheng and Chiu [9]. For the hardened state, compressive strength was tested.

TABLE 1 – CHEMICAL AND PHYSICAL CHARACTERIZATION OF MKS.

Chemical compost	**MK1 (%)**	**MK2 (%)**
SiO_2	65.63%	47.50%
Al_2O_3	26.30%	43.08%
Fe_2O_3	3.93%	7.06%
TiO_2	0.73%	1.62%
Outros	3.40%	2.36%
Blaine fineness [cm²/g]	18610	15670
Specific mass [g/cm³]	2.57	2.57

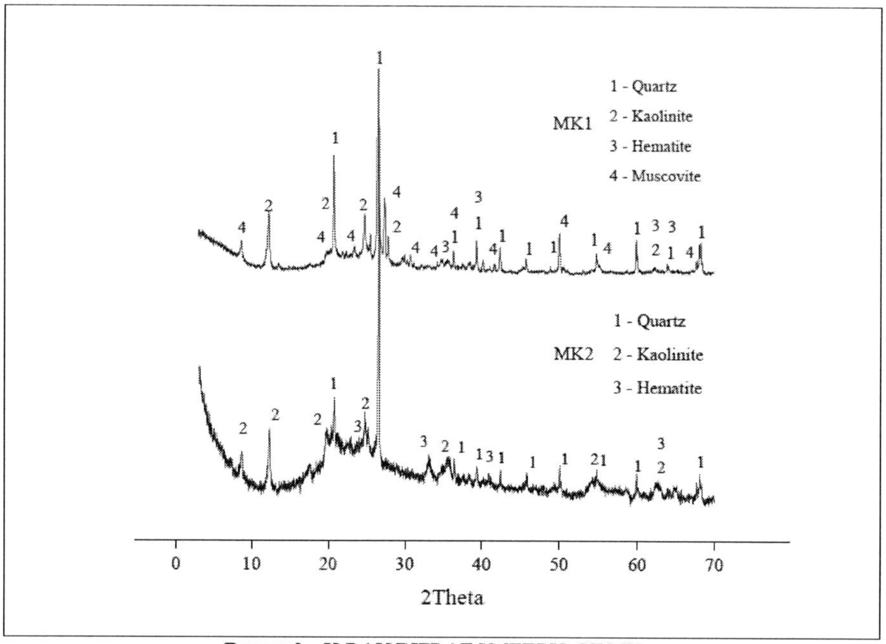

FIGURE 2 – X-RAY DIFRATOMETRY OF MKS.

Results and Discussion

Pozzolanic Activity by Chapelle Modified

To evaluate the reactivity of MKs, Modified Chapelle method was used, which measures the pozzolanic activity index. Its result informs the content (in milligrams) of calcium hydroxide (Ca(OH)2) fixed per gram (g) of material. The results are shown in Figure 3.

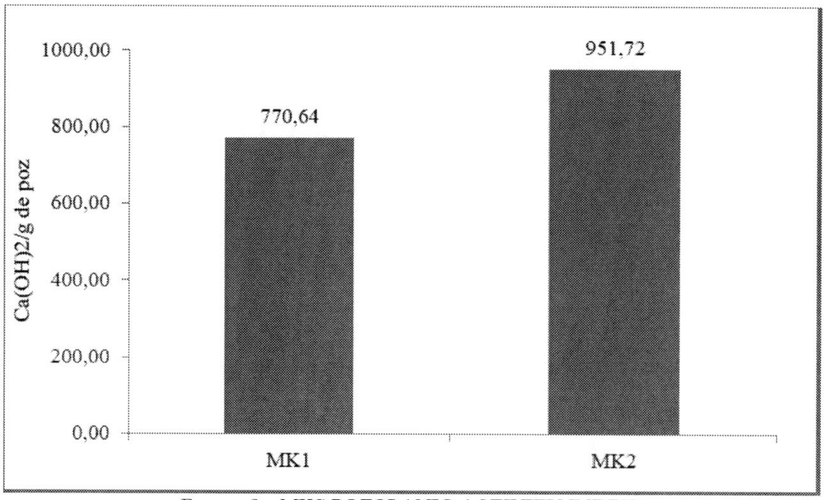

FIGURE 3 – MKS POZOLANIC ACTIVITY INDEX.

Figure 3 shows that the pozzolanic activity index of MK2 surpasses that of MK1 by 23.5%; therefore, it can be stated that MK2 is more reactive than MK1.

It should be noted that the modified Chapelle test does not accurately simulate the conditions of the alkaline attack of sodium hydroxide on the MK particles that occurs on geopolymerization reaction, since calcium hydroxide is used. However, if Ca(OH)2 is fixed by MK, there is an indication that this material is pozzolanic, with a disorganized molecular structure and reactive potential.

This assertion is corroborated by the diffractograms shown in Figure 2. Directly, in the MK1 diffractogram it can be seen a large amount of crystalline peaks and absence of the characteristic amorphous halo, characteristics of non or less reactive materials. The opposite is shown in the MK2 diffractogram, crystalline peaks of lower intensity and presence of amorphous halo, characteristics of more reactive materials.

The particle size of the material also influences its chemical reactivity [10], [11]. The finer, the larger the surface area and consequently the greater its reactive capacity. Curiously, MK1 is finer than MK2 (Table 1), even though it had lower reactivity by modified Chapelle and a predominantly crystalline diffractogram. Its high fineness is not sufficient to generate significant reactivity, since there is evidence of an inefficient burning during its manufacturing process (fact revealed by the presence of residual kaolinite peaks in the diffractogram), and presence of impurities on its composition (muscovite peaks [12]).

In Figure 4, the results of initial setting time of the geopolymer pastes are shown. Paste 1 had the MK1 as precursos; and paste 2, the MK2.

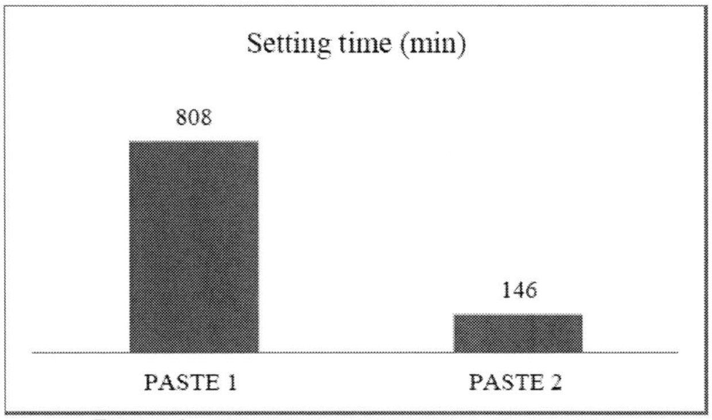

FIGURE 4 – SETTING TIME OF GEOPOLYMERIC PASTES.

It is noted that paste 2 has a shorter initial set time than paste 1, with a significant difference of approximately 11 hours. Since Provis & Devender [13] demonstrate that the reactive alumina has a significant importance on regiment of the initial kinetics of the geopolymer gel formation, on hardening of the pastes and the condensation reactions being chemically initiated by this compound, the phenomenon of the longest initial set time of MK1 can be explained by the fact that MK1 has a lower amount of Al2O3 (26,30% vs 43,08% of MK2 - Table 1). In addition, from this amount, only an amorphous part may be chemically available for reaction.

Traditional Portland cement has an average initial set time of 3 to 4 hours. The paste 2 has a similar time, which is an advantageous feature for possible future applications of geopolimeric cements.

Compressive Strength

In Figure 5, the compressive strength results of the geopolymers or geocements produced with the different MKs are shown.

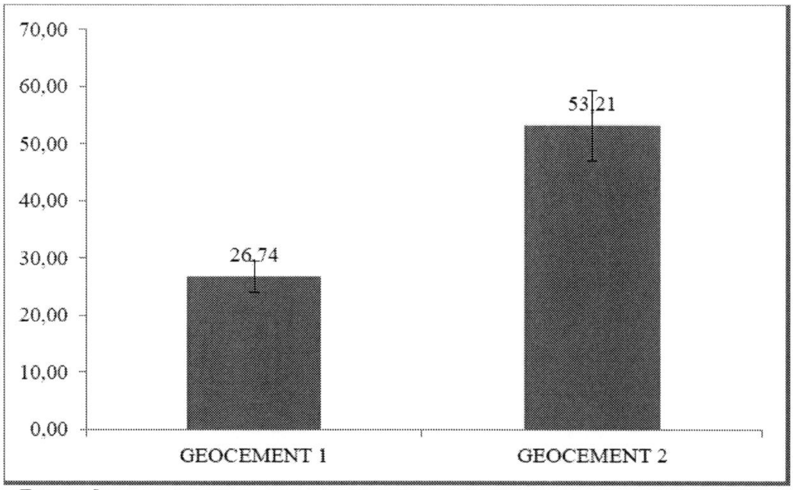

FIGURE 5 – COMPRESSIVE STRENGTH OF GEOCEMENTS PRODUCED WITH DIFFERENT MKS.

Through the analysis of Figure 5, it can be seen that the compressive strength of the geopolymer cements produced with MK2 were 98.99% higher than those produced with MK1. This fact is explained by the difference in reactivity between MKs.

Conclusions

It should be noted, as conclusions, that:

- The analyzed metakaolins have different chemical reactivities, with MK2 being more potentially pozzolanic than MK1 in 23.5%.
- Compressive strength of the geopolymer cements produced with MK2 were 98.9% higher than those produced with MK1.
- Setting times presented big differences, of the order of hours.
- Properties of geopolymer cements are significantly altered when using different MKs.
- With regard to applications of geopolymeric cements, it is necessary to make a previous analysis of reactivity of its precursor, for determination of molar ratios and dosages.

Bibliographic References

1. FELIX, A. S. et al.. Estabilização alcalina de adobes. In. Seminario Iberoamericano de Arquitectura y Construcción con Tierra, 15°, 2015, Cuencae. Equador. Anais.

2. GRANIZO, M. L; BLANCO-VARELA, M. T.; MARTÍNEZ-RAMÍREZ, S. Alkali activation of metakaolins: parameters affecting mechanical, structural and microstructural properties. Advances in Geopolymer Science & Technology. Journal of Materials Science. Springer. vol. 42, p. 2934–2943, 2007.

3. VASSALO, E. A. S. Obtenção de geopolímero a partir de metacaulim ativado. 2013. Dissertação (Mestrado em Construção Civil). Universidade Federal de Minas Gerais. Belo Horizonte.

4. DAVIDOVITS, J. 30 Years of Sucesses and Failures in Geopolymers Aplications. Market Trends and Potential Breakthroughs. Geopolymer 2002 Conference, Melbourne, Austrália, 2002.

5. DUXSON, P.; LUKEY, G.C.; DEVENTER, J.S.J. Physical evolution of Na-geopolymer derived from metakaolin up to 1000°C. Advances in Geopolymer Science & Technology. Journal of Materials Science. Springer. vol. 42, p. 3044–3054, 2007.

6. VILLA, C. et al. Geopolymer synthesis using alkaline activation of natural zeolite. Construction and Building Materials, vol.24, Issue 11, p. 2084-2090, 2010.

7. BORGES, P. H. R.; LOURENÇO, T. M. F.; FOUREAUX, A. F. S.; PACHECO, L. S. Estudo comparativo da análise de ciclo de vida de concretos geopoliméricos e de concretos à base de cimento Portland composto (CP II). Ambiente Construído, Porto Alegre, v. 14, n. 2, p. 153-168, abr./jun. 2014.

8. DAVIDOVITS, J. Geopolymer Chemistry and Applications. 3ª Edit. 2011.

9. CHENG, T. W.; CHIU, J.P. Fire-resistant geopolymer produced by granulated blast furnace slag. Mineral Engineering, vol 16, 2002.

10. TEMUUJIN, J; WILLIAMS, R. P.; RIESSEN A. van. Effect of mechanical activation of fly ash on the properties of geopolymer cured at ambient temperature. Journal of Materials Processing Technology, vol 209, Issues 12-13, 2009.

11. ALCANTARA, P. S. X. Blocos intertravados coloridos para pavimentação com incorporação de resíduos de cerâmica vermelha em prol da redução de pigmentos. 2015. Dissertação (mestrado) – Programa de Pós-Graduação em Engenharia Civil e Ambiental. Universidade Federal de Pernambuco. Caruaru.

12. DUXSON, P. The structure and termal evolution of metakaolin geopolymers. 2006. Thesis.

13. PROVIS, John L.; DEVENTER, Jannie S. J. van. Geopolymers – Structure, processing, properties and industrial appplications. 2009.

IC-NOCMAT 2019

18th International Conference on
Non-Conventional Materials and Technologies
"Construction Materials & Technologies for Sustainability"
(18th NOCMAT 2019)
24th – 26th July 2019
Nairobi, Kenya

THE INFLUENCE OF WOOD SHAVINGS TREATMENTS ON THEIR COMPATIBILITY WITH CEMENTITIOUS MATRIX

Mônica Cristina Pinto[1], M'hamed Yassin Rajiv da Gloria1, Romildo Dias Toledo Filho[1]

[1] Department of Civil Engineering, COPPE, Universidade Federal do Rio de Janeiro, Rio de Janeiro, RJ, Brazil;

ABSTRACT

This work aimed to study the effect of different wood shavings treatments on the removal of extractives for a better chemical compatibility with the cementitious matrix. The compatibility was accessed through determination of setting time of the matrix by the Vicat test. First of all, the wood shavings were immersed in water at temperatures of 25 ºC, 60 ºC, 80 ºC, 85 ºC, 90 ºC, and 95 ºC, respecting the wood to water ratio (w_d/w) of 1/10, by weight. Three immersion times for the lower temperatures were used: 2 hours, 24 hours and 48 hours for 25 ºC; 0.5 hour, 1 hour and 2 hours for both 60 ºC and 80 ºC. An immersion of 0.5 hour was used for the remaining temperatures. Then, cementitious pastes were fabricated with the residual waters and the setting time was determined by Vicat apparatus. The results showed that washing the particles in water at 80 ºC for 0.5 hour gave results closed to those for either greater temperatures or greater immersion times. Then, 80 ºC / 0.5 hour were chose to test w_d/w equal to 1/5 and 1/4. The subsequent results prove that the water is capable of solubilize more extractives, so it is possible to immerse more wood shavings in the less quantity of water.

Keywords: wood shavings, extractives, treatment, setting time, cementitious matrix.

INTRODUCTION

There is an increasing interest in the researching and developing of sustainable materials for civil construction. This is happening because of the global warming and the necessity of energy saving [1]. Bio-concretes associating inorganic binders and lignocellulosic aggregates have been shown as an alternative to this purpose [2]. Many authors have proved the feasibility of wood shavings as aggregates in the bio-concretes production [3, 4, 5, 6, 7]. Additionally, it is a consensus between them the need of pretreatments to improve the wood-cement compatibility. However, each one suggests a different way to do that.

The compatibility is related with the degree of cement setting after being mixed with water and wood particles. They are considered compatible if the presence of wood does not disturb the chemical processes of cement setting and hardening [8]. The main inhibitors of these processes are extractives (resins, fatty acids, terpenes, simple sugars, phenolic compounds) and hemicelluloses, which can be degraded by the high alkalinity of cement and release byproducts [9, 10]. However, it is important to emphasize that different retarding degree of cement hydration happens because of different components of wooden extractives or their quantities [11].

The compatibility between wood and cement can be assessed by different methods. The CA-factor is defined as the ratio of the amount of heat released from a mixture between wood and cement, and the heat released from a cement paste [8]. The Tmax (maximum temperature) and tmax (time to achieve this temperature) during the cement hydration are directly associated with the compatibility between cement and wood [11]. X-ray diffraction, thermal, spectroscopic and scanning electron analysis can be carried out to identify hydration products in mixes with different wood species [12]. Through Vicat test, the effect of water extractives on cement setting can be monitored [13].

There are two possible mechanisms of inhibition effect of wood on Portland cement [11]. First, the nucleation of Ca(OH)2 and CSH gel is delayed because the extractives make complexes with the metal ions in the cement solution, decreasing the concentration of Ca2+ and disturbing the solution equilibrium. Second, the hydration process is slowed because a thin adsorption layer of organic compounds is formed on the surface of cement grains [14].

In order to improve the compatibility between wood and cement, several methods are employed. Some authors suggested removing part of the extractives with hot water at 80ºC or 60 ºC, cold water, sodium hydroxide (NaOH) and calcium hydroxide (Ca(OH)2) [15, 16, 17, 18]. In the cold water and calcium hydroxide extractions, some researches have been using the wood-to-water ratio (wd/w) of 1/20 or 1/10 [19, 20, 21]. Others have cited the addition of 2% to 5% of cement setting accelerators as calcium chloride in the mixture [3, 4, 6, 8, 10].

In this work it is proposed to study the efficiency of different aqueous extraction methods (cold and hot water) to remove wood extractives to improve the compatibility between wood shavings and cementitious matrix. In addition, an attempt to optimize wd/w is also made.

MATERIALS AND METHODS

Materials

Cementitious materials

The matrix used in this study was designed in a previous work at NUMATS research center. It associated 60% of Brazilian Portland cement (CP V ARI), 30% of metakaolin and 10 % of fly ash. Material densities and their chemical composition are shown in Table 1.

TABLE 1: CHEMICAL COMPOSITION AND DENSITY OF THE MATRIX MATERIALS

Chemical elements	Portland cement	Metakaolin	Fly ash
CaO	72.638 %	-	2.290 %
SiO_2	13.224 %	41.858 %	50.963 %
Al_2O_3	4.904 %	48.606 %	34.194 %
Fe_2O_3	3.876 %	5.894 %	5.333 %
SO_3	3.355 %	1.259 %	1.593 %
K_2O	1.075 %	0.589 %	3.521 %
SrO	0.484 %	-	0.024 %
TiO_2	0.298 %	1.021 %	1.240 %
MnO	0.146 %	0.114 %	0.046 %
BaO	-	0.402 %	0.594 %
ZrO_2	-	0.194 %	0.111 %
NbO	-	0.038 %	0.006 %
ZnO	-	0.016 %	0.045 %
Ga_2O_3	-	0.009 %	-
Rb_2O	-	-	0.028 %
Y_2O_3	-	-	0.012 %
Density	3.09 g/cm^3	2.51 g/cm^3	1.89 g/cm^3

Raw material

Wood shavings were collected in a wood processing industry, located in Rio de Janeiro (Brazil). The residue was composed of at least four wood species (Manilkara salzmanni, Hymenolobium petraeum, Cedrela fissilis and Erisma uncinatum Warm) in random proportions. Due to their higher specific area, the particles passing through the 1.18 mm sieve were removed. Wood shavings general aspect and granulometric curve is shown in Figure 1 and Figure 2, respectively. Their apparent density and moisture content were 600 kg/m^3 and 15%, respectively.

FIGURE 1: WOOD SHAVINGS

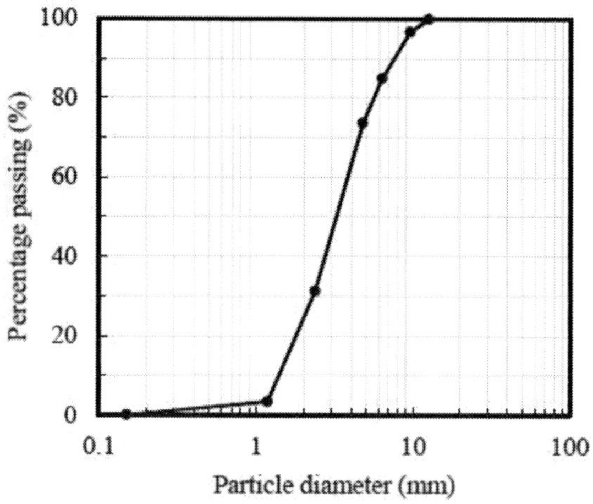

FIGURE 2: GRANULOMETRIC CURVE

Methods

Water extracts preparing

First, water extracts were prepared by immersing wood shavings in water with a wd/w of 1/10. The temperatures and immersion times considered are shown in Table 2.

TABLE 2: EXTRACTION TEMPERATURES AND TIMES

Temperature	Immersion times (h)		
25 °C	2	24	48
60 °C	0.5	1	2
80 °C	0.5	1	2
85 °C	0.5	-	-
90 °C	0.5	-	-
95 °C	0.5	-	-

After the results of Vicat test, one temperature and time of immersion were used to test wd/w of 1/5 and 1/4 as an attempt to optimize it. In this case, the particles were not only immersed, but they were also shaken with a spatula every 5 minutes to increase the release of extractives.

Mixing procedure

Water and solution of water extractives were weighed, respecting the water-to-cement ratio (w/c) of 0.40. The cementitious materials previously homogenized in a plastic bag and the water were added in a 5-l mixer. After 30 seconds of rest and 30 seconds of mixing at low speed, the mixer was stopped. Then, the paste was mixed again at high speed for 1 minute.

Cement setting monitoring via Vicat test

Using the mixing procedure described above, mixtures were prepared with each of the waters in Table 2 and distilled water (for the reference). The mixture obtained was then poured into a truncated cone-shaped mold (height = 40 mm, diameter 1 = 80 mm, and diameter 2 = 70 mm). Set monitoring was carried out according to NBR NM 65 Brazilian standard [22] with a manual Vicat apparatus illustrated in Figure 3. This equipment consists of a 1.13 mm diameter needle fixed on a 300 g moveable rod. In this test, when the material is fluid, the needle passes through the total height of the sample. When the paste starts setting, the needle penetrates the material until (4±1) mm from the mold base. The final setting time occurs when the needle penetrates only 0.5 mm.

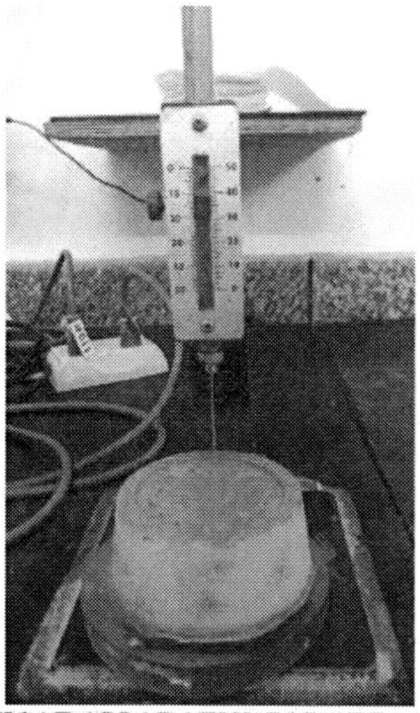

FIGURE 3: VICAT APPARATUS FOR SET MONITORING

RESULTS AND DISCUSSIONS

Figure 4 contains water obtained after immersing wood particles and used in this research to fabricate pastes. For water temperature of 25 °C, it can be observed a coloration difference between the three samples, due to the quantity of solubilized extractives per time. The darker is the sample, more extractives it contains. The same occurred at 60 °C, but the color difference is little. At 80 °C, the three samples had the same color, revealing that increasing the residence time above 0.5 hour do not solubilize additional amount of extractives. Observing the colors of water for the temperatures of 80 °C, 85 °C, 90 °C, and 95 °C, for the immersion time of 0.5 hour, it can be observed that all the samples have approximately the same color. It means that the increase in temperature did not affect, visually, the amount of solubilized extractives. The quantification and chemical composition of the removed extractives are being determined and will be informed in a coming publication.

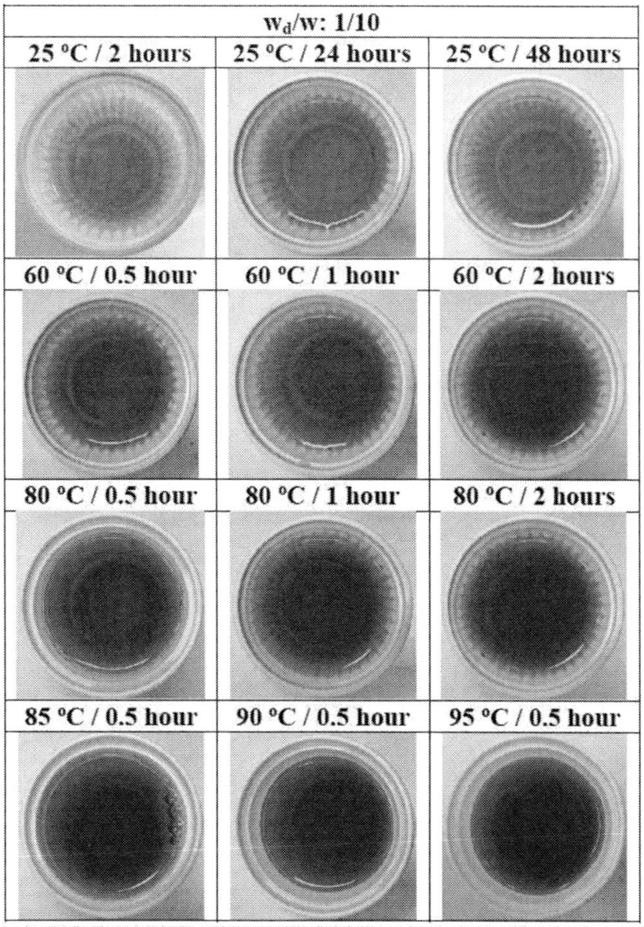

FIGURE 4: WATER OBTAINED AFTER WASHING WOOD SHAVINGS AT DIFFERENT TEMPERATURES AND RESIDENCE TIMES

As an attempt to optimize the wd/w in order to reduce the amount of water to be used in the wood shavings processing, it was also investigated the wd/w proportions of 1/5 and 1/4. The washing water was darker when the concentration of wood shavings was increased, as observed in Figure 5. It is an indication that the use of less water at this temperature and immersion time was capable of solubilizing the extractives present in the material.

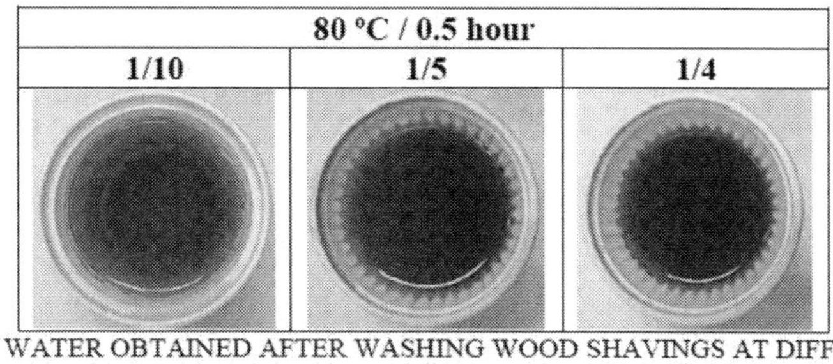

FIGURE 5: WATER OBTAINED AFTER WASHING WOOD SHAVINGS AT DIFFERENT w_d/w

Figure 6 shows the setting time measurement determined by Vicat apparatus for the matrix mixed with deionized water (reference) and with wood residual water. The initial and final set times of the reference occurred at 01:22 hour and 02:41 hours. When the matrix was mixed with residual water, the setting was delayed. For 25 ºC / 2 hours, the amount of solubilized extractives delayed the initial setting time about 01:01 hour and for 25 ºC / 24 hours it delayed 01:23 hour. When the immersion residence was increased to 48 hours, the initial setting time was not modified. The final setting time was disturbed almost in the same proportion. For 25 ºC / 2 hours, the final setting time was 01:00 hour higher than the reference. For 25 ºC / 24 hours and 25 ºC / 48 hours it was around 01:15 hour.

Similar behavior for the initial setting time was observed at 60 ºC. For 60 ºC / 0.5 hour, the initial setting time was 01:20 hour higher than the reference, for 60 ºC / 1 hour and 60 ºC / 2 hours it was around 01:33 hour. The final setting time was 01:22 hour higher for 60 ºC / 0.5 hour and 60 ºC / 1 hour and 01:37 hour for 60 ºC / 2 hours.

For 80 ºC / 0.5 hour, solubilized extractives delayed the initial setting time in 01:48 hour and for 80 ºC / 1 hour it delayed 01:58 hour. When the immersion residence was increased to 2 hours, the initial setting time was the same as 80 ºC / 1 hour. Final setting time had the same behavior. It was 01:35 hour higher for 80 ºC / 0.5 hour and 01:52 hour higher for the other two samples.

85 ºC / 0.5 hour delayed the initial setting time in 01:56 hour. 90 ºC / 0.5 hour and 95 ºC / 0.5 hour had almost the same result. The final setting time was also almost the same value for these three samples.

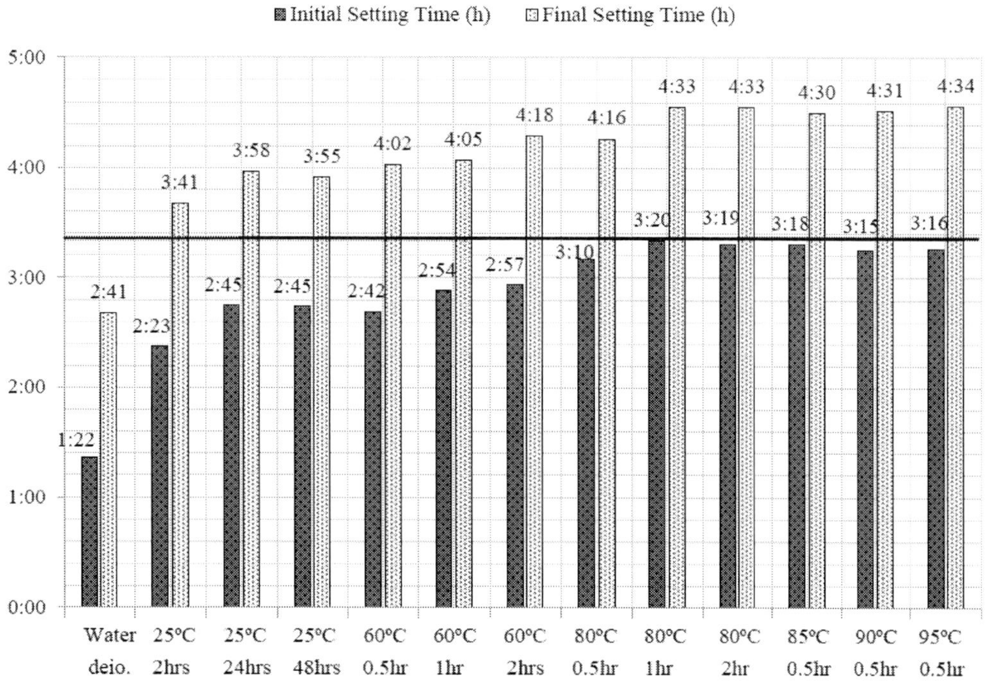

FIGURE 7: RESULTS OF INITIAL AND FINAL SETTING TIME FOR $w_d/w=1/10$

In Figure 7, it can be observed that setting times of 1/5 and 1/4 samples were delayed 02:45 hours and 03:03 hours, respectively, from the reference. The final setting time was also modified in 02:58 hours and 03:35 hours. This is one more prove that the water is capable of solubilizing more extractives, so it is possible to immerse more wood shavings in the same quantity of water.

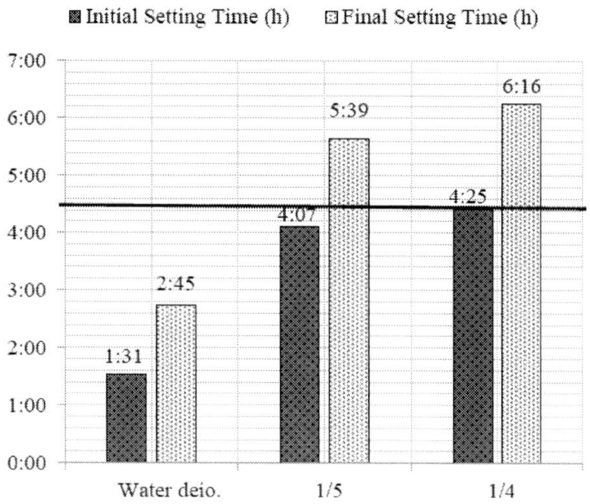

FIGURE 7: RESULTS OF INITIAL AND FINAL SETTING TIME FOR DIFFERENTS w_d/w

CONCLUSIONS

From the tests carried out in the present research, the following conclusion can be obtained:

- For the wd/w = 1/10, washing wood particles in temperature of 25 °C solubilized only part of the extractives. The initial setting time was increased by 01:01 hour when the residence time was 2 hours and 01:23 hour when the residence time was 24 or 48 hours in relation to the reference.
- Increasing the temperature to 60 °C results in a rise in the initial setting time. Only 0.5 hour of residence was responsible for an increase of 01:20 hour in the initial setting time. Changing the time to 1 hour, initial setting time increases 01:33 hour in relation to the reference.
- When 80 °C was used, the initial setting time increased even more. 0.5 hour of residence was responsible for a rise of 01:48 hour. Changing the residence time to 1 or 2 hour increased the initial setting time in 01:58 hour.
- Using temperatures above 80 °C did not increase the amount of solubilized extractives.
- Changing wd/w resulted in more extractives in the water, so the initial setting time was highest for 1/5 and 1/4.

ACKNOWLEDGEMENTS

The authors gratefully acknowledge the Conselho Nacional de Desenvolvimento Científico e Tecnológico, CNPq, (Brazilian National Science Foundation), for financial support for this work.

REFERENCES

1. Amziane S, Collet F. Bio-aggregates based building materials. Rilem State-of-the-Art Reports. 2017;23.

2. Florentin Y, Pearlmutter D, Givoni B, Gal E. A life-cycle energy and carbon analysis of hemplime bio-composite building materials. Energy and Buildings 2017;156:293-305.

3. Latorraca JVF. Eucalyptus spp. na produção de painéis de cimento-madeira. Curitiba: Universidade Federal do Paraná 2000; 191p. PhD Thesis.

4. Pimentel LL. Telhas onduladas à base de cimento Portland e resíduos de Pinus caribaea. Campinas: Universidade Estadual de Campinas 2000; 67p. Master Thesis.

5. Govin A, Peschard A, Fredon E, Guyonnet R. New insights into wood and cement interaction. Wood Research and Technology 2005;59(3):330-335.

6. Santos RC, Mendes LM, Mori FA, Mendes RF. Aproveitamento de resíduos da madeira de candeia (Eremanthus erythropappus) para produção de painéis cimento-madeira. Cerne 2008;14(3):241-250.

7. Gloria MYR. Desenvolvimento e caracterização de painéis sanduíches de concreto com núcleo leve e faces em laminados reforçados com fibras longas de sisal. Rio de Janeiro: Universidade Federal do Rio de Janeiro 2015; 115p. Master Thesis.

8. Jorge FC, Pereira C, Ferreira JM. Wood-cement composites: A review. Holz als Roh- und Werkstoff 2004;62(5):370-377.

9. Vaickelionis G, Vaickelioniene R. Cement hydration in the presence of wood extractives and pozzolan mineral additives. Ceram-Silik 2006;50(2):115-122.

10. Pereira C, Jorge FC, Irle M, Ferreira JM. Characterizing the setting of cement when mixed with cork, blue gum, or maritime pine, grown in Portugal – I: temperature profiles and compatibility indices. J Wood Sci 2006;52(4):311–317.

11. Na B, Wang Z, Wang H, Lu X. Wood-cement compatibility review. Wood Research

2014;59(5):813-826. 12. Dong Z, Yu Y, Song P, Ma L, Lu F. Effects of two types of waste wood species on the hydration characteristic of Portland cement. Journal of Advanced Concrete Technology 2016;14:13-20.

13. Diquélou Y, Gourlay E, Arnaud L, Kurek B. Impact of hemp shiv on cement setting and hardening: Influence of the extracted components from the aggregates and study of the interfaces with the inorganic matrix. Cement & Concrete Composites 2015;55:112-121.

14. Miller DP, Moslemi AA. Wood-cement composites: Effect of model compounds on hydration characteristics and tensile strength. Wood and Fiber Science 1991;23(4): 472-482.

15. Beraldo AL, Carvalho JV. Compósito Eucalyptus grandis – cimento Portland. Scientia Forestalis 2004;65:150-161.

16. Castro V, Araújo RD, Parchen C, Iwakiri S. Avaliação dos efeitos de pré-tratamento da madeira de Eucalyptus benthamii Maiden & Cambage no grau de compatibilidade com cimento Portland. Revista Árvore 2014;38(5):935-942.

17. Iwakiri S, Silva LS, Trianoski R, Bonduelle GM, Rocha VY. Avaliação do potencial de utilização da madeira de Schizolobium amazonicum "Paricá" e Cecropia hololeuca "Embaúba" para produção de painéis cimento-madeira. Cerne 2012;18(2):303-308.

18. Iwakiri S, Trianoski R, Cunha AB, Prata JG, Hara M, Bila NF, Luis RCG, Araújo RD. Propriedades tecnológicas de painéis cimento-madeira produzidos com partículas de eucalipto. Revista de Ciências Agroveterinárias 2015;14(3):217-223.

19. Quiroga A, Marzocchi V, Rintoul I. Influence of wood treatments on mechanical properties of wood-cement composites and of Populus Euroamerica wood fibers. Composites 2016;84:25-32.

20. Lima AM. Influência de três tipos de tratamentos físicos na resistência à compressão de compósitos cimento-madeira à base de resíduos de serraria. Belém: Universidade Federal do Pará 2005; 112p. Master Thesis.

21. Beraldo AL, Bartholomeu A, Batista A, Fagundes A, Rolim M, Segantini A. Viabilidade de Fabricação de Compósitos Resíduos de Madeiras e Cimento Portland (CBC). In: Workshop reciclagem de resíduos como materiais de construção civil, São Paulo 1996; 77-82.

22. ASSOCIAÇÃO BRASILEIRA DE NORMAS TÉCNICAS. NBR NM 65: Cimento Portland – Determinação do tempo de pega. Rio de Janeiro: ABNT, 2003. 4 p.

18th International Conference on
Non-Conventional Materials and Technologies
"Construction Materials & Technologies for Sustainability"
(18th NOCMAT 2019)
24th – 26th July 2019
Nairobi, Kenya

BAMBOO REINFORCED CONCRETE: LESSON LEARNED, PROHIBITIONS AND OPPORTUNITIES

Kent A. Harries[1], Hector Archila[2], David Trujillo[3], Sebastian Kaminski[4], Edwin Zea Escamilla[5]

[1] University of Pittsburgh, USA;

[2] University of Bath, UK;

[3] Coventry University, UK;

[4] Arup, London, UK;

[5] CCRS University of Zurich, Switzerland;

ABSTRACT

The use of small diameter whole-culm (bars) and/or split bamboo (a.k.a. splints or strips) has often been proposed as an alternative to reinforcing steel in reinforced concrete. The motivation for such replacement is typically cost and the drive to find more sustainable alternatives in the construction industry. Although bamboo is a material with extraordinary mechanical properties, this paper will summarise the reasons that for most load-bearing applications, bamboo-reinforced concrete is an ill-considered concept: having significant durability, strength and stiffness issues. Additionally, it is argued that bamboo-reinforced concrete does not possess the environmentally friendly credentials often attributed to it. Finally, the authors identify applications in which bamboo reinforcement may prove an acceptable alternative to steel provided durability concerns can be addressed.

Keywords: bamboo, bamboo reinforcement, bamboo-reinforced concrete, durability, life cycle assessment

INTRODUCTION

The use of small diameter whole-culm (bars) and/or split bamboo (a.k.a. splints or strips) has often been proposed as an alternative to relatively expensive reinforcing steel in reinforced concrete. The motivation for such replacement is typically cost and the drive to find more sustainable alternatives in the construction industry. Although bamboo is a material with extraordinary mechanical properties, its use in bamboo-reinforced concrete is an ill-considered concept, having significant durability, strength and stiffness issues; additionally, it does not meet the environmentally friendly credentials often attributed to it [1].

BAMBOO REINFORCED CONCRETE

Published accounts indicate that the use of bamboo to reinforce concrete structures dates back a century in Southeast Asia. Early experimental studies of bamboo-reinforced concrete were conducted at MIT in 1914 [2], in Germany [3] and Italy [4] in the 1930's, the United States following the Second World War [5, 6], and Colombia [7]. These studies used either bamboo bars (whole-culms of small diameter) or splints (semi-round strips).

Much early interest in bamboo-reinforced concrete is attributed to the US Navy and their interest in rapid [re-]construction in Southeast Asia following the Second World War. Glenn [5] highlighted the relatively poor behaviour of bamboo reinforced concrete, identifying issues including: a) high deflection, low ductility, brittle failure, and reduced ultimate load capacity when compared to steel-reinforced elements; b) bonding issues associated with excessive cracking and swelling of bamboo; and, c) the need for using asphalt emulsions. Glenn prescribed 3 to 4% bamboo reinforcement ratio and an allowable bamboo stress of 20 to 28 MPa calibrated to control deflections. Brink and Rush [8] also promulgated an allowable stress approach for designing bamboo-reinforced concrete comparable to the contemporary approach [9] for steel-reinforced concrete; they recommended an allowable bamboo tensile stress of 28 MPa, a bond strength of 0.34 MPa, and for serviceability requirements, a bamboo modulus of 17.2 GPa. Geymayer and Cox [10] recommended a hybrid design approach in which a bamboo-reinforced concrete flexural element is designed as an unreinforced concrete member with a maximum tensile stress of $0.67\sqrt{fc'}$ (MPa units). To this, 3 to 4% bamboo reinforcement is added resulting in, they claim, a factor of safety on the order of 2 to 2.5. Recognising the unique and limited bond behaviour of bamboo, bond strength was prescribed to be 44 N per mm of reinforcing 'bar' circumference and that the embedment must exceed 305 mm [10]. This is a maximum bond stress of about 0.15 MPa. Using either approach, bond capacity will always control design. As a basis of comparison, a 25 mm diameter bamboo reinforcing bar embedded 305 mm can develop only between 3.5 kN [10] and 8.4 kN [8]. By contrast, a 9.5 mm diameter steel reinforcing bar in the same conditions can develop 29.4 kN.

A number of research papers describing bamboo-reinforced flexural members confirm the basic premise of the design methodology proposed by Geymayer and Cox [10]. Optimal ratios of longitudinal bamboo reinforcement ratios range from 3 to 5% from which the capacity of an otherwise unreinforced concrete beam is increased at least 2.5 times [11-16]. Ghavami [17] demonstrated the importance of providing at least minimum bamboo reinforcement and appropriate surface treatment to enhance bond reporting that beams with a 3% ratio of treated split bamboo reinforcement had four times the ultimate capacity of comparable unreinforced concrete beams. By contrast, bamboo-reinforced concrete with approximately 1.4% splints having no bond enhancement provided no improvement over the behaviour of unreinforced concrete [18, 19]. Similarly, bamboo-reinforced slabs having a reinforcement ratio of only 0.5% developed a single large crack and exhibited significant reinforcement slip [20].

Agarawal et al. [18] showed the significant beneficial effects of 'treating' bamboo splints with commercial epoxy-based adhesives in order to enhance bond. They reported average bond stresses (from pull-out tests) on the order of 0.13 MPa for plain bamboo splints (a value echoing the recommendation of Geymayer and Cox [10]) and values as high as 0.59 MPa (350% increase) when epoxy adhesive was used to coat the splints. This behaviour translated to improved flexural response. Similarly, Ghavami [21] reports a 430%

increase in the value of bond strength for epoxy-coated bamboo splints (2.75 MPa) embedded in concrete over uncoated splints (0.52 MPa). Ghavami also conducted tests with an asphalt and sand coating which resulted in a bond strength of 0.73 MPa. Agarawal et al. report that 8% untreated bamboo was necessary to result in flexural behaviour similar to that of a steel-reinforced concrete member having a reinforcing ratio 0.89% whereas epoxy-coated splints required a reinforcing ratio of only 1.4 % to achieve similar behaviour; implying a 470% improvement in behaviour when the splints were coated.

Pull-out bond tests of splints [10] and round culms [22] having varying embedment lengths are a more realistic manner of determining bond properties. Both [10] and [12] conclude that the average bond stress decreases as the embedment length increases, and that this decrease is significantly more pronounced than is observed in [isotropic] steel reinforcing bars. Such a reduction can be explained by the greater effects of shear lag and the poor transverse material characteristics of the anisotropic bamboo. Bamboo splints, which have no pronounced deformations (thus relying mostly on friction to transfer stress), exhibit a lower bond stress than round culms for which the nodal protrusions provide some degree of mechanical interlock with the surrounding concrete. Geymayer and Cox [10] concluded that bamboo splints had an effective bond length, beyond which further increases in embedded length had no effect on available capacity; from this they established their bond strength recommendation.

All known studies that address bond of bamboo in concrete identify shrinkage of untreated, green or pre-soaked bamboo, and swelling cycles resulting from variations in moisture in the concrete as being detrimental to bond. As a result, most studies recommend coating bamboo in a moisture barrier although sealing inadequately seasoned bamboo into a watertight environment has the potential to exacerbate decay.

There are few known studies specifically addressing the durability of bamboo embedded in concrete. Nonetheless, there is considerable literature addressing the durability and treatment of different biomass materials (occasionally including bamboo) in cementitious materials. Gram [23] represents perhaps the first significant study in this regard. Recent and very thorough reviews are provided by [24] and [25].

Portland cement concrete is a highly alkali environment which provides a passivating environment for embedded steel reinforcement. In contrast, alkali treatments are often used to break-down the cell structure of lignocellulosic materials such as wood, hemp, flax and bamboo [26] in order to retrieve, expose or treat their fibres. Such treatment may improve bond with polymeric resins in composite materials but are clearly undesirable in the case of bamboo bars used in bamboo-reinforced concrete. Hosoda [27] reports a 50% loss of bamboo tensile capacity following one-year conditioning in a high alkali water bath; after three years, the bamboo retained only 30% of its initial strength. Hemicellulose is reactive with calcium hydroxide ($Ca(OH)2$) present in cement paste [28-31] leading to crystallisation of lime in the biomass pores. Lignin is soluble in hot alkali environments [23] as is the case during cement hydration, and potentially when the concrete is exposed to direct sunlight in a tropical environment. Reducing alkalinity whether using ternary cements [32] or through carbonation [33] were found to only partially mitigate the degradation of biomass. Lignocellulosic materials in hydrated cement are also embrittled by mineralisation associated with cations (primarily Ca2+) in the concrete pore water [34]. Water absorption is a critical durability concern for biomass of any kind embedded in a cementitious matrix [25]. Water absorption and hygrothermal cycling result in volumetric change of the embedded biomass leading to interfacial damage and micro- and macro-cracking. These effects increase permeability, driving the deleterious processes described previously.

Biological attack is arguably the most critical concern for bamboo. Embedment in concrete is not believed to be sufficient to protect bamboo from insect – especially termite – attack. Termites can pass into cracks as small as 0.8 mm [35]. Bamboo-reinforced concrete is likely to exhibit such cracks from temperature, shrinkage and/or load effects. Thus, bamboo reinforcement requires chemical treatment through its entire wall thickness to mitigate insect attack [36, 37]. Fungal attack (rot) requires aerobic conditions and a moisture content typically exceeding 20% [38]. Bamboo that is fully or partially embedded in concrete is vulnerable to rot because concrete (or mortar) is porous and moisture is easily transported through capillary action [39] and through existing cracks. Surface or 'paint-on' treatments are generally not considered to provide sufficient protection against rot in timber [38, 40, 41] or bamboo [42]. Except in cases in which the concrete remains dry throughout its service life, decay is possible even when the bamboo is coated in a bituminous or epoxy coating.

ISSUES ASSOCIATED WITH BAMBOO REINFORCED CONCRETE

Concrete reinforced with bamboo, rather than conventional steel reinforcement exhibits a different behaviour and therefore needs to be designed using different paradigms. Fundamental differences between bamboo and conventional reinforcing steel are as follows [1]:

1. Bamboo is essentially an elastic brittle material; this limits the 'allowable' stress that may be utilised with bamboo based on the margin of safety desired.
2. The characteristic values of longitudinal tensile modulus and strength of bamboo are typically a tenth that of steel. As a result of the low modulus, serviceability considerations (i.e., deflections and crack control) are significant and typically govern design despite the low allowable strength.
3. The coefficient of thermal expansion of bamboo is: a) different from that of steel and concrete, which are, themselves, similar; and b) is almost an order of magnitude greater in the transverse direction than in the longitudinal direction affecting compatibility with the surrounding concrete, significantly impacting composite bond behaviour.
4. Unlike steel, bamboo is dimensionally unstable and requires some form of treatment to resist moisture transmission. Due to anisotropy, dimensional stability is not uniform in longitudinal and transverse directions, also potentially affecting bond behaviour.

5. Although not affected by corrosion, bamboo is susceptible to various degradation mechanisms associated with exposure to varying hygrothermic conditions and a high-alkali environment. There is no published or industry guidance that suggests that embedding bamboo into concrete will protect it against rot, even if it is coated with a water-proofing product.

One of the reasons steel reinforced concrete has been such a successful material is that its ductility allows engineers to safely design statically indeterminate structures by making use of the lower bound theory of plasticity. The absence of ductility in bamboo-reinforced concrete implies that not only is it inadequate for seismically active regions, it is inappropriate for statically indeterminate structures. In addition, there are other practical issues that hinder the use of bamboo as a reinforcing material in conventional construction. These include:

6. The anisotropic nature of bamboo makes hollow bars prone to crushing or splitting during transportation, handling and erection; bamboo bars must be handled with additional care not required for steel bars.
7. There is no known research addressing methods of splicing or the behaviour of splices in bamboo reinforcing bars. Like steel, bamboo bars are practically limited to about 6 m in length; thus splicing will be necessary in some applications.
8. There is no known research addressing the anchorage (beyond bond development) of bamboo in concrete. Whereas steel bars are easily bent, it is not believed to be practical to bend bamboo bars in a manner appropriate to provide anchorage in concrete. Thus, the only practical anchorage for bamboo bars is straight bar development.
9. Utilising on the order of 4% reinforcement ratio, bamboo-reinforced concrete will have congested bar details. This congestion, and the variability in bamboo bars, leads to the recommendation that, in order to facilitate adequate consolidation of the concrete, bamboo bars should be placed with a spacing of at least 3 bar diameters. This limit may result in concrete sections being larger than is strictly required to satisfy strength design considerations. As a result, a greater amount of concrete will be necessary to meet the load carrying requirement of the functional unit, which compromises the environmentally friendly credentials often attributed to it.
10. Bamboo bars will float in concrete. This requires bars to be tied in place to resist uplift. With the larger number of bars present, this may be a cumbersome requirement.
11. In addition to through-thickness treatment for protection from insect and fungal attack, pre-treatment of bamboo with special coatings to enhance bond and/or the use of waterproof membranes in ground-supported slabs are laborious and require expensive and complex application systems. This is counter to claims that bamboo-reinforced concrete is a sustainable, local and low cost alterative in developing regions.

12. Unlike steel, that when properly confined can be relied upon to contribute as 'compression reinforcement', the poor transverse properties of bamboo make it ill-suited for use in compression zones, including columns.
13. Bamboo is known to creep under the effects of sustained loads limiting the sustained tensile force that can be practically resisted.
14. The behaviour of bamboo at elevated temperatures or in fire conditions is unknown. Bamboo properties degrade above 50oC [43]. The glass transition temperatures of lignin and hemicellulose (the primary components of the bamboo matrix) range from 97 – 171oC and 140 – 180oC, respectively [44]. It is likely that the behaviour of bamboo reinforcement under fire conditions is inferior to that of steel.

There is a common belief that bamboo reinforcement for concrete is a ''green'' or ''sustainable'' alternative to steel reinforcement. However, life cycle assessment (LCA) of both systems used as reinforcement of a prototype three-bay portal frame shows that the production of the bamboo reinforced structure will have emissions almost twice those resulting from the production of the same structure reinforced with steel [1]. This increase is attributed to the considerably greater amount of concrete necessary to meet the load carrying requirement of the prototype structure. The emissions savings achieved by replacing steel with bamboo reinforcement are surpassed by the emissions from the additional concrete and transportation [1, 45]

Bamboo reinforced concrete is an ill-considered concept. More importantly, bamboo reinforcement – if used safely – is not an environmentally friendly or sustainable alternative to steel. Bamboo-reinforced concrete must be designed to remain uncracked; the presence of bamboo reinforcing is intended to impart a degree of ductility to the section and may impart some post-cracking reserve capacity in the event of an overload that results in cracking. This post-cracking behaviour is only possible if there is sufficient bond between the bamboo and concrete. It has been shown that some bond-enhancing surface treatments are sufficient to impart the bond capacity required. Nonetheless, the required 'uncracked' design increases concrete member dimensions and has a 'knock on' effect resulting in increased formwork and foundation requirements. Additionally, the poor durability and bond characteristics of bamboo require through-thickness treatment and additional surface treatment of bamboo reinforcement, respectively. Such treatments, as described in the literature, are labour intensive, costly, and often utilise materials of known toxicity or which have handling restrictions associated with workplace health and safety. Vo and Navard [24] draw a very prescient conclusion in this regard: "A large proportion of [the methods used to overcome issues of biomass durability when embedded in concrete] are effectively helpful in easing the concrete preparation and leading to better final materials. However, most of them, if not all, have little practical value since they are either impossible to be implemented because of the use of chemicals which are not environmentally-friendly or much too expensive."

POTENTIAL PRACTICAL USES OF BAMBOO AS REINFORCING MATERIAL

While bamboo-reinforced concrete is impractical in primary structural members [1], certain related applications may be practical provided issues of durability, dimensional stability and bond between bamboo and concrete are addressed.

1. Small cane or bamboo splints may be an alternative for crack control reinforcement for slabs on grade (slabs cast on the ground) provided at least 3% bamboo is used. Such slabs are designed to remain uncracked and/or are provided with control joints to facilitate controlled cracking. Slabs on grade are known to absorb moisture from the ground; thus this must be mitigated in order to avoid the potential for rot of the embedded bamboo.
2. Light cement bamboo frame (LCBF) panels, known colloquially as bahareque construction [46], are well established. LCBF construction is a modern technique utilising composite shear panels constituted of a wall matrix of bamboo or metal lath nailed onto a bamboo framing system, plastered with cement or lime mortar render. This method works well because the stresses in the wall matrix are very low. Provided the bamboo is treated against insect attack and kept dry through good design, the lifespan of the system is expected to exceed 30 years.
3. Small culm or bamboo splints have been proposed as reinforcement for hollow-core masonry construction in non-seismic environments [47]. Such bamboo-reinforced masonry may provide a means of strengthening otherwise unreinforced masonry. This requires further study and the application suffers from some of the same issues affecting bamboo-reinforced concrete described in this paper.
4. Javadian et al. [48] have proposed the use of a heat-treated, densified engineered bamboo composite for concrete reinforcement. The resulting composite strips have a reported tensile strength of 295 MPa and a modulus of 37 GPa. To be used as concrete reinforcing bars, the composite strips are coated with epoxy resin and sand is broadcast onto this as a means of enhancing bond. Bond capacities were reported to be about 80% of comparable steel reinforcement bond strength. Such engineered bamboo composite reinforcing bars hold promise for overcoming some of the obstacles associated with using bamboo as concrete reinforcement. To the authors' knowledge no LCA or similar comparison with steel has been made to document assertions of "sustainability". Nonetheless, it is clear that the additional processing, energy and the resins used on their production will have a significant impact on environmental impact and cost.
5. Finally, Bamboo-fibre reinforced concrete has been proposed and demonstrated by multiple researchers [49-51]. The nature of fibre reinforcement for concrete is quite different from conventional discrete bar reinforcement and beyond the scope of this summary.

CONCLUSION

Although widely proposed as a 'sustainable' alternative to the use of steel reinforcement in concrete, bamboo reinforced concrete is an ill-advised concept. Mechanically, bamboo is unable to contribute to reinforced concrete behaviour in a meaningful manner requiring bamboo-reinforced elements to remain essentially uncracked. This results in concrete member sizes considerably greater than required if conventionally reinforced and an effective prohibition on the design of indeterminate (continuous) structures (which itself results in design efficiency). Additionally, for a variety of reasons, bamboo – without considerable treatment to resist the alkali environment, rot and biological attack – is unlikely to be found to be durable in a concrete environment. Taken together, it is highly unlikely that any economy or sustainability benefit results from safely designed bamboo-reinforced concrete.

ACKNOWLEDGEMENTS

This review was developed when the lead author was Leverhulme Visiting Professor at the University of Bath in academic year 2018. The support of the Leverhulme Foundation is gratefully acknowledged.

REFERENCES

[1] Archila, H., Kaminski, S., Trujillo, D., Zea Escamilla, E. and Harries, K.A. (2018) Bamboo Reinforced Concrete – a Critical Review, Materials and Structures 51:102. https://doi.org/10.1617/s11527-018-1228-6

[2] Chow, H.K. (1914) Bamboo as a material for reinforcing concrete. Massachusetts Institute of Technology. http://hdl.handle.net/1721.1/81516.

[3] Datta, K. (1936) Versuche Uber die Verwendung von Bambus im Beton bau [Experiments on the use of bamboo in concrete construction], Der Bauingenieur (BerlinWilmersdorf), 17(1).

[4] De Simone, D. (1939) Norme, Materiali e Strutture per la Limitazione dell'impiego del Ferro nelle Construzioni, Annali dei Lavori Publlici, 77(12), 1189-1223. English translation by A. Holme, Substitutes for Steel in Reinforced Concrete, Royal Engineers Journal (Chatham-Kent), 54 (June 1940), 237-244.

[5] Glenn, H.E. (1950) Bamboo reinforcement in Portland cement concrete, Engineering Experiment Station, Clemson Agricultural College, Clemson, South Carolina, Bulletin No. 4, May 1950.

[6] Smith, E.F. and Saucier, K.L. (1964) Precast concrete elements with bamboo reinforcement, U. S. Army Engineer Waterways Experiment Station, Vicksburg, Mississippi, Technical Report No. 6-646, May 1964.

[7] Hidalgo López, O. (1974) Bambú, su cultivo y aplicaciones en fabricación de papel, construcción, arquitectura, ingeniería, artesanía [Bamboo, its cultivation and applications

in papermaking, construction, architecture, engineering, crafts]., Cali, Colombia, https://searchworks.stanford.edu/view/164053.

[8] Brink, F.E. and Rush, P.J. (1966) Bamboo Reinforced Concrete Construction, US Naval Civil Engineering Laboratory Report, Port Hueneme, CA, 18 pp

[9] American Concrete Institute (ACI) (1956) ACI 318-56 Building code requirements for reinforced concrete. Detroit, MI.

[10] Geymayer, H.G. and Cox, F.B. (1970) Bamboo Reinforced Concrete, Journal of the American Concrete Institute 67(51), 841-846.

[11] Rahman, M.M, Rashid, M.H., Hossain, M.A., Hasan, M.T. and Hasan, M.K. (2011) Performance Evaluation of Bamboo Reinforced Concrete Beam, International Journal of Engineering and Technology 11(4), 142-146.

[12] Khare, L. (2005) Performance Evaluation of Bamboo Reinforced Concrete Beams, MSc Thesis, University of Texas at Arlington. 100 pp.

[13] Kankam, J.A., Ben-George, M. and Perry, S.H. (1988) Bamboo-Reinforced Concrete Beams Subjected to Third-Point Loading, ACI Structural Journal 85(1), 61-67.

[14] Subrahmanyam B.V. (1984) Bamboo reinforcement for cement matrices, New Reinforced Concrete, Surrey University Press, Guildford, England, 141–194

[15] Indra, I. G. (1982) Bamboo as Expedient Reinforcement for Concrete Beams, MSc Thesis, University of Ottawa.

[16] Youssef, M. A. R. (1976) Bamboo as a Substitute for Steel Reinforcement in Structural Concrete, New Horizons in Construction Materials 1, 525-554, Envo Publishing Co., Lehigh Valley.

[17] Ghavami, K. (2005) Bamboo as reinforcement in structural concrete elements, Cement and Concrete Composites 27, 637-649.

[18] Agarwal, A., Nanda, B. and Maity, D. (2014) Experimental investigation on chemically treated bamboo reinforced concrete beams and columns, Construction and Building Materials 71, 610-617.

[19] Sevalia, J.K., Siddhpura, N.B., Agarwal, C.S., Shah, D.P. and Kapadia, J.V. (2013) Study on Bamboo as Reinforcement in Cement Concrete, International Journal of Engineering Research and Applications, 3(2), 1181-1190.

[20] Terai, M. and Minami, K. (2012) Research and Development on Bamboo Reinforced Concrete Structure, 15th World Conference on Earthquake Engineering, Lisbon.

[21] Ghavami, K. (1995) Ultimate load behaviour of bamboo-reinforced lightweight concrete beams, Cement and Concrete Composites 17, 281–288.

[22] Sakaray, H., Togati, N.V.V.K., Reddy, I.V.R. (2012) Investigation on Properties of Bamboo as Reinforcing Material in Concrete, International Journal of Engineering Research and Applications, 2(1), 77-83.

[23] Gram H. (1983) Durability of natural fibres in concrete. Stockolm: Swedish Cement and Concrete Research Institute. http://www.cbi.se/objfiles/1/HansErikGramavh_-1704081873.pdf.

[24] Vo, L.T.T and Navard, P. (2016) Treatments of plant biomass for cementitious building materials – A review, Construction and Building Materials 121, 161–176.

[25] Pacheco-Torgal, F. and Jalali, S. (2011) Cementitious building materials reinforced with vegetable fibres: A review, Construction and Building Materials 25, 575–581

[26] Pickering, K.L., Efendy, M.G.A. and Le, T.M. (2016) A review of recent developments in natural fibre composites and their mechanical performance. Composites Part A: Applied Science and Manufacturing, 83, 98–112.

[27] Hosoda, K. (1942). Bamboo reinforced Concrete, Syukyosya Syoin, Japan.

[28] Toledo Filho, R., Scrivener, K., England, G. and Ghavami, K. (2000) Durability of alkali-sensitive sisal and coconuts fibres in cement mortar composites. Cement and Concrete Composites 22, 127–143.

[29] Ramakrishna, G. and Sundararajan, T. (2005) Studies on the durability of natural fibres and the effect of corroded fibres on the strength of mortar. Cement and Concrete Composites 27, 575–582.

[30] Kriker, A., Bali, A., Debicki, G., Bouziane, M. and Chabannet, M. (2008) Durability of date palm fibres and their use as reinforcement in hot dry climates. Cement and Concrete Composites 30, 639–48.

[31] Roma, L., Martello, L. and Savastano, H. (2008) Evaluation of mechanical, physical and thermal performance of cement-based tiles reinforced with vegetable fibers. Construction and Building Materials 22, 668–74.

[32] John, V., Cincotto, M., Sjotrom, C., Agopyan, V. and Oliveira, C. (2005) Durability of slag mortar reinforced with coconut fibre. Cement and Concrete Composites 27, 565–74.

[33] Tonoli, G., Santos, S., Joaquim, A. and Savastano, H. (2010) Effect of accelerated carbonation on cementitious roofing tiles reinforced with lignocellulosic fibre. Construction and Building Materials 24, 193–201.

[34] Jorge, F.C., Pereira, C. and Ferreira, J.M.F. (2004) Wood-cement composites: a review, Holz Roh Werkst 62, 370–377.

[35] University of Tennessee Agricultural Extension Service (2001) Subterranean Termite Control, PB1344-3M-4/01. http://trace.tennessee.edu/utk_agexdise/15/. (Accessed November 2017).

[36] Liese, W. and T. K. H. Tang (2015) Preservation and Drying of Bamboo, in Bamboo. The Plant and its Uses, Vol. 10, W. Liese and M. Köhl, Eds. Springer, pp. 257–297.

[37] Kaminski S., Lawrence A., Trujillo D. and King C. (2016a) Structural use of bamboo. Part 2: Durability and preservation, The Structural Engineer, October 2016, 38–43.

[38] Ridout B. (1999) Timber Decay in Buildings: The conservation approach to treatment, E. & F.N. Spon, Abingdon, UK.

[39] Martys, N. and Ferraris, C. (1997) Capillary transport in mortars and concrete, Cement and Concrete Research, 27(5), 747-760.

[40] FPL (2010) Wood handbook—Wood as an engineering material. General Technical Report FPL-GTR-190: U.S. Department of Agriculture, Madison, WI, USA

[41] Wood Protection Association (2012) Manual: Industrial Wood Preservation – Specification and Practice, 2nd Edition, Wood Protection Association Castleford, UK

[42] Liese W. and Kumar S. (2003) INBAR Technical Report 22: Bamboo Preservation Compendium, INBAR, Beijing, China.

[43] Correal, J.F. (2016) Bamboo design and construction. Chapter 14 in Nonconventional and Vernacular Construction Materials: Characterisation, Properties and Applications, (Harries and Sharma, editors) Woodhead (Elsevier) Publishing Series in Civil and Structural Engineering No. 58.

[44] Youssefian, S. and Rahbar, N. (2015) Molecular Origin of Strength and Stiffness in Bamboo Fibrils. Nature/Scientific Reports. 5 (11116)

[45] Zea Escamilla, E, and Habert, G. (2013) Environmental impacts from the production of bamboo based construction materials representing the global production diversity, Journal of Clean Production, 66, 117-127.

[46] Gonzalez, G. and Gutierrez, J. (2005) Structural performance of bamboo 'bahareque' walls under cyclic load, Journal of Bamboo and Rattan, 4(4), 353-368.

[47] Moroz, J.G., Lissel, S.L. and Hagel, M.D. (2014) Performance of bamboo reinforced concrete masonry shear walls, Construction and Building Materials 61, 125-137.

[48] Javadian, A., Wielopolski, M., Smith, F.C. and Hebel, D.E. (2016) Bond-behavior study of newly developed bamboo-composite reinforcement in concrete. Construction and Building Materials, 122, 110–117.

[49] Terai, M. and Minami, K. (2012) Basic Study on Mechanical Properties of Bamboo Fiber Reinforced Concrete, IABSE 2012, Cairo.

[50] Ahmad, S., Raze, A. and Gupta, H. (2014) Mechanical Properties of Bamboo Fibre Reinforced Concrete, 2nd International Conference on Research in Science, Engineering and Technology, March 21-22, 2014, Dubai.

[51] Brindha, M., Khan, S., Narayanan, S.S., Kumar, A.M.S. and Viviek, V. (2017) Properties of Concrete Reinforced With Bamboo Fibre, International Journal of Innovative Research in Science, Engineering and Technology, 6(3), 3809-3812.

IC-NOCMAT 2019

18th International Conference on
Non-Conventional Materials and Technologies
"Construction Materials & Technologies for Sustainability"
(18th NOCMAT 2019)
24th – 26th July 2019
Nairobi, Kenya

INFLUENCE OF PORCELAIN TILE POLISHING RESIDUES AS A SUPPLEMENTARY CEMENTITIOUS MATERIAL IN CONCRETE

Humberto Mycael Mota Santos[1], Erika Pinto Marinho[1], Ana Cecília Vieira da Nóbrega[1]

[1] Postgraduate Program in Civil and Environmental Engineering, Agreste Academic Center, Federal University of Pernambuco, PE, BR;

ABSTRACT

Porcelain tile polishing residues (PPR) are generally generated in the porcelain tile production process, where much of it is disposed of in landfills. This paper aims to produce sustainable concrete partially replacing Portland cement by these residues, as supplementary cementitious material (0-50%, by cement weight). The properties of the concrete in the fresh and hardened state, including some indicators of durability, were analyzed to 30 MPa concretes. They are dosed using the absolute volume method by the American Concrete Institute (ACI), adapted by the Brazilian Portland Cement Association (ABCP), using a water consumption of 195 l/m^3 to obtain an initial 40-60 mm. Also, the cement consumption was kept constant for all the mixtures, to evaluate the real interference of the PPR. As more cement content was replaced, the slump dropped, but it could be attenuated with the use of superplasticizer. The replacement of cement by PPR up to 40% presented a viable solution for both sustainability and the development of more resistant and durable concrete structures.

Keywords: Concrete, porcelain tile polishing, residues, supplementary cement material.

INTRODUCTION

The polishing porcellanato residues (PPR) are usually generated at the porcelain production process. It has a reduced grain size, white and moistened powder. Around 200 ton of residue, each month is produced in a medium enterprise in the northeast part of Brazil [1, 2], as well as around 1,000 ton each week in a large company in the southern region [3].

The PPR is rich mainly in silica, followed by alumina, and magnesium oxide (provided by the abrasives) [3-8]. In general, it is composed of crystalline phases [3, 4, 6-11], that is not interesting for the pozzolanic process, but some authors highlight their amorphous composition [3, 6-10, 12-17], and can be used as additional cementitious material depending on its reactivity, contributing to the increase of strength and durability of the cementitious materials.

These residues are well indicated to reuse in concrete because they do not need considerable additional energy to crush or sieve, as well as they already had an approximately adequated granulometric distribution to be used as supplementary material in cementitious systems. There is the possibility to develop structural artifacts, as well as non-structural ones [2].

In this scenery, previous publications highlighted the positive influence of the PPR in concrete: 1) a filler [1, 18, 19]; 2) reducing the permeability and porosity [3, 7, 8, 10, 20], consequently, improving the durability [3, 8, 20]; 3) as nucleation centers [3], improving the growth for cement hydration [3, 18]; 4) as later pozzolanic activity, in the case of some percentage of amorphous silica [9, 12, 13, 21]; 5) increasing the mechanical compressive strength [2, 3, 8, 19]; as well as, 6) reducing the use of Portland cement [3].

On the other hand, in the fresh state, the slump reduced when the amount of residue increased [2, 18, 19, 22]. Their high superficial area increases the water needed on it. On top of that, the irregular shape of the PPR particles difficult the rolling motion effect of the concrete [5, 23, 24]. However, this problem can be solved through an adequate dosage, determining to what extent the PPR can be incorporated in order to avoid problems, or using a superplasticizer additive (SP), which improves the dispersion of the particles.

In this way, this article aims to analyze the influence of the partial substitution of Portland cement by polishing porcellanato residues (PPR) in concrete in the contents of 10%, 20%, 30%, 40%, and 50%.

EXPERIMENTAL PROGRAM

Raw Materials

It was adopted the Brazilian type CPV ARI Portland cement [25] (equivalent to ASTM type III and high early strength [26]; or EN CEM I 52.5 according to EN 197-1 standard), having compressive strength at 28 days of 37 MPa. The chemical composition and the main physical characteristics of the cement used comply with the Brazilian Standard (NBR5733:91 [27]). The porcellanato polishing residues (PPR) were collected from a company specialized in producing ceramic tiles (including porcellanato) located at João Pessoa, Paraíba state, Brazil. The polished powder material was oven-dried for 24h at 100°C, grounded in a ball mill (20 balls, 3h, 20 liters capacity) and sieved in 0.075 mm (# 200). A commercial superplasticizer solution based on a modified polycarboxylate was employed, with density from 1.07 g/cm³ to 1.1 g/cm³ and solid content between 28.5% and 31.5%. Rounded coarse aggregate (0/19 mm) and sand (0/4.8 mm) were used. They were previously oven-dried for 24h at 100°C. Their physical and chemical characteristics are given in Table 1, as well as the particle size curves are exposed in Figure 1. Table 1 also gathered the physical characteristics of all the used raw materials.

It can be observed in Table 1 that the PPR has a smaller actual and specific mass than the cement, with a higher surface specific area by the BET (Braunauer-Emmet-Teller) method, requiring more water in the system and smaller average diameter. In Figure 1, it can be observed that the PPR has a particle distribution similar to that of cement, but it confirms that the PPR is thinner.

TABLE 1 – PHYSICAL PROPERTIES OF THE RAW MATERIALS.

Material	Density (g/cm³)	Specific gravity (g/cm³)	BET (m²/g)	Avarage diameter (μm)	Maximum diameter (mm)	Fineness
Cement	0.91	3.15	1.90	7.89	-	-
PPR	0.43	2.48	2.79	4.32	-	-
Sand	1.39	2.58	-	-	2.57	2.40
Gravel	1.48	2.63	-	-	19	7.03

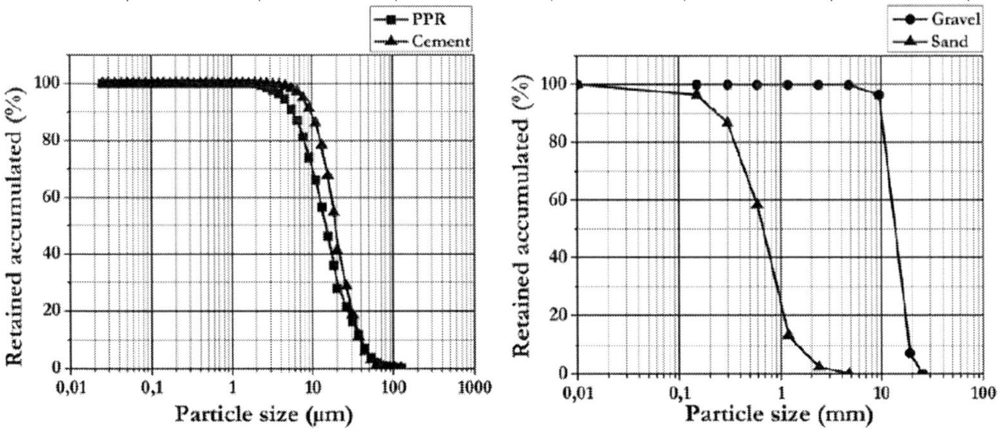

FIGURE 1 – GRANULOMETRIC DISTRIBUTION OF THE PPR, CEMENT, GRAVEL, AND SAND.

The chemical characterization of the porcelain tile residue can be seen in Table 2. It was settled applying XRF (N2 atmosphere), using an EDX 720 of Shimadzu. Previously the chemical tests the PPR was sieved in 0.075 mm (# 200) and oven dried for 24h at 110°C and for a second time. The porcellanato powder is composed primarily of siliceous and aluminum, which guarantee the pozzolanic activity possibility or, at least, some reactivity with the cement compounds. The significant amount of magnesium is related to the abrasive usually used in the polishing process. According to ABNT NBR 12653 [28], this PPR has one of the requirements to be classified as a pozzolanic material, since the sum of SiO2+Al2O3+ Fe2O3 = 84.532% ≥ 70%, as well as the amount of SO3 and Na2O are less than or equal to 4% and 1.5%, respectively.

TABLE 2 – CHEMICAL COMPOSITION OF THE PPR.

Oxides (wt. %)	SiO_2	Al_2O_3	Fe_2O_3	CaO	SO_3	MgO	K_2O	LOI
PPR	64.32	19.23	0.94	1.71	0.10	10.32	2.80	-

The mineralogical composition of the PPR was determined by X-ray diffraction (CuKα, λ = 1.5418 Å, 40 kV/40 mA, 10° to 70° (2θ) in 2.4°/min reading time) using an X-ray Bruker D8 Advance Davinci. The results show that the porcellanato polishing residues have some significant crystalline phases (Figure 2): mullite (17°, 26°, 31°, 35°, 42° e 58°), quartz (21°, 39°, 41°, 46°, 50° e 55°), silicon carbide (34°, 37° e 60°) and magnesium oxide (38°). The presence of mullite and quartz are related to porcellanato composition, while the silicon carbide and magnesium oxide come from the abrasive polishing process, as already verified in XRF. Similar results were found in Nascimento [4] and Steiner [29]. Also, as can be seen in Figure 2, the PPR has amorphous phases, indicated by an extended and accentuated halo between 20° and 30°, which could be assumed that the PPR is a reactive material as can contain amorphous siliceous and aluminum-based phases.

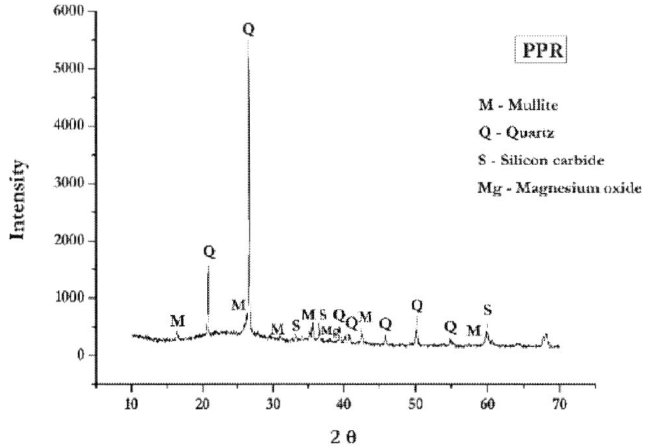

FIGURE 2 – MINERALOGICAL CHARACTERIZATION OF THE PPR.

The pozzolanic activity index (PAI) of the porcelain polishing residue was carried out as directed by ABNT NBR 5752 [30], in which a PAI equal to 79.80% was obtained. Thus, the PPR was not classified as pozzolanic because, according to ABNT NBR 12653 [28], the minimum PAI to classify the material as pozzolanic is 90%. However, as the pozzolanic activity index (PAI) near to the limit was reached, we tested the PPR as supplementary cementitious material, performing the compressive strengths at 91 days.

Concrete Blend Design

The mix design for the PPR based binders was determined according to the Portland Cement Brazilian Society (ABCP), according to Rodrigues [31]. The ABCP method is based on the method of the American Concrete Institute (ACI) 211, and the characteristics were adapted to Brazil. In the determination of the standard mixture, the characteristic compressive strength of the concrete (fck) required was 30 MPa. A standard deviation (Sd) was adopted in condition A, applicable to classes C10 through C80, as established by ABNT NBR 12655 [32], where the standard deviation has a value of 4 MPa. Finally, another parameter established was the trunk cone reduction between 40 and 60 mm. In total, eleven mixes were tested, using 195 l/m3 as the minimum operated water required to obtain an initial slump of 40-60 mm.

The standard mixture, concerning cement mass, was 1: 1.94: 2.62: 0.5 (cement: sand: gravel: a / c), with cement consumption equal to 390 Kg/m3. Besides, the cement consumption was kept constant for all mixes, to evaluate the real interference of the PPR. To keep the cement consumption constant, the amount of sand was altered according to each mix. On the other hand, the modification in the proprieties could be due to the alteration in the cement amount. Superplasticizer (SP = 1%, 2%, and 2.5%) was added due to the loss of workability when PPR was incorporated into the mixes. The PPR was used as supplementary cementitious material (10%, 20%, 30%, 40%, and 50%) replacing the cement partially. Therefore, the consumption of aggregates was determined, leading to the final mixture proportions presented in Table 3.

TABLE 3 – CONCRETE MIXTURES WITH CEMENT CONSUMPTION OF 390 KG/M³

	Cement	Sand	Gravel	PPR	W/C ratio	Superplasticizer
Reference	1.00	1.94	2.62	0.00	0.50	0.010
S10	0.90	1.92	2.62	0.10	0.50	0.010
S20	0.80	1.90	2.62	0.20	0.50	0.010
S30	0.70	1.88	2.62	0.30	0.50	0.020
S40	0.60	1.85	2.62	0.40	0.50	0.025
S50	0.50	1.83	2.62	0.50	0.50	0.025

Mixing and Concrete Casting Procedure

The six groups of concrete were prepared according to ABNT NBR 5738 [33] standard and following the Table 4 (100 x 200 mm). In the process was used a concrete-mixer with 400 liters of capacity working at a constant speed. The mixture procedure was carried out following this order: 50% of water (superplasticizer already diluted on it), 100% of gravel, 100% of sand, 100% of PPR (in the proportion defined for each mix) and the others 50% of the water. Mixing time procedure was 5 minutes. After 24h of casting, the samples were demolded and immediately full immersed in lime-saturated water until the tests.

Fresh and Hardened State Tests

In this work, all the mixes were tested in the fresh and hardened states. One bath to slump cone test (ABNT NBR NM 67 [34]). Three 10 cm x 20 cm specimens to compressive strength at 7, 28 and 91 days (ABNT NBR 5739 [35]. The compressive strength was settled using a hydraulic-type universal testing machine called UH-I Series, using a constant speed of 3 mm/min. Three 10 cm x 20 cm specimens to water capillarity at 28 days (ABNT NBR 9779 [29]).

RESULTS AND DISCUSSION

Fresh State Results

The slump values obtained using the cone test for all formulations are exposed in Figure 3. The mix proportion design adopted was to achieve a slump between 40 and 60 mm. These values were confirmed to the reference mixture. However, in order to minimize the PPR harmful interference in the mixtures workability, it was decided to additivity all the mixes with superplasticizer, once the casting process could be prejudiced using dry concretes. With this in mind, 1% of superplasticizer was used to the reference, S10, and S20. 2% to S30, as well as 2.5% to S40 and S50.

It should be highlighted that the workability decreased, even increasing the superplasticizer dose. However, all the concretes could be well-molded without a problem. The partial substitution of cement by PPR decreased the workability in 7.69% when 10% of cement was substituted by PPR; 46.15% for S20; 84.62% for S30; 88.72% for S40; and, 91.28% for S50.

The reduction in workability was also observed in Santos [5], in which PPR was added in mortars. The consistency index decreased from the reference value by up to approximately 20%. In Souza [7], it was also reported a decrease in workability in non-superplasticizer (from 50 mm to 5 mm) in samples with up to 40% PPR addition. It was only possible to add 50% of PPR when 0.50% of superplasticizer was used, having the reference mixtures and the addition of 50% of PPR rebates of 150 mm and 5 mm, respectively.

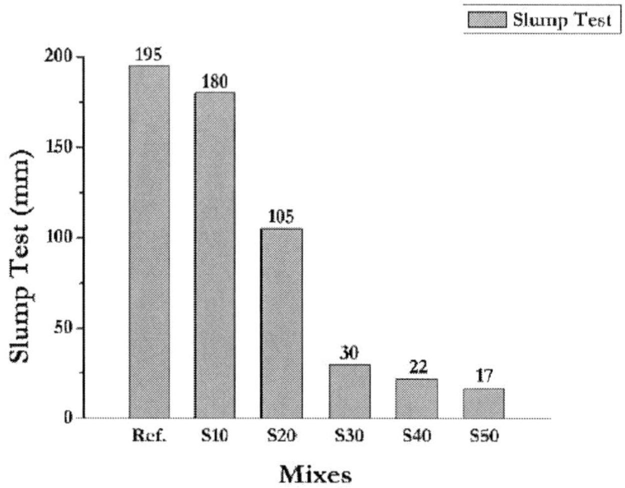

FIGURE 3 – SPREAD (MM) TO THE MIXTURES.

As obtained in the characterization of the PPR, its specific surface area is higher than that of cement. So, more water was needed to wrap the grains when this residue is incorporated. Also, Santos [5] analyzed the morphology of the porcelain polishing residue and found that its particles are irregular and angular, which can cause a reduction in workability, since the sliding process between the particles is prejudiced.

Hardened State Results

Compressive strength

The values of the average compressive strength at the ages of 7, 28 and 91 days for all mixtures are shown in Figure 4. As observed, the reference concrete reached values at 28 days higher than 30 MPa, certifying the dosage process adopted for a C30 concrete (f_{ck} of 30 MPa).

As expected for the mixtures with partial cement replacement, the compressive strength at 1- day decreases as the amount of PPR increases in the concrete, due to the reduction of sand in the matrix, altering the compactness dynamics of the systems. Although the consumption of cement remained constant in cubic meters, the matrix was physically altered.

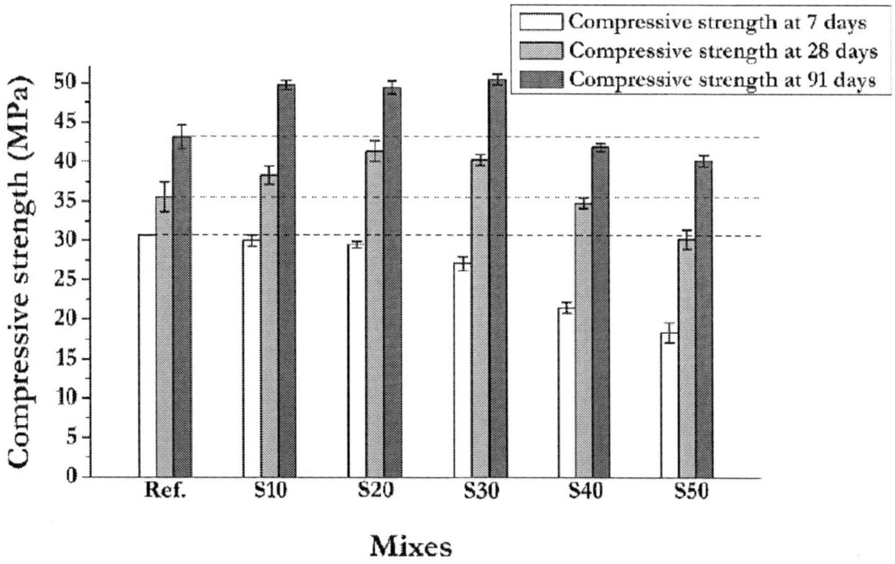

FIGURE 4 – COMPRESSIVE STRENGTHS AT ONE, SEVEN AND 28 DAYS.

The partial substitution of cement by PPR decrease the average compressive strength at 7-days in 2.29% when PPR replaced 10% of cement; 3.98% for S20; 11.61% for S30; 29.80% for S40; and, 40.12% for S50. It can be noted in Figure 5 that the compressive strength value for S10 and S20 at 7-days is relatively approximated to the reference one (taking into consideration the standard deviation bars), probably related to the fewer substitution rates associated to the filler action [3, 13, 20, 23].

At 28 days, however, higher average compressive strength values were reached for the concretes S10 (>7.45%), S20 (>16.16%) and S30 (>13.18%), when compared to the reference one. Positively, taking into consideration the standard deviation bars, S40 presented similar resistance behavior than the concrete without PPR. As expected, there was a decrease of around 15.15% when PPR replaced 50% of cement. Significant rises in compressive strength at 28 days were reported in [7, 36]. Resistances up to 44% higher were achieved in Souza [7] at 90 days when PPR was added 40% of PPR to the 1:2:3 (cement:sand: aggregate) concrete using superplasticizer. About the substitution percentage, Steiner, Bernandin and Pelisser [8] observed an improvement in resistance in the proportions of 10, 20 and 30% to the reference, where substitution greater than 30% tended to a reduction in resistance.

Similar behavior was also observed to the 91 days of curing with higher average compressive strength values for S10 (>15.15%), S20 (>14.52%) and S30 (>16.88%) than the reference concrete. In the same way, similar resistance behavior was achieved for the S40 concrete and the reference one. Jacoby and Pelisser [3] noticed that the higher ages resistance increased when the amount of PPR was elevated in the composition, showing the pozzolanic effect of the material.

The drop on the compressive strength for S50 concrete was 40.12% at seven days, 15.15% at 28 days, and just 6.95% at 91 days, confirming the reactivity of the PPR. Pacheco *et al.* [9] also found significant differences at early curing ages and smaller differences at long curing ages working with cement replacements by wastes in concrete. Therefore, both the filler effect and the PPR reactivity contribute to the compressive strength. Indeed, other researches also confirmed the PPR pozzolanic effect [3, 8, 10, 20, 24], as well as the filler effect [1, 18, 19]. The PPR in cement-based systems is well-known for its positive interference in compressive strengths.

Capillary absorption

The water absorption by capillarity, as expected, increase with the time. Also, it decreases as the amount of PPR increase in the mixture (Figure 5). Comparing the results with the reference sample, S10 and S20 reached higher absorption values (up to 72h with 48.25% and 18.42% respectively), while S30, S40, S50 show less capacity of absorption (-8.77%, -14.04%, - 17.54%). Similar performance was obtained by Breitenbach et al. [23], in which contents of 5% to 20% of PPR presented greater water absorption by capillarity, and for contents higher than 20% there was a considerable decrease. 30% of PPR reduced in 49.06% related to 20% of the residue.

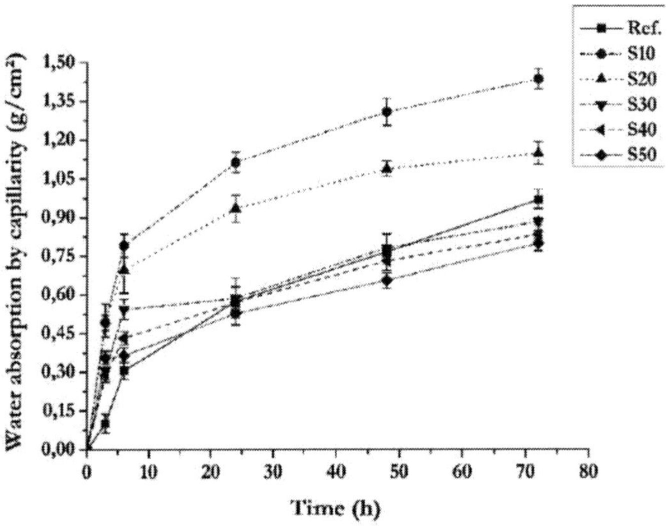

FIGURE 5 – CAPILLARY ABSORPTION AT 28 DAYS.

Probably, the tortuosity effect was responsible for reducing the capillarity absorption progress when more PPR was incorporated. Also, according to Breitenbach *et al.* [23], the water penetration is inversely proportional to the diameter of the capillaries; the smaller diameter of the particles of the PPR allowed the formation of pores with smaller and connected diameters, increasing the capillary absorption. The results agree with Silva [6] and Wang and Tian [13]. Both evaluated the mean pore diameter as the PPR was incorporated, where results obtained by Silva [6] showed that the average pore diameter decreased 55.18% for concrete with addition of 20% PPR, and the obtained by Wang and Tian [13], the reduction was 36.81% for mortars with 30% partial replacement of cement by PPR.

CONCLUSIONS

The results obtained have been promising for concretes with partial replacement of the concrete by the porcelain polishing residue (PPR).

- The PPR presented adequate physic-chemical characteristics for incorporation in cementitious materials. Although the pozzolanic activity index (PAI) did not reach the minimum of 90% required by NBR 12653, it was reactive.
- The partial replacement of the cement by the PPR decreased the workability of the concretes as the PPR levels increased, being the superplasticizer very efficient to solve this problem.
- The average compressive strength of the concretes decreased to 7 days in proportion to the increase of PPR contents. However, at 28 and 91 days, this behavior was altered, with the mixtures S10, S20 and S30 higher resistance and S40 similar resistance to the reference.
- Concerning the water absorption by capillarity, the mixtures S10 and S20 showed higher absorptions when compared to the reference, but with higher levels of PPR (S30, S40, and S50) the capillary absorption was lower than the reference.

REFERENCES

1. Nascimento, M.V.L.A., Estudos de blocos intertravados de concreto para pavimentação com incorporação de resíduo de porcelanto, in Programa de pós-graduação em engenharia civil e ambiental. 2016, Universidade Federal de Pernambuco: Caruaru-PE. p. 96.

2. Souza, P.A.B.F., Estudo do comportamento plásticos, mecânico, microestrutural e térmico do concreto produzido com resíduo de porcelanato, in Programa de pós-graduação em ciência e engenharia dos materiais. 2007, Universidade Federal do Rio Grande do Norte: Natal-RN. p. 230.

3. Jacoby, P.C. and F. Pelisser, Pozzolanic effect of porcelain polishing residue in Portland cement. Journal of Cleaner Production, 2015. 100: p. 84-88.

4. MVLA, N., Estudos de blocos intertravados de concreto para pavimentação com incorporação de resíduo de porcelanato, in Programa de Pós-graduação em Engenharia Civil e Ambiental. 2016, Universidade Federal de Pernambuco: Caruaru.

5. Santos, M.L.L.O., Aproveitamento de resíduos minerais na formulação de argamassas para a construção civil, in Programa de Pós-graduação em Ciência e Engenharia de Materiais (PPGCEM). 2008, UFRN: Natal.

6. Silva, G., Estudo do comportamento do concreto de cimento Portland produzido com a adição do resíduo de polimento do porcelanato, in Pós-graduação em Engenharia Metalúrgica e de Minas. 2005, Universidade Federal de Minas Gerais: Belo Horizonte.

7. Souza, P.A.B.F., Estudo do comportamento plástico, mecânico, microestrutural e térmico com concreto produzido com polimento de porcelanato, in Programa de pós-graduação em ciência e engenharia dos materiais. 2007, Universidade Federal do Rio Grande do Norte: Natal.

8. Steiner, L.R., A.M. Bernardin, and F. Pelisser, Effectiveness of ceramic tile polishing residues as supplementary cementitious materials for cement mortars. Sustainable Materials and Technologies, 2015. 4: p. 30-35.

9. Pacheco-Torgal, F. and S. Jalali, Reusing ceramic wastes in concrete. Construction and Building Materials, 2010. 24(5): p. 832-838.

10. Penteado, C.S.G., E.V. Carvalho, and R.C.C. Lintz, Reusing ceramic tile polishing wast in paving block manufacturing. Journal of Cleaner Production, 2016. 112(Part 1): p. 514- 520.

11. Wattanasiriwech, D., A. Saiton, and S. Wattanasiriwech, Paving blocks from ceramic tile production waste. Journal of Cleaner Production, 2009. 17(18): p. 1663-1668.

12. Cheng, Y., et al., Test research on effects of ceramic polishing powder on carbonation and sulphate-corrosion resistance of concrete. Construction and Building Materials, 2014. 55: p. 440-446.

13. Wang, G. and B. Tian. Effect of Waste Ceramic Polishing Powder on the Properties of Cement Mortars. in International Conference on Energy and Environment Technology. 2009.

14. Puertas, F., et al., Ceramic wastes as alternative raw materials for Portland cement clinker production. Cement and Concrete Composites, 2008. 30(9): p. 798-805.

15. Naceri, A. and M.C. Hamina, Use of waste brick as a partial replacement of cement in mortar. Waste Management, 2009. 29(8): p. 2378-2384.

16. Lavat, A.E., M.A. Trezza, and M. Poggi, Characterization of ceramic roof tile wastes as pozzolanic admixture. Waste Management, 2009. 29(5): p. 1666-1674.

17. Zhi-sheng, Q., Test research of effects of ceramical polishing powder on workability and compressive strength of concrete [in Chinese],. 2011, Northeastern University.

18. Silva, G.J.B., Estudo do comportamento do concreto de cimento Portland produzido com a adição do resíduo de polimento de porcelnato, in Programa de pós-graduação em engenharia metalúrgica e de minas. 2005, Universidade Federal de Minas Gerais: Belo Horizonte - MG. p. 107.

19. Penteado, C.S.G., E. Viviani de Carvalho, and R.C.C. Lintz, Reusing ceramic tile polishing waste in paving block manufacturing. Journal of Cleaner Production, 2016. 112, Part 1: p. 514-520.

20. Pelisser, F., L.R. Steiner, and A.M. Bernardin, Recycling of Porcelain Tile Polishing Residue in Portland Cement: Hydration Efficiency. Environmental Science & Technology, 2012. 46(4): p. 2368-2374.

21. ZHONG, M., D. SU, and Y. ZHAO, Hydration Activity of Ceramic Polishing Powder at Autoclaved Condition. Advanced Materials Research, 2012. 568: p. 392-395.

22. Santos, M.L.L.O., Aproveitamento de resíduos minerais na formulação de argamassas para construção civil, in Programa de pós-graduação em ciência e engenharia dos materiais. 2008, Universidade Federal do Rio Grande do Norte: Natal-RN. p. 163.

23. Breitenbach, S.B., et al., Adição de resíduo do polimento de porcelanato em argamassas de restauro à base de cal. Cerâmica, 2017. 63: p. 395-401.

24. Wang, G., et al., Pozzolanic Activity of Ceramic Polishing Powder as Cementitious Material. Materials Science Forum, 2011. 675: p. 135-138.

25. ABNT – Associação Brasileira de Normas Técnicas. NBR 5733 – Cimento Portland com de alta resistência inicial. Rio de Janeiro, 1991.

26. ASTM C150-07. Standard Specification for Portland Cement, Annu. B. ASTM Standars, 2007.

27. Wen, S. and D.D.L. Chung, Effect of stress on the electric polarization in cement. Cement and Concrete Research, 2001. 31(2): p. 291-295.

28. ABNT. Materiais Pozolânicos – Requisitos (Pozzolans — requirements). NBR 12653. Rio de Janeiro, Brazil; 2012.

29. Steiner, L.R., Efeito do resíduo do polimento do porcelanato como material suplementar, in Programa de Pós-Graduação em Ciência e Engenharia de Materiais. 2014, Universidade do Extremo Sul Catarinense - UNESC: Criciúma.

30. ASSOCIAÇÃO BRASILEIRA DE NORMAS TÉCNICAS (ABNT). Materiais Pozolânicos – Determinação de atividade pozolânica com cimento Portland – Índice de atividade pozolânica com cimento Portland – Método de ensaio. (Pozzolans – determination of pozzolanic activity with Portland cement — Test method). NBR 5752. Rio de Janeiro, Brazil; 2012.

31. Rodrigues, P., Parâmetros de dosagem de concreto. ET-67. 3 ed ed. 1998, São Paulo: IBRACON – Associação Brasileira de Cimento Portland.

32. ASSOCIAÇÃO BRASILEIRA DE NORMAS TÉCNICAS. Concreto de cimento Portland - Preparo, controle e recebimento - Procedimento. NBR 12655. Rio de Janeiro, 2006.

33. ABNT – Associação Brasileira de Normas Técnicas. NBR 5738 – Concreto – Procedimento para moldagem e cura de corpos de prova. Rio de Janeiro, 2015.

34. ABNT – Associação Brasileira de Normas Técnicas. NBR NM 67 – Concreto – Determinação da consistência pelo abatimento do tronco de cone. Rio de Janeiro, 1998.

35. ABNT – Associação Brasileira de Normas Técnicas. NBR 5739 – Concreto – Ensaios de compressão de corpos-de-prova cilíndricos. Rio de Janeiro, 2007.

36. Brigoline, G.J., Estudo do Comportamento do Concreto de Cimento Portland Produzido com a Adição do Resíduo de Polimento do Porcelanato, in Engenharia Metalúrgica e de Minas. 2005, Universidade Federal de Minas Geriais: Belo Horizonte.

SOCIAL, ECONOMIC & CULTURAL DIMENSIONS

IC-NOCMAT 2019

18th International Conference on
Non-Conventional Materials and Technologies
"Construction Materials & Technologies for Sustainability"
(18th NOCMAT 2019)
24th – 26th July 2019
Nairobi, Kenya

AFFORDABILITY AND SUSTAINABILITY OF NON-CONVECTIONAL CONSTRUCTION MATERIALS IN KENYA

Joseph Mwiti Marangua[1], Raine Isaksson[2], Wachira J. Muthengia[3], Thiong"o J. Karanja[4] and John M. Kinuthia[5]

[1] Department of Physical Sciences, Meru University of & Technology, Meru – Kenya;

[2] Department of Engineering Sciences, Uppsala University, Uppsala – Sweden;

[3] Department of Physical Sciences, University of Embu, Embu- Kenya;

[4] Department of Chemistry, Kenyatta University, Nairobi- Kenya;

[5] University of South Wales, Pontypridd- United Kingdom;

ABSTRACT

Globally, there is growing demand for low cost, ecologically friendly and technologically suitable construction materials. This is driven by the need for sustainable building materials that could be defined as affordable and with low carbon footprint. Housing plays an important role within sustainable development agenda. Affordable housing is therefore a key priority in most developing countries. Non-convectional construction materials are emerging as a viable option for low cost and eco-friendly materials in Kenya. Kenya is endowed with vast deposits of diversified natural resources, from which different kinds of non-convectional construction materials can be produced. Therefore, there is need for research that focuses on development, application and management of these materials. Materials such as, sugar cane baggasse, rice husks, pineapple wastes, banana stalks, sorghum stalk, bamboo, fired bricks, elephant grass, wheat straw ashes, lime, soil and clay among others are potential non-convectional materials for construction applications. The study aimed at identification and of selected non-convectional materials in Kenya. The potential materials were qualitatively discussed in terms of availability, usability, acceptability of the material and the service life of the expected structures. The most frequently mentioned non-conventional materials available in Kenya were identified and reviewed based on the defined criteria with the purpose to establish a shortlist for the most promising materials. These were studied in more detail resulting in an assessment highlighting opportunities and challenges for building sustainability. Preliminary results indicated that traditional building materials could be sustainable while having a low level for acceptability.

Keywords: Affordable; Eco-friendly; Low carbon footprint; Non-convectional; Sustainable materials; Wastes

INTRODUCTION

The world population is currently (2018-2019) growing at a rate of around 1.07 % per year. As the world population increases, the demand for housing increases proportionately. In addition, there is an increasing demand for decent and affordable housing in the world towns and cities that have grown at a rapid urban growth. The dramatic problems linked to it include; huge shelter needs, emergence of squatter settlements and slums, and deficient infrastructure in many low-income settlements. The shortage of decent housing is a problem that many Kenyans, especially the low income households who comprise more than 70% of residents in the urban areas, are confronted with[1]. The housing deficit in Kenya stood at 2 million in 2012 and continues to grow at the rate of over 200,000 units a year. Convectional building materials in most developing countries, like Kenya are beyond reach to a majority of the population[2]. This has resulted in the proliferation of informal settlements in urban areas with 60% of the population living in informal settlements[3]. Families live in overcrowded homes typically with one room and no adequate ventilation. These families are at risk of diseases such as malaria, respiratory infections and jigger (worm) infestation among others. Therefore, there is an important need to identify and develop alternative sustainable and affordable building materials to meet the current housing crisis[3]. There is growing demand for low cost, ecologically friendly and technologically suitable construction materials[4]. This is driven by the need for sustainable building materials that could be defined as affordable and with low carbon footprint. Housing plays an important role within sustainable development agenda. Affordable housing is therefore a key priority in most developing countries. Non-convectional construction materials are emerging as a viable option for low cost and eco-friendly materials in Kenya[3]. Kenya is endowed with vast deposits of diversified natural resources, from which different kinds of non-convectional construction materials can be produced. Therefore, there is need for research that focuses on development, application and management of these materials. Materials such as such as sugar cane baggasse, rice husks, pineapple wastes, banana stalks, sorghum stalk, bamboo, fired clay bricks, elephant grass, wheat straw ashes and mud among others are potential non-convectional materials for construction applications. The study aimed at identification of selected non-convectional materials in Kenya. The most frequently mentioned non-conventional materials available in Kenya were identified and reviewed based on the defined criteria with the purpose to establish a shortlist for the most promising materials. These were studied in more detail resulting in an assessment highlighting opportunities and challenges for building sustainability.

RESEARCH METHODOLOGY

The research methodology adopted in this study was largely based on previous case studies, using a deductive approach and qualitative analyses. The data collected in this study was mainly on secondary sources. In essence, the nonconventional construction materials considered include, mud, grass, cow dung, wood, wood ash, banana leaves, bamboo, fired clay bricks, The potential of nonconventional materials was classified in terms of availability, usability, acceptability of the material and the service life of the expected structures. The research has identified and classified the sustainability credentials in Kenyan non convectional materials.

Findings and Discussions

The results obtained from the evaluation matrix are presented in Table 3.1

Material	Availability, Affordability, Usability, Acceptability, Environmental friendliness and Service Life
Rammed Earth	This consists of ramming slightly damp soil between stout formwork using heavy rammers. It makes fairly strong and durable walls and floors when it is made thick enough with properly prepared, stabilized soil[3]. The rammed earth technology has been used to construct houses in Kenya. This is because they are cheap to construct, and the raw materials are readily available. In Kenya rammed earth houses have been constructed in many places such as Banana Hill in Kiambu County among others.
Papyrus Reeds	Dry papyrus reeds have been used as a building material in Kenya. Reeds have been woven into mats for use as wall or ceiling panels, shade roofs among others. Thatch made from untreated reeds has been reported to last only 1 year but, if well laid, treated and maintained, it has been found to last for 5–10 years[5].
Grass	The use of thatched roofs is common in many places in Kenya. When well laid and maintained, it can last for 10–20 years or longer. A good-quality thatching grass must be fibrous and tough, with a minimum length of 1 metre. During construction, the grass chosen is selected on the basis that it has thin stems without hollows, a low content of easily digestible nutrients and the ability to withstand repeated wetting without decaying.
Cow dung	Cow dung has been used in construction of houses in rural Kenya especially by nomadic communities such as Maasai. To improve the durability , Cow dung is mostly mixed with clay and used for housing[6]. Cow-dung has been found to improve the plasticity of clays and acts as reinforcing agent reducing concentrated cracks that can lead to breakage within the raw bricks. Upon firing the dung fibres ignite, thus assisting in even firing of bricks and minimizing the development of high temperature gradients within the brick - a phenomenon which may lead to firing cracks. When the fibres burn out they leave cavities within the brick which reduce unit weight and improve thermal characteristics[7]. There is low acceptability of cow dung in its use in construction activities.
Sugar cane bagasse	Sugar cane baggasse is a multi-process by-product produced from the milling of sugarcane. Bagasse is the fibrous material remaining after removing the sugar, water, and other impurities from the sugarcane delivered to the mill. The bagasse is normally used as biofuel. The ashes of the fuel can be used as building material. It is not well exploited in its application as a construction material in Kenya. There are preliminary studies being conducted to use it as a pozzolana in production of blended cements.
Wood	Wood has been extensively used in construction in Kenya since and before Kenya‟s independence. Wood is mostly used to put up semi-permanent houses in Kenya. Wood is a commonly used construction material in many parts of Kenya because of its availability, reasonable cost, ease of working, attractive appearance and adequate life if protected

	from moisture and insects. However, stringent measures have been put in place in Kenya to protect forests against deforestation particularly in effort to compact the negative effect of climate change. As a construction material, wood may be the last option for consideration in regions where other materials should be considered first, simply on a conservation basis. Wood for building is available from many different species with widely varying characteristics. Some species are used in the form of small poles for light construction, while other species are allowed to mature so that timber may be sawn from the large logs. The species that produce small, inexpensive poles in rather short growing periods often grow in the fringes of agricultural land and can be used without danger to the ecology of the region.
Papyrus	Papyrus is an aquatic plant that has a woody, blunty triangular stems and grows up to about 15 ft in quietly flowing water up to three feet deep. The stem can grow up to six centimeters in width near the top, while the bottom can portion can be up to six inches in width[8]. It has a long history of being used as low cost construction material by communities living around the lake regions in Kenya. It is easily degraded by heat, insects and easily attacked by fire when used as a construction material. Therefore exhibit low service life. It is however considered as a eco-friendly construction material.
Sisal fibre /stem	Sisal fibre is one of the strongest natural fibres. It has traditionally been used as reinforcement in gypsum plaster sheets[9]. It is a dominant construction material along the Kenyan coastal region. Sisal fibres have been used in a mixture with other materials such as gypsum and cement. Sisal fibres have the ability to withstand degradation from bacteriological attack better than other organic fibres, but are attacked by the alkalinity of cement. However, research has been carried out to make sisal fibre, like other natural fibre composites, into reliable cement reinforcement for long-term use in exposed situations[9].
Straw	Baled straw, if supported by a framework of wooden poles, can be used to construct temporary walls. Straw has also been used as raw material for manufactured building boards. Straw and split bamboo can be cement plastered to permanent structures, such as vaults and domes, at low cost[9].
Bamboo	Bamboo is a perennial grass with over 550 species, found in most parts of Kenya. It contains a large percentage of fibre, which has high mechanical performance. However, the major limitation of bamboos as construction materials is that they exhibit low durability; they are highly flammable and easily prone to split. This usually prevents the use of nails. It has been reported that cutting a notch or a mortise in a bamboo drastically reduces its ultimate strength. The remedy is the use of nodes as places of support and joints, and the use of lashing materials (strings) in place of nails. Dry bamboo is extremely susceptible to fire, but it can be covered or treated with a fire-retardant material. The strength properties of bamboo vary widely with species, growing conditions, position within the culm, seasoning and moisture content. Generally bamboo is as strong as timber in compression and very much stronger in tension. However, bamboo is weak in shear, with only about 8 percent of

	compressive strength, whereas timber normally has 20–30 percent. It is used mainly in building construction, for wall poles, frames, roof construction, roofing and water pipes and, after splitting, to form flattened boards or woven wall, floor and ceiling panels. Bamboo without proper seasoning and preservative treatment will rot and be attacked by insects, particularly if used in moist locations, such as in earth foundations. Bamboo is readily available and affordable in Kenya. It exhibits low level of acceptance majorly due to its short durability since it is easily degraded by moisture and fire outbreaks. Socially, it is considered a construction material for the poor.
Fired Clay Bricks	Fired clay bricks have been readily prepared in many regions in Kenya due to the significant availability of clay deposits across the country. In addition, minimal technological skills are required for production of fired clay bricks hence they have a high level of acceptability and usability. Previously, fired clay bricks have been regarded as old fashioned and replaced by other perceived modern walling units within such as machine cut stones especially in urban areas in Kenya. Presently, the fired clay bricks are enjoying renaissance as novel construction materials due to their abundance availability, unique benefits such as improved aesthetics, low maintenance cost among others. However, prolonged, exposure of housing units made from fired clay bricks to wet conditions such as rainfall results in their degradation. Therefore, they exhibit short service life. Measures taken to improve their durability include blending of fired clay bricks with cement. In addition the firing of these bricks requires wood and that this could lead to deforestation.
Adobe or sun-dried soil (mud) blocks The	Adobe or sun-dried soil (mud) blocks are made from a mixture of clay-based soil, water and sometimes straw. Wet adobe is formed into bricks and then laid out to dry in the sun for several days. The greatest disadvantage of adobe is its vulnerability to water and rain. Adobe blocks have low level of acceptability since it is sometimes locally regarded as „the material of the poor", which may limit its application. Adobe blocks are used for the construction of walls, with wet adobe serving as the mortar. Methods of construction vary considerably by climate. Most homes have a coating of adobe on the outside to create a smooth surface, which can be painted. Adobe has been used in construction activities mostly in rural areas but in urban areas, the use of this sustainable material is not very common. Generally, the adobe blocks are environmentally friendly and exhibit medium service life[10].
Elephant grass	Elephant grass has is similar to bamboo, but with the difference that the stem is not hollow. The fibres of the grass can be used to partly or wholly replace the asbestos in net and corrugated roofing sheets. However, the sheets are more brittle and have a slightly lower strength than asbestos-cement sheets. There is low use elephant grass in construction activities since in majority of the cases, it is considered as fodder. It is however a potentially sustainable construction material.

Palm	In Kenya, the naturally drying palm leaves locally referred to as „Makuti" is mainly used as thatching construction especially along the Kenyan coastal area. Coconut palm *Cocos nucifera* is also used. It is a very eco-friendly or "green" building material as it is made from leaves that are naturally wilting on the palm tree and therefore harvesting does not affect the parent plant in any way. The individual pieces (or tiles) have to be weaved first. Locals in the villages usually do this during their spare time. „Makuti" thatching is affordable since the materials are obtained locally. When used in construction the „Makuti" thatching offer great thermal comfort in the houses due to the fact they are good insulators. They are however not very durable due to degradation due to fires, insect and rainfall. In addition, they, require frequent maintenance. A „makuti" roof is estimated to last 6 to 8 years and re-ridging and re-valleying in particular will be required several times during the lifespan of the thatch.
Coir waste	Coir is a by-product of coconuts. The husk is used for making coir mats, cushions and as fuel. It can be mixed with cement, glue or resins, either to produce low-density boards with good insulating and sound absorption properties, or to be compressed to make building boards[9]. It is also used as reinforcement in cement for making roofing sheets. The different materials can also be classified based on their use in house building. This division could be seen as foundation, walls and roof. The most critical part for being able to build a house is the wall. For this the three most promising materials are rammed earth, Adobe and fired bricks. From sustainability point of view considering both price and carbon emissions the best candidates are Adobe and rammed earth, both having the problem of low acceptability and short service life. The issue of service life has possible technical solutions whereas the low acceptability could be more of a challenge. In forthcoming research, the proposed criteria need to be operationalized in order to permit a quantitative comparison. Here, we propose creating a Likert scale of 1 to 5 for each criterion. The end points could be 1= Very bad and 5=Very good Availability could be a function of price and thereby Affordability, but could also be described as tons or m3 available per year. The higher the amounts of available material, the greater the opportunities for use are. As a start this parameter is proposed to be twofold describing tons available per year and the percentage of the country where the material is available. This information can be translated into the 1-5 Likert scale. Affordability relates mainly to walls that correspond to the main cost of the building. Here a value per harm concept could be used [11]. This would be expressed as m2 wall per price. This would include the main material, any mortar used and the labor cost/m2 wall. These values need to be converted into the 1-5 Likert scale. Usability could be seen to relate to the easiness of material use. In most cases an investment in a house is too much to bear at one time and financing therefore is a restriction for most people. This favor building materials that are cheap or that can be used for incremental house construction. This makes different types of bricks and blocks popular. The end points of the Likert scale could be 1= Expensive and difficult to use and 5=Affordable and easy to use. Acceptability could also be based on a Likert scale of 1 to 5 where the end points are 1= Not at all acceptable

and 5= Very acceptable. The environmental friendliness could also use the value per harm concept where the indicator would be m2/kg CO2. For buildings the effect on climate due to carbon dioxide (CO2) emissions seems to be the main environmental problem. The World Business Council for Sustainable Development (WBCSD) estimates that about 7% of man-made CO2-emissons come from cement only [12].

CONCLUSION

From the study it was concluded that elephant grass, wood, reeds, palm, sisal fibre, banana leaves and bamboo are commonly available and affordable in Kenya and majorly used for housing. They exhibit short service life due to degradation since they are highly susceptible to decay caused by fungi, insects and easily prone to destruction by fire. They have low level of acceptability due to the fact they are considered as construction materials for the „poor". The fired clay bricks, adobe (sun dried soil) and coir waste exhibit medium service life. They have low acceptability level. Generally, all the materials considered were found to be potential sustainable materials for housing construction in Kenya.

RECOMMENDATIONS FOR FURTHER RESEARCH

a) Development of affordable preservation methods for elephant grass, reeds, palm, straw, bamboo, or banana leaves to increase their life span when used in as construction materials.
b) Further research is required to establish the bioclimatic housing benefits on the use of elephant grass, wood, reeds, palm, sisal fibre, banana leaves, bamboo, wood, fired clay bricks, adobe (sun dried soil) and coir waste.
c) Life Cycle Analysis of the materials is needed to provide more information on the overall performance of the indentified non-convectional materials.
d) Explore on value addition options of the material to enhance their suitability, acceptability, durability and aesthetic value. There is need also to explore means of blending the materials with conventional materials to enhance them. For example, fire clays can be enhanced by having a coat that is weather resistant.
e) The documentation of flexible codes of best practice (as flexible standards) for use of most of these materials could start to create confidence in their uptake.

REFERENCES

1. Mwelu, J. Prepared By The PSUP Team Nairobi. 2015, 96.

2. Wachira, J.M.; Thiong"o, J.K.; Marangu, J.M.; Murithi, L.G. Physicochemical Performance Of Portland-Rice Husk Ash-Calcined Clay-Dried Acetylene Lime Sludge Cement In Sulphate And Chloride Media. Advances In Materials Science And Engineering 2019, 2019, 1–12.
3. Adegun, O.B.; Adedeji, Y.M.D. Review Of Economic And Environmental Benefits Of Earthen Materials For Housing In Africa. Frontiers of Architectural Research 2017, 6, 519–528.
4. Mwiti, M.J.; Karanja, T.J.; Muthengia, W.J. Properties of Activated Blended Cement Containing High Content of Calcined Clay. Heliyon 2018, 4, E00742.
5. Mpakati-Gama, E.C.; Wamuziri, S.C.; Sloan, B. The Use of Alternative Building Materials In Developing Countries: Addressing Challenges Faced By Stakeholders. Sri Lanka 2012, 11.
6. 6. Manette, N.; Oyawa, W.; Nyomboi, T. Potential of Quarry Dust and Cow Dung As Stabilisers For Black Cotton Soil Eco-Blocks For Housing. Civil and Environmental Research 2015, 9.
7. Biwott, J.K. A Study of Challenges Facing Construction of Affordable Decent Low-Cost Housing In Turkana County. 2018, 105.
8. Atieno, A.J.; Ed, B. Utilization and Conservation Of Papyrus Plants For Sustainable Livelihoods In Kusa Swamp, Lake Victoria, Kenya. 134.
9. Sen, T.; Reddy, H.N.J. Application Of Sisal, Bamboo, Coir And Jute Natural Composites In Structural Upgradation. International Journal Of Innovation 2011, 2, 6.
10. Imarert, R.O. ^ The Role of Small Scale Brick Producers and The Use Of Brick? In Rural Housing: 161.
11. Isaksson, R.B., Garvare, R. and Johnson, M. The crippled bottom line– measuring and managing sustainability. International Journal of Productivity and Performance Management, 2015 64(3), 334-355.
12. WBCSD (2018), Technology Roadmap Low-Carbon Transition in the Cement Industry, available at: www.wbscd.org.

18th International Conference on
Non-Conventional Materials and Technologies
"Construction Materials & Technologies for Sustainability"
(18th NOCMAT 2019)
24th – 26th July 2019
Nairobi, Kenya

APPLICATION OF ALTERNATE CONSTRUCTION TECHNIQUES INREHABILITATION OF URBAN SLUMS

Sridevi Changali[1], Rosie Paul[2], Shubha B. A.[3]

[1] Co-Founder & Principal Architect, Masons Ink, Bengaluru, India;

[2] Co-Founder & Principal Architect, Masons Ink, Bengaluru, India;

[3] Senior Architect, Masons Ink, Bengaluru, India;

ABSTRACT

Approximately 62 million people inhabit slums in India today. The living conditions and overall standard of construction is appalling. This paper aims at looking at this larger issue and scope out how alternate building materials and techniques could create a positive impact and become a change maker in building construction. This would be done through case studies of 2 urban in-sets in 2 slums located in Bengaluru, India designed and executed by the authors. The 2 selected case studies were successfully completed in November 2017 using 2 differing wall materials – fly ash bricks and Compressed Stabilized Earth Blocks. The roof employed a filler slab technique that reused the existing clay tiles from the older unit. The tools used to measure the impact would include – post occupancy surveys, thermal comfort, study of costs of construction and tools to analyze these projects against the four pillars of sustainability.

Keywords: Mud Construction, Alternate building materials, Impact assessments.

INTRODUCTION – THE RISE OF URBAN SLUMS IN INDIA

India has seen a steady growth of slums since independence in 1947. Up until 1950, slums were concentrated in and around industrial areas. During the next decade, the percentage of slums increased to 18% and saw a higher increase in the 80's. Cities being economic centers providing diverse employment opportunities attract a lot of rural-urban migration thus resulting in mushrooming of informal settlements in and around these city cores. If we were to list the issues and concerns relating to the growth of these settlements it would consist of various topics ranging from health and safety to physical infrastructure. The large masses of people living in these settlements are living in unsanitary conditions without basic needs; there is overcrowding, lack of clean drinking water, high levels of pollution and no access to other needs like schools, healthcare, recreation to name a few[1][2]. The census of 2011 also states that almost 1out of 6 city residents in India are slum dwellers i.e. a whopping 62 million people inhabit slums in India today[3]. So any impact to this sector will see a significant impact on a monumental scale. Therefore this paper attempts to look at how as practitioners in the building sector can create a positive impact on this larger issue.

SLUMS DEFINED IN THE INDIAN CONTEXT

What exactly constitutes a slum? As per the central government, the definition of a slum is as follows:"...are usually a cluster of hutments with dilapidated and infirm structures having common or no toilet facilities, suffering from the lack of basic amenities, inadequate arrangement for drainage and for disposal of solid wastes and garbage. These inadequacies make the living conditions in slums extremely suboptimal, unhygienic and results in usually higher incidence of air and water borne diseases for the dwellers."[4]

Improving built fabric in this context directly impacts 3 main segments:

1. Human Life
2. Building Construction
3. Environment

These aspects directly touch many aspects of the UN sustainable development goals that came into effect in January 2016 signed and adopted by 193 countries. Although in general, improvements aimed at improving built fabric in urban slums would impact or fulfill the aims of several of the 17 development goals, in specific, it would directly be applicable to Goal 11 that is tailored towards creating sustainable cities and communities. Some global facts are quite astounding – the number of people living in cities is projected to double by the year 2050, although cities occupy only 3% of land it is responsible for 75% of carbon emissions and lastly, 828 million people globally reside in slums and the number is rising [5]. So the fact remains that action needs to be taken and alternate and more environmental friendly solutions need to be innovated is pertinent.

Slums and Their Built Fabric

When we look at the built fabric in the Indian context, there are 3 types of housing categories, namely, kutcha house that is defined as a temporary structure, Pucca house that is considered as permanent and Semi-pucca that falls in the middle of these two definitions. The percentage of each is as follows (Refer FIGURE 1):

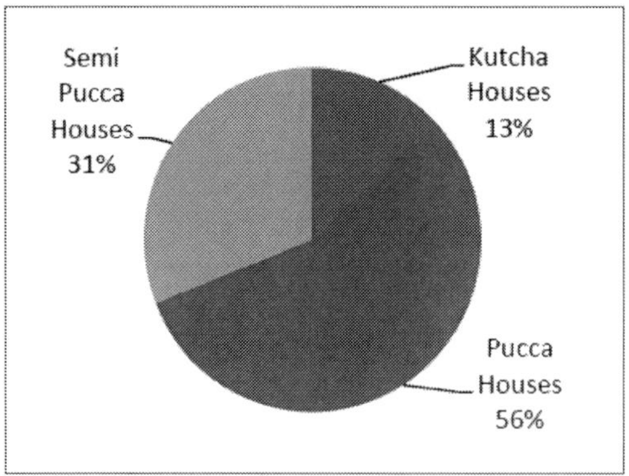

FIGURE 1 - TYPES AND PERCENTAGE OF HOUSING CATEGORIES

(Of the total estimated housing stock of 211.86 million, 117.63 million houses are pucca houses, 65.53 million units are semi-pucca houses, and the remaining 28.70 million units are kutcha houses.)[7]

Census 2011 has classified houses according to the construction material used for the roof, wall and floor of the structure. The census recorded building elements of 15.9 million households in the country (Refer FIGURE 2 & 3).

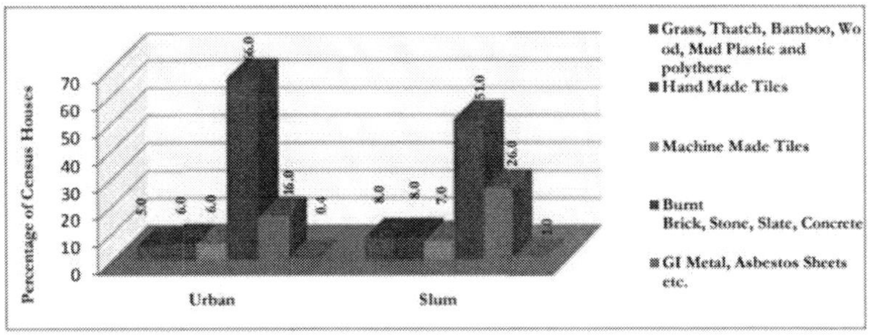

FIGURE 2 - CENSUS HOUSES BY PREDOMINANT MATERIAL OF ROOF -2011[6]

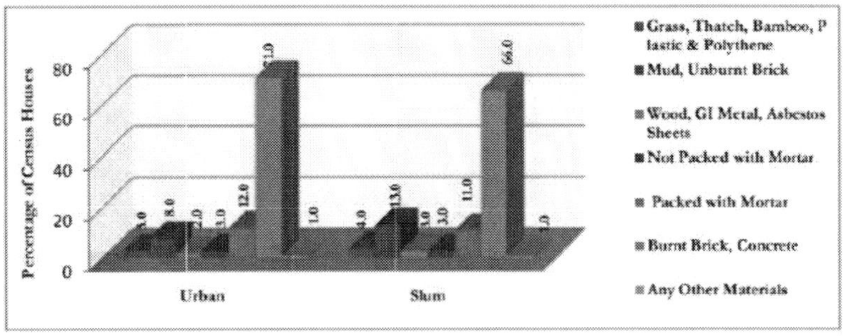

FIGURE 3 - CENSUS HOUSES BY PREDOMINANT MATERIAL OF WALL-2011[6]

The inference from the above figures is that the majority of households are built with burnt materials, for both walls and roof; which is a major constituent of embodied energy. It is also important to note the volume of the houses using unhealthy materials like GI sheets and asbestos sheets. The former increases thermal gain when used for both roofs and walls; while the latter has been proven to emit unhealthy fumes. Thus, it stresses on the need for alternative materials in the wake of sustaining natural resources for the future. The following tables state the quantity of embodied energy for various materials in wall and roof respectively. [8]

TABLE 1 - EMBODIED ENERGY FOR VARIOUS WALL MATERIALS [8]

Material	Embodied energy MJ/sqm of surface area	Embodied energy MJ/sqm of floor area	Percentage with base case (%)
Base case : Bricks	615	1139.9	
Flyash Lime Gypsum Blocks	242	448.8	39.37
Hollow concrete Blocks	193	358.39	31.45
Compressed Stabilised earth Blocks	195	361.64	31.75

TABLE 2 - EMBODIED ENERGY FOR VARIOUS ROOF MATERIALS [8]

Material	Embodied energy MJ/sqm of surface area	Embodied energy MJ/sqm of floor area	Percentage with base case (%)
Base case : RCC	847	847	
High Albedo Roof (Reflective tiles)	34	34	4
High Albedo Roof (White coating)	23	23	2.7
RCC Filler Slab	590	590	69.65

In this context, the following case studies will look at the application of alternate construction components like fly ash blocks, Compressed Stabilized Earth Blocks (CSEB) and filler slabs. In addition to that, the case studies will be analyzed with respect to the four pillars of sustainability.

PROJECT GRIHAKENDRA

Grihakendra aims at developing sustainable housing infrastructure and promoting healthy living conditions amongst the low income communities. There are three key institutions that are involved with this initiative. Selco Foundation, an NGO working towards developing sustainable living solutions for the underserved communities. Their expertise in understanding the living conditions and lifestyle of slums at the grass root level and desire to improvise their built environment set the project in motion. Fedina, another NGO working on empowering the economically weaker sections of Indian society manages the timeline and finances of the project. Lastly, Masons Ink studio, a design and construction firm practicing sustainable architecture and heritage conservation is the architectural plug in. Thus, the project brings together financial institutions, ground partners and technical professionals to realize sustainable housing solutions.

The first phase of the project was the construction of two grant houses in the year 2017 in select slums of Bengaluru in order to showcase the sustainable techniques and low cost construction solutions at hand, with an eye to establish community driven process that provides design and financial assistance to those wanting to elevate their living conditions. By 2020, sixty such dwellings are projected to be built on a subsidised loan that is made available to the beneficiary. Subsequently, in the coming years, this initiative hopes to bring a positive impact to the built and social fabric of Bengaluru's slums.

The design process involves active participation of the identified beneficiaries and their communities in pursuance of healthy and hygenic living conditions for the present as well as the generations to come. The process has been summarised with a flowchart below (Refer FIGURE 4).

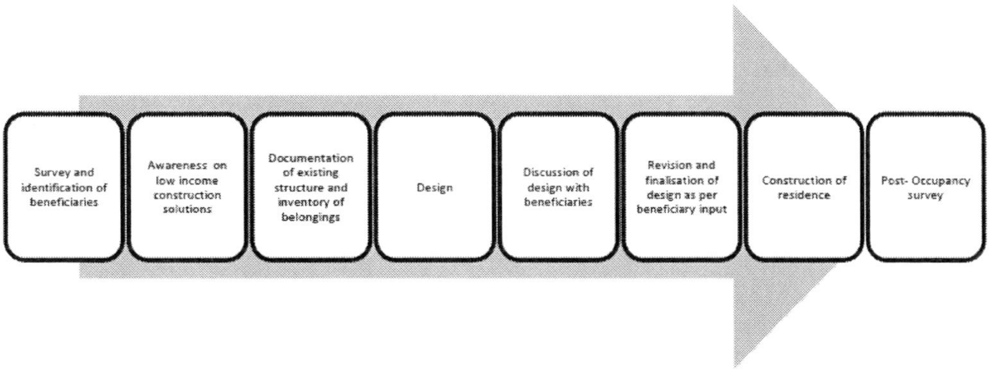

FIGURE 4 - DESIGN PROCESS

CASESTUDIES - GRANT HOUSES

Casestudy 1 - Ambedkarnagar, Old Byappanahalli , Bengaluru

Located in ward no 50, east of Bengaluru, Ambedkarnagar is a locality housing 1000 dwellings on plots measuring 14sqm, with an average of four inhabitants per dwelling. The basic infrastructure of the locality includes electricity, sanitation, communal drinking water connection and minor streets. A larger percentage of the households are built with fired bricks or concrete blocks 2-3 storey high with roofing combination of RCC and asbestos sheets. The plots have abutting walls on three sides, thus, the access to air and light is only through the front facade.

IMAGE 1 - THE AMBEDKARNAGAR LOCALITY

The identified beneficiary, Mrs Kuppamma, a senior citizen working as helper with the municipality owned a single storey fired brick residence with asbestos roofing. The layout of the dwelling included a kitchen at the rear end, a living as well as bedroom in the middle and a bathroom in the front. Upon inspection, it was observed that the building lacked a strong structural foundation and the impractical location of the kitchen had its share of drawbacks like lack of ventilation, lighting and storage space for cooking purposes which raise concerns on health and safety principles.

Design Process

As an outcome of the design process (refer FIGURE 4), the concept design revolved around healthy living conditions and spatial rearrangement of the existing furniture. The front of the dwelling was designed to house a staircase and an independent bathroom, the kitchen at the entrance along a window for proper ventilation and reduction of dependency on artificial light, lastly, the living as well as bedroom at the rear with storage niche and lofts. Since the locality has communal water connection, a 1000L underground sump was provided to store water. The concept design was presented to the beneficiary, who upon analyzing the design proposed a few changes. The household was provided with solar energy to meet daily electricity needs.

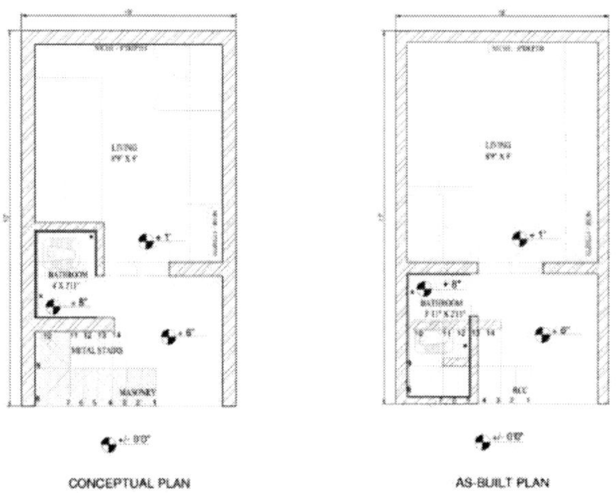

FIGURE 5 - CONCEPTUAL AND AS BUILT PLANS

Building Materials

The building elements of the dwelling include 3' deep SSM foundation; Fly ash blocks sized 40.6 x 20.3 x 15.2cm, RCC staircase and filler slab roofing system. The filler material for the roof and the railing for the staircase are reused products sourced from nearby demolition sites.

IMAGE 2 - THE BEFORE AND AFTER PHOTOGRAPHS OF THE DWELLING

Casestudy 2 - New Byappanahalli, Maruti Sevanagar , Bengaluru

Located in ward no 79, east of Bengaluru, New Byappnahalli is a locality housing nearly 450-500 dwellings on varied sized plots, with an average of five inhabitants per dwelling. The basic infrastructure of the locality includes electricity, individual drinking water connections and alleys. A larger percentage of the households are built with fired bricks or concrete blocks 2-3 storey high with roofing materials like Mangalore tiles, RCC and asbestos sheets.

IMAGE 3 - THE NEW BYAPPANAHALLI LOCALITY

The identified beneficiary, Mrs Shantamma, a senior citizen, owned a single storey three side open, dilapidated residence built with mud in the cob building technique with Mangalore tile roof on a plot measuring 19.5sqm. The layout of the dwelling was an open plan with living, bedroom and kitchen as connected spaces. Upon inspection, it was observed that the building lacked foundation, had water seepage issues during rains and was void of sufficient lighting.

Design Process

The beneficiary lived with a granddaughter and an abusive son. Her major requirement was to isolate the living quarters of her son citing safety. The concept design revolved reflecting healthy and safe living conditions. As a result, the front of the dwelling was designed to house living and kitchen for the women and rear with a bathroom, staircase and a room with an independent access for the son. The concept design was presented to the beneficiary, who upon analyzing the design, gave her consent to begin construction. During different stages of construction, objections were raised by the neighbors on position of access to the son's room as well as the main entrance. Due to repeated conflicts, the design was revised. The changed layout included the kitchen, staircase and the bathroom in the front and the living and son's room with an internal access in the rear.

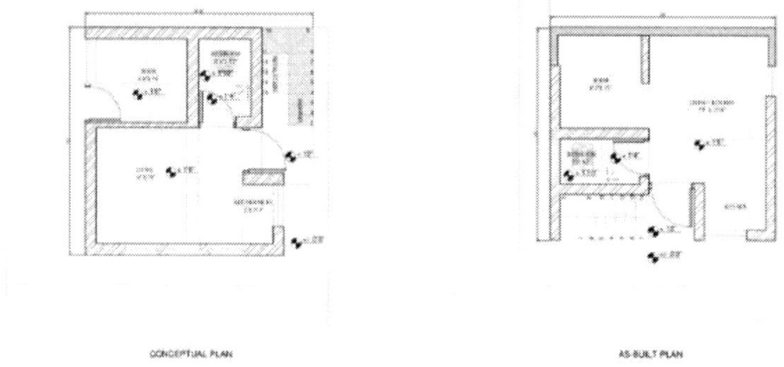

FIGURE 6 - CONCEPTUAL AND AS BUILT PLANS

Building Materials

The building elements of the dwelling include 3' deep SSM foundation, a combination of fly ash blocks sized 40.6 x 20.3 x 15.2cm and CSEB sized 30.48 x 15.2 x 20.3 cm, RCC staircase and Filler slab roofing system. The walls were a combination of materials due to the fact that CSEB's breathability is affected in case of abutting walls. It was substituted with fly ash blocks for one wall. The filler material for the roof was the Mangalore tile of the earlier dwelling and the railing for the staircase was sourced from a nearby demolished site. The idea of making CSEB blocks manually on site was not feasible due to unavailability of space for preparation and curing in the immediate vicinity, hence had to be sourced.

IMAGE 4 - THE BEFORE AND AFTER PHOTOGRAPHS OF THE DWELLING

POST OCCUPANCY SURVEY

A year into finishing of the structure, a post occupancy survey was conducted to analyze the merits and demerits of design in detail. Both the beneficiaries rated their dwellings based on parameters like lighting, ventilation, safety and thermal comfort. The details of which are listed in table 3.

TABLE 3: POST OCCUPANCY SURVEY

POST OCCUPANCY SURVEY OF GRANT HOUSES AT AMBEDKARNAGAR AND NEW BYAPPANAHALLI, BENGALURU											
Sl no	Questionnaire	Kuppamma					Shanthamma				
1	How would you rate the building Layout?	1	2	3	4	5	1	2	3	4	5
2	How would you rate the storage space in the building?	1	2	3	4	5	1	2	3	4	5
3	How would you rate privacy inside the building?	1	2	3	4	5	1	2	3	4	5
4	How would you describe the noise in the building?	1	2	3	4	5	1	2	3	4	5
5	How would you rate the average indoor temperature ?	1	2	3	4	5	1	2	3	4	5
6	How would rate incoming amount of light?	1	2	3	4	5	1	2	3	4	5
7	How would you rate your independency on artificial light?	1	2	3	4	5	1	2	3	4	5
8	Overall does the building meet your daily needs?	1	2	3	4	5	1	2	3	4	5
	Average	4					3.87				

Apart from the ratings, the buildings thermal performance as well as Interior Lux levels was recorded using surface thermometer and lux meter respectively. The results of which are attached as figures below (Refer FIGURE 7 and 8). The only drawback was non recording of pre demolition data so as to compare the before and after outcomes.

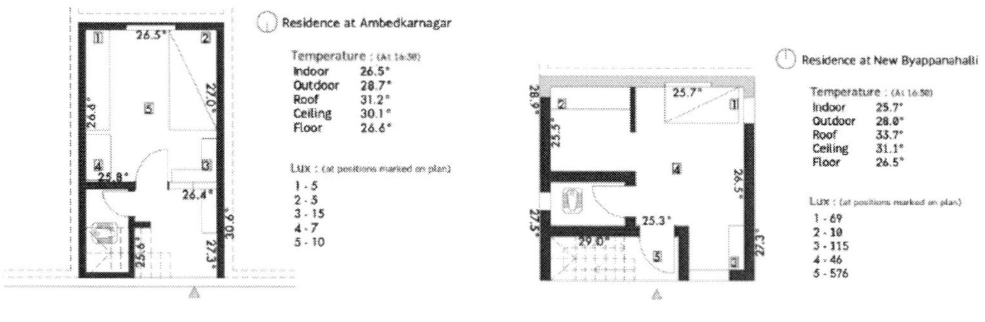

FIGURE 7 AND 8 - POST OCCUPANCY THERMAL AND LUX ANALYSIS

VERSUS

VerSUS is a self-evaluation tool to analyze completed projects against four pillars of sustainability, namely, environmental, cultural, social and economic. The tool bridges the gap between vernacular heritage and contemporary architecture achieved by identifying primary principles from the vernacular for the purpose of integrating those into the design and operation of the eco responsive buildings.[9]

Casestudy 1

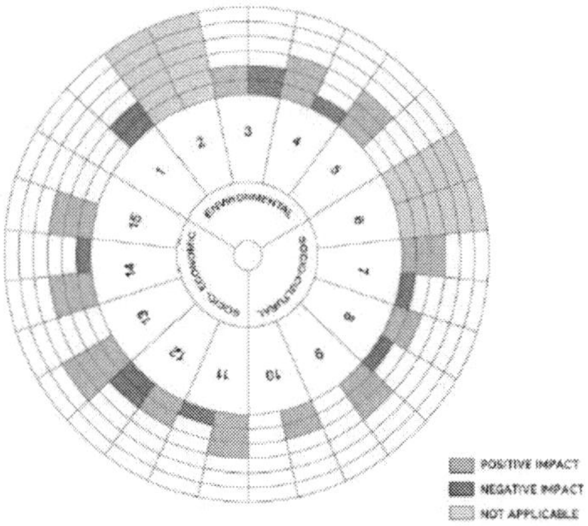

LEGEND	
1	TO RESPECT NATURE
2	TO BE APPROPRIATELY SITUATED
3	TO REDUCE POLLUTION AND WASTE MATERIALS
4	TO CONTRIBUTE TO HEALTH QUALITY
5	TO REDUCE NATURAL HAZARD'S EFFECTS
6	TO PROTECT CULTURAL LANDSCAPE
7	TO TRANSFER CONSTRUCTION CULTURES
8	TO ENHANCE CREATIVITY
9	TO RECOGNISE INTANGIBLE VALUES
10	TO ENCOURAGE SOCIAL COHESION
11	TO SUPPORT AUTONOMY
12	TO PROMOTE LOCAL ACTIVITIES
13	TO OPTIMISE CONSTRUCTION EFFORTS
14	TO EXTEND THE BUILDINGS LIFETIME
15	TO SAVE RESOURCES

FIGURE 9 - VERSUS ANALYSIS OF AMBEDKARNAGAR RESIDENCE

The analysis mirrors efforts taken to enhance construction practices and highlights contribution towards betterment of health quality. The transfer of construction cultures in terms of upskilling and involvement of local labors in sustainable construction tools has also been outlined as a positive factor. On the other hand, the tool helped analyze few shortcomings like investment in transportation costs for procurement of building blocks from distant sources along with discouragement of self-building.

Casestudy 2

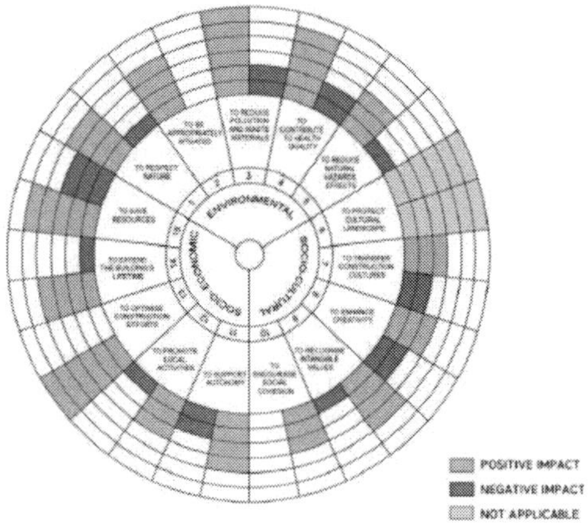

LEGEND	
1	TO RESPECT NATURE
2	TO BE APPROPRIATELY SITUATED
3	TO REDUCE POLLUTION AND WASTE MATERIALS
4	TO CONTRIBUTE TO HEALTH QUALITY
5	TO REDUCE NATURAL HAZARD'S EFFECTS
6	TO PROTECT CULTURAL LANDSCAPE
7	TO TRANSFER CONSTRUCTION CULTURES
8	TO ENHANCE CREATIVITY
9	TO RECOGNISE INTANGIBLE VALUES
10	TO ENCOURAGE SOCIAL COHESION
11	TO SUPPORT AUTONOMY
12	TO PROMOTE LOCAL ACTIVITIES
13	TO OPTIMISE CONSTRUCTION EFFORTS
14	TO EXTEND THE BUILDINGS LIFETIME
15	TO SAVE RESOURCES

FIGURE 10 - VERSUS ANALYSIS OF NEW BYAPPANAHALLI RESIDENCE

Similar to case study 1, the analysis reflected positive assets like encouragement of social cohesion and autonomy support by involvement of beneficiaries in design process. It also underlines the considerably low energy impact due to the usage of CSEB and fly ash blocks and the reuse of multiple building elements. Conversely, it indicates demerits like lack of attention to saving natural resources by demolition of existing structure instead of repairing it and retail procurement of CSEB instead of in-situ block making training.

CONCLUSION

There were innumerable learning's from the construction of the two grant houses in terms of planning, execution and social responsibility. The verSUS criteria being a holistic approach, helped reflect on the strengths and the shortcomings of the project Grihakendra and directed to improve upon the lacking principles in order to achieve a substantially sustainable building in an urban context. The principal learning is lack of thorough planning reflected in the total expenditure on the two dwellings USD 4045.87 (1$ = 71.09 INR) and USD 5523.36 (1$ =71.09 INR), respectively, against the initial budget of USD 3516.47 (1$ = 71.09 INR). The shortcomings, which were carefully assessed are now being amended in the first phase of reformation of Bengaluru slums. Professional training of identified local masons at COSTFORD, Kerala on sustainable building techniques and structured planning of the dwellings construction in order to save cost are some of the amendments. It has been observed that the study has been fruitful since the project has achieved the necessary economic sustainability in addition to healthy built environment.

REFERENCES:

1. Ministry of Home Affairs, India. CENSUS OF INDIA 2011 - CIRCULAR No. 8. New Delhi, 2010.
2. Bandyopadhya, Abir, and Vandana Agrawal. "Slums In India: From Past To Present". International Refereed Journal Of Engineering And Science (IRJES), Volume 2, no. Issue 4, 2013, pp. 55-59., Accessed 26 Feb 2019.
3. "The San Diego Union-Tribune". Sandiegouniontribune.Com, 2019, https://www.sandiegouniontribune.com/sdut-census-1-in-6-india-city-residents-livesin-slums-2013mar22-story.html. Accessed 20 Feb 2019.
4. "Sustainable Development Goals | UNDP". UNDP, 2019, https://www.undp.org/content/undp/en/home/sustainable-development-goals.html. Accessed 5 Feb 2019.
5. Ministry of Urban housing and Poverty alleviation. State Of Slums in India: A Statistical Compendium. Government of India, Delhi, 2013, pp. 14-24.
6. Ministry of Urban housing and Poverty alleviation. State Of Slums In India: A Statistical Compendium. Government of India, New Delhi, 2015, pp. 36-38
7. Nenova, Tatiana. Expanding Housing Finance To The Underserved In South Asia : Market Review And Forward Agenda. World Bank, 2010, p. 13, http://documents.worldbank.org/curated/en/393181468114238340/Expandinghousing- finance-to-the-underserved-in-South-Asia-market-review-and-forwardagenda. Accessed 24 Feb 2019.
8. Gupta, Nand Kishore et al. "Quantifying Embodied Energy Using Green Building Technologies Under Affordable Housing Construction". Open Journal Of Energy Efficiency, vol 02, no. 04, 2013, pp. 171-175, http://www.scirp.org/journal/ojee. Accessed 28 Feb 2019.

9. Moriset, Guilland H et al. Versus : Lessons From Vernacular Heritage To Sustainable Architecture. Craterre-ENSAG, 2014.

18th International Conference on
Non-Conventional Materials and Technologies
"Construction Materials & Technologies for Sustainability"
(18th NOCMAT 2019)
24th – 26th July 2019
Nairobi, Kenya

A STUDY OF THE SOCIO-ECONOMIC CULTURAL AND ENVIRONMENTAL IMPACT OF THE USE OF UNCONVENTIONAL BUILDING MATERIALS IN CONVENTIONAL BUILDING TYPOLOGIES

Rosie Paul[1], Sridevi Changali[2]

[1] Co-Founder & Principal Architect, Masons Ink, Bengaluru, India;

[2] Co-Founder & Principal Architect, Masons Ink, Bengaluru, India;

ABSTRACT

India is one of the fastest growing economies in the world. Subsequently, the size of the construction industry alone is projected to be 1 trillion USD by the year 2025 thus emphasizing the need for the use of alternate techniques in construction that are sustainable as well as cost effective. This paper aims to emphasize the benefits of using non-conventional materials with a special focus on earthen construction and other alternate construction techniques through a wider range of building typologies, thus showcasing its versatility. This will be done through case studies of buildings designed and executed by the authors ranging from farm houses to vocational training centers to Low Income housing. These projects will be analysed along the four pillars of sustainability namely – the environmental, the cultural, the social and economic impacts that the unique construction techniques and the construction processes have brought about. The techniques employed in these constructions ranges from Compressed Stabilized earth blocks, Rammed Earth, Nubian Vaults, Filler slabs, Traditional clay roofs etc. The different building typologies will be analyzed against a framework or matrix which highlight the positive impact these techniques have in the context of holistic sustainability.

Keywords: Mud Construction, Alternate building materials, Impact Assessments

INTRODUCTION

The construction industry is one of the biggest consumers of energy and raw materials, and a major contributor to greenhouse gas emissions. According to the UN Global annual status report 2017, buildings and construction together account for 36% of global final energy use and 39% of energy-related carbon dioxide (CO_2) emissions [1]. This becomes a huge concern especially in the developing nations where the percentage of unbuilt is much higher than the percentage of built which means more potential for construction. In the case of India, rapid urbanization and development brings about a construction requirement of 700 to 900 million square meters of commercial and residential space [2]. These alarming statistics emphasize the need to look into sustainable solutions for the future, as continuing in the current trend will have dangerous implications on the environment. The sustainable development goals by the UNDP have listed out 17 sectors to ensure future sustainable development. In the construction sector the relevant ones being - Sustainable cities and communities (goal 11), Responsible consumption and production (goal 12), Affordable and clean energy (goal 7), Industry infrastructure and innovation (goal 9) [3]. India is set to be home to the 3rd largest construction industry by the year 2025, hence there is an urgent need for a "green shift" that looks into less energy intensive and sustainable construction methods.

SUSTAINABLE DEVELOPMENT AND ITS FOUR DIMENSIONS

Sustainable development can be defined as the processes and pathways followed to achieve the long term goal of sustainability [4].

It involves the incremental changes and interventions that are implemented to reduce negative environmental impacts and help bring about social and economic equity. Since the construction sector is a major contributor to the development of any country and influences both socio-economy and the environment –it is important to look at sustainability in its holistic sense. Sustainability has 4 dimensions that are not independent of each other but are intertwined with the other, namely Environment, Society, Economy and Culture [4].

The Environmental dimension includes protection of bio-diversity, reduced embodied energy, reduced pollution and waste, controlled use of resources and use of local resources. Social Sustainability includes capacity building, promotes autonomy wherein the ustainable solutions respond directly to the social needs with health and safety being a primary concern. Economic sustainability looks into affordability and promoting the local economy. It takes in to account the economic viability of a project as well, in terms of operational costs and maintenance.

Cultural vitality is as important to a healthy sustainable society as the socio-economic and environmental factors. The Cultural dimension recognizes the importance of local building cultures and protection of cultural landscapes and identities [5].

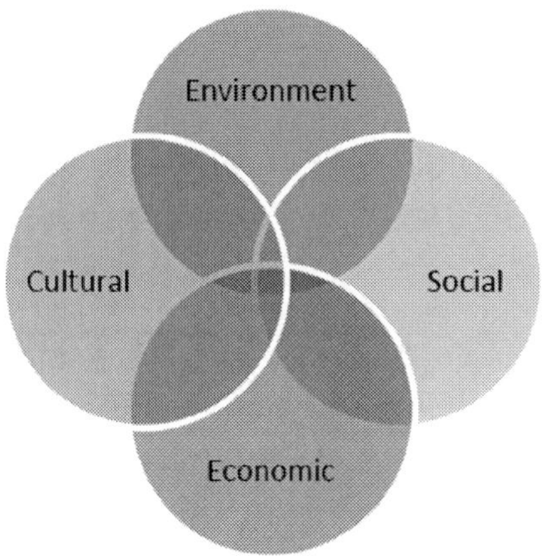

FIGURE 1 - THE FOUR DIMENSIONS OF SUSTAINABILITY

THE NEED OF NON- CONVENTIONAL MATERIALS IN CONSTRUCTION

India is the second largest producer of cement in the world with concrete being one of the most preferred construction materials in the industry. Paradoxically, the cement industry contributes to 5% of the global co2 emissions [6].This underlines the need to look at less energy intensive that have minimal impact on the environment. Using alternative building construction systems, methods and techniques can reduce environmental effect and embodied energy [7]. In this scenario, earthen construction could play a vital role in providing viable solutions to the future.

MUD AND ITS VERSATILITY

Mud if used appropriately can respond well to all four dimensions of sustainability. This has been further discussed in this paper using three case studies of varying scale and typology namely - a private residence, a vocational training centre and an affordable housing prototype that was built for an urban slum. These projects have been conceived and executed by the authors through their architecture studio Masons Ink, a design and construction firm specialised in sustainability, heritage conservation and capacity building. To study the projects effectively involves considering its impact on the socio-cultural fabric along with its environmental and economic benefits. This has been done using the VerSUS criteria.

VERSUS

It is a self-evaluation approach put together by an international team of multi-disciplinary professionals. The tool identifies the fundamental principles of sustainability from traditional vernacular architecture and integrates them to form 15 comprehensive principles along the Environmental, Socio- economic and socio-cultural axes to promote sustainable architecture. This provides a simplified method to assess if a project has rendered itself truly sustainable by positively impacting not one but all dimensions of sustainability. The VerSUS tool is more of a qualitative evaluation than a quantitative one. It provides as a means to study a completed project for both its positive and negative impacts on the social, economic, cultural and environmental fabric of its surroundings. The learnings from this evaluation in turn help future constructions to be rendered holistically sustainable.

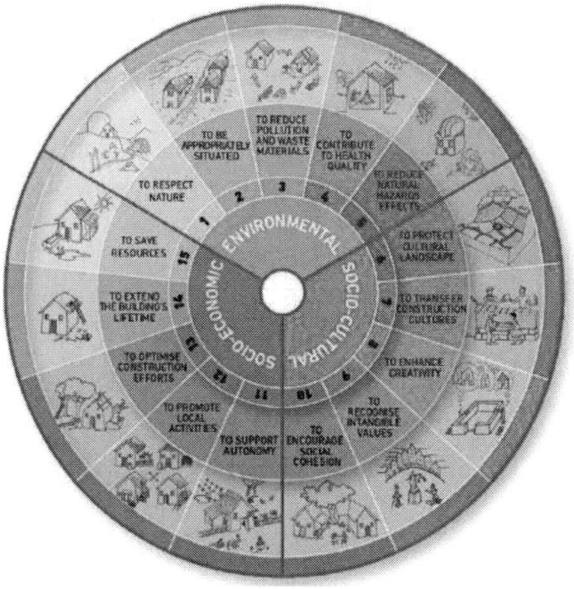

FIGURE 2 – THE 15 PRINCIPLES OF THE VERSUS WHEEL

CASE STUDY ANALYSIS ALONG THE VERSUS CRITERIA

Case Study 1 – Raa Maram

Raa Maram is a private residence that is located within a 25 acre organic farm near the Manchi reserve forest located in Denkanikottai Tehsil of Krishnagiri district in Tamil Nadu, India. It is a farmhouse whose design was inspired from the existing vernacular architecture of the place. The total built-up area of the project is 418sq.m and 92sq.m of semi-covered verandah spaces. Accessibility to the site was difficult that led to a different construction approach – looking into local labour and materials to avoid transportation.

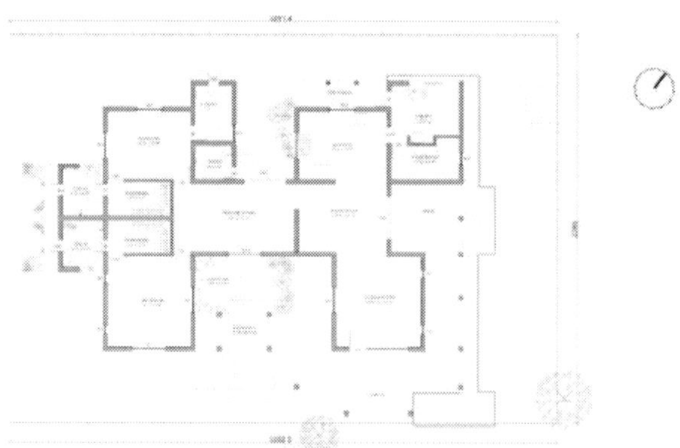

FIGURE 3 - RA MARAM PLAN

Environmental aspects

The project was designed to align with the principles of cradle to grave concept, minimizing the carbon footprint and embodied energy of the construction. The structure's material palette was entirely sourced from within the site, ensuring minimum transportation and transformation from the sourced material to the final product. The walls are mainly made of stabilized mud blocks wherein the soil was excavated and the blocks made on the site. The foundation and certain feature walls were built with local Sadarhalli stone, sourced and sized from within the site. The wood for the rafters and false ceilings were re-purposed from an abandoned factory further minimizing embodied energy of the material. Flooring is done with local stone tiles or coloured oxide.

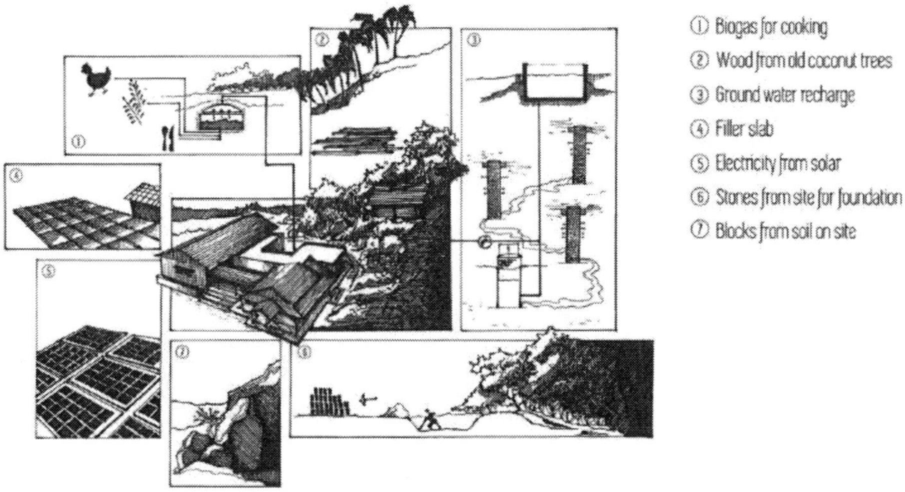

FIGURE 4- SKETCH SHOWING MATERIALS AND TECHNIQUES USED IN THE PROJECT

Socio-cultural aspects

The planning and layout of the building along with its built form has a very strong connect to the vernacular architecture of the region with sloped roofs, courtyards, verandahs etc. These features make the structure not only climate responsive but also culture- specific. Moreover the layout is also a reflection of local practices in the village. The verandahs allow for community discussions and drying of spices. The choice of material of stone and mud has been inspired from local building practices.

IMAGE 1- VERANDAH AND COURTYARD OF RAAMARAM

Socio-economic aspects

The use of mud as the primary construction material enabled local autonomy as the labour was trained in sustainable and cost-effective construction which contributed to skill-building of the human resource in the local aspect, earning them an additional livelihood. The local activities were further promoted as local carpenters and stone masons were used for the stone walls and the structural members for the roof.

IMAGE 2- VIEW OF THE HOUSE AND UPSKILLING OF MASON

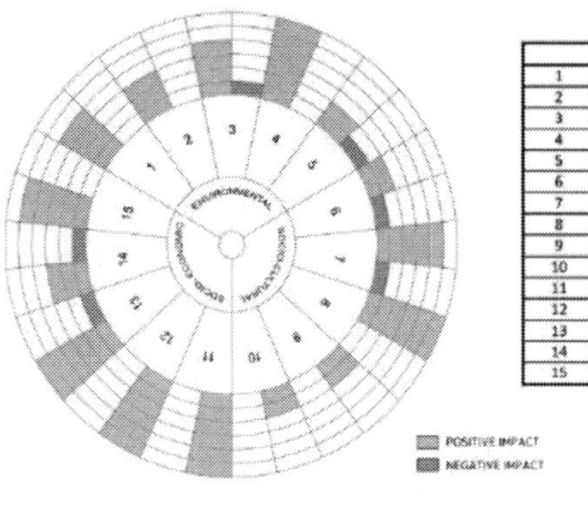

LEGEND	
1	TO RESPECT NATURE
2	TO BE APPROPRIATELY SITUATED
3	TO REDUCE POLLUTION AND WASTE MATERIALS
4	TO CONTRIBUTE TO HEALTH QUALITY
5	TO REDUCE NATURAL HAZARD'S EFFECTS
6	TO PROTECT CULTURAL LANDSCAPE
7	TO TRANSFER CONSTRUCTION CULTURES
8	TO ENHANCE CREATIVITY
9	TO RECOGNISE INTANGIBLE VALUES
10	TO ENCOURAGE SOCIAL COHESION
11	TO SUPPORT AUTONOMY
12	TO PROMOTE LOCAL ACTIVITIES
13	TO OPTIMISE CONSTRUCTION EFFORTS
14	TO EXTEND THE BUILDINGS LIFETIME
15	TO SAVE RESOURCES

FIGURE 5 - VERSUS ANALYSIS FOR RAAMARAM

The project when analysed through the VerSUS wheel shows that the construction had more of a positive impact to its surrounding environment and society. This helps to understand that inspiring from the vernacular, the use of local materials and labour can aid in making a building more sustainable. However as a result of evaluation though the VerSUS criteria has brought to limelight that the building lacked in certain aspects when looking at holistic sustainability. For example, looking at the socio-cultural aspect, though local materials were used, the ideal technique that should have been employed was an advanced form of cob as that was the traditional technique in the area. Construction time was higher due to training of locals, village festivals etc and it increased the construction cost of the building.

Case Study 2: Snehadaan

Snehadaan is a vocational training centre that houses a candle making unit, media centre and a candle retail store. It is located at Carmelaram in Bangalore, India and was commissioned by the Sneha Charitable Trust which imparts training and mentoring for people living with HIV. The vocational training centre has a total built up area of 185sq.m and is situated on a site of 10 acres that houses other facilities for the trust.

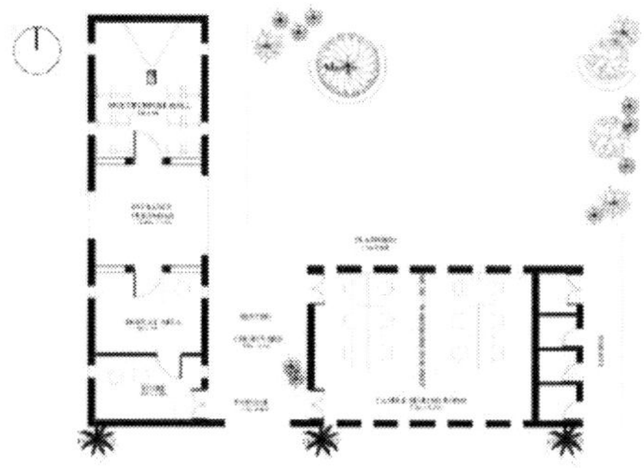

FIGURE 6 - SNEHADAAN PLAN

Environmental aspects

The large site provided an opportunity to source large quantities of soil which was one of the deciding factors for choosing to use mud as the primary construction material. The location identified for the structure had dense foliage and the building layout was done to follow the landscape and avoid the cutting of trees. The walls are of compressed stabilised mud blocks produced on the site. The windows and doors are from crate wood that was upcycled.

IMAGE 3- SHOWING THE CSEB BLOCKS AND FENESTRATION

Socio-cultural aspects

The design promotes social cohesion through common shared semi covered spaces promoting social activity. Mud being a healthy material to handle, the construction process was a participative one which encouraged community interaction and brought about joint ownership to the building.

Socio- economic aspects

Since the building was envisioned to be a vocational training centre the construction process was conceived as one that would add an additional vocation to the benefactors. They were trained in mud block making and about 65% of the blocks were made by the benefactors themselves. Affordability being one of the key factors that drove the project, led to some compromises in material selection - metal rafters were used for the roof which has a comparatively higher embodied energy as opposed to wood.

IMAGE 4- CSEB MAKING IN PROGRESS

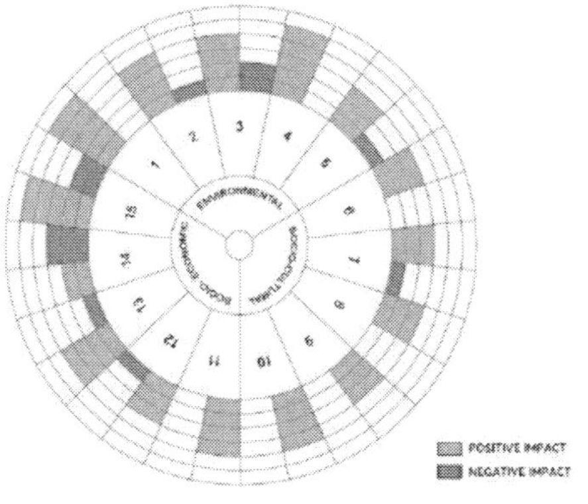

LEGEND	
1	TO RESPECT NATURE
2	TO BE APPROPRIATELY SITUATED
3	TO REDUCE POLLUTION AND WASTE MATERIALS
4	TO CONTRIBUTE TO HEALTH QUALITY
5	TO REDUCE NATURAL HAZARD'S EFFECTS
6	TO PROTECT CULTURAL LANDSCAPE
7	TO TRANSFER CONSTRUCTION CULTURES
8	TO ENHANCE CREATIVITY
9	TO RECOGNISE INTANGIBLE VALUES
10	TO ENCOURAGE SOCIAL COHESION
11	TO SUPPORT AUTONOMY
12	TO PROMOTE LOCAL ACTIVITIES
13	TO OPTIMISE CONSTRUCTION EFFORTS
14	TO EXTEND THE BUILDINGS LIFETIME
15	TO SAVE RESOURCES

FIGURE 7 - VERSUS ANALYSIS FOR SNEHADAAN

The VerSus wheel though overall positive, points out certain venues where the approach could have been more sustainable. For example, the use of certain energy intensive materials in construction such as the metallic rafters, ceramic tiles in the toilets etc. Certain construction details that were not followed during construction may lead to higher maintenance thereby increasing operational costs.

Case Study 3: Social Housing

This is a grant house built as part of the Griha Kendra project at New Byappanahalli, Bangalore, India. It is a locality housing nearly 450-500 dwellings on varied sized plots, with an average of five inhabitants per dwelling.The project aims at developing sustainable housing infrastructure and promoting healthy living conditions amongst the low income communities. This project is facilitated by 3 partners – Selco foundation - an NGO working towards developing sustainable living solutions for the underserved communities, Fedina – an NGO working on empowering the economically weaker sections of Indian society and Masons Ink. The identified beneficiary for this grant house is a senior citizen and owned a single storey - three side open, dilapidated mud residence with tiled roof on a plot measuring 19.5sqm. In such scenarios there are rarely any unbuilt areas on the plot. Since the conditions of the existing house were not meeting the basic comfort standards and needs of the dwelling existing house was demolished and a new one constructed.

IMAGE 5- BEFORE AND AFTER IMAGES OF THE GRANT HOUSE

Environmental Aspects

The decision to demolish the structure came after careful considerations of the possibilities of retaining the existing structure but since both the foundation and roofing conditions were of poor quality it was better to reconstruct. The building materials used were energy efficient such as mud blocks for the walls with open sides and fly ash for the wall adjoining the other building. A flat roof was proposed versus a sloping roof to allow for vertical expansion is possible in the future. A filler slab technique was proposed for the flat roof in order to reduce the volume of cement in the structure with the tiles from the old structure as the fillers.

IMAGE 6- FILLER SLAB AND SUNSHADE DETAIL

Socio-cultural aspects

The original material of the demolished house was mud was respected in the new construction. The design was a participative process which ensured that the output would respond well to the needs of the beneficiary. Also certain traditions were taken into consideration in order to promote cultural identity. For example there is always an area demarcated as the puja (religious space). The sun-shades towards the street have very minimal protrusion in order to facilitate funeral processions that go along the street that is 60 cm wide.

Socio- economic aspects

The reconstruction was done keeping the possibility of extension in the future wherein the foundations have been planned to incorporate vertical expansion. Numerous materials were reused – railings, tiles, old windows. The structure was built by masons from the same community thereby promoting local activities and providing autonomy.

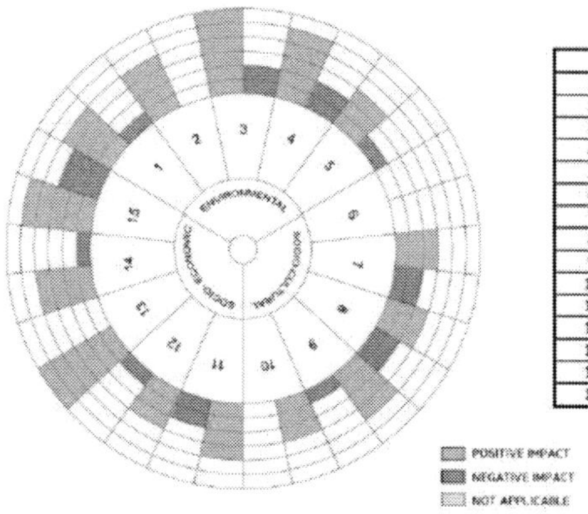

LEGEND	
1	TO RESPECT NATURE
2	TO BE APPROPRIATELY SITUATED
3	TO REDUCE POLLUTION AND WASTE MATERIALS
4	TO CONTRIBUTE TO HEALTH QUALITY
5	TO REDUCE NATURAL HAZARD'S EFFECTS
6	TO PROTECT CULTURAL LANDSCAPE
7	TO TRANSFER CONSTRUCTION CULTURES
8	TO ENHANCE CREATIVITY
9	TO RECOGNISE INTANGIBLE VALUES
10	TO ENCOURAGE SOCIAL COHESION
11	TO SUPPORT AUTONOMY
12	TO PROMOTE LOCAL ACTIVITIES
13	TO OPTIMISE CONSTRUCTION EFFORTS
14	TO EXTEND THE BUILDINGS LIFETIME
15	TO SAVE RESOURCES

FIGURE 8 - VERSUS ANALYSIS FOR THE GRANT HOUSE

The analysis highlights the contribution towards betterment of health quality by upgrading the levels of natural light and ventilation. On the other hand, the tool helps analyse few shortcomings like investment in transportation costs for procurement of the mud blocks due unavailability of space for production and storage of these mud blocks.

CONCLUSION

From the analysis of these 3 projects we come to understand the various aspects that need to be considered when thinking of holistic sustainability. The VerSUS criteria is a great tool that compliments the quantitative analysis of a completed project such as embodied energy calculations, carbon footprint calculations etc. It helps to evaluate the project on both its tangible and intangible impacts of its surroundings. In the case of the three projects mentioned in this paper, the common link that repeats through all 3 case studies is the use of mud as a construction material, this along with an environment-sensitive approach to design and the construction process. The relation between construction sector and environmental pollution is clear from the statistics mentioned in this paper. It is evident that a change in dynamic or thought process is necessary to create a positive impact. A re-analysis of the materials commonly used in present construction or the factors that lead to their selection is needed. This goes to prove that mud can indeed respond to the needs of today and the future as long as it is contextually appropriate. Throughout the world there are examples where this material has been used in varying typologies starting from quaint residences to high rise constructions, institutions and educational spaces and has brought a positive addition to that locale. It is important to recognise that a material that was used traditionally docs not necessarily mean that it is out-dated and irrelevant for the future. On the contrary it testifies itself through its resilient buildings that have stood the vagaries of time. It is also vital to understand that a particular material is ideal only when it's a local resource and availability is not a concern, hence it is always necessary to consider the sustainability principles and make the right choices right from the conception of a project.

REFERENCES:

1. Towards a zero emission, efficient and resilient buildings and construction sector. Global Status report. (2017). UN environment and international agency.
2. Mc Kinsey and company (2010). India's urban awakening: Building inclusive cities, sustaining economic growth.
3. UNDP. (2019). Sustainable Development Goals | UNDP. [online] Available at: https://www.undp.org/content/undp/en/home/sustainable-developmentgoals.html [Accessed 28 Feb. 2019].
4. UNESCO. (2019). Education for Sustainable Development. [online] Available at: https://en.unesco.org/themes/education-sustainable-development [Accessed 28 Feb. 2019].
5. Moriset, Guilland H et al. Versus : Lessons From Vernacular Heritage To Sustainable Architecture. Craterre-ENSAG, 2014.
6. M Andrew, R. (2018). Global CO2 Emissions from Cement Production. Earth System science data, (10), pp.195 - 217.
7. Embodied energy of common and alternate building materials and technologies. (2003). Energy and Buildings, (35), pp.129 - 137.

18th International Conference on
Non-Conventional Materials and Technologies
"Construction Materials & Technologies for Sustainability"
(18th NOCMAT 2019)
24th – 26th July 2019
Nairobi, Kenya

USE OF NON-CONVENTIONAL MATERIAL IN THE BUILDINGS OF THE ITAMARATI FARM, BEFORE AND AFTER THE ADAPTATIONS TO THE MODEL OF RURAL SETLEMENTS MANAGED BY INCRA.

Gisele Santos Estrella[1], Marney Pascoli Cereda[1], Luca Estrella Medeiros[2], Milene Santos Estrella[1]

[1] Dom Bosco Catholic University, Brazil;

[2] Anhanguera Uniderp University, Brazil,

ABSTRACT

The article compares the use of unconventional materials used in the remaining buildings of Itamarati Farm, in the municipality of Ponta Porã, state of South Mato Grosso, before and after assimilation of the property and adaptation for use in agrarian reform. The methodology consisted of visits and photographic records of the remaining buildings of the old farm, and of the new buildings, which are now part of the New Itamarati District. Interviews were conducted with officials of the involved and technical bodies, meetings with settled families,The theoretical reference was based on articles, books, historical documents. The analysis of the materials used in the area corresponding to the Itamarati Farm, later bought and occupied by the Agrarian reform National Institute (INCRA) as a federal settlement, showed, as expected, that all the buildings of the period were built with conventional material. Even the wood found in the employees' homes and in the church, which was little altered by occupation, However, in the same area, buildings of commercial and social use, housing and complementary buildings were identified, with an estimated result of 35% regarding the use of non conventional material. These buildings were erected or adapted in invasions that occurred at the beginning of the stage Itamarati I, since there was no care to prevent the uncontrolled occupation of the properties available in the purchase process. The survey showed that 85% of the families reused the materials of the houses and shacks during the improvements and annexes in their dwellings. Other unconventional material identified was cut eucalyptus in areas of encouraged planting, mainly used as pillars and beams. In the area of the New Itamarati District, there are still some non-inhabited buildings, built in haste by the inhabitants of the beginning of the occupation, who used eucalyptus shells for closing the walls and making doors and windows, classified with 50 to 65% unconventional.

Keywords: Non conventional materials, Agrarian reform, Rural buildings, Rural settlement.

INTRODUCTION

The advent of the industrial revolution in the seventeenth century, building materials known as conventional or industrialized (BARBOSA, [1]) began to be used and marketed, which are still today the most used in civil construction, but are also causing significant impacts to the environment (CASAGRANDE JR, [2]) .

Around the world, the construction industry is responsible for 50% of natural resources and 40% of energy inputs from all sources used to heat, light and ventilate buildings. These data consider the life cycle of the buildings, which includes, in addition to the energy consumption in the useful life of the buildings, also the one spent in the construction of the building materials themselves and in their deconstruction (TAVARES, [3]). The industrialized materials also require enormous amounts of energy in their manufacturing and handling processes (FRAANJE, [4]; BARBOSA, [5]) .

With rapid industrialization and urban development, the need for conventional materials has grown steadily. On a global scale, more than one ton of concrete per capita can be calculated (RADONJANIN et al., [6]). As an example, the volume of natural resources used by the construction industry, such as wood and iron, accounts for at least a third of the total consumed by society as a whole (SILVA, [7]).

The increase in the demand for constructions worldwide results in a similar increase in energy and the level of pollution that the construction industry causes, which in turn generates environmental pollution, but has also promoted research on non-conventional materials, with a more economical and renewable materials that can replace conventional industrial materials and also address issues of environmental preservation.

The use of non-conventional building materials, such as raw earth, is an alternative to environmentally-friendly bio-construction, as well as a healthier, more sustainable way of life (BAYER, [8]).

However, in spite of the large number of authors who present the advantages of the use of non-conventional building materials, the use of non-conventional materials is incipient and more concentrated in the urban area, where they are already valued and prioritized (PICUNO, [9]); UDDIN, [10]); GHAVAMI, [11]) . This urban use is even more contradictory when one considers the concept of nonconventional materials (FRAANJE, [12] ; PORTELA, [13]; GHAVAMI, [14]) as those already available in nature, a concept that is not always very clear.

In Brazil, conventional materials are defined by the Brazilian Association of Technical Standards, which are obligatory and establish legal certainty, and should be considered by the constructors and professionals as a state-of-the-art reference (SINDUSCOM, [15]). Currently, it was prepared by a Commission, the Brazilian standard of construction with adobe, awaiting its publication by the Brazilian Association of Technical Standards (ABNT), as Barbosa [16] pointed out. Other unconventional material, already widely used from technical standards in countries such as Colombia, Chile and Ecuador, is bamboo, which is being drafted the first Technical Standard through a Commission composed of professors from some universities and related companies (BARBOSA, [17]).

Another frequent issue in the literature is the frequency of use of unconventional materials in buildings of lower economic and cultural level (BARBOSA, [17] ; CASAGRANDE JR, [2]); CAIADO, [18]). In the rural area, unconventional materials should be more frequent, since the use of conventional materials would present additional difficulties in delivery due to logistics, distance and lack of information (TERRA, [19]; MULE, [20]; ZENID, [21]). For the characteristics described, the use of non-conventional materials should be more frequent in the field, especially in buildings are simpler and modest. Despite the advantages pointed out in the literature regarding the environment, the use of non-conventional materials is still little studied in the Brazilian rural environment. The evaluation of the occurrence and quantity of unconventional constructive material in buildings in rural areas may point out the difficulties that may interfere in the choice of these materials. To evaluate the use of non-conventional material, the remaining buildings of tha Itamarati Farm, which according to Terra [19]) date from 1975, were analyzed as well as those resulting from construction and adaptations for use as a rural settlement by INCRA, in the beginning of 2001 (TERRA, [19]), in the municipality of Ponta Porã, Mato Grosso do Sul.

METHODS/DATA

The study was carried out in the area located at 22° 32 '10 "S latitude 55° 43' 32" W longitude and altitude of 655m, corresponding to the former Itamarati Farm, by businessman Olacir Francisco de Morais. The area belongs to the Municipality of Ponta Porã, Mato Grosso do Sul, is now occupied by Itamarati I and II settlements. Both are close to the city of Ponta Porã, about 45 km, and the neighboring Paraguayan city of Pedro Juan Caballero.

In the same area, Itamarati Farm was set up with business planning, as the Company owend by Olacyr de Moraes planned all the necessary structure to carry out the planned activities, which were technically very diverse (Appendix I). After the purchase by the federal government, it was necessary to adapt from the National Institute of Colonization and Agrarian Reform Laws (INCRA) as a seat in the settlements (INCRA, [22]).

The areas studied, both settlements, are divided by BR MS – 164 (Federal Highway), which links the cities of Ponta Porã to Antônio João, and divides Itamarati settlements I from Itamarati II.

In the area of Itamarati I are the remaining buildings of Itamarati Farm, which are some residences. In Itamarati II settlement, the main buildings connected to the administration of the old farm, such as buildings such as schools, health center, hospital, churches (evangelical and catholic), airport, cafeterias, restaurant, supermarket and other facilities in the areas of agricultural production, common use, residences (Appendix I). For both construction systems, both for the company Itamarati Farm in 1973, and for the adaptation of the remaining buildings by INCRA from 2001, the difficulties in acquiring the building materials foreseen for the buildings would be the same once which occupy the same area. To obtain information necessary to explain the decisions taken regarding the use of non-conventional material, in addition to the visits made to the selected houses, a questionnaire with open and closed questions was elaborated and the specific literature consulted, as well as other documents.

The objective information was obtained through interviews carried out on visits to public agencies, when documents were obtained for consultation, as well as maps of the region surveyed. The field results were obtained during visits and interviews with residents, during which the study focus was observed and photographed, in order to identify the types of materials used in the site.

Criteria for the percentage evaluation of the presence of non-conventional building material

The criteria established to evaluate the percentage of use of the unconventional materials in the analyzed buildings was a scale of 0% (conventional material only) to 100% (only non conventional material), from the perspective of the author of this study, a civil engineer, wich worked on the area for ten years, provided technical assistance, and visited 85% settlements of the South Mato Grosso in Brazil.

In order to evaluate the percentage of unconventional material of the 370 existing residences (BELLÉ, [23]), a random sample of 90 houses in the Itamarati II and 15 houses in Itamarati I, corresponding to about 40% of the total, were randomly sampled.

Once unconventional construction materials were identified and quantified, they were described using information from the literature, as well as the responses of the applied questionnaire. In this work, the use of wood boards was considered as non conventional material, using vertical wood boards with joint joints (BROSLER, [24]), as well as non-conventional materials highlighted by Ghavami [25]) and Barbosa [1]).

RESULTS AND DISCUSSION

The state of Mato Grosso do Sul still has extensive areas under different crops, predominantly soybean, sugar cane, beef cattle and, more recently, eucalyptus (CAMELINI & CASTILLO, [26]). With a low population density of approximately 7 inhab / km2 (IBGE, [27]), in the 1940s the federal government saw a large "demographic void" in the state, with lands that could be integrated into the productive process (TERRA, [19]).

In the 1970s, Olacyr de Moraes was encouraged by the government to acquire an extensive area in the southern region of the state, with easy credit and subsidized interest, giving rise to Itamarati Farm (ALVES et al., [28]).

Itamarati Farm should be understood in two distinct stages, within the context of the Brazilian agrarian reform. It was the first acquired by the Federal Government of the group ITAÚ S.A. in the year 1992 with 50 thousand ha, at a cost of R$ 55 million (brasilian currency). Olacyr de Morais was the main owner of the property, already one of the largest landowners in the country and a major soy producer. The main purpose of the purchase was to implement the Itamarati I Settlement Project, which occurred in the year 2000. In the area of approximately 25 thousand hectares, approximately 1145 thousand families were settled (INCRA, [22]).

The second phase, which occurred only in the year 2004 to mid-2006, under the new perspective of the Terra Vida settlement program, had INCRA / MS - National Institute of Colonization and Agrarian Reform along with the government sanction federal government the commitment and execution of one of the largest agrarian reform projects in Latin

America, ie PA Itamarati II with the remaining 50% of the territorial acquisition of the property of the old landowner Olacyr, through the record disbursement of the public figures the amount of R$ 165 million (brasilian currency), whose objective and new method of agrarian reorganization had determined the inclusion of more than 1,776 thousand settled families.

Although they occupy the same area, the construction system of Itamarati Farm S / A Agropecuária's headquarters differs from the system used as a model for the National Institute for Colonization and Agrarian Reform (INCRA), since the company planned all the necessary structure to carry out the activities planned, which according to Terra, [19]), were technically very diverse.

The property of the farm had a social and productive infrastructure suitable for commercial and service activities, such as real estate with the capacity to meet the social demand for health, education, leisure and sports of its employees, and may be districts of the region . After the transformation of the farm into settlement, this area was destined to install public services and employees' homes (BELLÉ, [23]).

There were numerous properties with capacity for use and use, as well as their privileged characteristics of natural resources, such as residential, social, collective: 370 residences and the main headquarters, 2 ecumenical temples, 18 rooms for employees, 1 hospital, 2 schools, 1 poly gym with semi-Olympic swimming pool with 1 football field, poliesportiva court, sand court (LUCENA et al., [29]).

For the construction of these buildings, the company would have faced difficulties in the purchase and transportation of construction materials, however, without going through resource constraints. It is also necessary to take into account the complexity of the works, which required the presence of engineering companies of different modalities.

Subsequently, the same area was invaded by interested in occupying the available buildings of Itamarati Farm, and at the same time, it was necessary to adapt some of the buildings to the needs due to management.

New conceptions and use of the rural dwellings of the Itamarati Farm and after the implantation of the Itamarati I and II settlements into the BRAZILIAN Agrarian reform.

The analysis describes the utilized percentages of non-conventional materials used in the old Itamarati Farm buildings, evaluates the restructuring and adaptations of the old constructions, and proceeds to the analysis of the new buildings located in Itamarati I and II.

In the rural area, unconventional materials should be more frequent, since the use of conventional materials would present additional difficulties in delivery due to logistics, distance and lack of information (TERRA, [19]; ZENID, [21]).

Itamarati I and II settlements were implemented differently in time, but with the same function, residence and production for the people who registered. However, they differ greatly in the presence of remaining buildings of Itamarati Farm. The Itamarati I is located along and alongside the BR-164, it does not count on buildings destined to the high officials, the fact of not having enclosure could have propitiated the spontaneous occupation, in some existent buildings as well as constructions improvised by the residents who came in search of a rural plot.

According to the account of an ancient resident, Mr. P.V. , at the beginning of the establishment of the settlements, beginning of the year 2000, the reason why Itamarati I had more commercial movement, real estate and the flow of new residents, was that besides the Itamarati II area was later acquired, Itamarati II was blocked and controlled by a guardhouse and INCRA employees worked and lived in the place, so there was more control, tano for the invasion in the buildings as to the spontaneous occupation by the settlers or even new residents looking for a rural plot to start agricultural production and settle there.

The origin of the settlers was very varied. Besides those originating from the countryside, they were settled from the urban areas of the surrounding cities such as Campo Grande, Dourados and others (TERRA, [19]). A differential in relation to other settlements was the participation of former Itamarati Farm employees in the distribution of lots, organized as an association (AMFI). Another difference was the participation of the four most important social movements in Mato Grosso do Sul, FETAGRI, MST, CUT, FAF, which greatly favored local occupation.

Another feature of the Itamarati I settlement was to attract other residents than just those interested in agrarian reform. This spontaneous agglomeration settled in the area of the settlement offering the settlers the services not foreseen by INCRA, mainly of commerce such as sale of fuel for the motorcycles, that are the most common own conduction in the settlements, to restaurants and hotels. For this there was invasion and occupation of real estate that were more or less intensely adapted for the new use. As this movement was intense and uncontrolled in the early years of the process, the possibility of using non-conventional materials is greater, due to the hurry and pressing need.

At present this unplanned agglomeration was elevated to the condition of a District due to the population increase, which required new buildings. In 2015, with around 16 thousand inhabitants, the Itamarati I settlement was elevated to the New Itamarati District through Law Project 02/2015, approved by the Ponta Porã city council, ranked as the fifth most populous in the state and the 38th in economic movement (PELLIN, [30]). The District has the areas of Itamarati I and II.

This nucleus has about 700 properties, 400 of them built at the time of the farm and about 300 others built after the Itamarati II settlement. There are, in the locality, dozens of commercial establishments, police station, school, nursery, health post, post office, headquarters of the

Reform Agrarian National Institute (INCRA) (recently closed) Post, State Agrarian agency (AGRAER) and many other services (Appendix I).

Buildings of common use

Itamarati I

Although much altered in its layout, there are still buildings remaining in the core of the former Itamarati Farm, mainly concentrated in the area of Itamarati II, as a medical station and school, which are in operation and in good repair.

Planning Itamarati Farm did not prevent some situations out of control. At the height of Itamarati Farm's soybean production in the mid-1980s, about 20,000 people, including employees from the most diverse areas and their families, formed the famous called "Bugreiro" which was a collection of shacks made of tarpaulins, which was located outside of the headquarters of the Farm, where the providers of itinerant services were installed in such temporary residences (CORREIO DO ESTADO newspaper, [31]).

Figure 1 shows the proper knowledge of the use of non-conventional material. This building is located at the entrance of the area of the buildings of the old farm in Itamarati II and is used for the organization, security cabin, control and sale of transport, sales of products, etc. It was an initiative of a resident, who locally sells garapa cane broth, seasonal fruits, salted, and other products of family farming. It also serves as a stopping point. The building was built in the middle of 2018, 75% of non-conventional materials such as reuse were used: in the closing of the walls metal tiles and wood panel, eucalyptus pillars in the structure of beams and pillars, reuse metal door.

FIGURE 1- Security Cabin

Figure 1 - Recent building for sale of products, garapa cane broth, bus stop, built at the entrance to Itamarati II settlement, in the municipality of Ponta Porã, MS.

The buildings, in the core of the farm, are masonry, with slab and are well preserved. The building of the District Townhall was constructed by the Townhall of Ponta Porã city using only conventional materials.

Another building in the Itamarati New District (Figure 2) stands out for the use of untreated eucalyptus pillars on the balcony of the restaurant, 25% of non-conventional materials such as structures and closure in eucalyptus.

FIGURE 2 – Restaurant building under construction

Figure 2 - Construction of the restaurant with 35% of unconventional materials, pillars and untreated eucalyptus beams.

Some other buildings, such as inside the old farm's Fuel Station, invaded the building at the beginning of the settlement and expanded it with 35% of non-conventional materials, such as eucalyptus structures (beams and pillars).

An example of the oldest trades is the Mix Restaurant, which was initially built with unconventional materials on the wood-framed balcony. After the renovation with the closing of the balcony, it was used of conventional materials in 75%.

The Hotel, formerly used as employee accommodation, was initially provided by INCRA to the current owner, and expanded by 25% of woodwork on the expanded balcony. On the other hand, the Agricultural Products Store was built by the owner.

The current gas station, located at the entrance of the District, where pet gasoline is marketed, was built with wood reuse material and consists of a space of 1.5 x 1.5m made of unconventional materials and tile of reuse.

Itamarati II

In Itamarati II, another building of common use is the headquarters of the Cooperative, formerly used as a market and butchery, today represents an important economic role for the residents of this nucleus. The building was built at the time of installation of Itamarati Farm, with conventional materials and is in good condition. It is of masonry with tiles of the type of fiber cement, commonly called "canaletão". The other rooms are also used for meetings and meetings of the members. It is located in front of the old INCRA station and former headquarters of the farm. In this area the access roads are asphalted and well maintained.

The multi-sport gym, with a swimming pool, is now in a precarious state of conservation, due to mismatches by the management of the sub-municipality of Ponta Porã, as some residents reported. This gym is still today the target of disagreements among the residents of this nucleus. The family that takes care of the place is a former employee of Itamarati Farm, and built for his family a house adjoining the gymnasium, the construction in masonry, which is in poor condition. In the same area besides the gym, there is the semi olimpic swimming pool, also of the time of the entrepreneur, near to the State School Nova Conquista.

Schools

The research shows that there are now four schools in the settlements: a municipal Nova Conquista in Itamarati II, which serves approximately 400 students, state schools: José Edson Domingos dos Santos, located in the group belonging to CUT, near the urban nucleus, in the Itamarati and serves around 900 students in the three periods, the Nova Escola Itamarati Núcleo Urbano, which works in the same space as the old farm school, considered the second largest school in the state of Mato Grosso do Sul serving approximately 1700 students and finally the Carlos School Pereira da Silva, located in the MST group (Itamarati I), attending about 700 students in two periods and is currently operating a small nursery in the urban nucleus, also functioning in two periods.

The building of the largest school located in Itamarati II, State School Nova Itamarati, received the renovation of the front facade and expansion with 100% conventional materials. Initially, in 1975, the school was inaugurated as the first one on the old farm and its construction was made with 75% of non conventional materials with the use of wooden boards as closure in the walls and foundation masonry in the foundation, the roof was with tiles of French type clay (telha francesa type).

Differences between the occupations

Spontaneous occupation is a term originally used by INCRA field workers to explain some of the agrovilas formed in each nucleus of the groups of families seated with those of the occupied areas in the old Village of the Dryer (term that originated from the silo for drying and storage of soybean grains) name still used by the residents, nowdays New District Itamarati (old settlement Itamarati I).

In relation to Itamarati II, in addition to its purchase by INCRA, it was carried out in 2004, its occupational development continued in another rhythm. From the beginning, its main entrance was always controlled by a guardhouse managed by INCRA employees and until 2008, the larger buildings belonged to the managing bodies like AGRAER, INCRA and City Hall. In this way, its growth was slower. Unlike Itamarati I, Itamarati II, stands out today for the real estate offer.

The following describes some differences of the remaining buildings in the settlements and the use of the unconventional materials.

Itamarati I already shows some development resulting from some improvements such as basic health units, schools and a powerful trade. However, the great local demand is still the infrastructure: streets that are constantly flooded, residences for which there is still a sewage network, street lighting and water.

Health Offices

In the settlements Itamarati I and II, are attended by 4 health posts. Currently the city of Ponta Porã has built or renovated the buildings. Only the health center of AMFI, located next to the church, the building is of the time of the old Farm and with conventional materials. The buildings at the Canãa Group's health centers, Geraldo Garcia (CUT I) and the District Nova Itamarati (located at Itamarati II) were built only with conventional materials and are in excellent condition.

Residences

The type of housing was also contemplated in the interview script, since the settlements have different dwellings. It was verified that in the nucleus of the old farm (in Itamarati II), we found wooden houses, of bricks and we observed that this distinction of adopted materials followed a hierarchy in the social classes of the employees.

In this sense, the rural house portrays the resistances and transformations that occurred outside the rural area. Through the situation of the house, its construction material, its aesthetics and functionality, it is possible to identify important aspects about what this rural house is, what its characteristics are (SILVA, [32]).

However, in the buildings built at the time of Itamarati Farm, conventional civil construction materials were used primarily in high-level buildings such as the Fazenda headquarters, in homes for employees such as doctors, engineers and team leaders. On the other hand, buildings that were used in 35% to 50% of unconventional materials were used for employees as technicians, cooks, field assistants, and other functions below the high hierarchy.

While the buildings of collective use, remaining Itamarati Farm are constructions with conventional material, the residences present greater variation in types of construction materials used.

As housing, 370 houses were considered, considering Itamarati I and II, which are occupied by employees and teachers who work in the settlements. The offices of INCRA and Agrarian Development and Rural Extension Agency (AGRAER) functioned in the old farmhouse, but today it is home to the subprefecture of Ponta Porã and AGRAER. At the site there are several rooms that are given to agencies such as the Brazilian Service of Support to Micro and Small Businesses, SEBRAE, National Rural Apprenticeship

Service, SENAR and other universities both for face-to-face courses, distance courses, trainings and other meetings under the organization of AGRAER employees.

The residences are of two types, temporary residences and permanent residence of the employees. Due to the characteristics described, the use of non-conventional materials should be more frequent especially in buildings that are simpler and more modest (Kumar, [33]). However, the analysis of the buildings built during the Itamarati Farm period shows that conventional building materials were used primarily in high-level buildings, such as in the Fazenda headquarters and in the homes of some employees, such as doctors, engineers and team leaders. The 62 masonry and slab dwellings in the Itamarati II urban nucleus have never been modified or expanded and are in a good state of conservation, were built with conventional materials until today are inhabited. Residents are employees of the Health Post, doctors, dentists, nurses, nuns and teachers. These houses were built of masonry and slab (0% of unconventional material) or wood with masonry foundation. In these houses the wood is conventional and treated, it complies with NBR 7190. The occupation of the houses followed a criterion that was established according to the hierarchy of the employees, the greater the hierarchy, the better the houses that occupied, with masonry, slab and with closing in bricks, concrete slabs and French type clay tiles.

Among the dwellings visited, the Itamarati II had a higher number of vacant dwellings due to the commercial flow of the settlements. Among them, 12 were not in good condition, they were built in 35% of non-conventional materials.

In contrast, to buildings with only conventional material, the wooden houses are in poor conservation status, have not been enlarged and 12 are uninhabited. Constructed with walls closing, wooden roof structure and French clay type tiles. The windows are metal and wooden doors. In some other houses the windows are made of wood. It was observed that the dwellings that were invaded are constructed with 100% of conventional materials. As an example, the building in front of the Cooperative, has been invaded and is inhabited and in a good state of conservation.

Three types of houses were identified in the Itamarati I and II urban centers: a) old farm houses, built of masonry, slab and French clay tiles, 75% to 100% of conventional materials; b) old houses of the Farm, built from foundation of masonry of basement and masonry up to 1m height and closure on walls, frames, wooden roof structure and clay tiles 50% of non-conventional materials; c) brick houses built after the laying of the settlements, without slab, covering of clay tiles or fiber cement, 25% of non-conventional materials.

The subprefeitura of the municipality of Ponta Porã does not provide to the local population, a sectoral planning of the existing buildings of the old Farm, which aims at a greater control of the technical management, efficient and adequate to the maintenance, use and occupation of the same ones. In the random sample of 62 houses in Itamarati II and 15 houses in Itamarati I, it was evidenced the abandonment of a good part of these houses, in both nuclei. There are no technical surveys so that the good state of conservation of these houses is preserved and to provide a social function for each dwelling or even for building without use.

The process of maintaining the buildings in the settlements in question depends, above all, on the social function and the use made of them. Of the 142 dwellings occupied by former employees, in the Itamarati I and Itamarati II centers, 107 are in good repair, built in 35% of non-conventional materials, and only 35 have been expanded. These extensions were made due to the need to be adapted for commercial activity, such as restaurants, pizzeria and beauty salon. In these extensions, 25% of non-conventional materials such as untreated eucalyptus were used in columns and beams.

In 25% of the houses that were expanded in these settlements, the conventional materials were used as masonry closure in the internal environments and wood for the structure in the balconies of the trade.

Regarding the state of conservation, it was observed that in the Itamarati II nucleus there are a greater number of houses uninhabited and in a precarious state of conservation. This is due to the fact that in the Itamarati II core there are fewer commercial buildings and people flows. In Itamarati I, in turn, there is a greater turnover of consumers and different types of commercial establishments.

In the interviews, residents reported that, there are still in the nucleus some housing for which INCRA itself did not establish the social function or even the right to live. Of the 42 buildings in Itamarati II, 28 used 35% of non-conventional materials and are in poor conservation status. At the time of the old farm, the employees had the right to housing and these were in good condition, as reported by the representative of the Association of Former Residents of Itamarati Farm (AMFI).

The Municipal Government of Ponta Porã has not yet defined norms for the maintenance of the housing infrastructure, which are remaining buildings of Itamarati Farm, in order to guarantee them good conservation. INCRA has already transferred the sub-prefecture to the City Hall, but some services are slow to walk. In this way, there are many abandoned houses in the core of the old farmhouse.

Of the 24 existing buildings that occurred due to spontaneous occupation at the beginning of the formation of the Nova Itamarati District, called "Vila Dryer" (Itamarati I), 35% of non conventional materials were used, such as the closing of walls with untreated wood, pillars, beams and structure of native and exotic wood such as eucalyptus.

In the District, 02 buildings from the time of the beginning of occupation in the Vila area of the dryer were located, with the use of 75% of unconventional materials, closure of the walls with eucalyptus peel and Eternit tiles.

The analysis of the unconventional material used in the buildings, both in common use and in buildings of a commercial or residential nature, highlight the little use they had, except for a few adequate wood, according to ABNT standards, but not exceeding 35%. Only in two types of housing built at the time of settlement, some residents used the bark of eucalyptus, predominantly used in the region. Later on, INCRA was encouraged to plant lots (BELLÉ, [23]).

It is natural that there was a similar pattern in relation to the constructive material, in this first phase, since for the two constructive models, the material had to be acquired in specialized stores. Therefore, it should be taken into account that, in the same place, at the time of the implantation of Itamarati Farm or the construction model of INCRA, the same difficulties of attendance and delivery would be encountered.

Airport

The airport and airstrip are abandoned and the population uses it to hike. Go to the end of the text when you finish, along with everything that has no unconventional material.

CONCLUSIONS

The remaining buildings of the old farm built in Itamarati II with 75% of conventional material are 90% inhabited or invaded, in contrast to buildings built with 50% of non-conventional materials, 70% are inhabited and the rest are uninhabited or invaded. The city council of Ponta Porã does not yet provide a strategic plan for the conditions of use of the dwellings in the urban center in Itamarati II. However, the buildings remaining in Itamarati I, do not find any uninhabited because the flow of people is larger and the real estate value is also higher in relation to the Itamarati II settlement. This fact was due to the fact that, from the outset, the Itamarati I occupation was freer, that is, at the time, there was a guardhouse that controlled the entrance to the Itamarati II, which made the invasions impossible more in the entire area in Itamarati I.

The analysis of the materials used in the area corresponding to Itamarati Farm, later occupied by INCRA as a settlement, showed, as it was expected that all the buildings of the time were built with conventional material. Even the wood found in the employees' houses and in the church, which was little altered by occupation, is processed wood, which complies with ABNT NBR 7190 standards, which probably explains the good conservation. However in the study of Brosler [24]), the closure in wood boards is considered as unconventional materials.

However, in the same area, buildings of commercial use, housing and social use, a percentage of unconventional material up to a maximum of 35% were identified, in this case buildings erected or adapted when at the beginning of the occupation phase, when not care was taken to prevent the uncontrolled occupation of the properties made available in the purchase process. In this case the unconventional material identified was cut eucalyptus in areas of planting encouraged by INCRA in 2011 (BELLÉ, [23]). The untreated eucalyptus wood was mainlyused as pillars and beams. In the area of the New Itamarati District, some non-inhabited buildings, built in haste by the inhabitants of the beginning of the occupation, can be seen, using eucalyptus bark for closing the walls and making doors and windows.

It was concluded that in the areas where there was greater planting of exotic species, such as eucalyptus, there was a higher incidence of use, both in the enlargement of residences and in other types of buildings, such as the restaurant and other cases. The government should encourage programs on the use of these materials so that settled families can establish greater autonomy in the construction of their houses and start their agricultural production, so loga are installed in their rural lot. Technical guides should be disseminated using technical standards from Brazil and other countries such as Peru, Colombia, Chile and others in order to propagate this cultural knowledge and make it efficient and accessible to all.

REFERENCES

1. BARBOSA, N. P., GONÇALVES J. da S., GHAVAMI K. Proposta de Norma Brasileira de Construção com Adobe. Centro de Tecnologia. Universidade Federal Paraíba. 2005.

2. CASAGRANDE JR, E. F. Princípios e Parâmetros para a Construção Sustentável. Universidade Feevale. Novo Hamburgo, RS, 2009. Disponível em: http://www.cbcs.org.br/. Acesso em: 11 ago. 2009.

3. TAVARES, S. F. Metodologia de análise do ciclo de vida energético de edificações residenciais brasileiras. 2006. 225f. Tese – Doutorado em Engenharia Civil. Universidade Federal de Florianópolis – UFSC, 2006.

4. FRAANJE, P. J. Building with nature making use of renewable resources, grown structures and growing constructions. In: International Symposium Beyond Sustainability, Eindhoven University of Technology, Netherlands, 28/29 Sptember, 2000.

5. BARBOSA, N. P., GONÇALVES J. da S., GHAVAMI K. Proposta de Norma Brasileira de Construção com Adobe. Centro de Tecnologia. Universidade Federal Paraíba. 2005.

6. RADONJANIN, V.; MALESEV, M.; MARINKOVIC, S.; MALTY, A. Green recycled aggregate concrete. Construction and Building Materials. 2013.

7. SILVA, V.G. Avaliação do desempenho ambiental de edifícios: qualidade na construção. n. 25. São Paulo: SindusCon, 2000.

8. BAYER, A. P. Proposta de Diretrizes para o Desenvolvimento da Arquitetura em Terra no Rio Grande do Sul, a partir da Interpretação de Estratégias Uruguaias. 2010. Dissertação (Mestrado em Engenharia Civil) Programa de Pós-Graduação em Engenharia Civil, UFRGS, Porto Alegre, 2010.

9. PICUNO, P. Use of traditional materials in farm buildings for a sustainable rural environmental. University of Basilicata. Potenza, It. International Journal of Sustainable Built Environment. Volume 5; May, 2016 pp 451-460.

10. UDDIN, N. A study on the traditional housing technology of Bangladesh. University of Chittagong, Bangladesh. Indian Journal of Tradicional Knowledge. Vol. 7(3). July, 2008.

11. GHAVAMI, K.; BARBOSA, P. N.; MOREIRA, L. E. Bambu como Material de Engenharia. In: Avaliação de Desempenho de Tecnologias Construtivas Inovadoras: Conforto Ambiental, Durabilidade e Pós Ocupação. 2017. p. 305-348.

12. FRAANJE, P. J. Building with nature making use of renewable resources, grown structures and growing constructions. In: International Symposium Beyond Sustainability, Eindhoven University of Technology, Netherlands, 28/29 Sptember, 2000.

13. PORTELA, M. O. B.; GOMES, J. M. A. Os danos ambientais resultantes da extração de argila no bairro Olarias em Teresina - PI. Universidade Federal do Maranhão. II Jornada Internacional de Políticas Públicas Mundialização e Estados Nacionais: a questão da emancipação e da soberania. São Luís – MA, agosto 2005.

14. GHAVAMI, K. Materiais e Tecnologias não Convencionais para o Século XXI. Pontifícia Universidade Católica do Rio de Janeiro, RJ, Brasil. 2014.

15. SINDUSCOM/MG Sindicato da Indústria da Construção Civil no Estado de Minas Gerais. Catálogo de Normas Técnicas - Edificações. Belo Horizonte: Sinduscon-MG/CBIC, 2017.

16. BARBOSA, N. P., GONÇALVES J. da S., GHAVAMI K. Proposta de Norma Brasileira de Construção com Adobe. Centro de Tecnologia. Universidade Federal Paraíba. 2005.

17. BARBOSA, N. P. Considerações sobre materiais de construção convencionais e não convencionais. Universidade Federal da Paraíba. 2015.

18. CAIADO, A. R. Contribuição ao Estudo da Rotulagem Ambiental dos Materiais de Construção Civil. Tese (Doutorado). Universidade de São Paulo/USP. 2014.

19. TERRA, A. Reforma agrária por conveniência e/ou por pressão? Assentamento Itamaraty em Ponta Porã – MS: "o pivô da questão". 2009. Tese (doutorado) - Universidade Estadual Paulista, Faculdade de Ciências e Tecnologia (UNESP). SP: Presidente Prudente, 2009.

20. MULE, B., M. Factors affecting the adoption of appropriate building materials and technologies (abm&ts) programme in North Eastern province of Kenya. Project Planning and Management Dept. University of Nairobi, 2012. 103p.

21. ZENID, G. J.; NAHUZ, M. A. R.; MIRANDA, M.J.A.C.; FERREIRA, O. P.; BRAZOLIN, S. Madeira: uso sustentável na construção civil. Secretaria do Verde e do Meio Ambiente do Município de São Paulo. SindusCon-SP – Sindicato da Indústria da Construção Civil do Estado de São Paulo; COMASP – Comitê de Meio Ambiente,

Segurança e Produtividade do SindusCon-SP. IPT – Instituto de Pesquisas Tecnológicas do Estado de São Paulo S. A. 2016.

22. Instituto Nacional de Colonização e Reforma Agrária. Available in:<www.incra.gov.br >

23. BELLÉ, M.F. L. Impactos sobre o desenvolvimento regional decorrentes do assentamento Itamarati, Ponta Porã (MS): 2001-2010. Dissertação (mestrado). 149f. Campo Grande, 2011.

24. BROSLER, T. M.; BERGAMASCO, S.M.P.P. Novas concepções da moradia no meio rural: uma reflexão a partir de assentamentos rurais no estado de São Paulo. Universidade Estadual de Campinas-SP. 2016.

25. GHAVAMI, K. Materiais e Tecnologias não Convencionais para o Século XXI. Pontifícia Universidade Católica do Rio de Janeiro, RJ, Brasil. 2014.

26. CAMELINI, J. H.; CASTILLO, R. Logística e competitividade no circuito espacial produtivo do etanol no Brasil. Boletim Campineiro de Geografia. AGB Campinas, v.2, n.2, 2012.

27. Instituto Brasileiro de Geografia e Estatística – IBGE- Available in: <https://cidades.ibge.gov.br/brasil/ms/panorama>

28. ALVES G. L; LESSA, M. F. Impactos sobre o desenvolvimento regional decorrentes do assentamento Itamarati, Ponta Porã (MS): 2001-2010. 150p.

29. LUCENA, P.L.; FIGUEIREDO, L. F.; SCHIERHOLT, M.I. MARIANI, M.A.P. ; LUCENA, R.M. Novo modelo de assentamento para reforma agrária- programa Terra Vida, o estudo de caso do plano de assentamento Itamarati II. XLV CONGRESSO DA SOBER. "Conhecimentos para Agricultura do Futuro". UEL. Londrina – PR , 2007

30. CORREIO DO ESTADO. Available in:< www.correiodoestado.com.br>

31. SILVA, A. N. Currículo e práticas mono/multi/interculturais e a produção de identidades e diferenças na Escola Estadual do Campo Nova Itamarati. Tese (Doutorado). Campo Grande. Universidade Católica Dom Bosco. 2014.

32. KUMAR, A.; DEKA, A.; SINHA, R. Rural Housing in India: Status and Policy Challenges. Lokashraya Foundation. Editorial Board: Prof. Ashok K. Pankaj; Shri R.N.P. Singh. New Dehli, 2016.

Appendix I

Quantity	Type of building (use and occupation)
Residencial	
01	Main office: masonry, built with exposed brick and reinforced concrete structure, 02 floors, roof with fiber cement roofing tiles.
224	Masonry houses
63	Wooden houses
08	Semi-detached masonry houses
Common usage	
06	Masonry guard
04	Cantinee
01	Restaurant
01	Supermarket
02	Hospital
01	Leisure areas with gym
01	Laundry service
01	Garagem
01	Parking place
02	Tower metalic made
01	Gas station
01	Masonry rail
02	Churches
01	
Agricultural Production	
01	Hangar , track with asphalt
O2	Equipments rooms
01	Airplane washer
01	Central radio and lighthouse
03	Masonry offices
03	Masonry warehouses
08	Sheds of metal structure
02	Sheds
02	Wooden tanks
03	Silos, trench type
18	Mansonry silos
05	Wooden shovels
02	Water wheels
06	Reservoirs
02	Masonry tanks
01	Saloon for slaughter
08	Artesian wells
06	Masonry modules for confinement
02	Semi Artesian wells
02	Pavilions

01	Poultry
01	Timber Swine
02	Wooden Sheds for sheep
01	Wooden pallet
08	Masonry waarehouses
02	Railway scales
10	Dryer of soyan (masonry)
04	Processing units buildings
27	Pivots
26	Tanks for dilution
26	Sheds
01	Factory of animal foods
16	Dams
08	Flight builts

Source: INCRA [19].

18th International Conference on
Non-Conventional Materials and Technologies
"Construction Materials & Technologies for Sustainability"
(18th NOCMAT 2019)
24th – 26th July 2019
Nairobi, Kenya

INFLUENCE FACTORS AFFECTING THE ADOPTION OF NON CONVENTIONAL MATERIALS IN THE RURAL DWELLINGS, SOUTH MATO GROSSO, BRAZIL

Gisele Santos Estrella[1], Milene Santos Estrella[1], Luca Estrella Medeiros[2]

[1] Dom Bosco Catholic University, Brazil;
[2] Anhanguera Uniderp University, Brazil,

ABSTRACT

Although the use of non-conventional building materials is encouraged for several reasons, including environmental and so on, their use is still incipient. The research analyzes the factors that influence the decision on the use of nonconventional material (MNC) in rural areas, based on the Itamarati I and II settlements, in the municipality of Ponta Porã, Mato Grosso do Sul, managed by the National Institute of Colonization and Agrarian Reform (INCRA). In the construction model adopted after the occupation of the lot, the settler receives a minimum plant that will have to execute on a joint effort basis, while the purchase of the material is made by INCRA. Sampling took into account the variables that could influence this decision, considering as main factor the existence of MNC in the place and the spatiality, which includes the location of the lot. Other variables selected from the literature were technical and social training, on-site dwelling time, the level of technical knowledge about construction, the characteristics of rural construction, involvement with social movements, and previous experience with non-conventional constructive material. The information was complemented with questionnaire applied and interviews with settled families. The quantification of the MNC used in the buildings was made on a visit to the lot on a scale of 0 to 100% for the extremes. The results showed that MNC were poorly present in the main residence, as was expected, where only eucalyptus logs and their processed wood were identified, both as conventional and nonconventional material (closure of walls) due to their planting in an area of the settlements. The distance from the lot in relation to the New District Itamarati, where there are stores of construction material, influenced the percentage of use, but not the other factors evaluated. Also, as expected, the other buildings, for which the residents do not receive INCRA material, presented a higher percentage of MNC, which varied from 35 to 65%. Bamboo, which is an easy-to-apply material, was only identified in complementary buildings (OE) such as chicken and vegetable sheds, as well as reuse material. It is recommended to encourage the planting and use in the rural buildings of eucalyptus and bamboo, in the Public policies of rural dwellings, besides informing the rural settlers about the appropriate use of nonconventional materials, meeting the world technical standards.

Keywords: Non conventional material; rural dwelling; rural settlement.

Introduction

The process of building a rural dwelling after conquesting the land, can be analyzed in the interweaving of the present with the past and with the future pretensions (BROSLER et al. [1]), and in a certain way ends the cycle of camping and opens the settler, in the case of land reform settlements. Da Matta [2] considers that the house, as an element of study, transcends the question of its construction, portrays internal aspects experienced as a Terra of cultural representation of man. According to Brosler and Bergamasco [3] represents a set of resistances, as well as the transformations of which it is surrounded. Although in the recent past the use of materials collected in the immediate vicinity with the advent of the industrial revolution in the seventeenth century began to be used and marketed the building materials that are now known as conventional or industrialized (BARBOSA, [4]), which are, to date the most used in construction, but also those that cause significant impacts to the environment (CASAGRANDE JR, [5]). The choice of materials to be used in the construction of a residence is a variable that depends on who will reside in the property, but in the specific case of housing programs carried out by governments, this does not always occur. In the study by Amado et.al [6] in East Timor, the housing project should be adapted to family habits so that there is effectiveness in the housing program and provide security, quality, adaptability and adaptation to family needs. However, Scott et al. [7] pointed out in Ireland a disconnection between the public policy spheres, marked by a tense relationship between rural communities and regulatory planning, so that the authors conclude that any housing policy for Rural areas must take into account the social, economic and cultural aspects of rural life, not just the environment and landscape criteria. Another problem in housing programs is the selection and use of building materials. In Brazil, conventional materials are mandatorily defined by the Brazilian Association of Technical Standards (ABNT), as provided for in laws and regulatory instruments, in a way that provides technical isonomy, which results in an indispensable reference in this regard. They are also one of the pillars of legal certainty, and should be seen by the builders and professionals as a state-of-the-art reference (SINDUSCOM, [8]). These materials make the construction sector worldwide responsible for the consumption of 50% of natural resources, as well as 40% of energy inputs from all sources used to heat, light and ventilate buildings. These data consider the life cycle of buildings, which includes not only energy consumption in the useful life of the buildings, but also that spent on the construction materials in the work itself and the deconstruction of these (Barbosa, [4]) is equivalent to that of a modest family in the period of one month. Rapid industrialization and urban development accelerated the use of conventional materials. At the global scale, more than one ton of concrete per capita is estimated (RADONJANIN et al., [9]) and the volume of wood and iron accounts for about one-third of the total consumed by society as a whole (SILVA, [10]). Civil construction draws much of the planet's natural resources consumed by mankind. In this sense, the use of non-conventional building materials, such as raw earth, is an alternative to environmentally-friendly bioconstructions, as well as healthier and more sustainable ways of life (BAYER, [11]). To address the lack constructive material, Ugochukwu et al. [12], recall the importance of using local materials available, such as land use and traditional architecture, noting that this type of building can provide thermal comfort, durability and economy. Uddin [13] analyzed the traditional constructive processes used in several areas of Bangladesh to highlight the importance of maintaining the constructive technology with adobe, emphasizing that at the time, there was still the ability of construction workers to execute it. The past knowledge of these practices be documented, so that they do not get lost at once. However, the author pointed out the importance of technical follow-up to the use of traditional constructive practices, encouraging the elaboration of an illustrated booklet to facilitate the use of practices with non-conventional materials found locally and based on culture, as well as the use of Technical Norms, when existing, and laboratory testing in order to make better use of these materials. Along the same lines, Gichunge [14] recommended that governments and other stakeholders in economic housing projects that aim to reduce the housing deficit and the cost, establish strategies to promote them such as projects, effective government participation in resources and recommend revised technical standards to address indigenous constructions and materials available locally. However, in spite of this, the use of local materials is infrequent, a fact explained by Araújo [15] for cultural reasons. Other difficulties are highlighted in Brazil by Barbosa [4] by government agencies towards everything that is not conventional, such as lack of political will, ignorance and intentional ignorance, and the lack of technicians to use them correctly. For several reasons and despite the high number of authors who present the advantages of the use of unconventional building materials (BROSLER, [16]; OLIVEIRA, [17]); VERDE GHAIA, [18]), this use is still incipient and curiously more found in the urban area, where are already valued and prioritized (PICUNO, [19]; UDDIN, [13]; GHAVAMI, 2017[20]), something that becomes contradictory, if one takes into account the concept of non-conventional materials such as those available in nature (FRAANJE, [21]; GHAVAMI, [22]). In the rural area, therefore, it would be expected that nonconventional materials would be more used in the construction of houses, given that one of the difficulties that is expected in these localities is precisely the purchase and delivery of conventional materials, of the distance (TERRA, [23]; ZENID, [24]). In the literature it is also mentioned the frequency with which the unconventional materials are used in buildings of lower economic and cultural level (CASAGRANDE JR, [5]; CAIADO, [25], BARBOSA, [4]). It is explained the difficulty to implement the use of non-conventional materials also by prejudice. Some materials, such as earth, bamboo, straw and vegetable fibers, are rooted in the minds of the vast majority of people as "poor material." This prejudice was perceived in the research results of Vilpoux, Estrella and Cereda, [26] that resulted in the settler considering the use of bamboo in the rural residence as cheap and not durable. Therefore, the study focuses on the methodology of rural housing implemented by the Institute of Colonization and Agrarian Reform (INCRA) for the establishment of settlements, where each lot is considered a unit of family agriculture and demands benefits from all spheres of government. The material used for construction in rural buildings, focusing on unconventional materials, is a subject that has not yet been approached, which is why it was chosen for the present research, in Itamarati settlements I and II. The largest Brazilian agrarian reform settlement is managed by the National

Institute for Colonization and Agrarian Reform (INCRA), which since 2000 has established a total of 2,835 families. This number by itself guarantees greatvariability, because according to Terra [23] the settlement is a place where different biographies meeting themselves and also where new processes of social interaction and identity begin, a privileged place. It means that access to land, a geographically located place, makes both work and residence possible, and transforms the "landless" into "landless" people (INCRA, [24]). Another advantage highlighted in the chosen place is that, besides rural is also remote, implanted in Mato Grosso do Sul, on the border between Brazil and Paraguay. In this model, the settlers receive a basic floor plan of 45.00 m2, as well as the necessary construction material, purchased by the agency itself through a bidding process under Law 8666/93 and made available to the settlers to build their residences in scheme of mutirão. Therefore a great uniformity of constructive material is expected in the main residence, since they were acquired in specialized stores. For this reason the research also analyzed the other buildings as the places for the storage of the production of the familiar agriculture, deposit for tools, implements and small agricultural machines, corral, besides gardens and animal enclosures for animal husbandry like chicken coop, because, contrary to what happens with the main residence, INCRA does not foresee the purchase of construction material other than in notices or specific projects. In this case the settled family would be more dependent, which could stimulate the use of nonconventional material. These arguments are the basis of the proposed research, which aims to evaluate the use of non-conventional material in the Itamarati I and II settlements in Ponta Porã, MS.

Methods

The research described is a result of the integration of the data obtained in the bibliographic and field research, being characterized as a quantitative exploratory approach.

The study site

The study was carried out at the Itamarati I and II settlement, located in the municipality of Ponta Porã, Mato Grosso do Sul, at coordinates 22° 32 '10 "S latitude 55° 43' 32" W longitude and 655m altitude, as verified in Figure 1 which situates the Itamarati I and II settlements in relation to the state of Mato Grosso do Sul and Paraguay and the New Itamarati District, a disordered agglomeration that houses the commerce stimulated by settled families. The BR164 cuts both settlements longitudinally. Figure 1 shows the geographical location of the study area of Itamarati I and II settlements and of the Nova Itamarati district of the BR BR164 highway in the municipality of Ponta Porã, MS, highlighting the cities in the vicinity and the border with Paraguay.

FIGURE 1 – GEOGRAPHIC LOCATION OF THE STUDY AREA OF THE ITAMARATI I AND II SETTLEMENTS, NOWADAYS NOVA ITAMARATI DISTRICT.

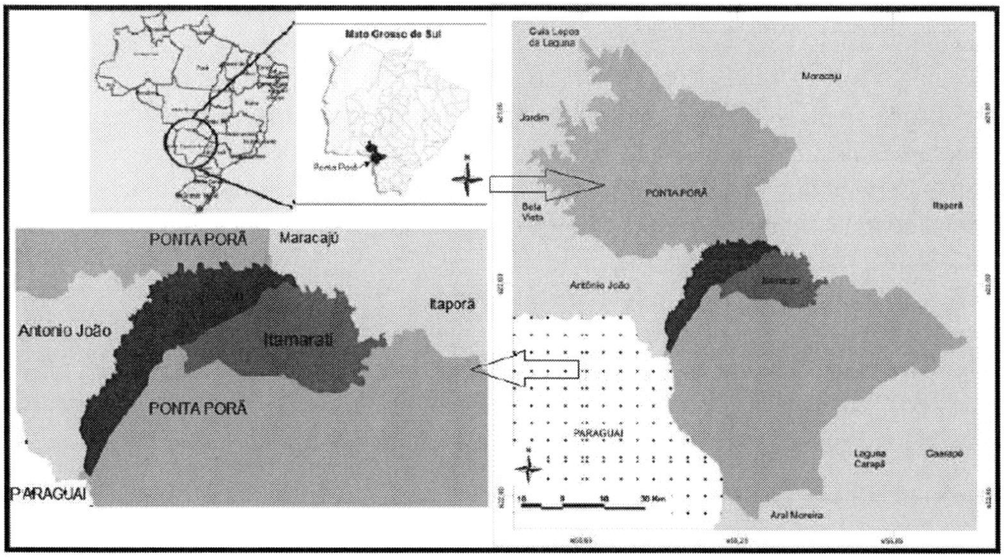

Source: Adaptation of the authors of the map provided by Terra [23] .

The settlements were implemented in two stages: Itamarati I in May 2001 and Itamarati II in May 2004, both under the management of the National Institute for Colonization and Agrarian Reform (INCRA). The division of the lots was made previously occupying the area in a total of 2,835 households. In Itamarati I, 1,143 families were settled, distributed among social movements in the state of South Mato Grosso: Central Unica dos Trabalhadores (CUT) with 280 families, Association of Residents of the Former Farm (AMFFI) with 150 families, Federation of Agricultural Workers (FETAGRI) with 395 families, Landless Workers Movement (MST) with 320 families. In Itamarati II, 1,692 families were settled. In 2015, the Itamarati Settlement was elevated to the District category of the city of Ponta Porã by the City council in 2015. According to IBGE, in 2012, the Itamarati I and II settlements had, in 2012, an estimated population of 15,867 inhabitants that are distributed over more than 50 thousand hectares (IBGE, [25]).

Construction system and non-conventional materials

In this present research the conventional materials evaluated will be the most commonly used in a rural dwelling and defined by the Technical Standards (ABNT) such as clay bricks (eight holes or solid brick), clay tiles, metallic, fiber cement, steel bars, cement, aggregates, among others. Civil construction draws much of the planet's natural resources consumed by mankind. In this sense, the use of non-conventional building materials appears as a more environmentally friendly alternative, as well as healthier and more sustainable ways of life (BAYER, [11]). In this same alignment, a group of people, scientists, professionals and professors in the field of engineering and architecture and other related, dedicated to the study and development of the sciences of materials and unconventional technologies met to found the Brazilian Association of Materials and

Non-conventional technologies (ABMTENC) at PUC-Rio. Professors Dr. Khosrow Ghavami and Normando Perazzo Barbosa are pioneers in research in this area, the first one to come from PUC-Rio, the second from the Federal University of Paraíba. This same association emphasized that the sustainable use of new technologies and innovative uses of non-conventional materials such as earth, bamboo, adobe, and taipa should be used in compliance with world technical standards and stimulate research and studies for Brazilian Standards. To provide low-cost buildings with adequate thermal comfort, aesthetics, durability and safety (BARBOSA, [4]). In this present research the following non-conventional materials will be considered for the criteria: 1. Wood: wood boards (BROSLER, [16]), eucalyptus costaneira (JUÍZO, [26]), eucalyptus (pillars and beams); 2. Earth: use of adobe, mud, adobe bricks with bamboo, straw (GHAVAMI, [20]); 3. Bamboo: in coverings, walls, frames; 4. Reuse materials: Metallic tiles used as walls and cover, old doors, pet bottles as cover, tires in enclosure, among others. Sampling of lots, the criteria established to evaluate the percentage of use of the nonconventional materials in the main dwelling and in other buildings are explained below.

Data collection

The objective information was obtained through interviews from a questionnaire carried out on visits to public agencies, when documents were obtained for consultation, as well as maps of the region surveyed. In addition, the study consisted of consulting the websites of these agencies to obtain details of each technical standard used during the purchase and delivery of the materials, the construction process of the dwellings and the use of conventional and non-conventional materials in the budget worksheets. To evaluate the percentage of use of the non-conventional materials in the analyzed buildings, it was a scale of 0% (conventional material only) to 100% (non-conventional material only). It was tried to take into account the most common non-conventional materials used in rural dwellings and highlighted in theoretical references such as pillars and beams of eucalyptus, costaneira, wooden tubs, untreated wood, bamboo, mud or adobe walls closure or with reuse materials such as metal sheets and other materials. In this analysis, wood boards were considered as non-conventional material (BROSLER, [16]).

Dimensions and Variables used in the research

After evaluating the criteria established in the previous item, we adopted the Dimensions and variables for the analysis of the influence factors that led the families to use non-conventional materials in the works in their rural dwellings. For the decision making of the choice of influence factors, the variables cited in the literature were considered as influencing the decision to use nonconventional material or not. Therefore, it was highlighted as decisive the premise that the decision of the use depends in the first place on the spatial location of the lot and the local availability of unconventional material, in quantity and quality adequate to replace the conventional material. In addition to these two variables considered as priorities, other information was collected using a questionnaire with open and closed questions. The other variables, as well as the bibliographic references that recommend them, are listed below.

Spatial Dimension: refers to the location of rural plots and their importance, was highlighted by Terra [23]; Mule [27]; Brosler [1]. Specifically in relation to the study site, the batch positions were considered in relation to the main locations where conventional construction material could be acquired, including delivery. The District of Nova Itamarati, a cluster formed along the BR164, is located in front of the settlements was considered the main place to find shops specialized in construction material, from its informal formation. Second was the city of Ponta Porã and its Paraguayan neighbor, Pedro Juan Caballero, both approximately 45 km, by the State Highway MS-164 that crosses the studied area. The hypothesis to be proven is that as a first decision, due to the necessity and urgency of constructive material, it would be possible to acquire the material close to the lot, which would prioritize the Nova Itamarati District, which began to be organized in 2000, but the unavailability of buying with option of delivery, the option to collect non-conventional material would be reinforced. To ascertain this hypothesis, the research established 4 sampling points in relation to the distance from the New Itamarati District, obtaining the points (A) from 8 to 14 km from the District, towards the Maracaju (B) exit from 14 to 20 km in direction to the exit to Ponta Porã, (C) up to 8km from the Nova Itamarati District (D) 8 to 14 km distance from the District, towards the Dourados exit, indicated in Figure 3 the sampling points in Itamarati I and II settlement for the visits to the lots necessary to investigate the hypothesis of the influence of spatiality on the use of unconventional material in the main residence and in the other buildings

FIGURE 3: SAMPLING POINTS IN THE ITAMARATI I AND II SETTLEMENT ESTABLISHED FOR VISITS TO THE LOTS NECESSARY TO INVESTIGATE THE HYPOTHESIS OF THE INFLUENCE OF SPATIALITY ON THE USE OF UNCONVENTIONAL MATERIAL IN THE MAIN RESIDENCE AND IN OTHER BUILDINGS.

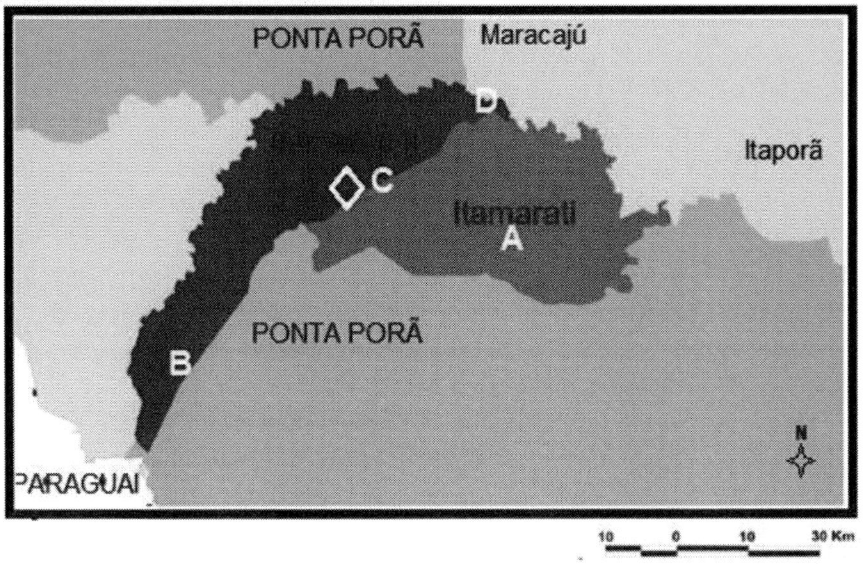

Source: Adapted by the authors (TERRA, [23]).

Legend: Sampling points in relation to the Nova Itamarati district: (A), (B), (C) and (D), marked on Itamarati I and II settlements maps; District Nova Itamarati. Following the established criteria, at each point (A), (B), (C) and (D), 30 lots were selected, in order to obtain a total of 120 lots visited. Sampling included lots in the areas of the 3 social movements and one of the former association of officials of Fazenda Itamarati (AMFFI).

To explain the results obtained, the literature consulted presents some suggestions. The following dimensions were adopted: Spatial (MULE, [27]) represented by the distance of points of sale of conventional building materials; (PENNA et al. [28]), Time of ownership of the rural plot (TERRA, [23]) and level of technical knowledge of the settlers (SCOTT et al. [7]), represented by three variables: (HONADLE [29]); The delay in the execution of the expansion or construction work, represented by the variables delay in the delivery of the materials, insufficient quantity in the work, lack of hand-ofthe- art equipment (GICHUNGE [14]); SILVA [30] and AMADO et al. [6] (Table 1), which is based on the results obtained in this study, and the results obtained in this work are summarized in Table 1, BELLÉ [30]; MULE [27] and knowledge about nonconventional constructive material (BARBOSA [4], GHAVAMI [20]; PICUNO [19]; LAWANIA et al. [32]; VILPOUX, ESTRELLA, CEREDA et al. [26] and FATTAH et al. [33]) One more dimension, the availability of unconventional material was included, to be logical, although no specific references were found. Once the dimensions that could influence the use of non-conventional material in the main residences and other buildings on the site were defined, questions were elaborated that were included in the questionnaire in relation to non-conventional materials. Additional information was obtained in interviews, guided by a script of semi-structured questions. In the interviews with technical personnel of the public agencies it was possible to have access to documents and maps of the researched region. The measurement of the use of non-conventional materials, both in the main residences and in the other buildings was done by direct observation of the buildings, recorded in photographs, followed by the following criterion of 0%, equivalent to only conventional material up to 100%, equivalent to only unconventional material. Once unconventional construction materials were identified and quantified, they were described using information from the literature, as well as the responses of the applied questionnaire, to explain the results obtained.

RESULTS AND DISCUSSION

The region of Itamarati Farm, throughout its history, has undergone the process of redistribution of lands. In the past, part of the territory occupied by the Matte Laranjeira Company became the National Agricultural Colony of Dourados (CAND), which in turn was acquired by Olacyr Moraes in the formation of Fazenda Itamarati (BELLÉ, [31]) It should be noted here that Olacyr de Moraes, in the decade of 1970, was encouraged by the government to acquire the farm, had easy credit and subsidized interest, participating in the process of change occurred in the economic and social relations of production of national agriculture. (ALVES et al. [34]). This fact should be understood in two distinct historical stages, within the context of the Brazilian agrarian reform: the first was marked by the purchase of the Federal Government in 1992, at the average cost of R$ 55 million with the ITAÚ SA group as the main owner of part of the property, Olacyr de Morais, one of the largest landowners in the country as a soy producer. The main objective was the implementation of Itamaraty I settlement, in year of 2000, with an average area of approximately 25 thousand hectares of the total of 50 thousand hectares belonging to the total area of the property, establishing more than 1145 thousand families, under the bases of the old Cartesian process of settlement (INCRA [24]). The second stage of implementation of Itamaraty II occurred only in 2004 until mid-2006, under the new perspective of the Terra Vida settlement program. It had as active members INCRA / MS - National Institute of Colonization of Agrarian Reform / MS, which, next to the sanction of the Federal Government, committed itself to execute one of the largest agrarian reform projects in Latin America, that is, the Settlement Itamaraty II , with the remaining 50% of the territorial acquisition of the property of the former landowner Olacyr, through the record disbursement of the public figures, the amount of R $ 165 million, whose objective and new method of agrarian reorganization had determined the inclusion of more than 1,776 thousand settled families (INCRA [24]). The Itamaraty II Settlement Development Plan followed a different model from that proposed in the area of Itamaraty I - the TERRA VIDA program. It comprised an area belonging to each family was divided into Family Home and Corporate Area. The first would be a small area, where families would develop activities according to their vocation: an orchard, a vegetable garden, a small animal husbandry, and could finance one or more activities by PRONAF, for representing the support program development of the federal government. The other fraction, called the Corporate Area, would serve to develop collective activities, whose objective would be to generate income for the distribution among the families and, also, the accomplishment of activities that could support the production in the familiar places. The corporate area was demarcated around the irrigation pivots (INCRA [24]). By the constructive system adopted by the INCRA, theoretically all the settlers counted on conventional constructive materials sufficient to build their own houses, within the Tecnical brasilian standards. The observations made during the visits in the 120 rural lots (small lands) visited allowed us to state that among the main households sampled, none came to use only conventional or even 65% unconventional material, since the highest percentage of use was 50%. On the other hand, when the residents were considered to be receiving conventional material for the main residence of INCRA, the results showed that there were still 21 residences of the 120 sampled lots that used only conventional material. Under the denomination of other buildings (OE), the buildings in the rural environment were considered to shed or raise domestic animals, vegetable gardens, warehouses of work materials. In the literature, it is possible to find some definitions as Brosler [1] defines in his work the name of ranch as the constructions related to the agricultural production. Kumar et al. [35] defines as an additional requirement for shed animals / chicken coop for rural houses. When the other buildings are included in the analysis, a similar number (20) of constructions is observed only with use of conventional material. However, the use of unconventional material in this type also did not reach 100%, but was higher than for the main residences, with 5 cases of 65% use. The most commonly used non-conventional materials were reuse such as metal tiles for closing walls, shredded tire tiles, reused wooden or metal doors, wooden boards, among others.

Spatial Dimension

The Table 1 below summarizes all the results obtained for the percentage of unconventional materials for main residences and other buildings for the spatiality dimension, adopted as the distance to the Nova Itamarati District.

TABLE 1 - PERCENTAGE OF UNCONVENTIONAL MATERIALS FOR MAIN RESIDENCES (R) AND OTHER BUILDINGS (OE), FOUND IN THE 120 LOTS DRAWN BY SAMPLING POINT IN THE ITAMARATI I AND II SETTLEMENTS, IN RELATION TO THE SAMPLING POINTS ESTABLISHED FOR THE SPATIALITY DIMENSION, SUCH AS DISTANCE TO THE NEW ITAMARATI DISTRICT.

	Quantity (%) of no conventional materials found												
	0		25		35		50		65		100		Total
place	R	OE	R	OE	R	OE	R	OE	R	OE	R	OE	R
A	4	3	15	4	6	13	5	9	0	1	0	0	30
B	9	8	8	2	7	9	6	9	0	2	0	0	30
C	3	5	16	5	9	9	2	10	0	1	0	0	30
D	5	4	7	2	17	15	1	8	0	1	0	0	30
Total	21	20	46	13	39	46	14	36	0	5	0	0	120

Subtitle: R – residence; OE – other buildings (such as production building, storehouse)

The results showed that 25% of use had higher results, due to the great ease of use of untreated eucalyptus, as woodwork in both pillars and beams and in general coverage.

Taking into account the availability of eucalyptus planted in the rural settlements Itamarati I and II. The percentage of use of the materials found in 50% referred to the use of both eucalyptus, reuse materials (doors and windows), as well as canvas, considered in the research as non-conventional material. The unconventional materials were found in garages with tarpaulin and the covered porch with reuse tile. In the study of Mule [27], which correlated locality as one of the influence factors for the use of construction materials and transport costs, in this study, spatial location did not influence the predominance of constructions with conventional materials. The spatial location was not an influence factor for the use of non-conventional materials, however, it was observed that if there is unconventional material available in the place, its use is verified when it is available. As a notable example, in both areas of the settlements, there is eucalyptus plantation and 70% of the lots visited are used. In some areas, such as A, we see bamboo, but in 4 lots visited. Only in 2 lots, bamboo was used in one fence and the other on the roof of a chicken coop. It is possible to observe that, of the 30 lots near the exit to Ponta Porã, in the settlement Itamarati (B), 6 used in 50% the unconventional materials like untreated eucalyptus in the pillars of the balconies and in the extension of the kitchen and reuse materials as the coverage used in the enlarged bathroom, made of tire. In contrast, of the 30 lots located near the exit to Dourados (D), 17 were used 35% of the unconventional materials in the enlarged houses, as the reuse of some materials as, the eucalyptus not treated in the pillars as in the structure balcony and bathroom coverage. On the walls of the kitchen that was enlarged, on one of the walls was used the demolition brick and reuse wood. In the location of the lots at the Itamarati I (A) funds, it was observed that 15 of the rural lots used 25% of nonconventional materials such as untreated eucalyptus pillars on the porch and roof structure, such as beams. Five of the lots used 50% of the materials with predominance of untreated eucalyptus and reuse materials such as tarpaulin on the garage cover and reuse windows, as can be seen in Table 3. None of the lots visited had 100% of unconventional material, but also no residences with 65% of nonconventional material, but considering only the residences only, in comparison a low number was observed, but similar to only conventional material. Profile similar to A and C in residences for 25% of nonconventional material. In the residences located at points A, B and C were similar in 35% of 6 to 9 residences. In counterpoint in the sampling D that presented 17 residences with 35% of nonconventional materials.

According to Mule [27] and Brosler [1], the spatial dimension represents the cost of a transport in greater distance, therefore the user would pay more expensive the freight of conventional material according to the distance of the delivery. Considering this reflection, one would expect a greater number of use of non-conventional materials in the most distant places of the Nova Itamarati District, where it would be possible to buy in houses specialized in construction materials and in the smallest distances, to obtain even the delivery without cost. This dimension would be even more valued, since in Itamarati settlements I and II, due to lack of collective transportation, the main means of locomotion are motorcycles that are not suitable for transportation of heavy materials and in a quantity quantity (NOGUEIRA [31]).

The analysis of the results shows that the spatial dimension, as suggested by Terra [23]; Mule [27] and Brosler [1] explains the higher occurrence of lots with 25% percentages of unconventional material in residences located at the most distant point A but it does not explain the equally large occurrence at point C, nor at point D at 35%.

The results obtained about the other buildings, Brosler [1] defines in her work the name improvement as constructions related to the agricultural production. Kumar et al. [5] (2016) defines as an additional requirement for shed animals / chicken coop for rural houses. In his research he identifies the percentages to which these buildings are twinned in residence or away from the main residence. It concludes that in 62.1% of the cases they had these buildings, attached or not to the main house. However, in 22.1% of the cases, they were twinned to the main house. He stressed that any rural housing policy or program, the Government should provide more assistance to this type of building, since it is an integral part of the main residence. In this present research, when the families were approached about the environment that stayed the most in the house or that they would like to extend always was the balcony. In 64% of the balconies it becomes a work area for families. Brosler [1] corroborates that the ranch, defined in the study, becomes the first construction carried out in the lot. This is replaced by the house - made of wood or masonry - built in another locality. Thus, the function of the ranch is modified, serving as a shed to store agricultural machinery, equipment, tools and supplies, as well as serving as a place to carry out activities related to agricultural production, such as the preparation of agricultural products for commercialization. In contrast to the study by Kumar (2016), Martins (2009) emphasizes that the location of the shed is defined in relation to the place of the house, being close according to the logic of the activities and the intensity of its use.

At the sampling point B, eight buildings (OE) with only conventional material were distinguished from 3 to 5 buildings with only conventional material at points A, C and D. The largest difference occurred at sampling points A and D with 35%, and 13 buildings were found at sampling point A and 15 at sampling point D.

The analysis of the results showed that the spatial dimension, as suggested by Terra [23]; Mule [27] and Brosler [1] explained the higher occurrence of lots with percentages of 35 to 50% of unconventional material in other buildings located in points (A and B).

However, at point D only more than 35% of unconventional material was found in other buildings. For percentages of 50% of unconventional material the points B and C presented similar numbers of 10 and 11 respectively. For 65% of nonconventional material the points A, B, C, D presented similar but smaller profiles (from 1 to 2). It was demonstrated that there was a greater predominance of the use of the nonconventional materials in the improvements. It is concluded that this use originated from the owner, since INCRA does not provide resources for other buildings, except through projects, which has now been more scarce Technical and social capacity of the settlers

In addition to the aspect of the material used in the construction, another modification occurs with respect to the use of the own workmanship in the construction. This activity no longer appears as part of the functions of these farmers, much less through solidarity relations such as the mutirão. When carried out by the farmer himself, this activity is done due to the knowledge he acquired in phases prior to the settlement, such as work with civil construction, in its insertion in the urban environment.

The study of the social capacity of the settlers includes the influence of social movements (SCOTT et al. [7]; PENNA et al. [28]), the period of possession of the rural plot and the level of technical knowledge of the settlers (HONADLE [29]). It was considered that the greater the participation in social movements, the time of possession in the rural lot and the greater technical knowledge would be the percentage of unconventional use in the homes as in other buildings.

The results showed that the percentage of 25% to 50% of the use of non-conventional materials predominated in the families that live in the settlements between 11 and 13 years old. However, a 25% result was found in families who have lived for more than 14 years. The lowest value found was 50% in 4 families that have lived for more than 14 years. While families living in settlements for more than 16 years have excelled in using 25 to 35 percent of unconventional materials. The families that lived less time used a smaller percentage between 5 to 9 families. In particular, 19 families used 25% of non-conventional materials as shown in Table 1.

Honadle [29] proposes that public administration disclose and release the necessary funds to stimulate development in rural areas and suggests that Rural Development is not independent within the National Department of Agriculture but be included in major health, education, housing, environment and other substantive areas to meet the real needs of the rural community through information on the use of non-conventional materials. According to Terra [23]; Mule [27] and Brosler [1], the spatial dimension represents the cost of a transport in greater distance, therefore the user would pay more expensive the freight of conventional material according to the distance of the delivery.

Considering this reflection, one would expect a greater number of use of non-conventional materials in the most distant places of the Nova Itamarati district, where it would be possible to buy in houses specialized in construction materials and in the smallest distances, to obtain even the delivery without cost. This dimension would be even more valued, since in Itamarati settlements I and II, due to lack of collective transportation, the main means of locomotion are motorcycles that are not suitable for transporting heavy materials and in a quantity quantity (NOGUEIRA [36]).

The analysis of the results shows that the spatial dimension, as suggested by Terra [23], Mule (2012) and Brosler [1] explains the higher occurrence of lots with 25% percentages of unconventional material in residences located at the most distant point A , but it does not explain the equally large occurrence at point C, nor at point D at 35%.

Technical and social capacity of the settlers

In addition to the aspect of the material used in the construction, another modification occurs with respect to the use of the own workmanship in the construction. This activity no longer appears as part of the functions of these farmers, much less through solidarity relations such as the mutirão. When carried out by the farmer himself, this activity is done due to the knowledge he acquired in phases prior to the settlement, such as work with civil construction, in its insertion in the urban environment.

The study of the social capacity of the settlers includes the influence of social movements (SCOTT et al. [7]; PENNA et al. [28]), the period of possession of the rural plot and the level of technical knowledge of the settlers (HONADLE [29]). It was considered that the greater the participation in social movements, the time of possession in the rural lot and the greater technical knowledge would be the percentage of unconventional use in the homes as in other buildings.

The results showed that the percentage of 25% to 50% of the use of non-conventional materials predominated in the families that live in the settlements between 11 and 13 years old. However, a 25% result was found in families who have lived for more than 14 years. The lowest value found was 50% in 4 families that have lived for more than 14 years. While families living in settlements for more than 16 years have excelled in using 25 to 35 percent of unconventional materials. The families that lived less time used a smaller percentage between 5 to 9 families. In particular, 19 families used 25% of non-conventional materials. Honadle [29] proposes that public administration disclose and release the necessary funds to stimulate development in rural areas and suggests that Rural Development is not independent within the Department of National Agriculture but is included in major health, education, housing, environment and other substantive areas to meet the real needs of the rural community through information on the use of non-conventional materials.

Characteristics of the main rural building

Through the analysis of the constructive materials and their relations with the way of life of these farmers, in this study, we analyzed new conceptions of the rural dwelling allied to the use of the nonconventional materials, to the cultural knowledge brought by the different origins of the settled families. Through the conquest of the land, the return of the family to the rural area also brings fragments of a culture external to its tradition of family farmer, represented in the reconstruction of its reproduced living space to the family (BROSLER [1]). The rural house, therefore, depicts the resistances and transformations caused before the exterior. Through the situation of the house, its construction material, its aesthetics and functionality, it is possible to identify important aspects to reflect on what is currently the rural house, what its characteristics, what has changed and where to go. Its projection in space and time in the consolidation of the house, in the process of conquering the house itself after the conquest of the land, can be analyzed in the interweaving of the present with the past and with the future pretensions (BROSLER et al. [1]).

The project of the houses was adapted to the familiar habits in which a central patio is integrated where the activities and the workshops are realized. Extensions consist of adding dormitories around, usually after each child gets married. The structures of the houses served three sectors: social, economic and environment. Many meetings and workshops were held with the local community to obtain indicators and parameters related to lifestyle, family income, bioclimatic characteristics to ensure thermal comfort in homes. The author suggests that a post-construction survey be conducted to ascertain the effective success of the program (AMADO et al. [6]). Of the 120 families, 36.7% answered that they spend most of their time inside the house, and asked which environments are the most answered: 30.8% stay in the kitchen, 25% stay on the porch, 13.3 stay in the living room, 7.5% stay in the rooms and 23.3% stay out of the house. The environments that most would like to expand are: 62 families responded to the balcony; 20 families gave preference to the kitchen; 16 chose the room; 15 the bathroom and 7 families the living room.

In contrast, of the 30 families located near the New District of Itamarati, D, 17 used 35% of the unconventional materials in the extended dwellings, such as the reuse of some materials such as eucalyptus not treated in the pillars as in the covering structure the balcony and bathroom. On the walls of the kitchen that was enlarged, on one of the walls was used the demolition brick and reuse wood.

In the location of the lots at the Itamarati I, Point A funds, they showed that 15 of the rural lots located in the Itamarati funds used 25% of unconventional materials such as untreated eucalyptus pillars on the porch and roof structure, such as beams. Of the 5 lots used 50% of the materials with predominance of untreated eucalyptus and reuse materials such as tarpaulin on the garage cover and windows of reuse wood, as shown in the graph.

The spatial location B, intermediate distance in relation to the BR164, towards the exit to Ponta Porã exit, was characterized by the use of materials in 25% and 35% obtained results as 7 lots, such as the use of untreated eucalyptus materials in covering structures for pillars and internal beams in kitchen and bedroom environments, as well as reuse of demolition doors and solid bricks for masonry closing in the kitchen and wood burning stove, as shown in Table 1.

Spatial location C - Data obtained show that 16 of the 30 families used 25% of conventional materials in the dwellings. The most commonly used materials were untreated eucalyptus wood on the pillars and beams of the balconies, reuse materials such as wooden boards to close the walls and demolition bricks on the walls on the balconies.

Other Buildings

In the analysis of this dimension were not included the buildings coming from resources of projects elaborated by the INCRA, like for example production of the silkworm. In order to better contextualize, INCRA implements production projects within family agriculture, through specific Programs such as the National Program for Strengthening Family Agriculture (PRONAF). There are many types of PRONAF (s), such as: Forestry; Agroecology; Eco; Eco / Dendê / Rubber; Woman; Young; Semi-arid; More Food; Agribusiness; Costing; Landfill (Banco do Brasil, 2019). To do so, the resource is available for these types of buildings, coming from Banco do Brasil. In this study, it was defined as other buildings (OE), used for the production of family agriculture, which were built by the family settled.

Continuing the analyzes within this variable, we anchor some definitions in the researched literature. The Brazilian Standard (NBR) 146531, defines as necessary building, being the result of work or service performed in a good and that can not be removed without destruction, fracture or damages. In this norm is defined the good as something that has value, susceptible of use or that can be object of right, that integrates a patrimony. According to Brosler [1] the name improvement or ranch is cited as constructions related to agricultural production. Kumar et al. [35] defines as an additional requirement for shed animals / chicken coop for rural houses. In his research he identifies the percentages to which these buildings are twinned in residence or away from the presidential residence. It concludes that in 62.1% of the cases they had these buildings, geminada or not to the main house. However, in 22.1% of the cases, they were twinned to the main house. He stressed that any policy or program of rural housing, the Government should provide more assistance to this type of building, as it is an integral part of the main residence.

In this present research, when the families were approached about the environment that stayed the most in the house or that they would like to extend always was the balcony. In

64% of the balconies it becomes a work area for the families, as another home space, these reports agree with Kumar et al. [35].

In most of the houses visited, 48%, it was observed that the other buildings served primarily as the temporary construction of the main residence for the initial installation of the family to start the production of family agriculture. The most used materials are the less conventional ones as reuse materials: metal roofing, improvised wood, roofing with tile reuse. These results agree with Brosler [1], in which he emphasized that the ranch becomes the first construction carried out in the lot by the newly installed family in its individual lot. This is replaced by the house - made of wood or masonry - built in another locality. Thus, the function of the ranch is modified, serving as a shed to store agricultural machinery, equipment, tools and supplies, as well as serving as a place to carry out activities related to agricultural production, such as the preparation of agricultural products for commercialization. In contrast to the study by Kumar [35]; Martins [37] emphasizes that the location of the shed is defined in relation to the place of the house, being close according to the logic of the activities and the intensity of its use.

CONCLUSIONS

The present study showed that the adopted methodology allowed the analysis of the dimensions based on the researched literature.

In the Itamaraty I settlement, both materials and construction processes used with nonconventional materials (masonry, adobe, bamboo and bamboo) were not adopted since they received from INCRA conventional materials for the rural dwellings in the lots As for the other buildings, for which they did not receive materials or even resources for their acquisition, it was possible to note the presence of unconventional material such as eucalyptus, reused material and even bamboo in chicken fences, taking into account that INCRA provides resources primarily for family housing and for other buildings are few projects for buildings related to the production of family agriculture.

This study brought a reflection on the transformations arising from a process of ruralurban-rural mobility, registered from the settlers of agrarian reform. These are new conceptions regarding rural dwelling, observed through the visits to the 120 plots of the Itamarati I settlement, in the municipality of Ponta-Porã / MS, and the analysis of the answers obtained in the interviews with the families living there, focusing on the unconventional materials used in constructions and constructive dwellings and their relations with the way of life of these farmers. In this sense, the aim was to contribute to the improvement of techniques for the use of non-conventional materials, promoting use initiatives, such as the adoption of adobe brick, industrial waste (such as ceramic waste) and others; in the construction of housing that meet structural and aesthetic requirements. The environmental importance becomes relevant, either by the reduction of industrial waste or by the reuse of waste from the industries producing red ceramics (OLIVEIRA [17]. The adoption of other standards in the construction of the settlers' housing was observed, which is reinforced not only by their passage through urban space, but, mainly, by the industrialization process that intensified in the civil construction market, disseminated by the means of mass communication. The traditional techniques of construction are diversified and known by the settlers, knowledge that serves only as an alternative in a situation of extreme necessity (SILVA [10]).

The renunciation of the use of these unconventional techniques is evidenced by the common desire, by the settled individual, to be able to construct or finish a masonry house with its own and government resources. However, it was possible to perceive that, even if masonry construction proves to be overwhelming, the seated subject, in general, manifests remnants of the elements that constitute a way of living and being in the field, observable in the provisional character of the construction of his House. The study sought to establish possible relationships between constructions, possibilities and techniques for the use of building materials, spatial location of lots, the various social movements represented there, the type of resource received by the funding agencies, and the socioeconomic and cultural factors that led families to choose by certain type of material used.

The results showed that there was no significant interference related to the location of the lots and the use of non conventional and conventional materials, however, revealed the need for actions that clarify and make feasible the adoption of non conventional materials in the construction practices, in order to expedite the obtaining and occupation by the families of the dwellings in their individual lots. It is explained the difficulty to implement the use of non-conventional materials also by prejudice. Some materials, such as earth, bamboo, straw and vegetable fibers, are rooted in the minds of the vast majority of people as "poor material." This prejudice was perceived in the research results of Vilpoux, Estrella and Cereda [26] that resulted in the settler considering the use of bamboo in the rural residence as cheap and not durable. In addition, the data obtained point to the need for the creation of new, more agile and speedy administrative instruments, from the public bodies managers and financiers, in order to provide decent housing, with conditions of thermal comfort, that meets the technical (architectural and complementary) projects that have been previously elaborated and followed during the construction stages.

In addition, actions to encourage the use of bio-construction techniques or the use of non-conventional materials in areas of agrarian reform settlements, with specific resource lines for rural housing made with non-conventional materials, are considered necessary. Technical Standards.

REFERENCES

1. BROSLER, T. M. Materiais não convencionais na construção civil: presente, passado e futuro no processo de conhecimento dos assentados de Mogi Mirim-SP. Campinas, SP: [s.n.], 2011.

2. DAMATTA, R. A casa e a rua. Espaço, cidadania, mulher e morte no Brasil. 4 ed. Rio de Janeiro: Guanabara Koogan, 1991.

3. BROSLER, T. M.; BERGAMASCO, S.M.P.P. Novas concepções da moradia no meio rural: uma reflexão a partir de assentamentos rurais no estado de São Paulo. Universidade Estadual de Campinas-SP. 2016.

4. BARBOSA, N. P. Considerações sobre materiais de construção convencionais e não convencionais. Universidade Federal da Paraíba. 2015.

5. CASAGRANDE JR, E. F. Princípios e Parâmetros para a Construção Sustentável. Universidade Feevale. Novo Hamburgo, RS, 2009. Available in em: http://www.cbcs.org.br/. Acesso em: 11 ago. 2009.

6. AMADO, P. M. RAMALHETE, I. AMADO, A. R.; FREITAS, J. C Frontiers of Architectural research. Vol. 6. Issue 1. March, 2017. Pp 74-88.

7. SCOTT, M.; MURRARY, M. Housing Rural Communities: Connecting Rural Dwellings to Rural Development in Ireland. Journal Housing Studies. 2009

8. SINDUSCOM/MG Sindicato da Indústria da Construção Civil no Estado de Minas Gerais. Catálogo de Normas Técnicas - Edificações. Belo Horizonte: Sinduscon-MG/CBIC, 2017.

9. RADONJANINI, V.; MALESEV, M.; MARINKOVIC, S.; MALTY, A. Green recycled aggregate concrete. Construction and Building Materials, Vol. 47; pp 1503-1511, 2013.

10. SILVA, V.G. Avaliação do desempenho ambiental de edifícios: qualidade na construção. n. 25. São Paulo: SindusCon, 2000.

11. BAYER, A. P. Proposta de Diretrizes para o Desenvolvimento da Arquitetura em Terra no Rio Grande do Sul, a partir da Interpretação de Estratégias Uruguaias. 2010. Dissertação (Mestrado em Engenharia Civil) Programa de Pós-Graduação em Engenharia Civil, UFRGS, Porto Alegre, 2010.

12. UGOCHUKWU, I. B.; IWUAGWU, B.C. Local building materials: affordable strategy for housing the Urban poor in Nigeria. Procedia Engineering Journal. International Conference on Sustainable Design, Engineering and Construction- ICSDEC 2015. Elsevier, Chicago, USA. p. 42-49.

13. UDDIN, N. A study on the traditional housing technology of Bangladesh. University of Chittagong, Bangladesh. Indian Journal of Tradicional Knowledge. Vol. 7(3). July, 2008.

14. GICHUNGE, H. Factors that contribute to the cost of provisions of low cost housing in Nairobi, Kenya. p12. Dr. H. GICHUNGE, Kenya. International Conference on Spaatial Information for Sustainable Development Nairobi, Kenya. 2-5 october, 2001.

15. ARAÚJO, H. G. Manualização de construções em adobe. UFC - Universidade Federal do Ceará. Centro de tecnologia, Departamento de engenharia estrutural e construção civil. Fortaleza, 2009.

17. OLIVEIRA, L. S.; BARBOSA, N. P. Estabilização de blocos de terra crua através de ativação alcalina. (Dissertação de mestrado). Universidade de João Pessoa. Paraíba.83 p.

18. VERDE GHAIA. Consultoria online. Disponível em: https://www.consultoriaiso.org/nbrs-regulam-materiais-na-construcao-civil. Acesso em: 4 de dez. 2018.

19. PICUNO, P. Use of traditional materials in farm buildings for a sustainable rural environmental. University of Basilicata. Potenza, It. International Journal of Sustainable Built Environment. Volume 5; May, 2016 pp 451-460.

20. GHAVAMI, K.; BARBOSA, P. N.; MOREIRA, L. E. Bambu como Material de Engenharia. In: Avaliação de Desempenho de Tecnologias Construtivas Inovadoras: Conforto Ambiental, Durabilidade e Pós Ocupação. 2017. p. 305-348.

21. FRAANJE, P. J. Building with nature making use of renewable resources, grown structures and growing constructions. In: International Symposium Beyond Sustainability, Eindhoven University of Technology, Netherlands, 28/29 Sptember, 2000.

22. GHAVAMI, K. Materiais e Tecnologias não Convencionais para o Século XXI. Pontifícia Universidade Católica do Rio de Janeiro, RJ, Brasil. 2014.

23. TERRA, A. Reforma agrária por conveniência e/ou por pressão? Assentamento Itamaraty em Ponta Porã – MS: "o pivô da questão". 2009. Tese (doutorado) - Universidade Estadual Paulista, Faculdade de Ciências e Tecnologia (UNESP). SP: Presidente Prudente, 2009.

24. ZENID, G. J. A madeira na construção civil. Instituto de Pesquisas Tecnológicas do Estado de São Paulo S.A.-I PT. São Paulo.

25. CAIADO, A. R. Contribuição ao Estudo da Rotulagem Ambiental dos Materiais de Construção Civil. Tese (Doutorado). Universidade de São Paulo/USP. 2014.

26. VILPOUX, O.; ESTRELLA, G. S.; CEREDA, M. P.; FOIZER, G. Percepção de assentados no Mato Grosso do Sul em relação às moradias de bambu. In: DRUMOND, P. M; WIEDMAN, G (Orgs.). Bambus no Brasil: da biologia à tecnologia. 1. ed. Rio de Janeiro: ICH, 2017.p. 335-348.

27. MULE, B. M. Factors affecting the adoption of the appropriate building materials and Technologies. (ABM&Ts) Programme in North Eastern Province of Kenya. University of Nairobi.

28. PENNA, C. ROSA, M. C. Estado, movimentos e reforma agrária no Brasil: reflexões a partir do INCRA. Universidade de Brasília (UnB), Brasília. DF. Lua Nova, São Paulo, 95: 57-85, 2015.

29. HONADLE, B. W. Rural development policy in the United States: a critical analysis and lessons from the "still birth" of the rural collaborative investment program. Vol. 42, 2011. p 56-69.

30. SILVA, C. M. G. da. Habitação Rural: uma luta pela cidadania. 2014. 152f. Dissertação (Mestrado). Universidade de São Paulo – Faculdade de Arquitetura e Urbanismo, São Paulo, 2014.

31. BELLÉ, M. de F.L. Impactos sobre o desenvolvimento regional decorrentes do assentamento Itamarati, Ponta Porá (MS): 2001-2010. 149f. Dissertação (mestrado) – Universidade Anhanguera - Uniderp. Campo Grande, 2011.

32. LAWANIA, K. K., BISWAS, W. K. Achieving environmentally friendly building envelope for Western Australia's housing sector: A life cycle assessment approach. International Journal of Sustainable Built Environment. Sustainable Engineering Group, Curtin University, Australia. Vol. 5(3). April, 2016, pp 210-224.

33. FATTAH, A. K; AL-TAMIMI, K.; HAMWEYAH, W.; IQBAL, F. Evaluation of sustainable concrete produced with desalinated reject brine. Department of Civil Engineering, College of Engineering American University of Sharjah, United Arabian Emirates. International Journal of Sustainable Built Environment. 2017.

34. ALVES, G. L.; BELLÉ, M. F. L.; SOUZA, C. C. Impactos sobre o desenvolvimento regional decorrentes do Assentamento Itamaraty, Ponta Porã (MS): 2001 -2010. Albuquerque: revista de História, v. 5 n. 9 p. 9-33, jan./jun. 2013, MS: Campo Grande.

35. KUMAR, A.; DEKA, A.; SINHA, R. Rural Housing in India: Status and Policy Challenges. Lokashraya Foundation. Editorial Board: Prof. Ashok K. Pankaj; Shri R.N.P. Singh. New Dehli, 2016.

36. NOGUEIRA, J. T. Fronteiriços: Diferentes territorialidades no assentamento Itamarati I–MS. 2011. 137f. Dissertação de mestrado. Universidade Federal da Grande Dourados. Mato Grosso do Sul. 2011.

37. NOGUEIRA, J. T. Fronteiriços: Diferentes territorialidades no assentamento Itamarati I–MS. 2011. 137f. Dissertação de mestrado. Universidade Federal da Grande Dourados. Mato Grosso do Sul. 2011.

IC-NOCMAT 2019

18th International Conference on
Non-Conventional Materials and Technologies
"Construction Materials & Technologies for Sustainability"
(18th NOCMAT 2019)
24th – 26th July 2019
Nairobi, Kenya

SOIL BUILDING BLOCKS STABILIZED WITH WHEAT STRAW AND CEMENT

Benhaoua, W.[1], Grine, K.[1], Kenai, S.[1]

[1] Geomaterials and civil engineering laboratory, University of Blida1, Algeria;

ABSTRACT

Stabilized earth blocs have been used in many countries as an alternative local construction materials with environmental and economic advantages. However, durability especially under humid environments could hinder their use. Stabilization using cement and natural fibres could enhance their durability and risk of cracking. The present paper deals with a laboratory investigation into the effect of adding a proportion of cement and/or naturel fibres (wheat fibres), on the behaviour of a local silty soil under compression, wetting/drying cycles, shrinkage and upon its aptitude to water absorption (capillary absorption) and thermal conductivity. Unconfined compressive strength (UCS), shrinkage, wetting and drying, capillary absorption and thermal conductivity tests were performed on both, untreated soil samples and soil samples stabilised with cement and/or wheat fibres. The results show that the cement has more effect on increasing the strength, the durability, the thermal conductivity and decreasing the capillary absorption and the shrinkage. However, when mixed with cement, the natural fibres lead to a decrease in the thermal conductivity, an increase in the capillary absorption and to a further reduction in the shrinkage.

Keywords: soil, wheat fibers, cement, durability, thermal conductivity.

INTRODUCTION

The world's population living in houses made of earth material is estimated at 30% [1]. The great advantage of using earth as a building material is its availability and low environmental impact. Earth material manufacture needs little energy and hence its use lowers cost as compared to conventional products such as cement and fired brick. From a technical point of view, traditional earthen constructions have particularly unfavourable characteristics like the low resistance to compression and lack of durability. Several methods have been developed in order to eliminate these disadvantages. Soil stabilization using chemical additives such as cement and lime alone or combined to a reinforcing materials such as natural fibres are the more common methods used to improve earth materials durability. Various research have been conducted to study the engineering properties and durability using chemical treatment [2, 3]. What has been reported from these studies is that addition of cement increases the compressive strength and the resistance to the wetting/drying cycles (weight loss) and decreases the capillary water absorption and the shrinkage of the stabilised materiel [2-7]. However, the thermal conductivity increases with increasing cement content [8, 9]. Extensive studies have been carried out on the utilisation of natural and industrial fibres in soil stabilisation. In terms of durability it has been reported that the total absorption and the swelling, increase with increasing the fibres content. However, the thermal conductivity and the shrinkage of the stabilised block decrease [10-15].

This paper describes an investigation into the effect of adding a proportion of cement and/or natural fibres (wheat fibres) to a silty soil on compressive strength and durability, in particular the wetting/and drying cycle and capillary water absorption test. Particular attention is given to the reduction of shrinkage and thermal conductivity. The aim is to find a mean of stabilising a local silty soil to be used as a block building material.

EXPERIMENTAL INVESTIGATIONS

Materials

The soil used in this study is a silty soil obtained from a site located at 30 km west of the city of Algiers. This soil was extracted at a depth of about 6 m. The physical and mechanical properties of the soil obtained from laboratory tests are given in Table 1.

The cement used in this study is a composite Portland cement type CEMIIA 42.5.

The fibres used in this study are natural commercially procured fibres (wheat straws). They have a diameter of 1 to 4 mm and a length of 10 to 40 mm.

TABLE 1 – Physical and mechanicals properties of the soil

Basic characteristics	**soil**
Color	red
Depth (m)	6
Sand (%)	30
Silt (%)	45
Clay (%)	25
Liquid limit (%)	30.2
Plastic limit (%)	16.9
Plasticity index (%)	13.3
Optimum water content (%)	10
Maximum dry density (kN/m^3)	19.8
Unconfined compressive strength (kPa)	372

Laboratory tests procedures

A series of laboratory tests of unconfined compressive strength, wetting/drying cycles, shrinkage, water absorption (capillary absorption) and thermal conductivity were conducted. Extensive combinations of cement and/or natural fibres (wheat fibres) were used for stabilization of the silty soil. The percentages by dry weight of wheat straw were 0%, 0.5%, 0.75% and 1% while those of the cement were 0%, 4%, and 8%. Specimens were prepared by compacting virgin and mixed samples in a Proctor mould at the modified optimal (OPM) water content, extruded in a standardised mould, covered in plastic film sheet to prevent moisture changes and left to cure respectively for 7 and 28 days before testing.

Unconfined compressive strength

The unconfined compression test was performed in accordance with the French standard (NF P94-077), using the triaxial apparatus. Three tests on a 38 in diameter and 78mm in length specimen were performed for each virgin and stabilised samples. The unconfined compressive strength value retained is the average of the three.

Wetting and drying test

The test consists in applying saturation and drying cycles to the prepared samples according to the American standard (ASTM D 559), after a curing period of 7 and 28 days. Each cycle consists in immersing all the specimens in water for a period of 5 hours then removing and drying them in an oven at a temperature of 71oC for a period of 42 hours. This is followed by a weighing operation of the samples to assess any loss in mass. A total of 12 cycles were performed for each sample in order to complete the test.

Capillary water absorption test

Capillary absorption test was conducted according to the French standard (XP P 13-901).The test consists in immersing the lower part of the cylindrical specimens at a 5 mm constant headwater tank for a period of time (t). The water absorption coefficient (Ca) corresponds to the rate of absorption after 10 min. It is given by the following formulae

$$Ca = \frac{W}{A * t^{0.5}}$$

With (W) is the mass of water in grams, absorbed by the block during the test; (S) the immersed area in centimetre squared and (t) the time of immersion of the block in minutes.

Thermal conductivity

The thermal conductivity (λ) was measured using the commercial CT-meter device (Fig. 1). The apparatus consists of a probe containing a resistive wire inserted between two samples. The samples were subjected to a heat flux delivered in the form of Joule effect by the resistive wire. The evolution of temperature measured in the course of time by a thermocouple, also contained in the probe at a known distance from the hot wire, makes it possible to determine the thermal conductivity of the material.

Fig.1. CT-meter device.

Shrinkage

The shrinkage was obtained following the method given in the French standard (XP P 13-901).The test consists in following the dimensional variation of the earth block sample due to water evaporation with time.

RESULTS AND DISCUSSION

Unconfined compressive strength

As can be seen from figure 2, an increase in in the unconfined compressive strength (UCS) value is clearly obvious with increasing the percentage of cement within the soil sample for both periods of cure, 7 and 28 days. However, higher values in strength are obtained with increasing the curing time. These results agree with previous studies reported in the literature by different researchers concerning soils treated with cement [16 -18].This increase in strength is mainly attributed to the active components of the Portland cement. Indeed, by hydration of the Portland cement, active compounds such as (tricalcium silicate, bicalcium silicate and tricalcium aluminate) crystallize, leading in creating bonds between the soil particles and hence transforming its structure to harder material. Whereas, when adding wheat fibres to the soil, a small increase in strength compared to the one obtained by adding Portland cement is obtained by the samples mixed with a proportion of 0.5 %. However, this gain decreases with increasing the proportion of the fibres from 0.5 % to 0.75% and up to 1%. This trend in the strength values is obvious for both cure periods. One the other hand, when mixing together a proportion by weight of cement (4% and 8%) and fibre wheat (0.5%, 0.75% and 1%) to the soil, the results show that no gain in strength is obtained. Indeed, results from figure 2 show that the unconfined compressive strength values are almost the same for samples stabilised with 4% of cement only and those mixed with both 4% of cement and wheat fibres at 7 and 28 days cure. Similar trend in the strength values is shown by the samples stabilised with 8% of cement and those mixed with a proportion of wheat fibres of 0.5, 0.75 and 1% for both curing periods. The gain in strength is mainly due to the action of cement.

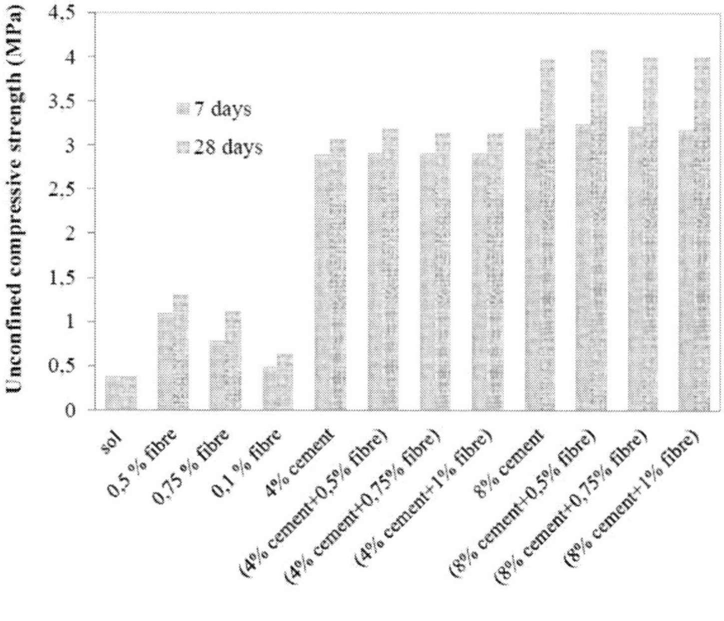

Fig.2. Influence of stabilizers on unconfined compressive strength.

Wetting and Drying Test

The results of the wetting-drying test are shown in Figure 3. The results show that samples stabilised with a proportion by dry weight of cement and fibres survived the 12 cycles despite the aggressiveness of this test. What can be highlighted during the realisation of the test is that, the virgin samples and the ones reinforced with the fibres alone did not complete any cycle. They collapsed just after the first saturation phase, before the completion of the 1st cycle.

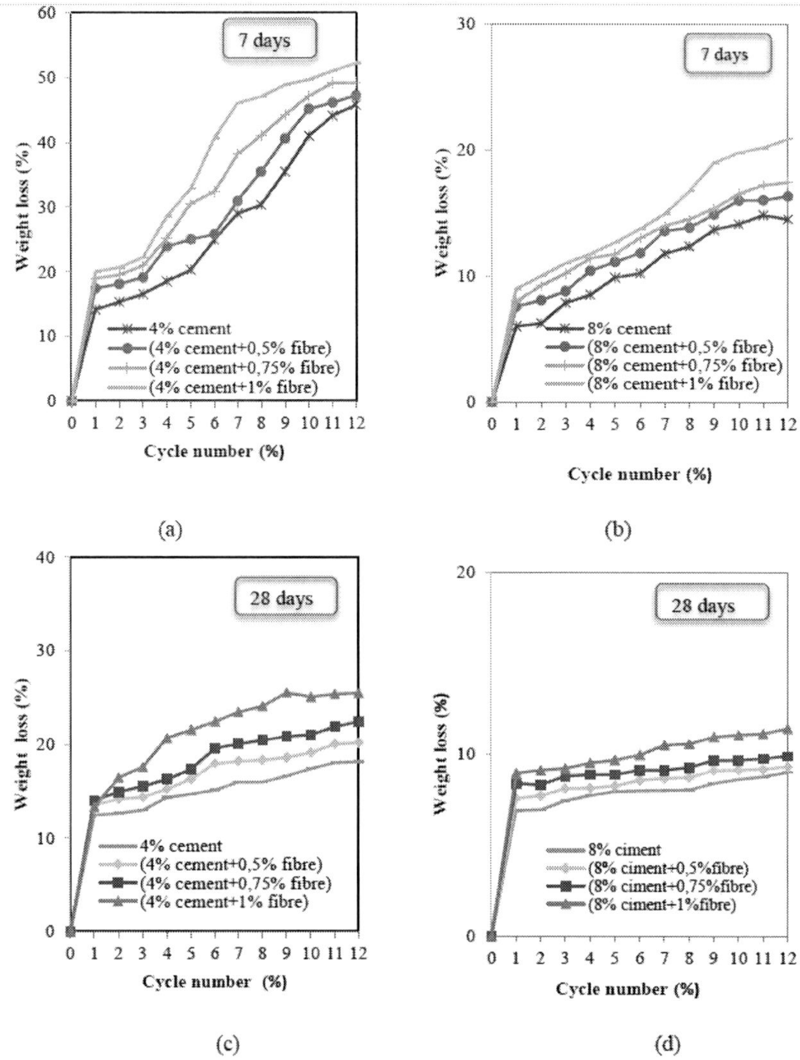

Fig.3. Loss of mass (wetting/drying test) of block soil as a function of cement and fibres.

As can be seen from (Figure 3 (a, b)), an increase in the weight loss with increasing the cycle's number is well obvious for samples stabilised with cement and those mixed with cement/and fibres wheat cured for 7 days. For both samples, cemented at (4%, 8 %) and those cemented at (4 %, 8%) and mixed with 0.5, 0.75 and 1% of fibres wheat, there is a sharp increase up to a certain value, followed by a progressive increase in the weight loss values with incising the percentage by weight of the fibres and the cycle's number. However, lower values in the weight loss are shown by the samples stabilised with 8 % of cement and the ones mixed with different proportion of fibres, (Figure 3, b). On the other hand, with increasing the cure period, for both samples cemented at (4 %, 8 %) and those cemented and mixed with different proportion of fibres, a further decrease in the weight loss with increasing cycle's number and the percentage of fibres is observed. In addition, the sharp increase part of the curves values are followed by a progressive increase tending towards an almost constant value in masse loss with increasing the percentage the fibres and the cycle's number, (Figure 3 (c,d)). Indeed, this trend is more pronounced with samples cemented at 8 % cement and mixed with different proportion of fibres, (Figure 3,d). The values of the weight loss for all the samples are less than 10 %. Samples stabilised with cement alone show more resistance to wetting/a drying cycles. Addition of fibres to the cement stabilized samples has a detrimental effect on weight loss.

Capillary water absorption

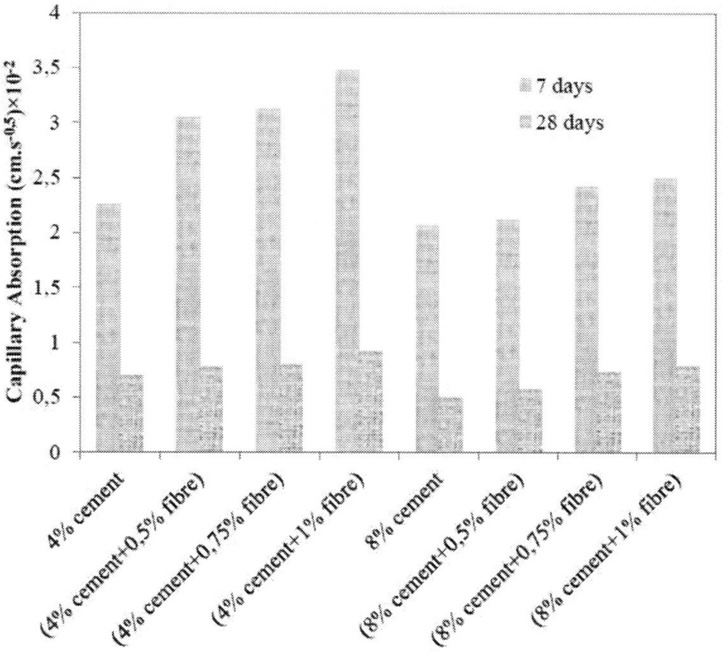

Fig.4. Capillary water absorption of block soil as a function of cement and fibres.

The effect of change in the cement and fibres content on the capillary absorption of the soil block is illustrated in Figure.4. The results show that the capillary absorption decreases significantly with increasing cement content. This, is due to the formation of cementation compounds as a result of the cement hydration, filling the voids of the soil and thus obstructing the water flow. However, the addition of wheat fibres increases the capillary absorption compared to the samples stabilized with cement alone. This behaviour can be explained by the fact that the addition of fibres generates voids in the soil-cement matrix which facilitates water penetrating the sample and by the higher water absorption capacity of the fibres. Generally, the capillary absorption decreases significantly with increasing cure period. Indeed, after 7 days cure, the silty soil stabilised with 8% of cement show an absorption value around 2.06%. At 28 days, the value is around 0.5% corresponding to a reduction of about 75%. This reduction in water absorption at 28 days can be attributed to the development of the cement hydration reaction with increasing times.

Thermal conductivity

Figure 5 illustrates the evolution of the thermal conductivity of the soil blocks with the cement content and the percentage of the wheat fibres. The results show that the thermal conductivity of the soil-cement blocks increases with increasing cement content. These results can be explained by the fact that the addition of cement brings fines to the initial material that clog the pores created at the proximity of large particles. The material obtained after mixing is less porous than the original material. The cement does not only stabilize the material but reduces its pores. These results agree with the ones reported in the literature [8, 9, 19]. In addition, the results obtained in this study indicate that thermal conductivity decreased with increasing fibres content within the cemented samples. The presence of fibres in the matrix generates voids, which leads to an increase in porosity and a decrease in density. The lower conductivity makes the material suitable insulation material for walls especially in hot climate regions to reduce energy consumption for cooling.

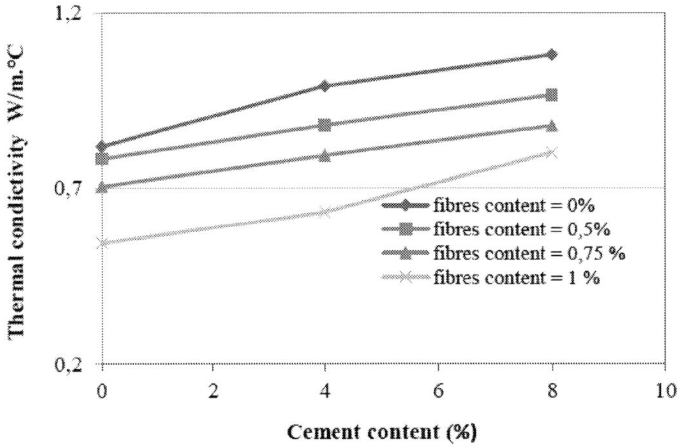

Fig.5. Thermal conductivity of block soil as a function of cement and fibres.

Shrinkage

Figure 6, illustrates the effect of adding cement/and fibres on shrinkage. The results show that the shrinkage increases as drying increases with time for all samples. This increase in shrinkage is more pronounced in the first four days then tends to stabilise to become constant after 14 days. What can be also noticed, is that the shrinkage decreases with increasing both the proportion of cement/and wheat fibres. However mixing together a proportion of cement with a proportion of fibres wheat, this lead to a further reduction with increasing the percentage of the fibres within the mixture. This fact is more attributed to the action of the fibres which reduce the shrinkage as the fibres tend to oppose the deformation [15, 19]. The addition of (0.75% fibres and 8% cement) decreased the shrinkage of a stabilised block by a proportion of 61.91% compared a proportion of 17.32% for a stabilised block using cement alone. The reduced shrinkage lowers the risk for cracking for blocks or rendering made of cemented soil with wheat fibres.

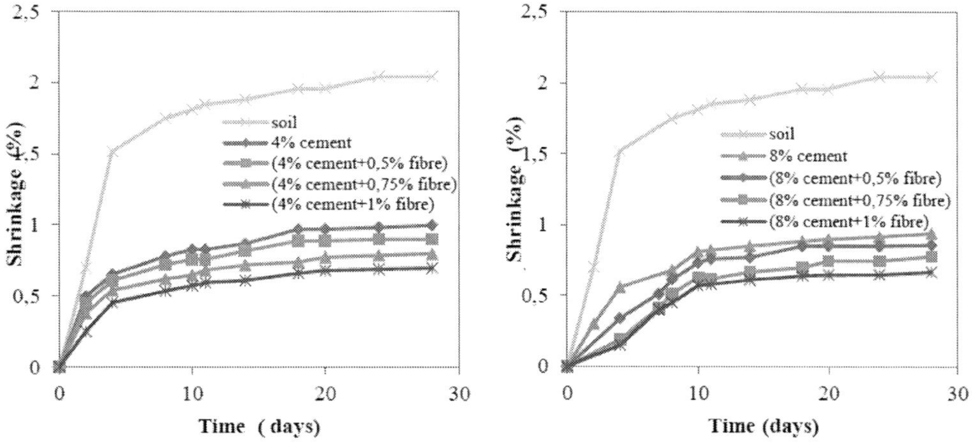

Fig .6. Shrinkage of earth blocks as a function of cement and fibres.

CONCLUSIONS

This paper aims to evaluate through a laboratory study the effect of adding a proportion of cement and/or natural fibres (wheat fibres), on the behaviour of a local silty soil under compression, wetting/drying cycles, shrinkage and upon its aptitude to water absorption (capillary absorption) and thermal conductivity. Based on the results obtained, the following conclusions can be drawn:

- The unconfined compressive strength increases with increasing cement contents. Moreover, the addition of fibres has no effect on the strength of the cement stabilized blocks. The gain in strength is more attributed to the action of the cement rather than to the fibres.
- Addition of a proportion by dry weight of cement to the silty soil significantly improves the durability in term of capillary absorption, wetting/drying at 7 and 28 days. However, addition of wheat fibres to cement-sol mixture has a negative effect as it increases weight loss and capillary absorption.
- The addition of cement increases the thermal conductivity. However the addition of fibres shows an opposite effect and could be used a good insulationg material to save energy.
- Mixing a proportion of cement and fibres wheat together with the silty soil decreases its shrinkage and hence its risk of cracking.
- A combined stabilisation of soil by cement and wheat fibres could improve both compressive strength, resistance to water and reduces both shrinkage and thermal conductivity and hence an environment friendly building material.

References

1. Houben, H., Guillaud H. Traité de Construction en terre, CRATerre, Éditions Parenthèses, Marseille, 1989.

2. Kerali, A.G. Destructive effects of moisture on the long-term durability of stabilized Soil blocks, PhD thesis. University of Warwick, 2000.

3. Walker, P. Characteristics of pressed earth blocks in compression, 11th International brick/block masonry conference, Shanghai, China, 14-16 October 1997, p.1-10.

4. Fopossi, A.J., Mutuku, R.N., Ngapgue, F. Effect of stabilizers on water absorption of compressed earth blocks made from mangu soil. International Journal of Engineering Sciences & Emerging Technologies, 2014. p. 490-495.

5. Guettala, A., Abibsi, A., Houari, H. Durability study of stabilized earth concrete under both laboratory and climatic conditions exposure. Construction and Building Materials, 2006. p. 119-127.

6. Bahar, R., Benazzoug, M., Kenai, S. Performance of compacted cement-stabilised soil. Cement & Concrete Composites , 2004. p. 811–820.

7. Gavigan, D., Goggins, J., Mccabe, B. Strength and durability performance of stabilized soil block masonry units, Global Thinking In Structural Engineering: Recent Achievements. IABSE Conference, Sharem EL Sheikh, Egypt, 2012.

8. Boussaid, S., El-Bakkouri, A., Ezbakhe, H., Taib, A., El-Bouardi, A. Comportement thermique de la terre stabilisée au ciment. Revue Française de Génie Civil, 5(4), 2001, p. 505-511.

9. Saidi, M., Cherif, A.S., Zeghmati, B., Sediki, E. Stabilization effects on the thermal conductivity and sorption behaviour of earth bricks. Construction and Building Materials, 2018, p. 566–577

10. Prabakar, J., Sridhar, R.S. Effect of random inclusion of sisal fibre on strength behaviour of soil. Construction and Building Materials. 2002. p.123–131

11. Yi Cai, Bin Shi, Charles W.W. Ng, Chao-sheng Tang. Effect of polypropylene fibre and lime admixture on engineering properties of clayey soil. Engineering Geology, 2006, p. 230–240

12. Sharma V, Bhanu M. Marwaha, Hemant K. Vinayak. Enhancing compressive strength of soil using natural fibers. International Journal of Sustainable Built Environment, 2016, p.141–155.

13. Taallah B, Guettala A, The mechanical and physical properties of compressed earth block stabilized with lime and filled with untreated and alkali-treated date palm fibers. Construction and Building Materials 104, 2016. p.52–62.

14. Ashoura A, Korjenicb A, Korjenicc S, Wuda-Benha W. Thermal conductivity of unfired earth bricks reinforced by agricultural wastes with cement and gypsum. Energy and Buildings. 2015. p.139–146

15. Bouhicha, M., Aouissi, F., Kenai, S. Performance of composite soil reinforced with barley straw. Cement & Concrete Composites, 2005. p. 617–621

16. Shooshpasha, I., Shirvani, R.A. Effect of cement stabilization on geotechnical properties of sandy soils. Geomechanics and Engineering, Vol. 8, No. 1, 2015, p. 17-31

17. Haralambos, S. Compressive strength of soil improved with cement. Proceedings of the Contemporary Topics in Ground Modification, Problem Soils, and Geo-Support, 2009. p. 289-296.

18. Akpokodje, E.G. The stabilization of some arid zone soils with cement and lime. Q. J. eng. Geol. London, 1985, p.173-180.

19. Zakham, N., El-Rhaffari Y., Ammari A., Cherraj M., Bouabid H., Gueraoui K., Samaouali, A., Mzerd, A. Influence of cement content on the thermal properties of compressed earth blocks (CEB) in the dry state. MATEC Web of Conferences 149, 010590. 2018.

18th International Conference on
Non-Conventional Materials and Technologies
"Construction Materials & Technologies for Sustainability"
(18th NOCMAT 2019)
24th – 26th July 2019
Nairobi, Kenya

THE POLICY, LEGAL AND ADMINISTRATIVE FRAMEWORK FOR ADOPTION OF INNOVATIVE BUILDING TECHNOLOGIES IN KENYA

Edwin K. Kaburua[1], Mugwima Njugunab[1], Titus Kivaac[1]

[1] Penn State University, University Park, State College, Beaver, Pittsburgh, PA, USA;

ABSTRACT

Innovative building technologies (ibt) are often exhibited as a panacea for mitigating housing delivery shortfalls in the light of the ever growing construction costs. the regulatory framework has however been often cited as a challenge to this probable solution of affordable housing. this paper examines the legal and administrative framework that bears on adoption of ibt in construction industry of kenya. to achieve this, the paper describes the existing legal framework influencing adoption innovative building technologies beginning with a broad perspective of the constitution of kenya in relation to housing as a basic human right. a review of relevant legilations that bear on innovative building technologies reveal some critical handicaps to adoption of ibt. the paper also appreciates considerable policy effort by the government to ease the barriers through the housing policy, proposed construction industry policy and proposed planning and building regulations. The paper relies on desktop study of relevant laws and policies together with relevant reviews on their bearing on ibt adoption. the main conclusion is that the policy and legal framework for innovative building technologies is elaborate but with segments of outdated provisions that serve to inhibit ibt adoption. these include the building code that put emphasis on material composition. this is however addressed in the proposed planning and building regulations which lay emphasis on performance. the administrative framework is liberal and uncoordinated. There is thus need to actualize the elaborate policy provisions through establishment of institutions with specific mandate to implement or oversee implementation of ibt policy provisions.

Keywords: Innovative building technologies, legal framework, housing policy, administrative framework

INTRODUCTION

The provision of adequate and affordable housing remains a challenge to most countries, particularly those in the developing world. Diverse strategies are pursued by various governments worldwide in order to address the problem; key among them is alternative building materials and technologies. It is understood that most of the buildings constructed using conventional building materials like natural stone, steel and mortar are unaffordable to a majority of people in the developing world, thus necessitating the development and adoption of alternative, relatively cheap, decent and durable on site produced materials (1). However, majority of populations remain steeped in traditional construction methods. Despite the high demand for housing especially in the city and the affordability challenges, adoption of alternative building technologies is rather subdued.

Most developers in Kenya stick to stone and cement, and there is not widespread use of any alternative building material (2).

The tepid adoption of IBT may be blamed on several factors but at the bottom line, a sound policy, legal and administrative support for innovative building technology (IBT) is paramount as the free market economy may otherwise not allow for significant technical innovation. This is more so because potential investors and users of the technologies are averse to risks associated with new technologies. Factors which contribute to this need for Government intervention include; the costs or disadvantages of adopting new technology, Government's need to protect inventors' rights and allow innovators to reap the benefits that offset the costs and risks of innovation, market size and integration influence where Government can facilitate the introduction of new technology in small or poorly integrated markets (3). Policy drawbacks and support for IBT can be found in various legislations and policies promulgated by the Government although the object of these policy documents are often broad as they seek to address wider goals. This paper identifies and discusses the various policy and legal documents that support or inhibit adoption innovative building technologies and the institutions that play a role in influencing adoption.

BEST PRACTICES IN INSTITUTIONAL AND LEGAL FRAMEWORK FOR INNOVATIVE BUILDING TECHNOLOGIES

International practices in development and promotion of Innovative Building Technologies and their experiences, successes or failures can serve to guide and inspire the development of the same in Kenya. Various countries have embarked on a range of interventions geared towards increasing supply of low cost housing and particularly through building materials and technologies. These interventions include; establishment of agencies which promote appropriate technologies, Establishment and facilitation of research centres, incentive mechanisms, and procurement methods that is founded on performance specifications and Design and Build procurement method, propagation of supportive policies among others.

1. Establishment of Agencies

In India there are two notable agencies whose mandate revolves on promoting Innovative Building Technologies. These agencies have distinct objectives in regard to enhancement of technology adoption;

a) The building materials and Technology promotion council (BMTPC) of India whose mission was to bridge the gap between research and development and large scale application of new building material technologies. Other objectives of BMTPC include packaging of proven innovative technologies for the benefit of entrepreneurs. Arising from this among other initiatives, India has been able to solve the problem of housing to masses through provision of house at affordable prices by adopting appropriate and cost-effective technologies (4).

b) The Housing and Urban Development Corporation (HUDCO) of India was set up by the Government of India to accelerate the pace of housing and urban development in the country. Apart from the financing operations, HUDCO offers consultancy services, promotes research and studies and help propagate use of local building materials, cost-effective and innovative construction technologies.

Through HUDCO, the Building Centre Movement was launched in 1988. The Building Centres play a significant role in the construction of cost effective livable, durable, aesthetic houses by using cost effective, appropriate, intermediate and sustainable technologies and also generate substantial employment opportunities by training not only in the conventional technologies but also in the cost effective building material trades. The building centres have developed into an effective construction delivery system for affordable housing and also contributing to employment and income generation at local levels. Apart from trainings, the centres have introduced significant technologies in production of building components and construction of houses & buildings and the range of cost savings has been from 15 to 40% over conventional costs, a contribution that was viewed as a best practice (UNCHS, 1998).

2. Incentive mechanisms for promotion of technology by European Countries

Another approach that can be borrowed from developing countries in technology promotion is the incentive mechanism that seeks to promote sustainable housing. The national building codes of member countries lays emphasis on energy efficiency aspect as the countries need to adapt their existing building codes to comply with the EU regulations. Instruments and initiatives that have being proposed, developed and implemented by EU municipalities to comply with the EU targets include; financial and fiscal incentives which can be in form of taxes, charges, subsidies and grants, Economic incentives which can be in form of pricing policy that ensure customers have incentives to conserve scarce resources, promotion and awareness incentives, which can be in form of offering education, advisory services, technical assistance, in-house advice and education, demonstration projects. (5) In the local situation, these incentives can be applied to encourage adoption of innovative building technologies.

3. Promotion of Building Technology Adoption in United States

There is no U.S. government agency with explicit responsibility for representing or encouraging enhancement of the construction industry but rather, new technology is promoted by government agencies through direct purchase for mission-oriented applications or through promulgation of policies aimed at broader social goals like environmental protection (3).

The direct purchase approach take the form of programmes such as the 'percentage-of-construction-cost set-aside for art' which guarantee a funding stream for public art projects which are usually sited in, on, or adjacent to the building or project being constructed. Lemer (1992) opines that such models have demonstrated success in dealing in the short term with a fragmented environment while building in the long term a coalition of interest. Other programmes of federal activities intended specifically to advance building technology include the Army's CPAR program, the Department of Energy's Building Technologies programs, and the activities of the National Science Foundation (NSF). These programs act through contracts and grants for specific studies, typically for projects of relatively short duration.

Lemer further observes that Procurement methods that encourage new technology have sometimes been used by government agencies. These include development of performance specifications and the solicitation of design-build proposals. Other approaches for promotion of technology that have received some support include the establishment of research centres focused on particular areas of building science and technology, regulatory policy or tax policy and technology transfer regulation, supporting education and communication in the building professions.

POLICY AND LEGAL INTERVENTIONS THAT BORDER ON IBT ADOPTION

There is no specific policy or legislative framework that has been promulgated to promotes adoption of Innovative Building Technologies in Kenya. Policy provisions that bear on Innovate Building Technologies can however be found in various policy documents that purpose to achieve other wider goals.

The Constitution of Kenya

Policy actions that stem from or have a linkage to the constitution are bound to merit wide public support and justify devotion of resources. The significance of IBT in relation to the constitution may be drawn from article 43 of the bill of rights where the right to accessible and adequate housing and to reasonable standards of sanitation is provided. Whereas this may not have a direct bearing to IBT, the affordability challenge that hampers provision of adequate housing using conventional technologies makes it imperative that the state has to seek alternatives that aim to overcome this hurdle. A further potential for influence of the Constitution in building technologies can be derived from the fourth schedule which delineates the functions of National government to include housing policy. County planning and development including Housing, and county public works is on the other hand in the realm of the County Governments functions. The constitution therefore sets a foundation for housing policy making and implementation of which Innovative Building Technologies is paramount.

The Kenya Vision 2030

The 2030 vision for housing and urbanization is an adequately and decently housed nation in sustainable all inclusive environment. The goal of the second medium term plan for vision 2030 was increase the annual production of housing units from 35,000 annually to over 200,000. The vision further aimed to achieve better development of and access to affordable and adequate housing.

The situation analysis of third medium term plan for vision 2030 acknowledges that the Housing sector promoted the use of Appropriate Building Materials and Technologies

(ABMT) through establishing 19 Housing Technology Training Centres at constituency level, training 300 community groups on use of ABMT and production of Expandable Polystyrene Panels (EPS).

Under the Programmes and Projects for 2018-2022, the third medium term plan targets to conduct Research and Development on Appropriate Building Materials and Technologies (ABMT): This will entail collaborative research, documentation and technology transfer on building and construction materials; training of 20,000 new trainees on ABMT; dissemination of appropriate building materials and technologies; and development of 20 ABMT Centres. The plan further targets to develop policy on lowering the cost of housing and improving accessibility to affordable mortgage.

Housing Policy for Kenya

The Sessional Paper Number 3 of 2004 on housing policy for Kenya recognizes the problem of urban housing as characterized by an acute shortage in the number of dwellings, overcrowding in the existing housing stock as well as the existence of sub-standard human settlements such as extensive slums and squatter settlements (6). The acute shortage in housing supply has pushed the cost of housing beyond the reach of majority of low income earners. The Sessional paper further observes that the majority of the people in urban areas do not own homes as the level of owner-occupancy has been declining. The Government of Kenya attributes this to; high cost of building and construction materials; inappropriate building and construction technologies; limited research on low cost building materials and construction technologies among other contributory factors (7). This formal acknowledgement creates an impetus for concerted efforts by both state and non-state agencies to devote resources in production and promotion of appropriate technologies.

The Policy lays the framework for Innovative Building Technologies as a strategy to provision of low cost housing right from the basic policy objectives. It sets out to encourage research and popularize the use of appropriate building materials that are locally available and low cost building technologies to reduce the cost of housing. It further acknowledges that the building materials and construction industry constitutes one of the most important sectors in Kenya's economy and that Some of the materials which are produced in some large-scale industries end up being costly due to high costs of production arising mainly from high electricity and transport costs. The anchoring of Innovative Building Technologies on the housing policy for Kenya is crucial in that it guides all important decisions and activities on appropriate IBT besides attracting devotion of resources on the area.

The specific input of the policy in relation to Innovative Building Technologies can be traced right from the policy statements for government action that ostensibly aim to increase the production of housing units through utilization of research findings as well as the use of innovative and cost-effective building materials and technologies. Out of the

16 policy statements, the following 10 statements are relevant to promotion of innovative technologies;

1. Increased funding for research on building materials and technologies.
2. Regular review of building materials taxes.
3. Establishment of a National Research Coordination Secretariat within the Ministry in charge of housing to coordinate and disseminate research findings.
4. Require all research actors to harness and document existing locally available building materials and technologies as well as disseminate this information to the users as appropriate.
5. Promote and encourage small-scale enterprises to engage in production and application of researched materials,
6. Promote the production of innovative building designs and traditional architecture that are cost effective and compatible with the use locally available and affordable materials
7. Stimulate the production and availability of conventional building materials like cement, steel and stones as part of the industrial policy,
8. Promote wider adoption and application of the revised Building By-Law's and Planning Regulations,
9. Encourage the public, private and voluntary sectors to utilize the research materials in their housing and other development programmes.
10. Promote intensified training in requisite skills and construction technologies through Youth Polytechnics, Women and Youth Groups, Community-Based Organizations and Appropriate Material Building Centre s and to encourage production and use of fire resistant building materials (6).

These policy statements provide a broad spectrum of action plans with emphasis on research and dissemination of research findings to promote wider application of innovative technologies. It is of essence that the implementation as well as the impact of these initiatives be evaluated for their effectiveness in addressing the housing delivery challenges in construction industry.

The Construction Industry Policy (Draft)

The Construction Industry Policy aims to create a well-coordinated and developed construction industry which will address the existing needs in order to contribute to sustainable socio-economic development. The policy identifies challenges that require to be addressed so that performance in the industry can be enhanced to include;

a) Low technological uptake and exposure levels of stakeholders to international best practices
b) Use of inappropriate construction material
c) Poor quality of works as a result of poor workmanship and use of substandard materials

Some of the policy statements made towards addressing the aforementioned challenges that relate to Innovative Building Technologies include; Creation of an enabling environment for the local construction industry players in order to boost growth of the industry, while encouraging international and regional players. This is informed by the recognition that the industry is still a net importer of construction materials, contractors, professional services and technology.

Policy actions identified for implementation in this regard include Promotion of the use of locally manufactured products in the industry; Creation of awareness and enforcement of quality standards in the industry, Nationalization and harmonization of all construction standards under the agency in charge of standards, Provide framework for standardization of informal local construction products and services, domesticate foreign construction and material standards designs to suit local conditions and Fast-track formulation and updating of construction regulations and standards on emerging technologies and best practices.

A REVIEW OF LEGISLATIONS THAT BEAR ON ADOPTION OF INNOVATIVE BUILDING TECHNOLOGIES

The Building Code of 1968

The building code which was anchored on the repealed Local Government Act was adopted in 1968. The building code has often been castigated as a barrier to use of innovative technologies. Whereas this law does not make direct provisions on IBT, it has significant bearing on the use of it. Part III of the code specifically deals with Building materials and stipulates that no person shall use or permit or cause to be used in the erection of a building any material which is not; of suitable nature and quality for purposes for which it is used, adequately mixed or prepared or applied, used or fixed in a proper manner so as to adequately perform the functions for which it is designed.

In defining the standards required of building materials, the code recommends conformity with British Standards or British Code of Practice or the provisions in the schedules of the Code. Section 36 however gives the council jurisdiction to decide whether any intended use of the material or proposed method of construction which is not specifically covered by the B.S, Code of practice or appropriate schedule of the building code as comparable therewith. (8)

The substance of the building code as it impinges on IBT comes into force when reference is made to the provisions in the third schedule. Rule 3 of the particular schedule provides that bricks or blocks used in any wall to which the rule relates shall be composed of burnt clay, stone, concrete or sand lime besides the other performance requirements. It further requires the solid material to be at least half of the volume of bricks or blocks. Rule 6 in the same schedule seeks to define the minimum thickness of external and party walls and proceeds to make elaborate provisions on the same but the minimum thickness is set to about 200mm. From the foregoing, it is apparent that whereas the code seeks to address the performance requirements for building materials, it also focuses on the physical properties of the materials and this process has potential to bring bias against materials that may meet the performance requirement but fail to meet prescribed physical properties. However, with the repeal of the Local government Act, it would appear that the Building code is adrift of the anchoring law and as such, it is inconsequential at the moment.

Draft Planning and Building Regulations of 2009

The draft Planning and Building Regulations were developed under the proposed Built Environment Bill that seeks to establish standards and practices in the built environment relating to building, maintenance and associated works and to make provisions for matters connected therewith. The regulations also sought to address the drawbacks noted in the Building code which include prescription of materials based on their physical properties. To address this, the draft regulations define the performance requirements of the materials with reference to the performance requirement for the particular building element where the material is used.

National Construction Authority Act

The National Construction Authority Act (2010) is an Act of parliament which provides for establishment of the National Construction Authority and connected purposes thereof. The influence of this law on IBT may be looked at in the context of some of the functions of the Authority as set out in Section 5 of the Act. Clause 2 Subsection (g) particularly spells one of the mandates of the agency to include; encouraging the standardization and improvement of construction techniques and materials. Whereas the authority may not have binding implications on the standardization and improvement of building materials and technologies, this function has direct implications on IBT. Other function outlined for the authority that could possibly be exploited by the authority to promote IBT include; promote and stimulate the development, improvement and expansion of the construction industry, undertake or commission research into any matter relating to the construction industry, promote quality assurance in the construction industry.

Public Health Act

It is critical that any building technology is not hazardous to the environment where it is applied. The Public health Act CAP 242 of the laws of Kenya sets out to make provisions for securing and maintaining health. Section 118(I) of the Act defines nuisance to include any public or any other building which is so situated, constructed used or kept to be unsafe, or injurious or dangerous to health. Building materials like asbestos roofing sheets was a popular roofing material but is no longer used largely due to health concerns nowadays. Summarily, it may be concluded that the Public Health Act endeavors to put checks in construction to avert nuisance wherein, construction materials and technologies are a major component.

Science, Technology and Innovation Act

The objectives of Science, Technology and Innovation Act is to facilitate the promotion, co-ordination and regulation of the progress of science, technology and innovation of the country; to assign priority to the development of science, technology and innovation; to entrench science, technology and innovation into the national production system and for connected purposes. The Act establishes the Kenya National Innovation Agency whose function entail the development and management the Kenya National Innovation System. The Act further sets us an Innovation fund whose object is to facilitate research for the advancement of science, technology and innovation. However, the general nature of the Act and the extensiveness of the Innovation and Technology field does not guarantee significant involvement of the agency in Innovative Building technologies

Standards Act – CAP 496

This Act of Parliament seeks to promote the standardization of specifications for commodities and establishes the Kenya Bureau of Standards as an agency whose functions include promotion of standardization in industry and commerce. In this regard, the Kenya Bureau of standards prescribes standards for various commodities including building materials but compliance to these standards applies to manufactured products. The Act also leaves the onus of obtaining certification on the producer of the material. It is thus apparent that naturally occurring materials and locally produced materials like the stabilized soil blocks do not bear the standard mark of quality but in reality, they may meet prescribed standards. (9)

The Public Procurement and Disposal Act (PPDA)

Section 60 of the PPDA provides that a procuring entity shall prepare specific requirements relating to the goods, works or services being procured that are clear, that give a correct and complete description of what is to be procured and that allow for fair and open competition among those who may wish to participate in the procurement proceeding. The Act requires that the technical requirements shall not refer to a particular trademark, name, patent, design, type, producer or service provider or to a specific origin unless there is no other sufficiently precise or intelligible way of describing the requirements; and the requirements allow equivalents to what is referred to. In this regards, the Act removes bias to use of Innovative Building Materials in public procurement. However, the conventional building designs and specifications oftentimes specify specific building materials even where there are suitable alternatives. For instance, the walling materials in most residential buildings are designed to be the conventional stones. This is in-spite of other alternatives like the expanded polystyrene panel, and stabilized soil blocks that could perhaps be equally viable or superior to the conventional stones.

INSTITUTIONAL FRAMEWORK GOVERNING INNOVATIVE BUILDING TECHNOLOGIES

As noted in the policy and legal framework, multiple institutions are involved with innovative building technologies at various levels in the country. These include the State department for Housing, the Country Governments, Government agencies like the National Construction Authority, the Kenya Bureau of Standards, Kenya Building Research Institute, and several non-governmental institutions. However, the activities of these agencies remain largely uncoordinated.

The State Department for Housing, Urban Development and Public Works

The fourth schedule of the constitution delineates the national and government functions and the housing policy function is apportioned to the National Government. The Executive order number 1 of 2018 in turn delineates the function of providing policy direction and coordination of all matters related to housing and urban planning and development to the State Department for Housing and Urban Development. Other functions for the department include Housing Policy Management, Development and Management of Affordable Housing; Management of Building and Construction Standards and Codes, Building Research Services among other functions. It is thus apparent that the department has great significance in formulation and implementation of policies on Innovative Building Technologies (10)

National Construction Authority

The National Construction Authority is established under the National Construction Authority Act and the mandate of the Authority is to regulate the construction industry and coordinate its development. The functions of the Authority as set out in Section 5 of the Act include; encouraging the standardization and improvement of construction techniques and materials. Whereas the authority may not have binding implications on the standardization and improvement of building materials and technologies, this function has direct implications on IBT. Other functions outlined for the authority that could possibly be exploited to promote IBT include; promote and stimulate the development, improvement and expansion of the construction industry, undertake or commission research into any matter relating to the construction industry, promote quality assurance in the construction industry. The Authority is also Formulating a Construction Industry Policy that provides a further avenue for influencing Innovative Building Technologies

National Housing Corporation (NHC)

National Housing Corporation is charged with the responsibility of providing subsidized housing and implementing government housing policies and programmes through tenant purchase, mortgages, rental and rural housing loans scheme. NHC was formed as part of Kenya's post-colonial housing policy underscoring the importance of providing decent shelter for all urban workers in the country. In its operations, the agency has directly been involved in production of innovative technologies and particularly the Expanded Polystyrene Panel that is produced at the Mavoko factory (11).

The National building Inspectorate

The National Buildings Inspectorate (NBI) is an agency under the State Department Housing and Urban Development and part of its mandate is to audit buildings for conformity with land registration, planning, zoning, building standards and structural soundness. (12) The buildings audit is critical to assessment of performance of Innovative Building Technologies during the life cycle of the building.

Kenya Building Research Centre

The Kenya Building Research Centre is domiciled under the State Department for Housing and Urban Development and is responsible for; dissemination of information relating to the building construction industry, carrying out research on building materials and systems to facilitate availability of innovative and cost effective building materials and technologies.

Kenya Bureau Standards

Kenya Bureau of Standards (KEBS) is a statutory body whose aims and objectives include preparation of standards relating to products, measurements, materials and processes and their promotion at national, regional and international levels; certification of industrial products; assistance in the production of quality goods; improvement of measurement accuracy and circulation of information relating to standards (13). One of the positive steps undertaken by KEBS in the line of enabling legislation is the enactment of guide standards on application of materials and technologies in 'Special Scheduled Areas' in the major urban centres of Nairobi, Mombasa, Nakuru, Eldoret, Kisumu and Embu (14).

The County Governments

The fourth of the Constitution spells the functions of County Governments to include county planning and development including Housing, and county public works. As such, the County Government has a big stake in influencing Use of Innovative Building Technologies through the various county laws that could be enacted in pursuant to these functions.

Academic Institutions

Academic Institutions establish units that conduct research and develop innovative building Technologies. This includes The Jomo Kenyatta University of Agriculture and Technology that established Sustainable Materials Research and Technology Centre. (SMARTEC) The centre conducts research, develop innovations, disseminate research findings/innovations through training/extension/consultancy activities, and produce/market appropriate construction materials/technologies. The Housing and Building Research Institute (HABRI) was established by the Government of Kenya as a project within the Architecture department in University of Nairobi. The body is however not in operation at present. The core mandate of HABRI was to undertake research on various aspects of low cost housing and community planning in both rural and urban areas (15).

Non-Governmental Organizations

Several non-governmental organizations are involved in promoting Innovative Building technologies. These include the United Nations Centre for Human Settlements UNCHS (Habitat), Action Aid Kenya, Practical Action group (Intermediate Technology Group) among others. The Un- Habitat is an international agency mandated by the UN General Assembly to promote socially and environmentally sustainable towns and cities with the goal of providing adequate shelter for all (16). In implementing its programmes, UNCHS (Habitat) focuses its efforts in the building-materials and construction sector in critical areas. These include; assisting member Governments in formulating and adopting implementable policies and strategies for a sustained growth of building materials sector; Creating increased awareness and knowledge among professionals and decision-makers on available, innovative and cost-effective technological options; Promoting or modifying building codes, regulations, standards and specifications to make them adaptable to local conditions among other initiatives (17) By focusing on these areas UNHCS approach has potential to promote adoption of innovative technologies through dissemination of information as well as addresses some of the factors cited as a hindrance to IBT adoption like the building laws. The major achievement of UNHCS is in organization of regional workshops in some of the participating countries aimed at supporting the countries in formulating standards and specifications for local low cost building materials (14).

The involvement of Practical Action group (previously identified as Intermediate Technology Development Group (ITDG) is in promoting adoption of Innovative technology through training, promotion and dissemination of alternative building technologies already developed. The group has particularly been cited to have been involved in promotion of Fibre Roofing Concrete tiles. The dissemination of technologies is done through seminars, workshops, participation in national agricultural shows and other demonstration forum and secondly, promotion of small private entrepreneurs (18).

Action Aid Kenya (AAK) is a non-governmental development agency which seeks to promote basic education in all forms. As set out in its mission, AAK works with communities to design, fund and manage integrated development programs. One of the approaches previously used by the organization through its technical support unit was, used to channel its assistance through primary schools in form of technical support and building materials. The local communities were expected to offer labor and to produce building materials. The assistance by AAK was also conditional in that the communities had to agree to use locally available materials like sun-dried bricks, burnt bricks, stabilized soil blocks, sisal cement tiles, etc., in their building programs (14).

CHALLENGES IN THE LEGAL AND ADMINISTRATIVE FRAMEWORK FOR INNOVATIVE BUILDING TECHNOLOGIES IN KENYA

There is no policy or legislation with a primary objective of spurring the development and use of Innovative building technologies in Kenya. Similarly, there is no single institution with a core mandate research, development and promotion of Innovative building technologies. Innovative Building technologies are thus seen as auxiliary functions to the multiple institutions and as such, this function could easily be ignored depending on the institution priorities. The institutions are also uncoordinated and the legal framework is comprised of some legislation that contains archaic provisions that detrimental to use of Innovative Building Technologies. Other relevant legislations are subdued in terms of providing a legislative framework that would safeguard adoption of Innovative Building Technologies. Specific challenges identified on relevant legislations and Institutions include;

1. Outdated and inadequate of legislative support

The relevance of Innovative Building Technologies has often been ignored in enactment of legislations that affect the choice of building materials. The building code of 1968 is prescriptive of materials based on their composition and physical size at the expense of performance specifications of the building element to which they are applied. The draft planning and building regulations seek to address this by prescribing performance specifications. These proposed regulations were however supposed to be anchored on the built environment bill which is not yet published.

The Standards Act focuses on standardization for manufactured commodities and this may likely be interpreted to mean naturally occurring building materials or any other material with no prescribed standards is inferior. This portends to bring bias against Innovative Technologies.

2. Lack of linkages and coordination among institution

Multiple institutions are engaged in diverse areas of Innovative Building Technologies including research and training, policy formulation and development. The institutions however remain largely uncoordinated and at times work at cross purposes. The fourth Schedule of the Constitution of Kenya vests the Housing Policy function to the national Government and devolves the functions of county planning and development including housing, and county public works. This separation means the County Governments have a big stake in housing policy implementation. Whereas the national government policies generally seek to promote innovative building technologies, this may not feature top in County Government priorities and the policy may just remain on paper.

The linkage between research and incubation of the technologies is weak and as such the transition between researched technologies and incubation of workable technology is poor. Whereas training institutions like the Jomo Kenyatta University of Agriculture and

Technology has established a center for sustainable materials research and technologies (SMARTEC), the other institutions often seek to develop and promote their own technologies with no attention to those already developed by other institutions. Conversely, the technologies developed by these institutions are not embedded in training of construction professionals within the academic institutions.

3. Generalities in Institutional Objectives

Institutions involved in Innovative Building Technologies often have other wider goals and as such, attention to development and promotion of IBTs could easily be overlooked.

The mandates of National Construction Authority that could be exploited to enhance Innovative Building Technologies include standardization and improvement of construction techniques and materials. This mandate is broad as it covers conventional and innovative technologies and as such, it does not guarantee significant attention to innovative building technologies. This holds also for other functions of the authority which include; promotion and stimulation of the development, improvement and expansion of the construction industry, undertaking or commissioning research into any matter relating to the construction industry, promote quality assurance in the construction industry are subject to varying interpretation. Similarly, other legislations like the Science and Innovation Act is general thereby leaving involvement of the Innovation agency in IBTs to its construal and discretion.

4. Inadequate Policy Implementation framework

The implementation framework to actualize the extensive policy objectives set out in the Housing Policy for Kenya and the draft construction Industry policy is lacking.

The local governments, who had the mandate of regulating and implementing the existing building code prior to the repeal of the local Government Act failed in many areas.

5. Financing IBT Projects – Risk associated

Financial Institutions consider IBT Projects to be risky and as such, they attach a high premium in terms of interest rate in financing IBT projects

6. Assimilation of IBT Training in academic institutions

Incorporation of Innovative building technologies in most tertiary institutions is slow and done in piecemeal.

CONCLUSIONS AND RECOMMENDATIONS

There is no specific policy or Act of parliament that gives a framework for use of Innovative building technologies but rather, the policy and legal framework can be found in bits from the hosing policy, the Vision 2030 and the draft Construction Industry Policy. Legislation on Innovative Building Technologies is also found in segments of various laws. The housing policy however gives considerable attention to promotion of Innovative building technologies but this support may be said to be lip service as it does not provide elaborate measures in terms of law and Institutional framework to actualize the provisions.

On the provisions pertaining to IBT in various laws, the building code has been noted to focus on the physical properties of the materials at the expense of material performance. The National Construction Authority Act is aloof on Innovative Building Technologies but this is understandably because of the general nature of its objectives. The Act however has great potential to influence IBT adoption depending on the level of operationalization of the related provisions therein. The provisions that direct bearing on Innovative Building technologies in Public health Act is remote but in any case, this Act server to ensure that any material used is not detrimental to public health.

The multiplicity of laws and regulations as well as multiple institutions with fragmentary roles in Innovative Building Technologies has been identified as major impediments to accelerated use of Innovative building technologies for housing in Kenya. The proposed Planning and Building Regulations are expected to broaden the range of acceptable construction materials beyond brick and mortar. Though cement will remain a vital component of any construction, locally available materials and latest building technologies could help offer faster and cheaper solutions for housing in the semi-permanent segment. It is hoped that there will be a reduction in the construction period, hastening the rate at which houses are completed.

The study thus recommends a framework for review of the policy and administrative framework as summarized in table 1 below. The framework further recommends actualization of the existing enabling/supportive provisions on IBT through establishment of institutions that will steer the implementation of objectives relating to IBT adoption. The institution would in turn have capacity to engage other agencies whose mandate has an influence on IBT adoption like the National Construction Authority.

TABLE 1- LEGAL AND POLICY AND ADMINISTRATIVE FRAMEWORK FOR ENHANCING ADOPTION OF IBT

	Policy/ Law/ Institutions	Relevant provisions on IBT	Recommendation
1.	The Kenya Vision 2030	a. Vision of adequately and decently housed nation in sustainable all inclusive environment b. Achievement in 2nd medium term; 19 Housing Technology Training Centres at constituency level, training 300 community groups on use of IBT and production of Expandable Polystyrene Panels c. 3rd medium term plan targets to conduct Research and Development on IBT and develop policy on lowering the cost of housing and improving accessibility to affordable mortgage	There is need to harmonize the medium term plans targets with subsequent target achievements. The 2nd term achievements are not anchored on specific 2nd term plan targets
2.	a. Housing Policy for Kenya b. Draft Construction Industry Policy c. National Construction Authority Act	Elaborate provisions that would enhance IBT adoption if actualized	Establish an institution with specific mandate of continuous review and implementing of the IBT policies
3.	Building code	Puts emphasis on Material composition The code stands docile/ in limbo following repeal of anchoring legislation	Adoption of Proposed Planning and Building regulations and anchoring them to relevant law
4.	Science Technology and Innovation Act	Framework for support of technology and innovation is generalized thus does not guarantee support of innovation in building technologies	Structuring of the Legislation to recognize distinguish and support the various priority agendas for research and Innovation
5.	Standards Act	Addresses standards in Manufactured commodities thus leaves out naturally extracted building materials	Expansion of scope to address naturally occurring building materials like coral stones, building sand and ballast
6.	PPDA	- Regulates Public Procurement including preparation of Technical specifications - The Act requires that technical requirements shall not refer to a particular trademark, name, patent, design, type, producer or service provider or to a specific origin unless there is no other sufficiently precise or intelligible way of describing the requirements; and the requirements allow equivalents to what is referred to.	- Particular Building Materials are often times specified in building plans or and in the Bills of Quantities and from the face of it, this goes against the Act. - Need for legal interpretation and if conventional designs contravene, professionals should be compelled to make specifications and designs that do not refer to particular building materials.
7.	Financing IBT	High risk associated with IBT results to relatively higher financing cost to IBT Projects	State Intervention mechanisms to guarantee favorable financing terms.
8.	Training on IBT	Poor assimilation of IBT in training curriculum	Structured engagement a body established to promote IBTs with trading institutions so as to adequately and continuously integrate IBT in curriculum
9.	State agencies involved in IBT	- Multiple agencies with similar but noncore mandates on IBT - Local Governments (Now County Governments) tasked with implementing building code	- Establish of an agency with core mandate of integrating IBT in construction Industry. Such agency to coordinate and leverage on state and non-state agencies involved in IBT - The mandate of implementing the proposed code of practice should be given to another authority which is solely responsible with dealing with buildings and housing as proposed in the Planning and building regulations

REFERENCES

1. HABRI. Low Cost Housing in Kenya. Nairobi Kenya : UON, 2007.

2. The ABC's of Affordable Housing in Kenya. Noppen, Aden Van. 2012.

3. Lemer, David R. Dibner and Andrew C. The Role of Public Agencies in Fostering New Technology and Innovation in Building. [Online] 1992. http/www.nap.edu/catalog/2070.html.

4. A. K. Jain, M. C. Paliwal. Adoption of Appropriate and Cost Effective. s.l. : World Academy of Science, Engineering and Technology, 2012.

5. Skinner, Dr. Ogenis Brilhante and Julia. Review of some incentive mechanisims being used by European Municipalities to promote sustainable housing. [Online] 2015. [Cited: 12 6, 2016.] https://www.ihs.nl/fileadmin/ASSETS/ihs/Marketing/prospective_students_pages/ECO_and_MFUI/Incentive.

6. G.O.K. Sessional Paper Number 3 on Housing Policy for Kenya. 2004.

7. —. The Ministry of Housing Revised Strategic plan 2008-2013. NAIROBI : Government Printer, 2010.

8. G.O.K. Building Code. Nairobi : Government Printer, 1968.

9. GOK. The Standards Act. 2013.

10. G.O.K. Housing and Urban. [Online] 2018. http://www.housingandurban.go.ke/our-functions/.

11. NHC. EPS Factory. [Online] 2009. http://www.nhckenya.co.ke/index.php?option=com_content&view=article&id=109&Itemid=68.

12. NBI. Natioanl Building Inspectorate. [Online] 2018. http://nbi.stratech.co.ke/about-us/.

13. ISO. Kenya (KEBS). International Standards Organization (ISO). [Online] 2016. http://www.iso.org/iso/about/iso_members/iso_member_body.htm?member_id=1854.

14. Towards populization of low cost building materials and technologies for urban housing in developing countries. Magutu, Jerry. 2015, International Journal of Scientific Research and Innovative Technology Vol. II, pp. 20-43.

15. HABRI. Low Cost Housing Techniques in Kenya. Nairobi : U.O.N, 2007.

16. UN-HABITAT. Mandate. UN-HABITAT. [Online] [Cited: 4 19, 2013.] http://www.unhabitat.org/content.asp?cid=2467&catid=1&typeid=24&subMenuId=0.

17. UNCHS. Workshop of the Network of African Countries on Local Building Materials and Technologies, Nairobi, Kenya. [Online] 1993. http://www.greenstone.org/greenstone3/nzdl;jsessionid=DBCFE95307AC8C817A11087CA08B8176?a=d&c=cdl&d=HASH70d5aa5c27a195065439a1.4&sib=1&p.a=b&p.sa=&p.s=ClassifierBrowse&p.c=cdl.

18. Practical Action. Technology challenging today's divided world. [Online] 2016. http://practicalaction.org/.

19. G.O.K. Draft National Building Regulations. 2010.

20. Jain, A K Paliwal, M C. Adoption of Appropriate and Cost Effective Technologies in Housing : Indian Experience. s.l. : World Academy of Engineering, Science and Technology, 2012.

MATERIALS WITH HIGH DENSITY / THERMAL MASS

18th International Conference on
Non-Conventional Materials and Technologies
"Construction Materials & Technologies for Sustainability"
(18th NOCMAT 2019)
24th – 26th July 2019
Nairobi, Kenya

CASE STUDY ON SMALL COMPANIES OF ORNAMENTAL STONE PROCESSING IN THE CITY OF MONTEIRO - PARAÍBA – BRAZIL

Thatiane Amanda da Silva[1], José Augusto Gomes Neto[2], Iracira José da Costa Ribeiro[1], Adri Duarte Lucena[1] Normando Perazzo Barbosa[2]

[1] Federal Institute of Paraíba, Monteiro, PB, Brazil;

[2] Federal University of Paraíba, João Pessoa, PB, Brazil;

ABSTRACT

The construction industry generates a large amount of pollutant waste in all constructive steps, this waste when improperly disposed of causes several environmental problems. The ornamental rocks are being used extensively in the buildings for coatings, furniture making and decoration objects. With the growth of this sector, the amount of waste generated increases the impact caused to the environment by the extraction of the raw material and the inadequate disposal of the waste generated. The present work had as objective to investigate the main forms of destination of the residues generated by the companies benefiting rocks in the city and Monteiro – Paraíba – Brazil. The visits were carried out in companies, the research team initially sought to know the main difficulties encountered in waste management through informal conversations with employees and owners, followed by the stages of production and final destination to know the types of waste generated and to analyze the forms of destination and the possible environmental problems generated. In the stages of processing of the rocks, two types of waste are generated: the residual dust and the flakes of stones of different sizes. According to CONAMA 307/2002 these wastes can be classified as class A materials that can be used in construction as aggregates. All wastes are deposited on land or sent to the landfill, causing disfigurement of the landscape, silting of water courses, and respiratory problems for the population. Workshops and lectures were held with city residents and students from the Federal Institute of Education, Science and Technology of Paraíba - Campus Monteiro to share knowledge about the environmental problems caused and to know techniques for reuse of waste. The situation of companies' sector of the beneficiary of ornamental rocks is worrying and requires immediate changes in waste management contributing to the sustainable development of the construction industry avoiding the increase of various environmental problems.

Keywords: Ornamental rocks. Waste. Reuse. Recycling. Sustainability.

INTRODUCTION

The main challenge of civil construction is to apply the concept of sustainability in its activities. The World Commission on Environment and Development known as the Brundtland Commission in April 1987 published the report Our Common Future which considers that sustainable development must meet the needs of the present generation without compromising future generations to meet their own needs. [1] (BRAZIL, 2018a).

The construction industry impacts the overall goals of sustainable development. This is proven by the international council of construction CIB that points to the construction industry as a sector that consumes more natural resources and uses energy intensively, generating considerable environmental impacts. There are also impacts associated with the generation of solid, liquid and gaseous wastes. It is estimated that more than 50% of solid waste generated by all human activities comes from construction. These environmental aspects, coupled with the quality of life that the built environment promotes, combine the relationships between construction and the environment. [2]

The Brazilian economy is characterized by a high level of waste of energy and natural resources. The reduction of these constitutes a true reserve of development for Brazil, as well as source of good business. When talking about the environment, there are unnoticed business opportunities or cost savings. As the environment is a potential resource poorly utilized, its inclusion in the business horizon can generate activities that provide profit or at least pay for energy, water, or other natural resources. Recycling waste, for example, is turning them into value-added products. Conserving energy, water and other natural resources is reducing production costs. [2]

The civil construction chain has a low tradition of innovation. However, a new, more eco-efficient construction market will require the continued development of a wide range of new products and processes that are appropriate to the different realities of the country. New markets must be created and the export trend intensified. [3]

The processes related to the extraction and processing of ornamental rocks promote a great economic growth in Brazil, but they have serious environmental consequences for the amount of waste generated, and its recycling and use in construction is a promising alternative that reduces environmental impacts and contributes to sustainable development of the sector. [4]

One of the sectors that improperly impacts the environment is the extraction of rocks as well as the beneficiation of the same. The extraction and processing of ornamental rocks contribute to the economic development of the country and the generated waste can be recycled and used in construction minimizing the environmental impacts and contributing to the sustainable development of the sector. [4] These wastes when misused can cause silting and turbidity of water resources, soil and groundwater contamination, air pollution, landscape disfigurement and damage to health. [5]

METHODOLOGY

The work was carried out in the city of Monteiro - Paraíba, Brazil. Initially the companies that work in the sector of ornamental stone processing were located, this activity was carried out to quantify the sites and to find partnerships for the development of the research. The difficulties encountered in stone management were analyzed. The positive and negative influences on society as a consequence of the place where each company is located.

With the purpose of knowing and analyzing the procedures performed in the treatment of rocks, the research team followed from the stage of purchase of the materials to the final disposal of the waste generated. The types of rocks commercialized in the region were identified, the types of residues generated at each stage of the processing of the pieces and classified according to the criteria established by the resolution of CONAMA 307/22. [6]

The team carried out a vast bibliographical review to improve the knowledge about techniques of ecologically correct disposal of residues from the beneficiary of ornamental rocks. Workshops and lectures were prepared for the population of the city of Monteiro-PB, a city in the surrounding area and the students of the Federal Institute of Science and Technology Education of Paraíba - Campus Monteiro, aiming to present the environmental problems caused by the waste and propose techniques ecologically and to exchange knowledge with the external community.

RESULTS AND DISCUSSIONS

In the city of Monteiro - Paraíba - Brazil there are two companies that benefit from ornamental rocks better known as marmorarias. The marble factory I (Figure 1) is located in the center of the city and performs all procedures for the processing of the pieces in the external area of the company's office, raw materials and equipment are also stored in the external area. However, the marble factory II (Figure 2) is located near the exit of the city, such company has a shed used to carry out the works and for storage of materials. Both marble mills buy the stones at the GRANFUJI and FUJI plants located in Campina Grande - Paraíba - Brazil. The pieces are bought according to the request of the clients, the marble houses exude the demand of Monteiro and neighboring cities. The two companies work only with marble and granite.

Figure 1 – Marble factory I

Source: Own (2018)

Figure 2 – Marble factory II

Source: Own (2018)

The stones are cut and polished according to the client's design specifications, in these steps two types of residual sludge residues are generated which when dried becomes a greyish powder similar to the cement visually (Figure 3). The second type of residue is the patchwork of stones in different sizes (Figure 4). The generation of this waste can vary according to weekly production, but on average 16 kg of powder and 1 m^3 of stone pieces are waste.

According to CONAMA Resolution 307/22 [6], these wastes are classified as Class A waste that can be reused and recycled in the construction industry.

Figure 3 – Residual dust

Source: Own (2018)

Figure 4 – Pieces of Stones.

Source: Own (2018)

The residual dust and the pieces of stones generated in the mortararia I are mixed with the household garbage routed to the municipal garbage dump and there they mix with different types of waste produced by the city. The owner has not shown any concern about the environmental risks caused by ways of disposing of the waste.

The marble factory II at the end of the week collects the Shed the two types of waste generated and plays in the near ground (Figure 5), the owners are aware that this form of disposal is generating serious environmental problems, but claims to have no knowledge of a technique ecologically correct.

Figure 5 – Disposal of waste on vacant lots

Source: Own (2018)

The workshops and lectures (Figure 6 and 7) made it possible to share the knowledge acquired during the research and aroused the interest of the external community for the development of techniques aimed at reducing the environmental impacts caused to the construction sector.

Figure 6 – Workshop

Source: Own (2018)

Figure 7 – Lectures

Source: Own (2018)

CONCLUSION

The improper disposal of waste from the processing of ornamental stones is a worrying situation that demands immediate solutions to avoid major environmental problems in the region such as the contamination of water courses and problems for fauna and flora and human health. First continue the development of lectures and workshops subsequently propose the reduction of the amount of waste generated amount of waste generated and then adopt reuse techniques for waste preferably in the construction industry itself in the construction industry itself and may become a new source of income in the region.

REFERENCES

1 BRASIL. A ONU e o meio ambiente. ONUBR – Nações Unidas no Brasil. 2017. Disponível em: < https://nacoesunidas.org/acao/meio-ambiente/>. Acesso em: 02 de agosto de 2018

2 BRASIL. Ministério do Meio Ambiente. Construção Sustentável. Disponível em: . Acesso em: 02 de agosto de 2018.

3 CONSELHO BRASILEIRO DE CONSTRUÇÃO SUSTENTÁVEL. A sustentabilidade é o grande desafio da atualidade e investir em preparo e conhecimento deve ser o foco da cadeia da construção civil. Disponível em: <http://www.cbcs.org.br/website/filiacao/>. Acesso em: 22 de outubro de 2018.

4 REIS, Alessandra Savazzini dos; ALVAREZ, Cristina Engel de. A sustentabilidade e o resíduo gerado no beneficiamento das rochas ornamentais. IV Encontro Nacional e II Encontro Latino-americano sobre Edificações e Comunidades Sustentáveis. 2007.

5 BARBOSA, J. F.; COSTA, V. S. da; LIMA, M. R. P. Avaliação da utilização de lama abrasiva gerada no beneficiamento de mármore e granito para a confecção de telhas de concreto. Revista Eletrônica de Materiais e Processos / ISSN 1809-8797 / v.8.1, p. 30 – 35. Espírito Santo, 2013

6 CONAMA - Conselho Nacional do Meio Ambiente, Resolução no 307, de 5 de julho de 2002. DOU 136, de 17 de julho de 2002, págs. 95-96.

18th International Conference on
Non-Conventional Materials and Technologies
"Construction Materials & Technologies for Sustainability"
(18th NOCMAT 2019)
24th – 26th July 2019
Nairobi, Kenya

RE-DESIGN PROPOSAL FOR THE THERMAL AND ENERGETIC EFFICIENCY OF A PRIMARY HEALTH CENTER IN TUCUMÁN, ARGENTINA.

Amalita Fernández[1], Silvia Beatriz Garzón[2]

[1] National University of Tucumán, CONICET, Argentina;

[2] National University of Tucumán, CONICET, Argentina;

ABSTRACT

The buildings destined to health establishments are among those that require greater consumption of energy for their operation. In Argentina, the "Norms of Organization and Functioning" of the National Quality Assurance Program for Medical Care demand high degrees of acclimatization. This necessity to maintain high environmental and health standards contribute to a greater energy demand; and, therefore, that the need arises to implement measures that generates very important energy savings that, in turn, make it possible to reorient resources towards health prevention and clinical care. This work proposes the architectural redesign for the thermo-energy efficiency of a Primary Health Care Center (CAPS) in the province of Tucumán, Argentina and the verification thereof, according to standards, in order to achieve its bioenvironmental adequacy through architectural and technological provisions that contribute to improve the comfort and health conditions of its users, to achieve the rational use of energy and the care of the environment. A methodological combination was used, through the following methods: case study, descriptive, analytical, deductive and correlational. The actions carried out were: 1) Recognition and geographical characterization of the study area; 2) Definition of bio-environmental guidelines and strategies; 3) Analysis of the prototype: architectural layout, technological and thermo-physical characteristics of the envelope; 4) Re-design proposal for its bioenvironmental adjustment; 5) Energy efficiency labeling. Likewise, the results reached in this respect and the conclusions in relation to the importance of applying and validating these guidelines and strategies in the design and production of new CAPS or in interventions in those already existing are exposed.

Keywords: bioclimatic adaptation, energy efficiency, energy labeling.

INTRODUCTION

If we start from the premise that architecture is a social work, the bioclimatic trend must be emphasized, because its spaces are directed to:

- The improvement the quality of life of the users from the point of view of the hygrothermal comfort.
- The integration of the architectural object into its context.

- Influencing the reduction of demand for conventional energy and the use of alternative energy sources, as a result of the ecological concept that frames this trend (Garzón, 2007).

The energy used in buildings represents a significant proportion of the total at regional, national and local levels. It is estimated that between 35 and 40% of all primary energy resources used in Argentina are destined to the conditioning of the built habitat. The application of bioclimatic strategies can contribute very positively and effectively to the growing energy and environmental crisis. (Evans, 2005).

Buildings used for health facilities are among those requiring greater use of energy to function (Chilean Energy Efficiency Agency, 2012). In Argentina, the "Rules of Organization and Functioning" of the National Program of Quality Assurance of Health Care (2004), demand high degrees of air conditioning, sterilization and control; this need to maintain high environmental standards contributes to high energy demand. Implementing energy efficiency measures would generate important savings that would allow reorienting resources to clinical care.

In the province of Tucumán there are 324 provincial health establishments administered by the Provincial Health System (Si.Pro.Sa.), of which 298 are primary health care centers. They are distributed geographically according to the number of inhabitants, these are: polyclinics, CAPS and sanitary posts (terms used by the officials, but not settled in any law). The Physical Resources area of the Si.Pro.Sa. is in charge of remodeling, maintenance and construction of new works. They are governed by the decrees presented in the "Rules of Organization and Functioning" of the National Programme for the Quality Assurance of Health Care; which contemplate minimum requirements of equipment and surfaces, as well as type of treatment of the surfaces and type of materials (for reasons of hygiene), type of facilities and the only thing mentioned related to the comfort is the temperature that the air conditioning systems must reach for specific premises (neonatology, operating rooms, intensive therapy unit, etc.).

The future presents us the need to confront the transformation of the establishments for the health, adapting them to the changes in the culture and to new paradigms in the architecture, which are not only scientific, but also human. Through design it is possible to improve the quality of life and even dignity at the moment of disease, it is the ethical responsibility of the architects to contribute to the construction of a safe habitat that dignifies the human being, especially when he/she is in a care center, as a patient, family member or medical and service personnel (Cerdés de Bello,2011).

Through a good proposal of architectural design, affordable, it can generate significant energy savings and consequently reduce the environmental impact. The objectives of this work are:

- To analyze the layout and architectural envelope of a prototype of a Primary Health Care Center (C.A.P.S.), in Banda del Río Salí, Cruz Alta, Tucumán, Argentina.

- To propose an architectural-bioclimatic redesign alternative of the prototype.
- To evaluate thermally and energetically the prototypes, making energy efficiency labelling, and comparing with the standard values.

METHODOLOGY

This work is done at different stages. In the first one, the climatic and geographical conditions of the region are defined. Then, the general strategies are determined and the particular guidelines are selected, for the bioenvironmental zone where the health center in analysis is located, according to IRAM Standard 11603:1996 "Bioenvironmental Classification of the Argentine Republic". In the third stage, an architectural and technological analysis of the State prototype is carried out in its present state and from its condition, a proposal of redesign is projected for its bioclimatic adjustment. Subsequently, the values of thermal transmittance for walls and ceilings in the summer and winter condition of the original prototype are calculated and compared with the IRAM Standard 11605:1996 of "Maximum values of thermal transmittance"; This proposes adjustments for the ceilings and walls in the composition and disposition of the materials to reduce the value of "K" for its improvement or to check with the minimum permissible. Finally, the thermal-energetic evaluation is completed with the labelling of both situations with the procedure indicated in the IRAM 11900:2017 Standard of "Energy Performance Index".

RESULTS

Identification and characterization of the work area

The study prototype, is located in La Banda del Río Salí, which is a city located in the department of Cruz Alta in the province of Tucumán, Argentina. The City, in addition to being the departmental head, is part of the agglomeration of the Gran San Miguel de Tucumán.

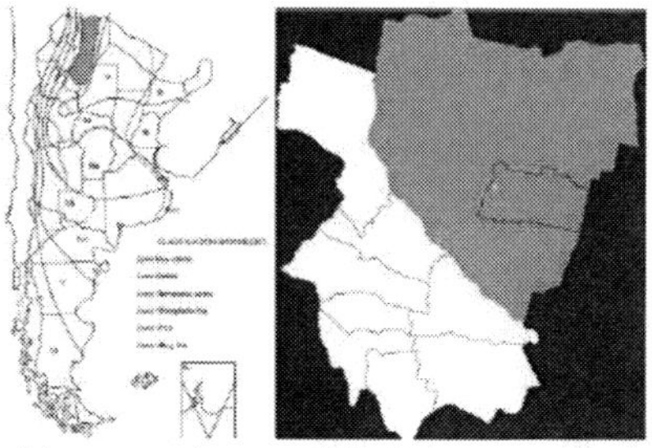

FIGURE 1: Bioclimatic Zoning of the Argentine Republic (IRAM 11603:1996); Geographical location of the study area.

The locality is in the zone Bio-Environmental II warm, sub zone B thermal amplitude less than 14 °c (Norma IRAM 11.603-2012).

CHART 1: Geographic information of the study area.

Bioenvironmental	Geographic area	Latitude	Length	Altitude
II b: warm with thermal amplitude <14°C	Banda del Río Salí City, Cruz Alta departament	26°50' S	65°09' O	432msnm

CHART 2: Climate information of the study area.

Summer			Winter		
Minimum average T. / HR máximum average	Average T. / avegare HR	Average máximum T./ Minimum average HR	Average Minimum/ Maximum average HR	Average T. / Average HR	Maximum average T./ Minimum average T.
20,76°C / 92%	26,08°C/ 75%	31,4°C/ 58%	9,84°C/ 87.2%	15,28°C/ 71%	20,72°C/ 54%

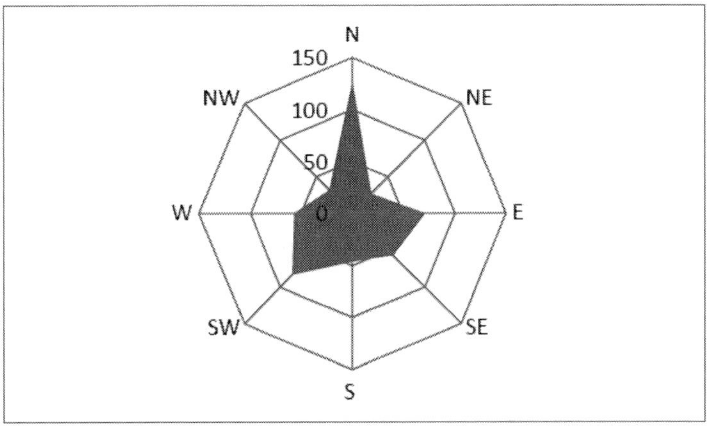

FIGURE 2: Direction of the wind

Determination of bio-environmental strategies and guidelines

In order to determine the Bio-environmental Strategies, the "Psychrometric" and "Olgyay" diagrams were used (Figure 2). From the maximum, medium and minimum average climatic data for winter and summer in the locality in analysis: values of temperature and relative humidity, of maximum, medium and minimum average values, segments were drawn on these diagrams. Thanks to the methodologies used, we obtained and considered those with higher percentages in relation to their removal from the comfort zone.

FIGURE 3: Psychometric and Olgyay Diagram.

From the psychometric diagram we obtained that the comfort zone corresponds only to the 4.46% annual and from the Olgyay diagram a 12% which shows an area of discomfort during most of the year, being necessary to adopt general strategies and particular design guidelines that provide maximum internal comfort with minimum energy cost.

It was determined that the main strategies to consider are: 1. Natural cooling strategy: Natural ventilation is a strategy to apply in summer period and allows to increase the feeling of comfort when it has an impact on people. The selected guideline for this case is "Cross ventilation".

2. Passive solar heating strategy: It is a strategy to apply in winter period; thus the design of the building should favor the capture, accumulation and distribution of the heat. In this case, the "direct gain through windows" guideline is chosen as appropriate; and "Protection of Openings" to control heat loss.

Analysis of the existing health prototype

Architectural Layout

The building is located in a house belonging to a residential complex, which underwent an enlargement and remodelling of one of the houses. It is implanted in a 9x40m terrain with an W-E orientation attached to a party wall belonging to the neighborhood. The premises are located around a main axis in the direction of the building, the covered area is of 180m2.

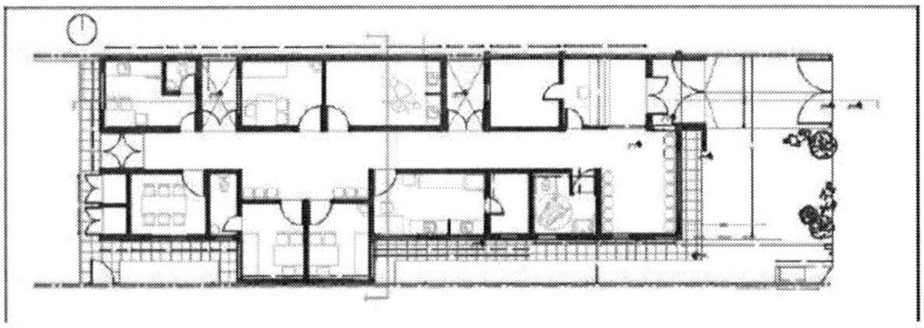

FIGURE 4: existing prototype floor

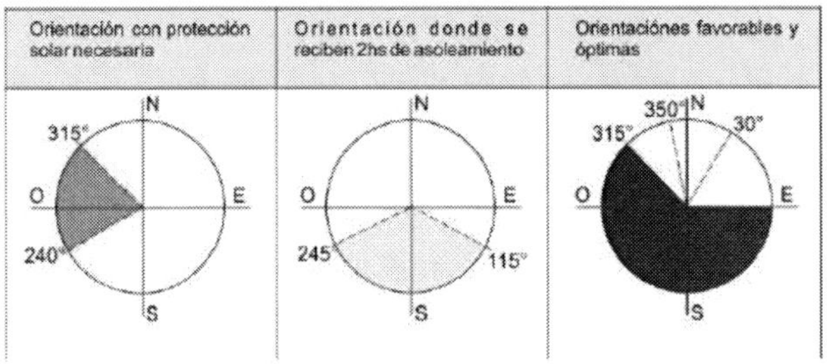

FIGURE 5: Recommended orientations

As for the orientations, no local is towards those recommended by IRAM for the use and control of the sun (Fig. 4) since it is attached to the party wall, not following that orientation; as well, refusing to the direction of the prevailing winds, not being able to take advantage of the strategy of natural ventilation.

Technological features of the envelope

The CAPS was built using a traditional system of point structural system of reinforced concrete columns and beams and exterior vertical enclosures of hollow ceramic brick of 0, 18m of thickness revoked in both sides. The windows are of white aluminium and common glass, the doors are also of white aluminium. The cover consists of a metallic structure with corrugated galvanized sheet color, with thermal insulation type Isolant with aluminium of 5 mm under the plate, a ventilated air chamber and a suspended ceiling of plaster with two inches of glass wool with kraft paper.

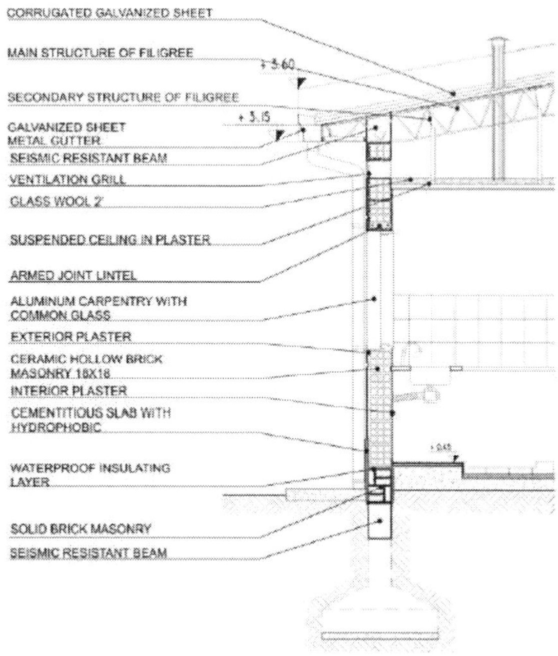

FIGURE 5: constructive detail

The coefficient of thermal transmittance K (W/m2 °c) was calculated for both elements and it was obtained that for the wall K = 1.77 W/m2 ° C and in the case of the cover K = 0.58 W/m2 ° C. The calculations made with the maximum permissible K values determined by the IRAM standards were compared for the three levels of hygro-thermal comfort:

Level A: Ecological.
Level B: Recommended.
Level C: Minimum.

CHART 4: VERIFICATION OF K STANDARDS

Verification K of norms for Walls						
Level A:	0.38	Level lB:	1.00	Level C:	1.85	Winter
It does not comply		It does not comply		It complies with the norms		

Verification K of norms for Cover						
Level A:	0.32	Level B:	0.83	Level C:	1.00	Winter
It does not comply		It complies with the norms		It complies with the norms		

Verification K of norms for Walls						
Level a:	0.45	Level B:	1.10	Level C:	1.80	Summer
It does not comply		It does not comply		It complies with the norms		

Verification K of norms for Cover						
Level A:	0.18	Level B:	0.45	Level C:	0.72	Summer
It does not comply		It does not comply		It complies with the norms		

From this it was determined, that the walls, meet only with level C, while on the deck is verified level B for winter.

Re-design of the sanitary prototype for its bioclimatic adjustment

For the proposal of improving the prototype, the main axis was maintained as in the original disposition, while maintaining the original structure of the house from which the prototype is made. We try to maintain the provisions and dimensions of the premises, since the original prototype is extremely functional. The building is detached from the North party wall, giving priority to this orientation for the main premises (medical offices and nursing). Windows of smaller dimensions are suggested, to avoid excessive gains during the summer, because for this type of function the large slots are unnecessary. Small eaves are added to the north, necessary to protect from the solar radiation in the summer, mobile protectors in the carpentry to avoid heat losses during the winter and to protect the less favorable orientations with deciduous native trees.

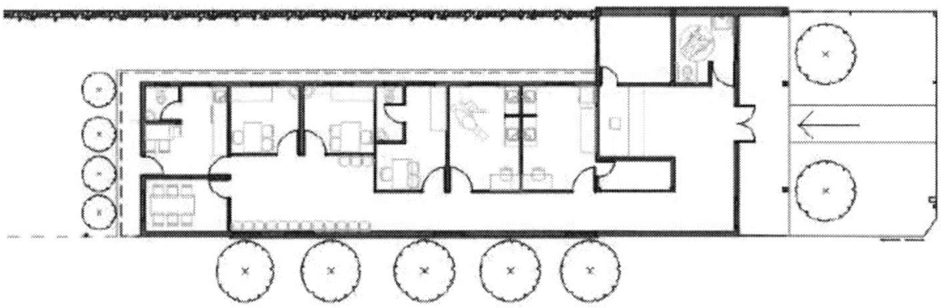

PICTURE 6: Porposal of re-design

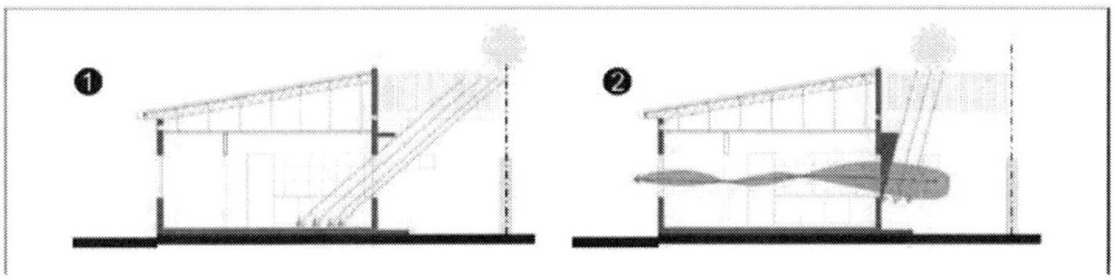

PICTURE 7: sunshine in winter (1) and summer (2)

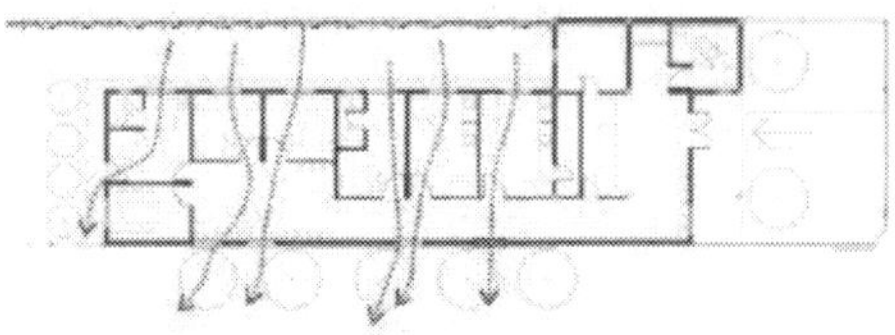

PICTURE 8: crossed ventilation

As for the enclosure, the sheet cover is maintained with ventilated air chamber, suspended ceiling and one more layer is added at 5cm of glass wool, achieving an insulation of 10cm. As for the walls, the masonry of hollow brick is completely dismissed, due to its low thermal-energetic efficiency and we used walls made with panels of expanded polystyrene of 8cm of thickness coated with structural plaster with an electro welded mesh. The calculated coefficient K (thermal transmittance) gives a value of 0.38 W/m² °c in walls and 0.34 W/m² °c for the cover. This way it is achieved that both belong to level A for winter and the cover reaches level B for summer.

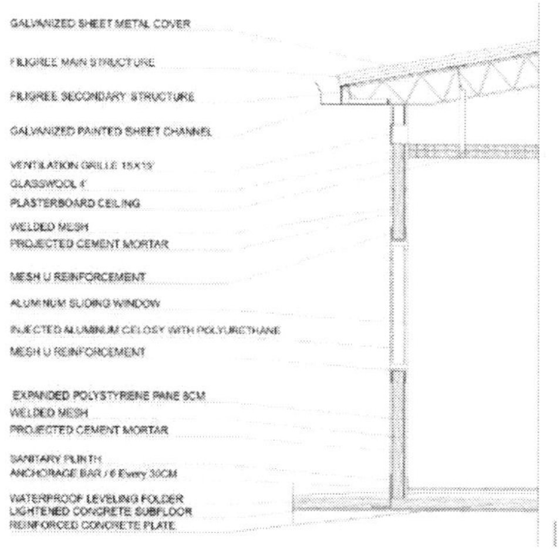

PICTURE 9: constructive detail

Verification K of norms for Walls						
Level A:	0.38	Level B:	1.00	Level C:	1.85	Winter
It complies with the norms		It complies with the norms		It complies with the norms		

Verification K of norms for Cover						
Level A:	0.32	Level B:	0.83	Level C:	1.00	Winter
It does not comply		It complies with the norms		It complies with the norms		

Verification K of norms for Walls						
Level a:	0.45	LevelB:	1.10	Level C:	1.80	Summer
It complies with the norms		It complies with the norms		It complies with the norms		

Verification K of norms for Cover						
Level A:	0.18	Level B:	0.45	Level C:	0.72	Summer
It does not comply		It complies with the norms		It complies with the norms		

Energy efficiency labelling of both situations

Finally, the labelling of energy efficiency was carried out, starting from the calculation of the IPE (Energy Performance Index kWh/m² year) for heating, cooling, hot water production and lighting. The same was elaborated in base of: IRAM 11900:2017 Standard - Computer Application for the calculation of Index of energy Benefits. Secretariat of State for Energy of the Province of Santa Fe and Secretariat of the Government of Energy, Presidency of the Nation. The State prototype ended being classified in class "E" (inefficient) with an IPE of 237 kWh/m² year, while in the re-design was significantly improved with an IPE 65 kWh/m² year.

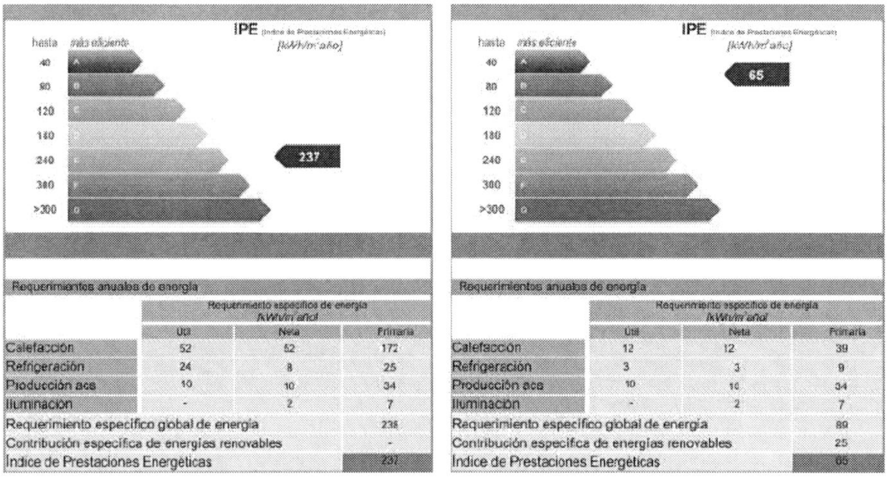

PICTURE 10: Labelling for both buildings.

CONCLUSIONS

Considering the conditions described in the State prototype, it was evident the need to propose alternatives of new solutions that allow us to improve the thermal comfort of the people who were in the building: patients and health professionals belonging to the community of the locality under study. In defining the bioclimatic strategies considered for the bio-environmental zone in "I1b-Warm" analysis, it was possible to select the design guidelines, so that the prototype responds to the thermal-energetic needs. The design strategies at site level in the field that were taken into account for the reformulation of the prototype, were the disposition northward of the main premises, for what the building was detached from the North party wall, thus achieving a better sunning and cross ventilation. It was proposed to maintain the plan of architectural needs, to redesign its form, to relocate functions and technological provisions in general, in order to allow the rational use of energy.

At the level of its technological resolution, neither the wall nor the original cover verified the recommended level "B recommended" of calculation of thermal transmission "K" (W/m² °c) established by the Standard IRAM 11.605:1996; with the improvement of the choice and disposition of the constructive elements in the redesigned enclosure, the calculation value of "K" was reduced; which allowed to verify the "A Ecological" level for the wall and "B recommended" for the roof; as it is indicated in the reference values of the IRAM 11,605 Standard. With respect to energy efficiency labelling in the case of the redesigned prototype, it was risen to level "B", improving three positions from the level "E" measured in the original prototype; according to the values established by IRAM 11,900:2017 Standards. The new enclosures that were studied up to the level of constructive details, were conformed in order to reduce the losses of heat and to increase profits in the critical time thus allowing to collaborate in the energy saving.

This work allowed us to devise a set of guidelines that can be applied in designs of new Primary Health Care centers and in remodeling of the existing ones that will manage to deepen in improvements for future reforms of buildings for the health, in general, and in the studied prototype, in particular. In this way, the results obtained contribute to the purpose of optimizing the conditions of comfort and thermo-energetic efficiency in buildings destined to health care in Tucumán, Argentina.

BIBLIOGRAPHY

1. AChEE. (2012). Guía de Eficiencia Energética para Establecimientos de Salud. Santiago de Chile, Chile: AChEE.

2. Cedrés de Bello, S. (2011). Tendencias en la arquitectura hospitalaria. Memorias de la Trienal de Investigación FAU 2011.(pp. 28-37) Caracas, Venezuela ISBN: 978-980-00-2654-0

3. Evans, JM. (2005) energía en el hábitat construido: panorama en Argentina. Los Edificios Bioclimáticos en los Paises de Ibero América Livro de Ponencias (pp. 97-104). Lisboa, Portugal: CYTED. ISBN No 972-676-200-6.

4. Garzón, B. (2007). Arquitectura bioclimática. Buenos Aires, Argentina. Nobuko.ISBN 978-987-584-096-6.

5. Secretaría de Estado de la Energía de la Provincia de Santa Fe. (2018). Aplicativo informático para el cálculo del índice de Prestaciones energéticas según normas IRAM 11.900.

6. Instituto Argentino de Normalización (IRAM) 11603. Clasificación Bioambiental de la República Argentina, 1996.

7. Instituto Argentino de Normalización (IRAM) 11601. Aislamiento térmico de edificios, métodos de cálculo, 2002.

8. Instituto Argentino de Normalización (IRAM) 11605. Acondicionamiento térmico de edificios, condiciones de habitabilidad en edificios, 1996.

9. Instituto Argentino de Normalización (IRAM) 11900. Cálculo de IPE, 2017.

10. Instituto Argentino de Normalización (IRAM) 11659. Etiqueta de eficiencia energética de calefacción para edificios. Instituto Argentino de Normalización, 2001.

11. Normativas del programa nacional de garantía de calidad aprobadas- Normas de Organización y Funcionamiento. Ministerio de Salud de la Nación, Buenos Aires. 30 de enero 2004.

IC-NOCMAT 2019

18th International Conference on
Non-Conventional Materials and Technologies
"Construction Materials & Technologies for Sustainability"
(18th NOCMAT 2019)
24th – 26th July 2019
Nairobi, Kenya

ABSORBENT MATERIALS FOR CLASSROOMS REVERBERATION CONTROL IN CLASSROOMS OF SAN MIGUEL DE TUCUMÁN. PARAMETERS DETERMINATION AND COMPARISON.

Leonardo Paterlini[1], Beatriz Garzón[1]

[1] San Miguel de Tucumán, Tucumán, Argentine;

ABSTRACT

This work aims at making proposals of architectural materials for the improvement of reverberation times (RT) in classrooms through a software. The specific objectives were to determine those sustainable and non-sustainable materials that could be used in classrooms for the optimization of the RT in schools. In this work, a methodological combination was used for carrying out different activities and obtaining results. In a first stage, the case study method was used to analyze the acoustics problematic of a school in San Miguel de Tucumán, Tucumán, Argentina. It is a special case because it belongs to the "program of 700 schools", at the national level. Also, we used the exploratory method for studying materials with possibility of including them to improve the acoustic conditions of classrooms. The descriptive method was also used to understand the totality of the buildings, their spaces and different characteristics. These methods involve tools and specific techniques and those that we use particularly to address and develop the subject matter should be mentioned: a) the RT is determined through the ArchRTCalculator software developed by the team according to the mathematical formula of Sabine. Starting from the evaluation of 4 alternatives of absorbent materials in a particular classroom, a comparison was made to determine which would be the most appropriate for this purpose and that could also be used.

Keywords: Habitat, Unconventional panels, Sugar agro-industry byproduct, Sustainability

INTRODUCTION

In 2004, the "National Program 700 schools" it was applied in Argentina, which had as main objective to solve the "unmet demand for Secondary Education (EM) and Basic General Education (EGB) for the next 5 years. Up to the month of March of 2018, the program has done 2370 schools across the country, and it is in the process of building and completing other 174 schools. In the province of Tucumán, 191 schools were completed. The construction of schools implied following technical specifications which were made by national bodies. Typologies, guidelines, forms, materials, lighting, etc were indicated in those specifications. In this documentation it does not exist any indication related to the acoustic materials that should be used. In addition, they lack of acoustic standards that should be completed to satisfy the educational function in the classroom. The proper acoustic conditioning for the teaching process - learning is vital. It should be mentioned that among the different tools that teachers use to teach, the spoken word is the most important.

There are currently many sustainable and non-sustainable materials to meet the Reverberation Time standards (RT) within the classroom. The general characteristics of the sustainable acoustic absorbent materials are that they are developed from fibers and other feedstocks that come from vegetables. Many of them are still in the evaluation stage, both for their acoustic characteristics, as well as for their resistance, durability, thermodynamics, etc. Some of these because of their aesthetic and acoustic features, could be used to achieve proper RT within the classrooms in Tucumán.

Starting from the evaluation of schools of the National Plan 700 Schools, in previous works, members of the work team of the Faculty of Architecture and Urbanism of the National University of Tucumán and of the Council of Scientific and Technological Researches (FAU-CONICET), belonging to the Group of Sustainable and Healthy Habitat (GHabSS), determined that there are acoustic deficiencies in the classrooms in schools that were built within the framework of the Plan 700 Schools. In view of this situation, FAU-CONICET developed the ArchRTCalculator software for the evaluation of RT in school spaces. This program allows us to make a quick evaluation and comparison when choosing acoustic absorbent materials.

OBJECTIVES

The main objective of this work is to make proposals of sustainable architectural materials for the improvement of the RT in classrooms through a software.

The particular objectives are to determine which are sustainable and non-sustainable materials that could be used in classrooms for the optimization of RT in schools through the use of the software ArchRTCalculator developed by the FAU-CONICET team.

METHODOLOGY

In this work, a methodological combination was employed. First, the case study method was used, which served to analyse the acoustic problems of a school. The Same, it is a particular case, since it belongs to the "Program 700 Schools". Also, the exploratory method was used, with which the appropriate materials were studied. The descriptive method served to understand all the buildings and their different characteristics. These methods involve specific tools and techniques and those that we use particularly to address and develop the topic in question should be mentioned. The RT was determined through the ArchTRCalculator software developed by the FAU-CONICET team, according to Sabine's mathematical formula. From the evaluation of all the materials in a classroom, a comparison was made to determine which would be most appropriate for its use.

ANÁLISIS DE LA ESCUELA CAMPO DE LAS CARRERA

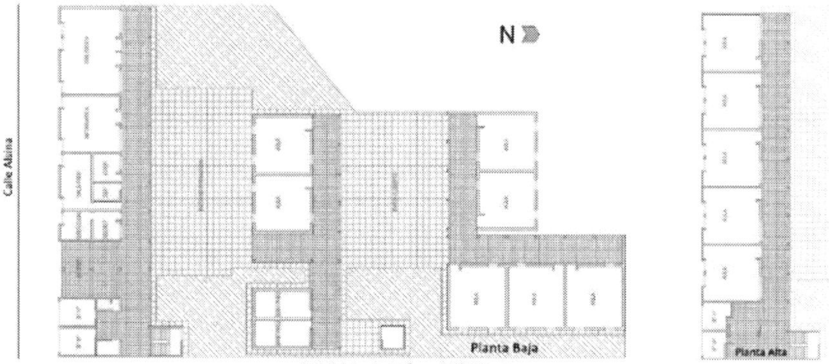

Graphic 1. School blueprint

The School Campo de las carreras is located in 1455 Alsina Street, in the city of San Miguel de Tucumán. It's both primary and secondary level. It is a 2-storey building in which, on the ground floor there is the administrative sector and classrooms, in the upper floor only classrooms.

DETERMINATION OF REVERBERATION TIME IN A STANDARD CLASSROOM

Graphic 2. School inside view. TR in a classroom.

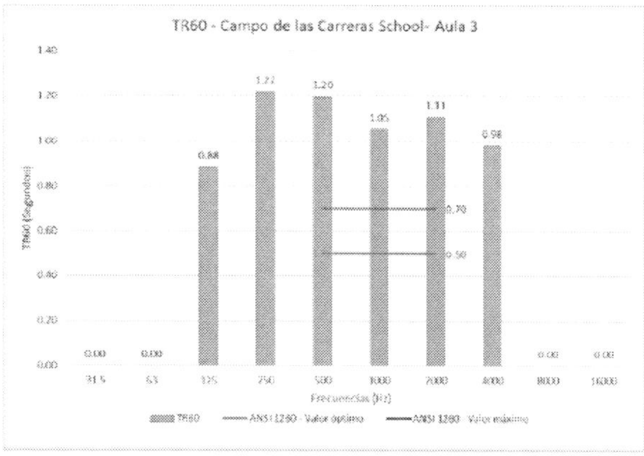

Graphic 3. TR in a standard classroom.

The reverberation time is the amount of time that a sound stays within an environment from the cessation of its sound source. High RT, means the extended permanence of the sound. In the case of the spoken word, the RT needs to be low in order to avoid the masking of words and syllables. For the situation of spaces where education is being imparted, it was stablished that the RT must be between 0.5 and 0.7 seconds. Outside these parameters, the intelligibility of the word is affected, thus the correct understanding of the message. This generates a detriment in the quality of education. This situation is more accentuated in younger students, who are learning their mother tongue.

As you can see, the RT is above the recommended level by ANSI 12.60 Standards, with special emphasis on the frequencies between 500 Hz and 2000 Hz, where the spoken word is located.

PROPOSALS FOR IMPROVING THE REVERBERATION TIME IN THE CLASSROOM

For the proper RT in the classroom, we propose 4 new acoustic absorbent materials that have sustainable characteristics.

Rockwooll Panels

These are mineral wool panels of volcanic rock with resistance to fire and water. It Is 100% natural and has good duration in time. For its use in schools, the existing suspended ceiling must be replaced in part. Of the nearly 54m2 suspended ceiling of plaster, 22 m2 should be replaced to obtain a suitable RT. This system would be suspended from the ceiling with a wooden structure to which perforated chipboard veneers would be riveted. Figure XXX shows the RT obtained from the intervention.

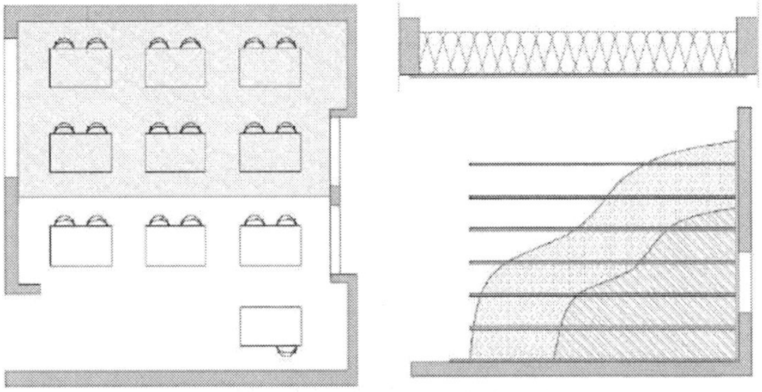

Graphic 4. Aula blueprint with the intervention

Graphic 5. Interior del aula con el rockwool panel. TR after intervention.

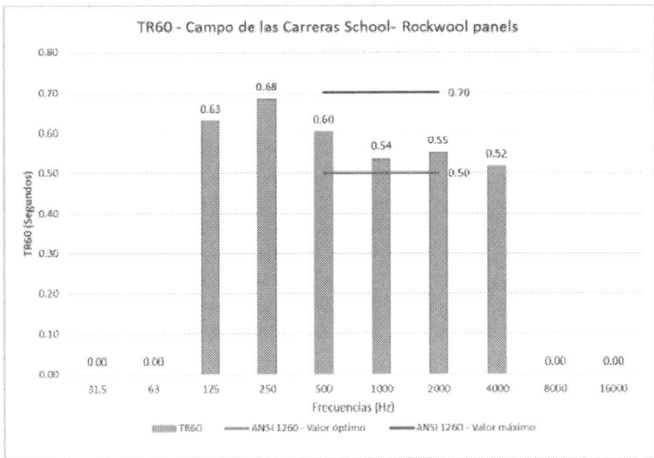

Graphic 6. TR after intervention.

Absorbent pleated

It Is a system formed by recessed sound-absorbing boards. This system uses ecological materials, as well as other technology for its operation. It is constructed with metal profiles that are hung from the existing slab. The wood can be fitted using a staples system. Above the boards, we add glass wool so that it will work together to solve the absorbent system. It is noted that when using glass wool, thermal insulation is given to the system. It is necessary to apply to the classroom 24 m2 to achieve a correct functioning but only in the area of the spoken word.

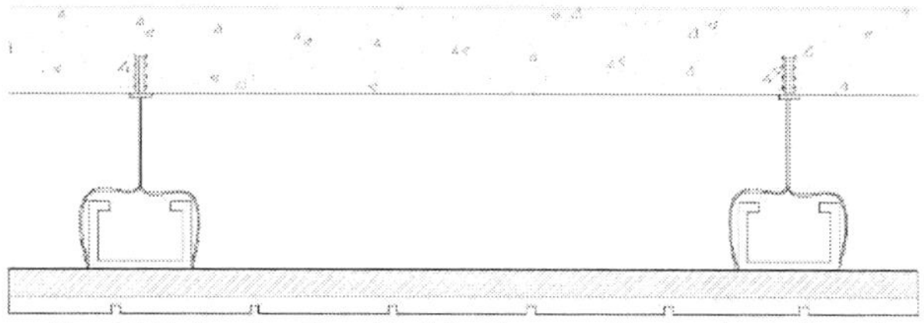

Graphic 7. Construction detail for the acoustic absorbent pleated.

Graphic 8. Interior view for the proposal.

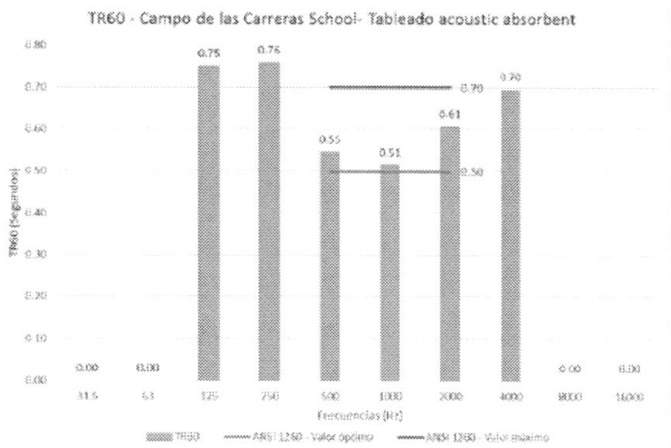

Graphic 9. TR after intervention.

NATURAL AGLOCORK PANELS

This system consists on panels of natural cork granules with glue type adhesives. All materials are biodegradable. It has characteristics of acoustic insulating, acoustic and barrier absorbent for noises generated by impacts or vibrations. It should be placed on another support system for operation in ceiling. It will be necessary to replace 24m2 of the existing ceiling by this system to achieve an appropriate RT.

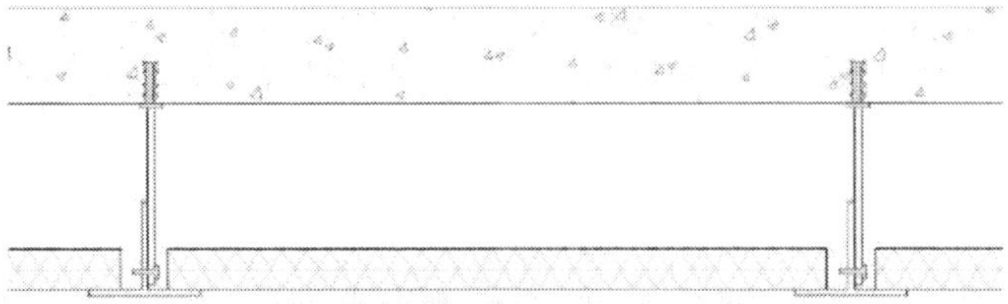

Graphic 10. Detail for aglocork panels

Graphic 11. Interior view for the intervention with aglocork panels.

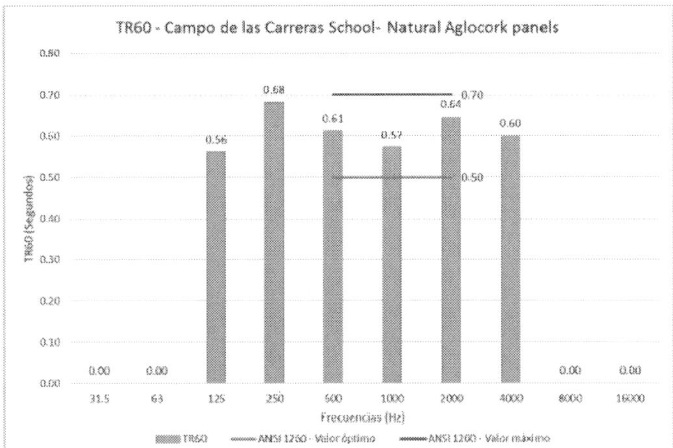

Graphic 12. RT after intervention

GeoPannels

They Are panels made from raw material of industrial waste. In Addition, in its production, there is a minimal production of traces of carbon, so they collaborate to reduce global warming. These are fibrous panels that allow clean cuts, without breakages. The system must have a clamping structure in addition to sheets with perforations, behind them will be placed the geopannel. We will need to replace 30 m2, and still because of the characteristics of its absorption curves, we could not solve the problem in a standard classroom in the city of Tucumán.

Graphic 13. Inside view - TR after intervention

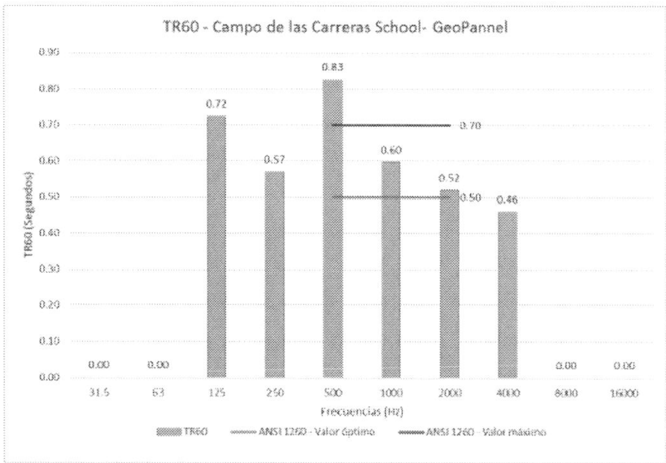

Graphic14. RT after intervention in the classroom.

CONCLUSIONS

From the use of new materials with acoustic absorbent properties, it is possible to improve the RT in the interiors of classrooms in Tucumán. The study of 4 alternatives, allows us to evaluate different options in order to generate an overcoming proposal that is appropriate in this field. We can observe that from these 4 proposals, the one with lower impact is the one with the Geopannels product, because this one, does not achieve in the case of study to generate in the area of frequencies of the spoken word, a correct RT. In this case, if we continue adding absorbent material to reach the comfort zone in the 500 Hz zone, the frequencies between 1000 Hz and 2000 Hz would be below the level recommended by ANSI 12.60 Standards, and the problem would continue. A similar situation happens with the Absorbent Pleated, which, while leaving the frequencies of the spoken word in the comfort zone, the most serious frequencies are still above. This would generate that if there were sounds with high frequency content, they would be reverberating, producing amounts of noise that could be annoying for school activities.

Another situation to take into account are the economic costs of each system; a study that will be pending for another future work.

REFERENCES

1. Argentina. Ministerio de Educación, Ciencia y Tecnología de la Nación. Repensar las escuelas. 2007. p. 5.
2. Pasch, V.; Miyara, F.; Yanitelli, M.; Cabanellas, S.; Miechi, P.; Accolti, E.; Marengo.
3. Instituto Argentino de Normalización y Certificación. NORMA IRAM 4062. Ruidos Molestos al vecindario. Primera y segunda parte. Agosto. 2015. Argentina.
4. Instituto Argentino de Normalización y Certificación. NORMA IRAM 4113. Acústica. Descripción, medición y evaluación del ruido ambiental. Parte 1. Parte 2. Junio. 2009. Argentina.
5. Instituto Argentino de Normalización y Certificación. NORMAS ISO 1996-2:2007-03 (E). Acoustics – Description, measurement and assessment of environmental noise. Part 2. Determination of environmental noise levels. 1996. United States of America.
6. Acoustical Society of America. NORMA ANSI/ASA S12.60/2010/Part 1. American National Standard Acoustical Performance Criteria, Design Requirements, and Guidelines for Schools, Part 1: Permanent Schools. 2010. United States of America.
7. Rodriguez, F. Acústica de Aulas. Arquitectura Investiga, 2012. p. 3.
8. Pasch, V.; Miyara, F.; Yanitelli, M.; Cabanellas, S.; Miechi, P.; Accolti, E.; Marengo Rodríguez, F. Acústica de Aulas. Arquitectura Investiga, 2012. p. 12
9. Paterlini, L. Garzón, B. Confort acústico en Aulas de Escuela en Tucumán, Argentina. Análisis, evaluación y concienciación. XXVIII Encontro da Sociedade Brasileira de Acústica. Porto Alegre – Brasil. Octubre. 2018.
10. Departamento de Tecnologías Audiovisuales Sección de Acústica de la Escuela Universitaria La Salle. Estudio para mejorar la insonorización de Aulas. Barcelona, 2004.

18th International Conference on
Non-Conventional Materials and Technologies
"Construction Materials & Technologies for Sustainability"
(18th NOCMAT 2019)
24th – 26th July 2019
Nairobi, Kenya

A DESIGN CONCEPT FOR USING BAMBOO IN LOW INCOME HOUSING IN KENYA

Shahryar Habibi[1], Esther Obonyo[2], Ali M. Memari[3]

[1] Visiting Postdoctoral Scholar, [2] Associate Professor, [3] Professor

[123] Department of Architectural Engineering, The Pennsylvania State University, University Park

ABSTRACT

Consistent with sustainability movement, bamboo has been identified as a green construction material. This paper discusses the development of a design concept for a typical bamboo house, including the layout configuration and the function/aesthetics properties that are important from architectural perspective. The main objective of this paper is to demonstrate an approach for using bamboo as a structural material for low-income housing. This paper presents a review of examples of vernacular architecture and building elements in Kenya. It then highlights the design of a bamboo-structure residential house based on bio-climatic design strategies. Bamboo housing can improve the financial stability and economic sustainability of low-income families. The development of bamboo housing systems can advance efforts directed at securing home ownership for low-income families through lowering the construction costs to levels that are within their budgets. The successful implementation of the proposed approach can help demonstrate the appropriateness of bamboo housing option for rapidly developing countries in East Africa. Bamboo is currently not being used as one of the main structural materials in the design and construction practice in the East African context. It is the contention of the authors that bamboo can be used as a suitable substitute for the more commonly used building materials in this context – concrete blocks, natural stones and burnt bricks. The discussion section of the paper includes a holistic comparison of different building materials, focusing specifically on how bamboo can address the sustainability-related issues such as: 1) the cost and use of cement-related concerns associated with concrete blocks; 2) the cost and destruction to land associated with the extraction of natural stones; and 3) the land use conflict, particularly with farming, reliance on wood fuel systems during the production of fired bricks, which contributes to air pollution and deforestation.

Keywords: Sustainability criteria, low-income housing, bamboo, cost-efficiency

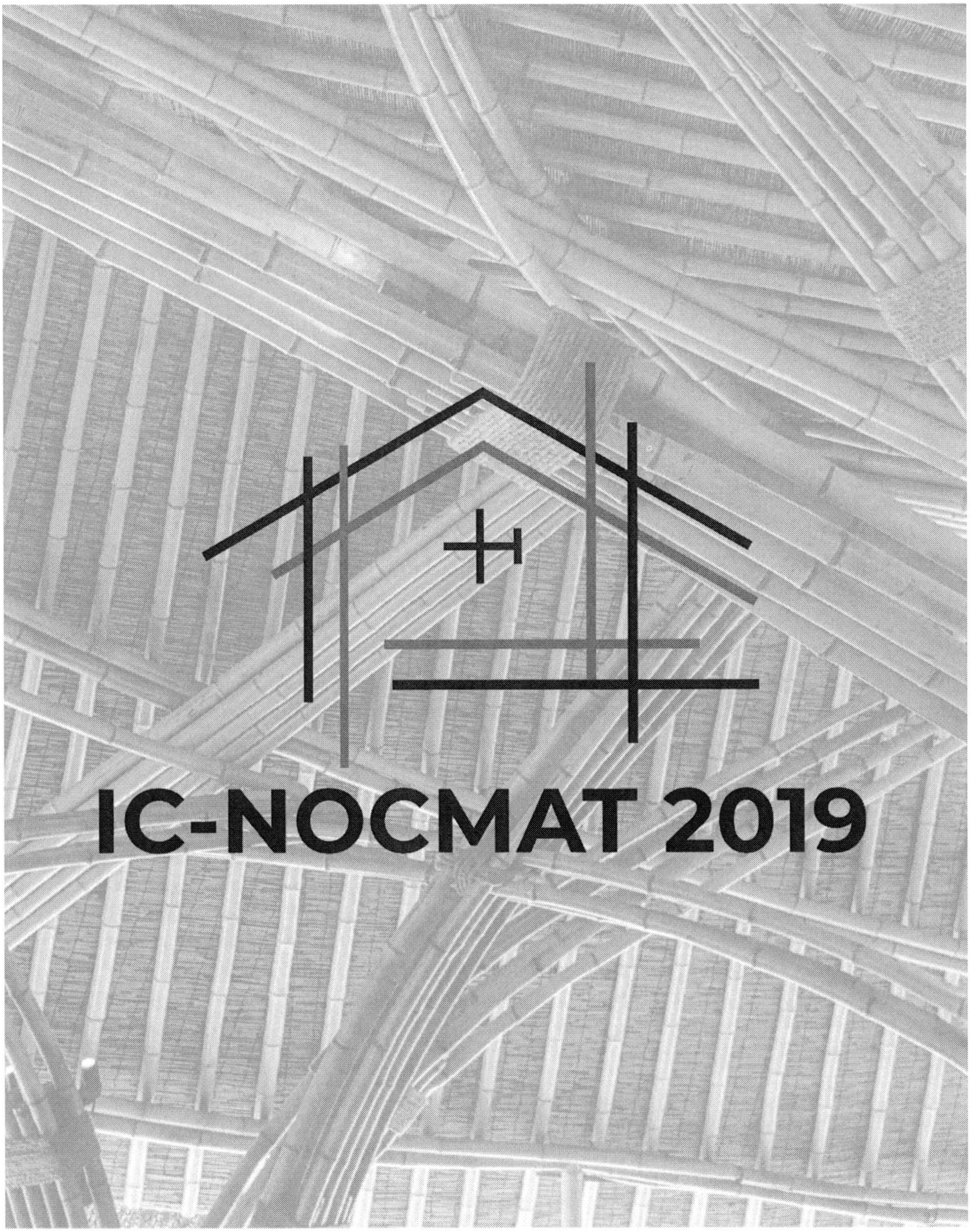

PPCEM
POST-GRADUATION PROGRAMM ON SCIENCE AND MATERIALS ENGINEERING FEDERAL UNIVERSITY OF PARAÍBA, BRAZIL

PUC RIO

SCIENTEC

abmtenc

PennState College of Engineering
ARCHITECTURAL ENGINEERING